Quinoa

Quinoa

Chemistry and Technology

Fan Zhu

School of Chemical Sciences, The University of Auckland, New Zealand

ELSEVIER

ACADEMIC PRESS

An imprint of Elsevier

Academic Press is an imprint of Elsevier
125 London Wall, London EC2Y 5AS, United Kingdom
525 B Street, Suite 1650, San Diego, CA 92101, United States
50 Hampshire Street, 5th Floor, Cambridge, MA 02139, United States
The Boulevard, Langford Lane, Kidlington, Oxford OX5 1GB, United Kingdom

ISBN: 978-0-323-99909-0

For Information on all Academic Press publications
visit our website at https://www.elsevier.com/books-and-journals

Publisher: Nikki Levy
Acquisitions Editor: Nina Bandeira
Editorial Project Manager: Katherine Esten
Production Project Manager: Paul Prasad Chandramohan
Cover Designer: Christian Bilbow

Typeset by MPS Limited, Chennai, India

Contents

About the author xiii
Preface xv

1. Development of quinoa grain as a sustainable crop 1
1.1 Introduction 1
1.2 Genetic diversity of quinoa 1
1.3 Cultivation of quinoa in different environmental conditions 4
1.4 Collaborations for quinoa development 5
1.5 Development of quinoa in different countries 6
1.6 Organic quinoa 7
1.7 Quinoa greens 8
1.8 Growing quinoa in outer space 9
1.9 Quinoa seed storage 9
1.10 Ecological and socioeconomic concerns of growing quinoa 10
1.11 Conclusions 11
References 11

2 Structure and nutrient composition of quinoa grain 17
2.1 Introduction 17
2.2 Quinoa grain structure 17
2.3 Chemical composition of quinoa grain 21
2.4 Interpretation and comparability of the data on proximate composition of quinoa 23
2.5 Proteins and amino acids of quinoa 24
2.6 Lipids of quinoa 26
2.7 Carbohydrates of quinoa 27
2.8 Minerals, organic acids, vitamins, and carotenoids of quinoa 28
2.8.1 Minerals 28
2.8.2 Phytic acid and organic acids 30
2.8.3 Vitamins 31
2.8.4 Carotenoids 32
2.9 Conclusions 32
References 33

3 Quinoa grain milling 39
3.1 Introduction 39

3.2 Quinoa milling processes 40
3.2.1 Dry milling 40
3.2.2 Wet milling 57
3.2.3 Alternatives to the conventional milling processes 60
3.2.4 Other factors affecting the quality of quinoa milling fractions 60
3.2.5 Milling devices 61
3.3 Effect of milling on chemical and physicochemical properties of quinoa milling fractions 61
3.3.1 Effect of milling on nutritional and chemical composition and particle size 61
3.3.2 Effect of milling on physicochemical and functional properties 63
3.3.3 Comparison between wet and dry milling of quinoa grain 63
3.4 Conclusions 65
References 65

4 Quinoa starch 69
4.1 Introduction 69
4.2 Quinoa starch isolation 69
4.3 Quinoa starch composition 70
4.4 Structure of quinoa starch 73
4.4.1 Shape and size of starch granules 73
4.4.2 Polymorphism 76
4.4.3 Chemical structure 76
4.5 Physicochemical and functional characteristics of quinoa starch 80
4.5.1 Thermal property 80
4.5.2 Swelling and solubility 80
4.5.3 Rheological properties 83
4.5.4 Retrogradation property 86
4.5.5 Starch digestibility 87
4.6 Conclusions 88
References 88

5 Quinoa dietary fiber and nonstarch polysaccharides 93
5.1 Introduction 93
5.2 Dietary fiber composition of quinoa 94
5.3 Isolation of fiber fractions and polysaccharides of quinoa 95
5.3.1 Structure of fiber polysaccharides of quinoa 97
5.4 Physicochemical properties of dietary fiber of quinoa 99
5.4.1 Thermal and rheological properties 99
5.4.2 Water- and oil-holding capacities 103
5.4.3 Glucose, cholesterol, and bile acid adsorption capacity 103

5.5 Biological properties of fiber polysaccharides and fractions of quinoa 104
5.5.1 Antioxidant activity 104
5.5.2 Inhibition on food digestion 104
5.5.3 Antidiabetic activity 109
5.5.4 Cholesterol and bile acid adsorption 109
5.5.5 Anticancer property 110
5.5.6 Gastro-protection 110
5.5.7 Immunomodulation 110
5.5.8 Fermentative properties and gut microbiota 111
5.6 Discussion 111
5.7 Conclusions 112
References 113

6 Quinoa protein 117
6.1 Introduction 117
6.2 Composition, fractionation, and structure of quinoa proteins 118
6.3 Quinoa protein isolates 118
6.4 Functional properties of quinoa proteins and peptides 119
6.4.1 Solubility 120
6.4.2 Gelling property 120
6.4.3 Emulsifying property 121
6.4.4 Foaming property 121
6.4.5 Water and oil-binding property 122
6.4.6 Thermal property 122
6.4.7 Digestibility 123
6.5 Biological properties of quinoa proteins and peptides 123
6.5.1 Antioxidant activity 128
6.5.2 Antihaemolytic activity 128
6.5.3 Antibacterial activity 129
6.5.4 Antihypertension property 129
6.5.5 Adipogenesis inhibition 129
6.5.6 Antidiabetic capacity 130
6.5.7 Anticancer capacity 130
6.5.8 Antiinflammatory 131
6.5.9 Safety of quinoa protein, hydrolysates, and peptides 131
6.6 Effect of modifications and processing on physicochemical, functional, and biological properties of quinoa protein and hydrolysates 131
6.6.1 Physical treatments 132
6.6.2 Chemical and enzymatic treatments 142
6.6.3 Interactions of quinoa protein and peptides with other components 143
6.7 Conclusions 144
References 144

7 Quinoa lipids **151**
7.1 Introduction 151
7.2 Extraction of quinoa lipids 151
7.3 Analysis of lipid content and fatty acid composition in quinoa grain 156
7.4 Composition of different types of lipids including neutral and polar lipids in quinoa 156
7.5 Triacylglycerol composition of quinoa lipids 157
7.6 Fatty acid composition of quinoa lipids 157
7.7 Unsaponifiable lipidic fraction of quinoa lipids 159
7.8 Melting properties of quinoa lipids 159
7.9 Effect of processing on quinoa lipid composition 160
7.9.1 Effect of thermal processing 160
7.9.2 Effect of aging and oxidation 160
7.9.3 Effect of germination 161
7.9.4 Effect of electron beam irradiation 162
7.10 Effect of lipids on physicochemical properties of quinoa flour 162
7.11 Quinoa seed oil 162
7.12 Oils of quinoa leaves 164
7.13 Conclusions 165
References 165

8 Polyphenols, saponins, and other bioactive components of quinoa **169**
8.1 Introduction 169
8.2 Composition of polyphenols and saponins in quinoa 170
8.2.1 Extraction of phenolics from quinoa 170
8.2.2 Phenolic acids and derivatives in quinoa 170
8.2.3 Flavonoids and derivatives in quinoa 178
8.2.4 Saponins in quinoa 179
8.2.5 Phytoecdysteroids, betalains, and other minor components in quinoa 182
8.3 Biological activities of polyphenols, saponins, and other bioactive compounds in quinoa 183
8.3.1 Phenolics-related biological activities 183
8.3.2 Saponins-related biological activities 193
8.3.3 Biological activities related to other bioactive compounds (phytoecdysteroids, betalains, and alkylresorcinols) 196
8.3.4 General discussion on biological activities of quinoa minor components 197
8.4 Effect of processing on chemical composition and biological activities of polyphenols and saponins in quinoa 198
8.4.1 Drying 198
8.4.2 Simulated gastrointestinal digestion 198

8.4.3 Germination and fermentation 198
8.4.4 Home cooking methods 207
8.4.5 Milling 209
8.4.6 Ultrasound, high hydrostatic pressure, and electron beam irradiation treatments 209
8.4.7 Ultrasound-assisted extraction and microwave-assisted hydrolysis 210
8.5 Conclusions 210
References 211

9 Health and nutritional effects of quinoa 219
9.1 Introduction 219
9.2 In vitro studies on health and nutritional effects of quinoa 219
9.2.1 Whole quinoa seeds 220
9.2.2 Quinoa extracts 220
9.2.3 Quinoa proteins 227
9.2.4 Quinoa polysaccharide 231
9.3 In vivo studies on health and nutritional effects of quinoa 232
9.3.1 Quinoa seeds and flours 232
9.3.2 Quinoa protein 246
9.3.3 Quinoa extracts 247
9.3.4 Quinoa polysaccharide 248
9.3.5 Quinoa phytoecdysteroid extract 248
9.4 Human trials on health and nutritional effects of quinoa 249
9.4.1 Metabolic improvement 249
9.4.2 Glycemic control 249
9.4.3 Plasma lipid profile improvement 250
9.4.4 Reduction of cardiovascular disease risk 250
9.5 Component-biofunction relationships of quinoa 251
9.6 Negative health effects of quinoa 252
9.6.1 Quinoa allergy revealed by clinical reports 252
9.6.2 Contaminated quinoa 256
9.6.3 Intestinal permeability 257
9.7 Research gaps on health effects of quinoa 257
9.8 Conclusions 259
References 259

10 Domestic cooking and sensory properties of quinoa 267
10.1 Introduction 267
10.2 Consumer preference for quinoa 268
10.3 Domestic cooking of quinoa 268
10.3.1 Boiling 269
10.3.2 Steaming 269
10.3.3 Baking and roasting 270

10.3.4 High-pressure assisted cooking 270
10.3.5 Combined cooking methods 270
10.4 Sensory properties of cooked quinoa 271
10.4.1 Taste 271
10.4.2 Aroma 274
10.4.3 Texture 274
10.4.4 Interplay among different sensory attributes of quinoa 275
10.5 Conclusions 276
References 276

11 Thermal, nonthermal, and biotechnological processing of quinoa 279
11.1 Introduction 279
11.2 Nonthermal processing and modifications of quinoa seeds and flours 283
11.2.1 Washing 283
11.2.2 High hydrostatic pressure treatment 283
11.2.3 Cold plasma modification 284
11.2.4 Ultrasound modification 284
11.2.5 Electron beam irradiation 285
11.3 Thermal processing of quinoa seeds and flours 285
11.3.1 Extrusion 285
11.3.2 Boiling, steaming, and roasting 292
11.3.3 Heating under pressure 294
11.3.4 Drying 294
11.3.5 Microwave heating 294
11.3.6 Popping 295
11.4 Bioprocessing of quinoa seeds and flours 295
11.4.1 Fermentation 295
11.4.2 Malting 309
11.4.3 Germination 310
11.5 Conclusions 310
References 311

12 Quinoa-based food product development 317
12.1 Introduction 317
12.2 Gluten-containing products with quinoa 318
12.2.1 Cupcake 318
12.2.2 Tarhana and kishk 318
12.2.3 Biscuit and cookie 329
12.2.4 Bread 330
12.2.5 Pasta and noodle 333
12.2.6 Meat-based product 333
12.2.7 Beer 334

12.3 Quinoa-based gluten-free products 334
12.3.1 Intermediate products including composite flour, gel, and extract 334
12.3.2 Film and coating 353
12.3.3 Sprout 353
12.3.4 Infant food 353
12.3.5 Snack food 354
12.3.6 Biscuit and cookie 355
12.3.7 Cake 356
12.3.8 Bread 357
12.3.9 Pasta 359
12.3.10 Tempe 360
12.3.11 Probiotic and fermented drinks and yogurt 360
12.3.12 Cheese and cream analog 361
12.3.13 Milk substitute and porridge 362
12.3.14 Fish- and meat-based products 366
12.3.15 Beer 367
12.4 Labeling and increasing price issues of quinoa-based gluten-free products 368
12.5 Conclusions 369
References 370

13 Food and nonfood applications of quinoa fractionated products 377
13.1 Introduction 377
13.2 Food and nutraceutical applications of quinoa protein 378
13.2.1 Functional food ingredient and food component interactions 378
13.2.2 Nutraceutical ingredients 378
13.2.3 Encapsulation of bioactive compounds 386
13.2.4 Emulsion stabilizers 386
13.2.5 Films and coatings 388
13.3 Food and nonfood applications of quinoa starch 391
13.3.1 Films and coatings 391
13.3.2 Applications in meat products 391
13.3.3 Production of nanoparticles 391
13.3.4 Encapsulation of bioactive compounds 396
13.3.5 Emulsions 398
13.4 Food, nutraceutical, and nonfood applications of quinoa dietary fiber and nonstarch polysaccharides 398
13.5 Food and nonfood applications of quinoa lipids 402
13.6 Food, nutraceutical and nonfood applications of quinoa saponin, polyphenol, and other extracts 403
13.6.1 Food and nutraceutical applications of saponins 403
13.6.2 Nonfood applications of saponins 403

13.6.3 Applications of compounds other than saponins in quinoa 411
13.7 Lack of comparative studies between quinoa-based fractionated products and products from other food grains 412
13.8 Lack of studies on the impact of quinoa fractionation on the environment 413
13.9 Conclusions 413
References 414

Index **421**

About the author

Fan Zhu is a senior lecturer of Food Science at the University of Auckland, New Zealand. He holds a PhD in food chemistry from the University of Hong Kong and was a postdoctoral fellow at University of Arkansas, United States and University of Guelph, Canada. Zhu has had the responsibility of teaching a range of food science courses related to food chemistry, food analysis, food grains, and food processing. He is author or coauthor of over 180 peer-reviewed journal articles. He has been doing research in the areas of food carbohydrate chemistry, grain science, antioxidants and functional foods, and water relations of foods and has worked extensively on quinoa grain.

Preface

Quinoa (*Chenopodium quinoa* Willd.) has become trendy due to its attractive health benefits. It also has great significance for food security in the age of climate change and sustainability. The processing properties of quinoa are similar to those of major cereals and pseudocereals. However, our understanding on quinoa chemistry and technology is much less than that for the major cereals such as wheat and rice. The available information on quinoa has been much scattered.

The present monograph aims to summarize the results of reported research and to present different aspects of chemistry, nutrition and processing technology of quinoa and its components in a series of reviews. The reference list of each chapter serves as a primary source of literature. The book also aims to discuss the research methodology that has been hindering rather than improving our progress toward understanding of quinoa. The book has practical relevance to industrial and food applications as well as to academic research. It is useful to farmers, crop breeders, food scientists and technologists, nutritionists, salespersons, and any individuals who have a special interest in quinoa.

Omissions and errors of the book are attributed to me. I welcome suggestions and corrections from readers for future improvement. Finally, I would like to thank my students, mentors, collaborators, and family who made this work possible.

Fan Zhu

Development of quinoa grain as a sustainable crop

1

1.1 Introduction

Quinoa (*Chenopodium quinoa* Willd.) is a pseudocereal of the "goosefoot" genus *Chenopodium* in the Amaranthaceae family (Fagandini Ruiz et al., 2021). Quinoa has its origin in the Andes of South America where it was "mother grain" for the Incas (Miller et al., 2021). Quinoa has attractive nutritional properties such as being gluten free and having more balanced amino acid composition than common cereals as described in Chapter 2 of this reference book. It has attracted great attention from consumers. The major component of quinoa grain is starch (Chapter 4 of the book). Thus the processing property of quinoa can be similar to that of common starchy food grains such as rice. Quinoa seeds can be milled into flour, which is used to create a range of food products (Chapter 13). The cultivation of quinoa crop has spread to other parts of the world such as the United States and China (Wieme et al., 2020; Yang et al., 2019). The year 2013 was named as the International Year of Quinoa by the United Nations. The world production quantity of quinoa has reached over 160k tons according to the Statistics Division of the Food and Agriculture Organization (FAOSTAT, 2022). However, this may be an underestimation due to the lack of data from quinoa-producing countries other than those of South America. Quinoa is cultivated in a range of climate conditions ranging from being very dry (e.g., 200 mm rainfall per year) to humid and from sea level to 4000 meters above the sea level (González et al., 2011; Jacobsen et al., 2003; Martínez et al., 2009). It can be grown on acidic, nutrient-poor soils with a range of pH (Jacobsen et al., 2003). Some quinoa genotypes can grow in high salt concentrations such as in seawater (Manaa et al., 2019; Zou et al., 2017). The crop in general has a high level of resistance against frost. Quinoa of some varieties at certain growing stage can survive at −8°C up to 4 h (Jacobsen et al., 2003). The hardy nature of quinoa allows the crop to be grown in a wide range of environmental conditions and as a potential candidate for space missions (Ponessa et al., 2022; Sellami et al., 2021). Quinoa has great importance in food security as we are facing the climate change and put sustainability as our priority. There is a high genetic diversity of quinoa. The genome of quinoa has been mapped (Jarvis et al., 2017). This information provides a basis for further development and utilization of the crop. This chapter describes the genetic variety and various aspects of the development of the quinoa grain for food production.

1.2 Genetic diversity of quinoa

Quinoa was a major food crop for the indigenous populations of South America before the Spanish conquest. Quinoa was domesticated by the indigenous people

Quinoa. DOI: https://doi.org/10.1016/B978-0-323-99909-0.00011-8

over 7000 years ago on the highlands of the Bolivian and Peruvian Andean area near Lake Titicaca (Bazile, Jacobson et al., 2016; Fagandini Ruiz et al., 2021; Miller et al., 2021; Ruiz et al., 2014). Ecuador also has a long history of quinoa cultivation. Genetic analysis showed that the Ecuadorian landraces have their origins from the highlands of the Bolivian and Peruvian Andes (Bazile et al., 2015). Comparative studies showed that genotypes from different regions including Bolivia, Peru, and Ecuador had different agronomic and phenotypic properties (Bazile et al., 2015). Even though there are many genotypes/varieties of quinoa reported from different regions, a lack of a definite list of quinoa varieties has created a degree of confusions among quinoa breeders (Andrews, 2017). The nomenclature of quinoa varieties with detailed background information remains to be standardized to facilitate the management of the genetic resources.

There is great diversity in quinoa properties including morphological characteristics (Andrews, 2017; Medina et al., 2010; Ruiz et al., 2014) (Fig. 1.1). The color of quinoa grains can be white, red, black, pink, purple, yellow, and gray (Yang et al., 2019) (Fig. 1.2). The color may be related to the taste of the grain. For example, white quinoa tends to be less bitter than black one (Yang et al., 2019).

There are differences in the agronomic traits of quinoa varieties collected from different regions near the center of genetic diversity in South America. For example, when compared to quinoa genotypes collected from Ecuador, varieties from Peru and Bolivia tended to grow shorter, mature earlier, and had less biomass accumulation (Bazile & Bertero, 2015). New quinoa varieties have been developed in some countries such as Denmark, the United Kingdom, the Netherlands, Canada, Peru, and Chile (Bazile et al., 2015). Quinoa varieties from different regions should be compared for the breeding of new varieties with improved quality and desired agronomic traits (Bazile, Jacobson et al., 2016).

Quinoa-related wild species of the genus *Chenopodium* may be studied as a source of genes for improving crop quality such as improved tolerance to harsh environmental conditions and pests (Fagandini Ruiz et al., 2021). The related species include *C. petiolare*, *C. incisum*, *C. hircinum*, *C. quinoa* ssp. *pallidicaule*, *C. ambrosioides*, *C. quinoa* ssp. *melanospermum*, and *C. carnosolum*. However, very little research has been done on the potential of these species for breeding programs and food production.

During the last two decades or so, different molecular markers have been developed to analyze the diversity of quinoa (Fuentes et al., 2009; Jarvis et al., 2017; Saad-Allah & Youssef, 2018; Salazar et al., 2019; Zhang et al., 2017). For example, retrotransposons are genetic components of plant genome and their position and copy number may be altered (Finnegan, 1989). Long terminal repeat (LTR) and non-LTR retrotransposons are the two forms of retrotransposons. Interprimer binding site (iPBS) was developed as a marker (Barut et al., 2020; Kalendar et al., 2010). Many different quinoa accessions collected from different regions have been analyzed. For example, Hossein-Pour et al. (2019) used iPBS to study 17 quinoa genotypes grown in Turkey. Romero et al. (2019) employed microsatellite markers to analyze 26 quinoa genotypes from Peru. Zhang et al. (2017) used insertion/deletion markers to analyze 129 quinoa accessions from a world collection. Barut et al. (2020) analyzed 96

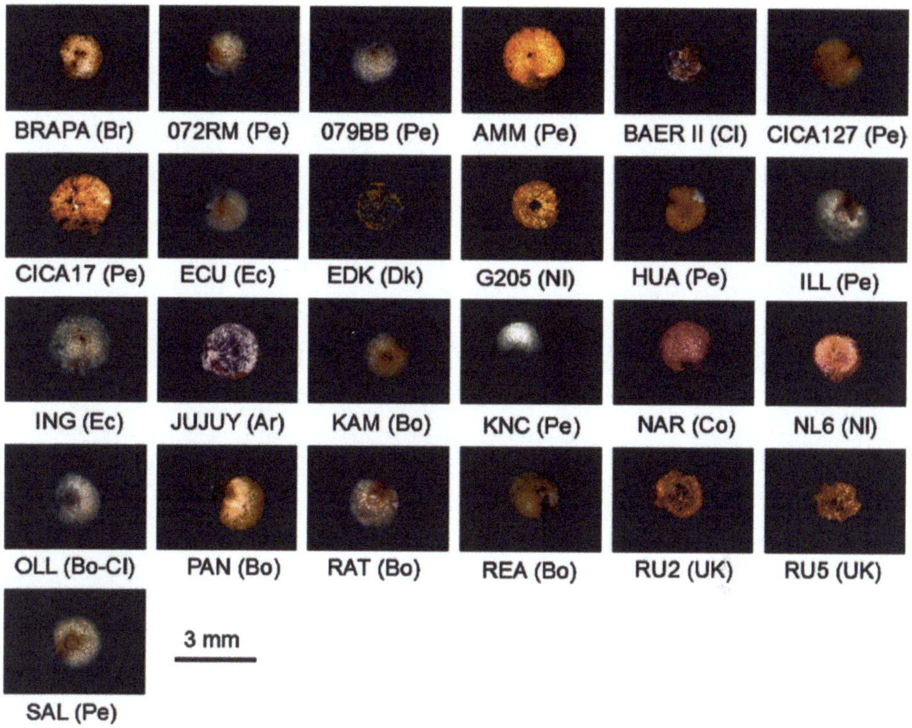

Figure 1.1 Diversity in grain morphology of 25 quinoa varieties collected from different geographic locations. Germplasm bank codes: BRAPA: 02-EMBRAPA, 072RM: 03−21 072RM-UNA-Puno, 079BB: 03−21−079BB-UNA-Puno, AMM: Amarilla de Maranganí, Cusco, BAER II: Baer-II-U. Concepción, CICA127: CICA-127-Cusco, CICA17: CICA-17-Cusco, ECU: ECU-420-INIAP, EDK: E-DK-4-PQCIP-DANIDA-UNA,G205: G-205−95-PQCIP-DANIDA-UNA,HUA: Huariponcho-CRIDER-Puno, ILL: Illpa-INIA-Puno, Ing: Ingapirca-INIAP, JUJUY: Jujuy-UNA, KAM:Kamiri-IBTA, KNC: Kancolla UNA, NAR: Nariño-INIA-Pasto, NL6: NL-6-PQCIP-DANIDA-UNA, OLL: Ollague, PAN: Pandela, RAT: Ratuqui-IBTA, REA: Real-IBTA, RU2: RU2-PQCIP-DANIDA-UNA, RU5: RU5-PQCIP-DANIDA-UNA, SAL: Salcedo-INIA-Puno. Geographical locations: Bo (Bolivia), Br (Brazil), Cl (Chile), Co (Colombia), Dk (Denmark), Ec (Ecuador), Nl (Netherlands), Pe (Peru), UK (United Kingdom) (Medina et al., 2010).
Source: Reprinted with permission from Elsevier.

quinoa accessions from different countries using iPBS-retrotransposon markers. Saad-Allah and Youssef (2018) found that inter simple sequence repeat (ISSR) and rapid amplified polymorphic DNA (RAPD) markers effectively distinguished different quinoa genotypes. Overall, these analyses from diverse studies showed that there is great molecular diversity among quinoa genotypes. Genotypes from Bolivia and Peru are more genetically diverse than those from other places such as Chile. A major issue is how to translate the information of this molecular diversity to improve the agronomic traits of quinoa for food production. For example, genotypes with resistance to preharvest

Figure 1.2 Different quinoa varieties cultivated in Inner Mongolia, China (Yang et al., 2019).
Source: Reprinted with permission under Creative Commons Attribution License.

sprouting remain to be developed (McGinty et al., 2021; Wu et al., 2020). Waxy and high amylose genotypes of quinoa have not been obtained yet. The genetic diversity of quinoa is reflected in the composition and properties of chemical components such as starch, proteins, and polyphenols (described in other chapters of this book).

1.3 Cultivation of quinoa in different environmental conditions

Systematic analysis of literature on the field trials of quinoa-growing performance showed that irrigation conditions (e.g., lack of irrigation or with saline) and sowing density and date are factors contributing to sustainable quinoa production in marginal lands and under abiotic stress (salinity and drought) or climate change (Benaffari et al., 2022; Murteira et al., 2022; Naheed et al., 2022; Walsh-Dilley, 2020). Murteira et al. (2022) showed that quinoa may be cultivated using marine and saline aquaponics under greenhouse conditions. Quinoa can also grow in saline conditions in the presence of heavy metal contamination. Naheed et al. (2022) analyzed the biochemical and physiological properties of quinoa grown in contaminated soils. They concluded that Ni-contaminated soils with a suitable level of NaCl salinity (150 mM) was suitable for growing quinoa. Quinoa can also be cultivated on less-drained and fine soils (Sellami et al., 2021). The applications and composition of fertilizers can influence the grain morphology and yield (Jorfi et al., 2022; Murteira et al., 2022). For example, in a field study, Jorfi et al. (2022) showed that high quinoa grain production was determined by the quinoa genotype and the use of suitable levels of P_2O_5 and $ZnSO_4$ of foliar application (Jorfi et al., 2022). The tolerance of quinoa may be increased using biostimulants (Benaffari et al., 2022). For example, Benaffari et al. (2022) showed that combined applications of arbuscular mycorrhizal fungi and vermicompost increased the adaptability of quinoa to drought conditions. Different regions have different soil and environmental conditions. This may lead to variations in the grain yield and quality for the same quinoa genotype. Overall, optimization of the agronomic practices could lead to reasonable production quantity of quinoa grains.

1.4 Collaborations for quinoa development

There have been global initiatives to facilitate the development of quinoa for cultivation in different regions. For example, seed exchange among the many parties participating in the quinoa studies should be encouraged (Fig. 1.3).

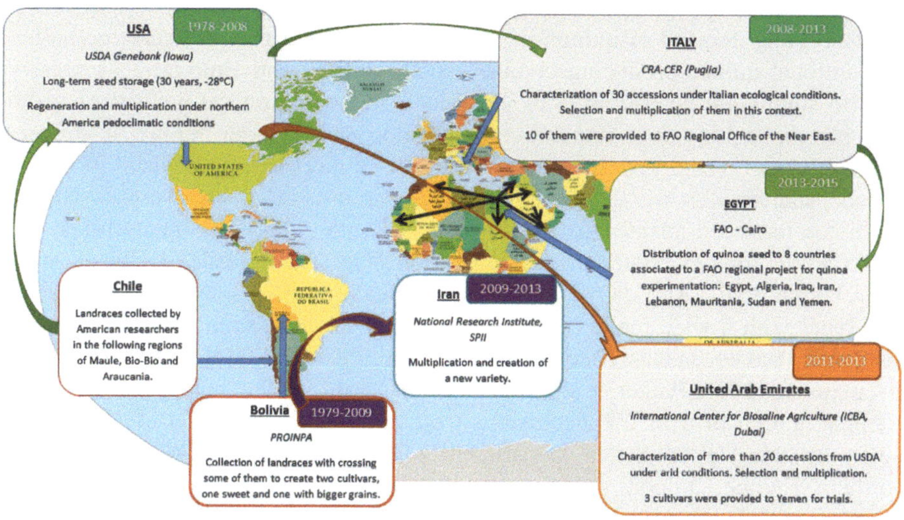

Figure 1.3 Seed transfers of quinoa from Andean nations to some new producers' nations worldwide (Bazile, Pulvento et al., 2016).
Source: Reprinted with permission under Creative Commons Attribution License.

Global Collaborative Network on Quinoa (gcn-quinoa.org) has been developed with more than 300 researchers representing over 75 nations sharing their experiences in adapting quinoa at local scale (Bazile et al., 2015; Murphy et al., 2016; Stanschewski et al., 2021). Major resources such as genetic pools, research facilities, and centers are found in countries where quinoa has been a major crop for generations. On the other hand, many countries such as African countries have a substantial gap in agronomic, nutritional, and sociocultural understandings about quinoa cultivation and consumption as a "new" food. Collaborative initiatives such as International Quinoa Research Symposium, Worldwide Consortium on Evolutionary Participatory Breeding in Quinoa, and Global Collaborative Network on Quinoa can provide communication forums for stakeholders such as researchers, breeders, growers, processors, distributors, consumers, policymakers, and all those interested from industries and academia in the development of quinoa. These stakeholders can promote knowledge exchanges and long-term exploitation of quinoa-related resources (Murphy & Matanguihan, 2015; Murphy et al., 2016).

The Quinoa Germinate Platform (http://germinate.quinoadb.org) has been created to manage quinoa data related to genotypes and phenotypes through an international initiative of collaboration (Stanschewski et al., 2021). This platform facilitates the standardization of quinoa dataset on a global scale. A guideline for environmental data collection and designing layouts with statistical significance for field trials has been proposed (Stanschewski et al., 2021).

1.5 Development of quinoa in different countries

The increasing demand of quinoa from the global market puts pressure on the land of traditional quinoa-producing countries such as Peru and Bolivia. This pressure leads to significant deterioration of local ecosystems and on the welfare of people from certain regions. It is imperative to develop and cultivate quinoa in different parts of the world to divert the pressure and also to meet to high demand for quinoa from consumers.

Quinoa has been adapted from its origin in South America to many other parts of the world for development and cultivation (Fig. 1.3) (Bazile, Jacobson et al., 2016; Hirich et al., 2021). Over the time, different quinoa genotypes have evolved. Genotypes with a wide range of agronomical and species adaption traits have been developed (Bazile, Jacobson et al., 2016; Bazile, Pulvento, et al., 2016; Murphy & Matanguihan, 2015; Präger et al., 2018; Wieme et al., 2020; Yang et al., 2019). For example, a total of over 200 accessions were obtained and taken back to be maintained by the United States Department of Agriculture (USDA) in 1980s. In Northern Europe, quinoa varieties that are day length neutral may be planted (Jacobsen, 2017). Harvesting of quinoa takes place there in early September. The yields of quinoa were found to be 1–3 tons/ha. The market price of the quinoa was relatively high (Jacobsen, 2017). In Southwestern Germany, European quinoa cultivars were grown with yields ranging from 1.73–2.43 Mg/ha (Präger et al., 2018).

One of the cultivars was sweet quinoa containing little amounts of saponins. In Italy, the Cereal Research Centre obtained 100 quinoa genotypes to cultivate, develop, and distribute within the Mediterranean region. In a systematic study, a total of 21 quinoa genotypes were subjected to field trial in over 20 countries outside the Andean region (Bazile, Pulvento, et al., 2016). Some of the genotypes had good adaption capacity and production stability to the environmental conditions of 9 new countries including Egypt, Yemen, Lebanon, Kyrgyzstan, Iran, Iraq, Tajikistan, Algeria, and Mauritania. The results showed that quinoa can be adapted to be grown in these countries with significant differences in geographic and environmental conditions (Bazile, Pulvento, et al., 2016). In mountainous areas of countries such as Nepal and Pakistan, quinoa has been tested for the potential in the security of sustainable nutrition (Adhikari et al., 2017). In Eastern and Southern Africa, quinoa evaluation trials have been systematically conducted in different countries including Djibouti, Ethiopia, Kenya, Somalia, South Sudan, Uganda, and Zambia (Maliro et al., 2021). Different quinoa varieties showed different adaptability to different ecological zones in these countries. These trials showed that seed viability needs to be significantly improved (Maliro et al., 2021).

Quinoa was introduced into China in the 1960s. It has been under significant development during the last decade (Shah et al., 2020; Yang et al., 2019) (Fig. 1.2). The production quantity and harvesting area have reached 20,000 tons and 12,000 ha, respectively. Quinoa cultivation has been conducted from an altitude of sea level to over 5000 m above the sea level (e.g., on the Tibetan plateau) in China. For example, a total of 15 quinoa accessions were planted and assessed for morphological and quality traits and for the adaption to the environmental conditions in the Northeastern part of China (Shah et al., 2020). The quinoa genotypes with higher grain yields had shorter and more compacted inflorescences and branches, whereas those producing more forage had more branches and intermediate plant height with thick stems in that part of China (Shah et al., 2020). A total of over 10 varieties have been registered by registration committees for production. Value-added quinoa products such as noodles, yogurt, and alcoholic beverage have been developed (Yang et al., 2019). In the quinoa development projects, genotypes with high performance were selected for future development and large-scale production. Overall, the new quinoa varieties selected during the plantation and breeding programs remain to be studied for their nutritional, sensory, biological, and chemical properties (Angeli et al., 2020).

1.6 Organic quinoa

The high demand of quinoa from the international market has driven the cultivation of the crop at low elevations such as near sea level. Such growing conditions has led to the attack on the crop by pests and the applications of pesticides (Cruces et al., 2021; Vilca et al., 2018). A significant number of consumers have become health and ecologically conscious. Thus there has been increasing interest in developing organic foods including organic quinoa (Jacobsen & Christiansen, 2016; Nosi et al., 2020; Yang et al., 2019). For example, organic quinoa production has been done in Peru (Cancino-Espinoza et al., 2018). A Life Cycle Assessment was done for the organic quinoa production there. The results showed that greenhouse gas emission of the organic quinoa product fell within the upper range of that for other organic products (e.g., rice and wheat), but was significantly lower than that for the production of animal-based products. The investment-to-protein ratio of quinoa was higher than that of most protein-rich food products. Overall, the production and consumption of organic quinoa are promising for commercial applications to meet the rising demand of organics from the market. However, the changing from traditional subsistence to industrial intensive farming tends to defeat the purpose of being environmentally sustainable (Cancino-Espinoza et al., 2018). The impact of organic quinoa farming on the total environment remains to be systematically studied.

1.7 Quinoa greens

The leaves, microgreens, and sprouts of quinoa have great potential to be developed as healthy foods (Adamczewska-Sowińska et al., 2021; Pathan & Siddiqui, 2022) (Fig. 1.4). These products have short growing durations with all year-round availability and may be cultivated in different settings in greenhouse, field, or high tunnel (Pathan & Siddiqui, 2022). Efforts have been made to optimize the amounts of bioactive compounds, biomass production, and nutritional quality of quinoa greens. For example, Adamczewska-Sowińska et al. (2021) analyzed the effects of harvest time, soil type, and sowing date on some nutrients and yield. Clay soil gave four times the yield of quinoa leaves with more $N\text{-}NO_3$, K, Ca, and Mg than sand soil. The leaves of quinoa sown in spring time had more $N\text{-}NO_3$ and K than those of quinoa sown in summer. Delaying harvest date decreased the contents of $N\text{-}NO_3$ and potassium in the leaves. The leaves with a height of 20–30 cm were suggested to be under the optimum conditions for their harvesting (Adamczewska-Sowińska et al., 2021). Nutritionally, quinoa greens are a source of dietary fibers, omega-3

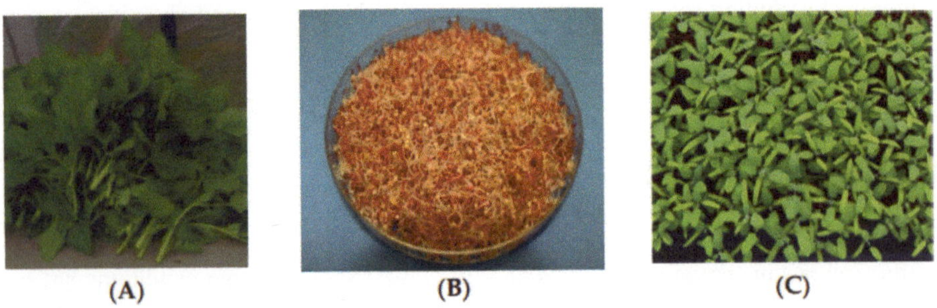

Figure 1.4 Quinoa leaves (A), sprouts (B), and microgreens (C) (Pathan & Siddiqui, 2022). *Source*: Reprinted with permission under Creative Commons Attribution License.

fatty acids, essential minerals, phenolics, proteins, and amino acids (Pathan & Siddiqui, 2022; Pathan et al., 2019; Villacrés et al., 2022). However, research and development on these quinoa greens has not been as extensive as for quinoa grains.

1.8 Growing quinoa in outer space

There is a need to prolong space missions for human exploration of outer space. The success of space exploration is partially dependent on the supply of fresh and nutritious foods (Carillo et al., 2020). Plant foods grown in a spaceship can support life, supply oxygen and food, and remove carbon dioxide. Quinoa showed resistance against a range of harsh conditions including UV-B radiation (González et al., 2009; Reyes et al., 2018). For example, Reyes et al. (2018) showed that various mechanisms of response of quinoa of a highland variety were involved depending on the dosage of UVB irradiation. González et al. (2009) showed that physiological and morphological responses of quinoa to UVB irradiation were dependent on the varieties. Overall, quinoa has the ability to adapt to extreme conditions as represented by its response to the UVB irradiation in these two studies. Quinoa has been tested for potential food production in space (Ponessa et al., 2022). Simulated extra-planetary experimental conditions were used in the tests including plasma radiation, low pressure, and cryogenic temperature. The final germination rates of quinoa reached values of up to 90% regardless of the extra-planetary experimental conditions. Plasma application to the quinoa seeds increased the germination rates. Cryogenic temperature and low pressure conditions had no effect on the final germination rates but delayed the germination process. These treatments affected the structural and mineral composition of the pericarp surface of quinoa seeds without changing the germination capacity. It was concluded that quinoa seeds had a high tolerance to the simulated extra-planetary conditions. Quinoa is a potential candidate for food production in space (Ponessa et al., 2022).

1.9 Quinoa seed storage

Quinoa seed storage is important to ensure long shelf-life (Bakhtavar & Afzal, 2020; Kibar & Temizel, 2021; Kibar et al., 2021). Quinoa seeds should be stored under suitable conditions (Arslan-Tontul, 2021). Bakhtavar and Afzal (2020) analyzed the suitability of dry chain technology in quinoa seed storage. Quinoa seeds were dried to reach a moisture content of 8%–14% before storage up to 18 months. The storage of the seeds was done using hermetic bags or conventional ones (with cloth, paper, jute, and polypropylene). The seeds with the lowest initial moisture content sealed in the bags of triple layers with low oxygen and moisture transmission rates had higher germination rates and lower moisture during storage, compared with those stored at other conditions. Quinoa seeds stored in the hermetic bag with the lowest initial moisture content had lower electrical conductivity, less malondialdehyde and reducing sugars, and more soluble sugars and higher α-amylase activity than those stored at the other conditions. The seeds sealed in traditional packaging materials gained the highest amount of moisture during the storage (Bakhtavar & Afzal, 2020). Kibar et al. (2021) and Kibar and Temizel (2021) studied the storage conditions on the nutritional and phenolic composition, in vitro antioxidant activities, and color properties of quinoa seeds. Temperatures (4°C–25°C) and durations (up to 360 days) of the storage were the variants. Overall, decreases in the concentrations of nutrients (except for minerals) were obtained during the storage. Increasing storage temperature and length further decreased these concentrations. The extents of changes in the chemical and nutritional properties and color of quinoa seeds were dependent on the storage conditions, quinoa varieties, and their interactions (Kibar et al., 2021). Overall, the storage methods and conditions are important to ensure the quality of quinoa seeds. Effective active and intelligent packaging methods may be developed for quinoa seed storage and quality maintenance.

1.10 Ecological and socioeconomic concerns of growing quinoa

Quinoa development has been a significant challenge to ecosystems of certain regions especially in developing countries. For example, a significant part of shrub land has been converted to farm land for quinoa cultivation in the highlands of Bolivia (Walsh-Dilley, 2020). The clearing of the land has left the soil without vegetation unprotected from wind erosion. The monoculture of quinoa leads to quinoa being grown in highland desert with declining crop yield. Many quinoa farmers have left the land to seek livelihood in other places such as in cities. This is a result of short-term interest of economic gain at the expense of the ecosystem. Such quinoa farming and cultivation are not sustainable. There is also growing concern on the impact of increasing quinoa production and rising price on the socioeconomic gap of farmers (Angeli et al., 2020; Bellemare et al., 2018; Kerssen, 2015). Sustainable repeasantization related to the quinoa boom can cause struggle for local farmers in Bolivia (Kerssen, 2015). Bellemare et al. (2018) showed that, in Peru, the increasing international trade of quinoa and crop price were not harmful to the welfare of quinoa farmers in Peru. However, Magrach and Sanz (2020) showed that the ever-increasing demand for quinoa from the global market would have significantly negative impacts on our environment with high carbon footprints. Increasing quinoa price drives increased applications of agrochemicals and land clearing in areas with fragile ecosystems. When the quinoa price makes a downturn to become below the production cost, quinoa farmers are forced to give up their dependence on quinoa farming (Magrach & Sanz, 2020). The COVID-19 pandemic has negatively affected the livelihood of small-scale farming of quinoa (Li & Chura, 2021). Management and policy practices aiming for sustainable quinoa production to reduce the ecological and socioeconomic impact should be developed. This requires significant political and ecological initiatives from governments and intergovernment cooperations.

1.11 Conclusions

There is great genetic diversity of quinoa on molecular and phenotypic levels. Such diversity allows quinoa to be adapted from its original growing environment to many different regions with significantly different agro-climatic and environmental conditions. Quinoa including organics and greens have been developed in different countries for food security and food production. Quinoa has potential as a space crop. Suitable storage conditions should be used to ensure the shelf-life of quinoa seeds. It is essential to develop collaborative network and initiatives among different stakeholders for quinoa production in different regions of the world. The genetic diversity in quinoa remains to be utilized to improve the agronomic performance and nutritional quality for food applications. Quinoa farming and cultivation should be sustainable, requiring government initiatives and regulatory farming policies.

References

Adamczewska-Sowińska, K., Sowiński, J., & Jama-Rodzeńska, A. (2021). The effect of sowing date and harvest time on leafy greens of quinoa (*Chenopodium quinoa* Willd.) yield and selected nutritional parameters. *Agriculture*, *11*, 405.

Adhikari, L., Hussain, A., & Rasul, G. (2017). Tapping the potential of neglected and underutilized food crops for sustainable nutrition security in the mountains of Pakistan and Nepal. *Sustainability*, *9*, 291.

Andrews, D. (2017). Race, status, and biodiversity: The social climbing of quinoa. *Culture, Agriculture, Food and Environment*, *39*, 15–24.

Angeli, V., Miguel, S. P., Crispim, M. D., Khan, M. W., Hamar, A., Khajehei, F., Graeff-Hönninger, S., & Piatti, C. (2020). Quinoa (*Chenopodium quinoa* Willd.): An overview of the potentials of the "golden grain" and socio-economic and environmental aspects of its cultivation and marketization. *Foods*, *9*, 216.

Arslan-Tontul, S. (2021). Moisture sorption isotherm and thermodynamic analysis of quinoa grains. *Heat and Mass Transfer*, *57*, 543–550.

Bakhtavar, M. A., & Afzal, I. (2020). Climate smart dry chain technology for safe storage of quinoa seeds. *Scientific Reports*, *10*, 12554.

Barut, M., Nadeem, M., Karaköy, T., & Baloch, F. (2020). DNA fingerprinting and genetic diversity analysis of world quinoa germplasm using iPBS-retrotransposon marker system. *Turkish Journal of Agriculture and Forestry*, *44*, 471–492.

Bazile, D., & Bertero, D. (2015). The dynamics of the global expansion of quinoa growing in view of its high biodiversity. In D. Bazile, D. Bertero, & C. Nieto (Eds.), *Book: State of the art report on quinoa around the world in 2013* (pp. 42–55). FAO/CIRAD.

Bazile, D., Bertero, D., & Nieto, C. (2015). *State of the art report of quinoa in the world in 2013*. Rome: FAO & CIRAD.

Bazile, D., Jacobsen, S. E., & Verniau, A. (2016). The global expansion of quinoa: Trends and limits. *Frontiers in Plant Science*, *7*, 622.

Bazile, D., Pulvento, C., Verniau, A., Al-Nusairi, M. S., Ba, D., Breidy, J., Hassan, L., Mohammed, M. I., Mambetov, O., Otambekova, M., Sepahvand, N. A., Shams, A., Souici, D., et al. (2016). Worldwide evaluations of quinoa: Preliminary results from post International Year of Quinoa FAO projects in nine countries. *Frontiers in Plant Science*, *7*, 850.

Bellemare, M. F., Fajardo-Gonzalez, J., & Gitter, S. R. (2018). Foods and fads: The welfare impacts of rising quinoa prices in Peru. *World Development*, *112*, 163–179.

Benaffari, W., Boutasknit, A., Anli, M., Ait-El-Mokhtar, M., Ait-Rahou, Y., Ben-Laouane, R., Ben Ahmed, H., et al. (2022). The native arbuscular mycorrhizal fungi and vermicompost-based organic amendments enhance soil fertility, growth performance, and the drought stress tolerance of quinoa. *Plants*, *11*, 393.

Cancino-Espinoza, E., Vázquez-Rowe, I., & Quispe, I. (2018). Organic quinoa (*Chenopodium quinoa* L.) production in Peru: Environmental hotspots and food security considerations using Life Cycle Assessment. *Science of the Total Environment*, *637–638*, 221–232.

Carillo, P., Morrone, B., Fusco, G. M., De Pascale, S., & Rouphael, Y. (2020). Challenges for a sustainable food production system on board of the International Space Station: A technical review. *Agronomy*, *10*, 687.

Cruces, L., de la Peña, E., & De Clercq, P. (2021). Field evaluation of cypermethrin, imidacloprid, teflubenzuron and emamectin benzoate against pests of quinoa (*Chenopodium quinoa* Willd.) and their side effects on non-target species. *Plants*, *10*, 1788.

Fagandini Ruiz, F., Bazile, D., Drucker, A. G., Tapia, M., & Chura, E. (2021). Geographical distribution of quinoa crop wild relatives in the Peruvian Andes: A participatory mapping initiative. *Environment, Development and Sustainability*, *23*, 6337–6358.

FAOSTAT. (2022). Food and Agriculture Organization Corporate Statistical Database. Available from: <http://www.fao.org/faostat/en/#data>.

Finnegan, D. J. (1989). Eukaryotic transposable elements and genome evolution. *Trends in Genetics*, *5*, 103–107.

Fuentes, F. F., Martinez, E. A., Hinrichsen, P. V., Jellen, E. N., & Maughan, P. J. (2009). Assessment of genetic diversity patterns in Chilean quinoa (*Chenopodium quinoa* Willd.) germplasm using multiplex fluorescent microsatellite markers. *Conservation Genetics*, *10*, 369–377.

González, J. A., Bruno, M., Valoy, M., & Prado, F. E. (2011). Genotypic variation of gas exchange parameters and leaf stable carbon and nitrogen isotopes in ten quinoa cultivars grown under drought. *Journal of Agronomy and Crop Science*, *197*, 81–93.

González, J. A., Rosa, M., Parrado, M. F., Hilal, M., & Prado, F. E. (2009). Morphological and physiological responses of two varieties of a highland species (*Chenopodium quinoa* Willd.) growing under near-ambient and strongly reduced solar UV–B in a lowland location. *Journal of Photochemistry and Photobiology B: Biology*, *96*, 144–151.

Hirich, A., Rafik, S., Rahmani, M., Fetouab, A., Azaykou, F., Filali, K., Ahmadzai, H., et al. (2021). Development of quinoa value chain to improve food and nutritional security in rural communities in Rehamna, Morocco: Lessons learned and perspectives. *Plants*, *10*, 301.

Hossein-Pour, A., Haliloglu, K., Ozkan, G., & Tan, M. (2019). Genetic diversity and population structure of quinoa (*Chenopodium quinoa* Willd.) using iPBS-retrotransposons markers. *Applied Ecology and Environmental Research*, *17*, 1899–1911.

Jacobsen, S. E. (2017). The scope for adaptation of quinoa in Northern latitudes of Europe. *Journal of Agronomy and Crop Science*, *203*, 603–613.

Jacobsen, S. E., & Christiansen, J. L. (2016). Some agronomic strategies for organic quinoa (*Chenopodium quinoa* Willd.). *Journal of Agronomy and Crop Science*, *202*, 454–463.

Jacobsen, S. E., Mujica, A., & Jensen, C. R. (2003). The resistance of quinoa (*Chenopodium quinoa* Willd.) to adverse abiotic factors. *Food Reviews International*, *19*, 99–109.

Jarvis, D. E., Ho, Y. S., Lightfoot, D. J., Schmöckel, S. M., Li, B., et al. (2017). The genome of *Chenopodium quinoa*. *Nature*, *542*, 307–312.

Jorfi, A., Alavifazel, M., Gilani, A., Ardakani, M. R., & Lak, S. (2022). Yield and morpho-physiological performance of quinoa (*Chenopodium quinoa*) genotypes as affected by phosphorus and zinc. *Journal of Plant Nutrition*. Available from https://doi.org/10.1080/01904167.2022.2035756.

Kalendar, R., Antonius, K., Smýkal, P., & Schulman, A. H. (2010). iPBS: A universal method for DNA fngerprinting and retrotransposon isolation. *Theoretical and Applied Genetics*, *121*, 1419–1430.

Kerssen, T. M. (2015). Food sovereignty and the quinoa boom: Challenges to sustainable re-peasantisation in the southern Altiplano of Bolivia. *Third World Quarterly*, *36*, 489–507.

Kibar, H., & Temizel, K. E. (2021). Kinetics of temperature and time effects on bioactive compounds and technological properties of quinoa varieties during storage. *Journal of Food Processing and Preservation*, *45*, e15297.

Kibar, H., Sönmez, F., & Temel, S. (2021). Effect of storage conditions on nutritional quality and color characteristics of quinoa varieties. *Journal of Stored Products Research*, *91*, 101761.

Li, F., & Chura, B. M. (2021). Quinoa and small-scale agriculture in times of COVID-19. *Anthropology Now*, *13*, 54–64.

Magrach, A., & Sanz, M. J. (2020). Environmental and social consequences of the increase in the demand for 'superfoods' world-wide. *People and Nature*, *2*, 267–278.

Maliro, M. F. A., Abang, M. M., Mukankusi, C., Lung'aho, M., Fenta, B., Wanderi, S., Kapa, R., Okiro, O. A., Koma, E., Mwaba, C., Isse, M. M., & Bazile, D. (2021). *Prospects for quinoa adaptation and utilization in Eastern and Southern Africa: Technological, institutional and policy considerations*. Addis Ababa: FAO.

Manaa, A., Goussi, R., Derbali, W., Cantamessa, S., Abdelly, C., & Barbato, R. (2019). Salinity tolerance of quinoa (*Chenopodium quinoa* Willd) as assessed by chloroplast ultrastructure and photosynthetic performance. *Environmental and Experimental Botany*, *162*, 103–114.

Martínez, E. A., Veas, E., Jorquera, C., San Martín, R., & Jara, P. (2009). Re-introduction of quinoa into arid Chile: Cultivation of two lowland races under extremely low irrigation. *Journal of Agronomy and Crop Science*, *195*, 1–10.

McGinty, E. M., Murphy, K. M., & Hauvermale, A. L. (2021). Seed dormancy and preharvest sprouting in quinoa (*Chenopodium quinoa* Willd). *Plants*, *10*, 458.

Medina, W., Skurtys, O., & Aguilera, J. M. (2010). Study on image analysis application for identification quinoa seeds (*Chenopodium quinoa* Willd) geographical provenance. *LWT-Food Science and Technology*, *43*, 238–246.

Miller, M. J., Kendall, I., Capriles, J. M., Bruno, M. C., Evershed, R. P., & Hastorf, C. A. (2021). Quinoa, potatoes, and llamas fueled emergent social complexity in the Lake Titicaca Basin of the Andes. *Proceedings of the National Academy of Sciences*, *118*, e2113395118.

Murphy, K. M., Bazile, D., Kellogg, J., & Rahmanian, M. (2016). Development of a world-wide consortium on evolutionary participatory breeding in quinoa. *Frontiers in Plant Science*, *7*, 608.

Murphy, K., & Matanguihan, J. (2015). Editor(s). *Quinoa: Improvement and sustainable production*. Hoboken, New Jersey: Wiley-Blackwell.

Murteira, M., Turcios, A. E., Calado, R., Lillebø, A. I., & Papenbrock, J. (2022). Relevance of nitrogen availability on the phytochemical properties of *Chenopodium quinoa* cultivated in marine hydroponics as a functional food. *Scientia Horticulturae*, *291*, 110524.

Naheed, N., Abbas, G., Naeem, M. A., Hussain, M., Shabbir, R., et al. (2022). Nickel tolerance and phytoremediation potential of quinoa are modulated under salinity: Multivariate comparison of physiological and biochemical attributes. *Environmental Geochemistry and Health*, *44*, 1409–1424.

Nosi, C., Zollo, L., Rialti, R., & Ciappei, C. (2020). Sustainable consumption in organic food buying behavior: The case of quinoa. *British Food Journal*, *122*, 976–994.

Pathan, S., & Siddiqui, R. A. (2022). Nutritional composition and bioactive components in quinoa (*Chenopodium quinoa* Willd.) greens: A review. *Nutrients*, *14*, 558.

Pathan, S., Eivazi, F., Valliyodan, B., Paul, K., Ndunguru, G., & Clark, K. (2019). Nutritional composition of the green leaves of quinoa (*Chenopodium quinoa* Willd.). *Journal of Food Research*, *8*, 55–65.

Ponessa, G. I., Such, P., González, J. A., Mercado, M. I., Buedo, S., González, D. A., Lalla, E., Freemantle, J., & Daly, M. G. (2022). Tolerance of high mountain quinoa to simulated extraplanetary conditions. Changes in surface mineral concentration, seed viability and early growth. *Acta Astronautica*, *195*, 502–512.

Präger, A., Munz, S., Nkebiwe, P. M., Mast, B., & Graeff-Hönninger, S. (2018). Yield and quality characteristics of different quinoa (*Chenopodium quinoa* Willd.) cultivars grown under field conditions in Southwestern Germany. *Agronomy*, *8*, 197.

Reyes, T. H., Scartazza, A., Castagna, A., Cosio, E. G., Ranieri, A., & Guglielminetti, L. (2018). Physiological effects of short acute UVB treatments in *Chenopodium quinoa* Willd. *Scientific Reports*, *8*, 371.

Romero, M., Mujica, A., Pineda, E., Ccamapaza, Y., & Zavalla, N. (2019). Genetic identity based on simple sequence repeat (SSR) markers for quinoa (*Chenopodium quinoa* Willd.). *Ciencia e Investigación Agrarian*, *46*, 166–178.

Ruiz, K. B., Biondi, S., Oses, R., Acuña-Rodríguez, I. S., Antognoni, F., et al. (2014). Quinoa biodiversity and sustainability for food security under climate change. A review. *Agronomy for Sustainable Development*, *34*, 349–359.

Saad-Allah, K. M., & Youssef, M. S. (2018). Phytochemical and genetic characterization of five quinoa (*Chenopodium quinoa* Willd.) genotypes introduced to Egypt. *Physiology and Molecular Biology of Plants*, *24*, 617–629.

Salazar, J., de Lourdes Torres, M., Gutierrez, B., & Torres, A. F. (2019). Molecular characterization of Ecuadorian quinoa (*Chenopodium quinoa* Willd.) diversity: Implications for conservation and breeding. *Euphytica*, *215*, Article number: 60.

Sellami, M. H., Pulvento, C., & Lavini, A. (2021). Agronomic practices and performances of quinoa under field conditions: A systematic review. *Plants*, *10*, 72.

Shah, S. S., Shi, L., Li, Z., Ren, G., Zhou, B., & Qin, P. (2020). Yield, agronomic and forage quality traits of different quinoa (*Chenopodium quinoa* Willd.) genotypes in Northeast China. *Agronomy*, *10*, 1908.

Stanschewski, C. S., Rey, E., Fiene, G., Craine, E. B., Wellman, G., Melino, V. J., Patiranage, D. S. R., et al. (2021). Quinoa phenotyping methodologies: An international consensus. *Plants*, *10*, 1759.

Vilca, F. Z., Moura Andrade, G. C. R., Zamalloa Cuba, W. A., & Tornisielo, V. L. (2018). Determination of pesticides residues in quinoa (*Chenopodium quinoa* Willd) using QuEChERS and LC-MS/MS. *Emirates Journal of Food and Agriculture*, *30*, 421–427.

Villacrés, E., Quelal, M., Galarza, S., Iza, D., & Silva, E. (2022). Nutritional value and bioactive compounds of leaves and grains from quinoa (*Chenopodium quinoa* Willd.). *Plants*, *11*, 213.

Walsh-Dilley, M. (2020). Resilience compromised: Producing vulnerability to climate and market among quinoa producers in Southwestern Bolivia. *Global Environmental Change*, *65*, 102165.

Wieme, R. A., Carpenter-Boggs, L. A., Crowder, D. W., Murphy, K. M., & Reganold, J. P. (2020). Agronomic and economic performance of organic forage, quinoa, and grain crop rotations in the Palouse region of the Pacific Northwest, USA. *Agricultural Systems*, *177*, 102709.

Wu, Q., Bai, X., Wu, X., Xiang, D., Wan, Y., Luo, Y., et al. (2020). Transcriptome profiling identifies transcription factors and key homologs involved in seed dormancy and germination regulation of *Chenopodium quinoa*. *Plant Physiology and Biochemistry*, *151*, 443–456.

Yang, X. S., Qin, P. Y., Guo, H. M., & Ren, G. X. (2019). Quinoa industry development in China. *Ciencia e Investigación Agraria*, *46*, 208–219.

Zhang, T., Gu, M., Liu, Y., Lv, Y., Zhou, L., et al. (2017). Development of novel InDel markers and genetic diversity in *Chenopodium quinoa* through whole-genome re-sequencing. *BMC Genomics*, *18*, 685.

Zou, C., Chen, A., Xiao, L., Muller, H. M., Ache, P., Haberer, G., Zhang, M., et al. (2017). A high-quality genome assembly of quinoa provides insights into the molecular basis of salt bladder-based salinity tolerance and the exceptional nutritional value. *Cell Research*, *27*, 1327–1340.

Structure and nutrient composition of quinoa grain

2

2.1 Introduction

Quinoa is a pseudocereal with small starchy dicotyledonous seeds (Ninfali et al., 2020; Ren et al., 2020). In most of the dicotyledonous plants including quinoa, nutritional components of the seed endosperm are depleted for its development. As a result, the reserve of energy is found in the cotyledons and in the perisperm of quinoa seed (Ninfali et al., 2020). The seeds of dicotyledonous plants including quinoa tend to have architectures different from those of monocotyledons including cereals (Ninfali et al., 2020). The quinoa plant produces oval-shaped, flat grains. These seed grains are very small with a diameter in the range of 1.0–2.6 mm and a 1000-grain-weight variation of 1.9–4.4 g (Graf et al., 2016; Koziol, 1993). The color of the grains ranges from pale yellow, red to dark (Graf et al., 2016). Quinoa grain has more lysine, an essential amino acid, than cereal grains. The quinoa seed does not contain gluten-type protein, which makes it a suitable source of foods for people with celiac diseases or wheat allergy. The seed has a relatively high nutritional quality of lipid fraction and is rich in minerals and some vitamins. The outer seed layer of quinoa may contain a significant amount of saponins. Saponins are bitter in taste and are traditionally considered as a type of antinutrients (Taylor & Parker, 2002). This chapter describes the physical structure of quinoa grains and the chemical composition of different nutrients. The knowledge summarized here provides a foundation for better understanding of other chapters of this book on the chemical, biological, and processing properties of quinoa.

2.2 Quinoa grain structure

The quinoa plant produces flat, oval-shaped seeds (like a flattened sphere) (see the front cover of the book). There are significant variations in the grain characteristics among different quinoa genotypes according to comparative studies (Abderrahim et al., 2015; Ghumman et al., 2021). For example, for six quinoa genotypes, the range of the grain density was 0.63–0.73 g/cm^3 (Ghumman et al., 2021). Präger et al. (2018) showed that 1000-grain-weights of four quinoa cultivars grown in southwestern Germany ranged from 1.2 to 3.3 g. Abderrahim et al. (2015) showed that among 13 types of colored quinoa seeds, the ranges of major and minor diameters were 1.78–2.52 and 1.66–2.37 mm, respectively. Botanically, the seed of the quinoa plant is achene. The seed color varies from being white, gray, pale yellow, to red and dark. Seeds of some genotypes have hints of pink and magenta hue. The seeds are held on the plant in five-lobed perianths. Abderrahim et al. (2015)

Quinoa. DOI: https://doi.org/10.1016/B978-0-323-99909-0.00001-5

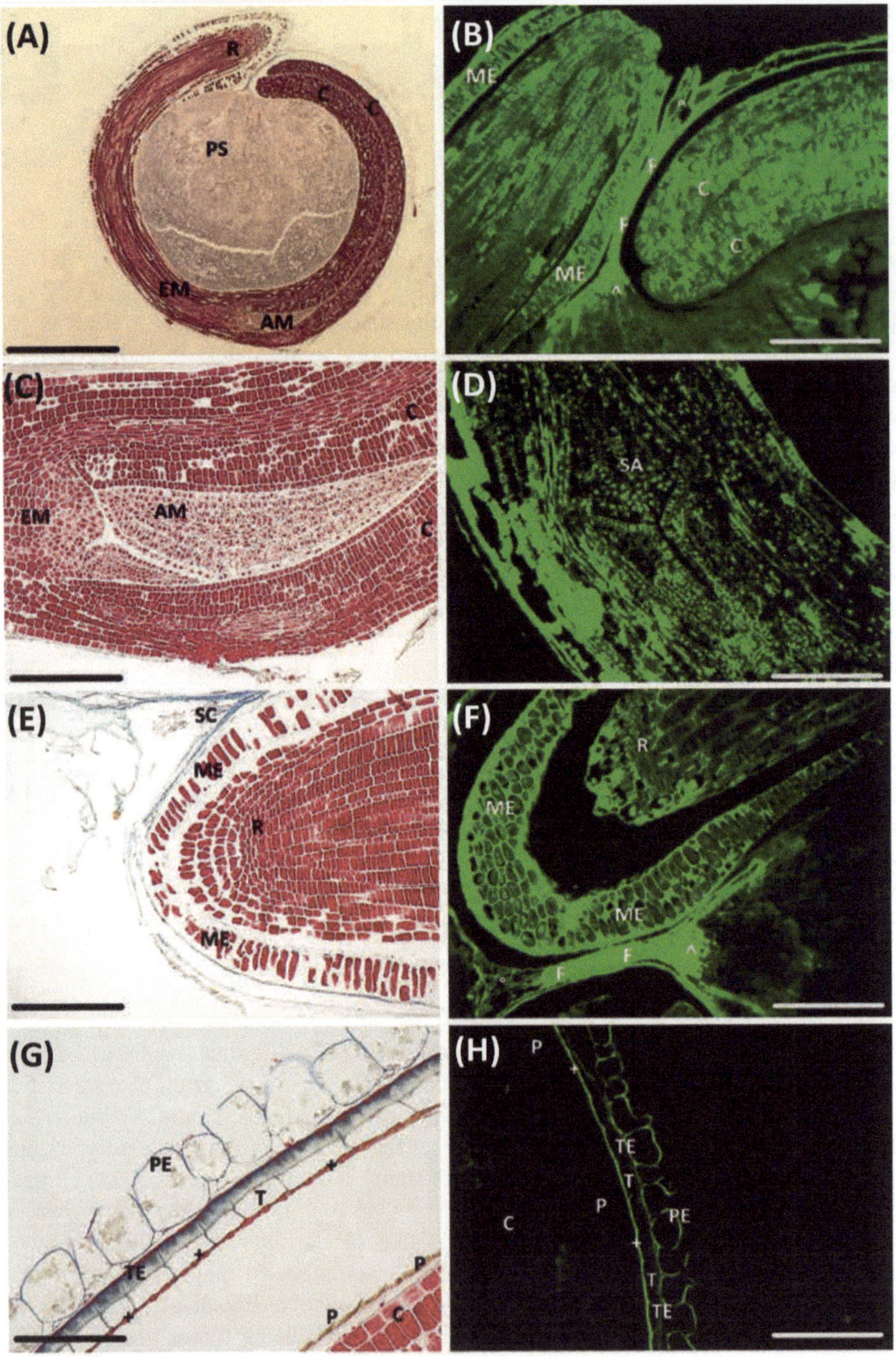
(A)
R
C
C
PS
EM
AM
(B)
ME
F
F
C
C
ME
(C)
C
EM
AM
C
(D)
SA
(E)
SC
ME
R
ME
(F)
R
ME
ME
F
F
(G)
PE
T
TE
P
P
C
(H)
P
TE
T
P
C
PE
T
TE

(*Continued*)

showed that, among 13 types of colored quinoa seeds, the fractal color values ranged from 51.0 to 71.8. The color of these quinoa seeds was significantly affected by the amounts and composition of betalains (Abderrahim et al., 2015). The structure where the seeds are attached to is loose, and the seeds can easily be removed by rubbing.

Like all other food grains, the structure of quinoa grain is multilayered (Fig. 2.1). The quinoa seed is encased in a pericarp, which is composed of two layers. The outermost layer is composed of papillose material (Prego et al., 1998). The pericarp contains saponins. Underneath the pericarp, the episperm (seed coat) may have different colors ranging from being transparent, gray, black, orange, red, brown, or white. It contains electron-dense bodies and starch granules (Prego et al., 1998; Varriano-Marston & DeFrancisco, 1984). These starch granules are not significantly degraded by the hydrolysis of amylases during the germination process (Varriano-Marston & DeFrancisco, 1984).

When a quinoa grain is sliced across its major axis, there is a center of perisperm that contains starch. The embryo with its radicle and two cotyledons is coiled around this perisperm (Fig. 2.2A and B). In cereal grains of the Poaceae family including maize, rice and wheat, most of the starch granules for developing embryos are stored in endosperm. In quinoa seeds, the endosperm is small with two layers around the structure of the hypocotyl-radicle axis (Fig. 2.2). The starch reserves in quinoa grain are found mostly in the perisperm compartment. Starch granules are found in polygonal cells with thin cell walls (Ninfali et al., 2020). The perisperm is surrounded by the embryo, which has radicle and storage cotyledons (Figs. 2.1 and 2.2). The perisperm has a thin wall and its volume is ~40% of that of the whole quinoa grain (Prego et al., 1998). The shoot apical meristem is detected in the middle of the embryo (Figs. 2.1 and 2.2). Procambium in the form of vascular bundles was observed in the radicle part (Ninfali et al., 2020).

The outer layers of the bran fraction of quinoa grain contain more lipids and proteins than the inner part of the perisperm. Some of the components including dietary fiber, minerals, polyphenols, and saponins are mostly found in the bran fraction. The quinoa embryo contains significant amounts of protein aggregates, lipid bodies, and various organelles. Globoid crystals in the protein aggregates have elements including magnesium and potassium. The proplastids contain deposits of phytoferritin (Prego et al., 1998; Varriano-Marston & DeFrancisco, 1984). Quinoa grain has a thinner seed coat than buckwheat and amaranth grains (Ninfali et al., 2020).

◀

Figure 2.1 Light microscopic images of transverse sections of a quinoa seed. A, C, E, G sections were stained with azan trichrome; B, D, F, H sections were unstained sections observed using fluorescence microscopy; ^, chalazal endosperm; +, peripheral endosperm; °, chalazal seed coat; AM, apical meristem; C, cotyledon; EM, embryo axis; F, funicle; ME, micropylar endosperm; P, protective sheath; PE, pericarp; PS, perisperm; R, radicle; SA, shoot apex; SC, seed coat; T, tegmen; TE, testa. Scale bars: (A) 200 μm; (B) 80 μm; (C) 55 μm; (D, F) 45 μm; (E) 43 μm; (G) 24 μm; (H) 50 μm (Ninfali et al., 2020).
Source: Reprinted with permission from the authors under Creative Commons Attribution Noncommercial License.

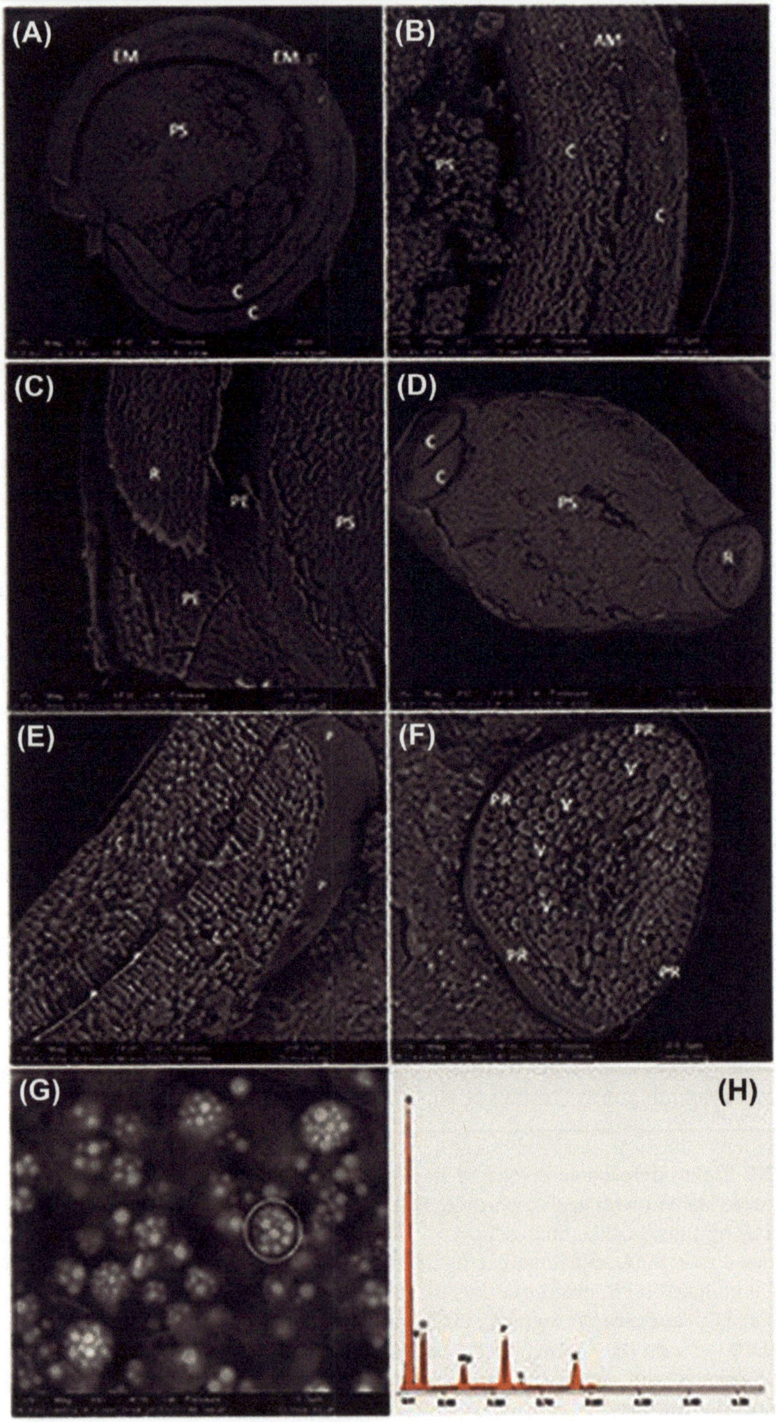

(*Continued*)

2.3 Chemical composition of quinoa grain

Carbohydrates, proteins, and lipids are the main components of quinoa seeds. The functional properties of these components significantly affect the physicochemical properties of quinoa food products. Starch is mostly found in the perisperm, whereas lipids and proteins are located in the embryo and endosperm (Ninfali et al., 2020; Prego et al., 1998). Quinoa grains tend to have more lipids and proteins and less starch as compared to major cereal grains of the Poaceae family including barley, wheat, rice, and maize (Satheesh & Fanta, 2018). There is significant diversity in the chemical composition of different quinoa genotypes (Craine & Murphy, 2020; Encina-Zelada et al., 2017; Filho et al., 2017; Graf et al., 2016; Li et al., 2021; Martín et al., 2014; Nowak et al., 2016; Präger et al., 2018). For example, among the 77 quinoa varieties, the moisture, protein, lipid, ash, and carbohydrate contents ranged from 9.2%–13.4%, 8.3%–11.4%, 5.4%–7.8%, 2.5%–4.1% and 78.5%–82.9%, respectively (Encina-Zelada et al., 2017). Quantitative differences in the relative ratio of the perisperm to the lipid- and protein-rich embryo may contribute to the differences in properties among different quinoa genotypes (Taylor & Parker, 2002). The difference in the moisture content of quinoa samples was most likely due to the difference in drying process. The differences in the chemical composition of different quinoa genotypes lead to differences in physicochemical and functional properties of quinoa flours (De Bock et al., 2022). This aspect is described in other chapters of this book such as Chapter 4.

Despite the genetic variations in the chemical composition of different quinoa varieties, the compositional differences between quinoa and other common food grains tend to be significant (De Bock et al., 2021; Filho et al., 2017; Nowak et al., 2016; Satheesh & Fanta, 2018; Valencia-Chamorro, 2003) (Table 2.1). For example, the protein content of quinoa tended to be higher than that of most cereal grains (Craine & Murphy, 2020; Valencia-Chamorro, 2003). The lipid content was higher than that of cereal crops (Valencia-Chamorro, 2003; Vera et al., 2019). Quinoa fell in the median range of dietary fiber contents when compared to cereal grains, whereas the ash content was higher than that of most cereals (Filho et al., 2017; Valencia-Chamorro, 2003; Zhu, 2020). The starch content of quinoa was lower than that of most cereals and was comparable to that of oats (Filho et al., 2017; Li & Zhu, 2018; Valencia-Chamorro, 2003). When compared to legumes, quinoa had

◀ **Figure 2.2** Environmental scanning electron microscopy (ESEM) analysis of sections of quinoa seed. (A) Transverse section; (D) longitudinal section; (B, C, E, F) cotyledon and root details; (G) high magnification of globoids inside cells of the root tip's cortical layer; the surrounded area represents the field for the analysis showed in H. (H) Elemental composition of the cell globoids. AM, apical meristem; C, cotyledons; EM, embryo axis; P, protective sheath; PE, pericarp; PR, protoderm; PS, perisperm; R, radicle; V, vascular bundles (Ninfali et al., 2020).
Source: Reprinted with permission from the authors under Creative Commons Attribution Noncommercial License.

Table 2.1 Chemical and nutritional composition of quinoa and some cereals and legumes (g/100 g dry weight) (Valencia-Chamorro, 2003).

Parameters	Quinoa	Barley	Maize	Rice	Wheat	Oat	Rye	Lupin	Soya
Protein	16.5	10.8	10.2	7.6	14.3	11.6	13.4	39.1	36.1
Lipid	6.3	1.9	4.7	2.2	2.3	5.2	1.8	7.0	18.9
Dietary fiber	3.8	4.4	2.3	6.4	2.8	10.4	2.6	14.6	5.6
Ash	3.8	2.2	11.7	3.4	2.2	2.9	2.1	4.0	5.3
Carbohydrates	69	80.7	81.7	80.4	78.4	69.8	80.1	35.3	34.1
kcal/100 g	399	383	408	372	392	372	390	361	451

kcal/100 g: 4 × (% protein + carbohydrates) + 9 × (% fat).
Source: Reprinted with permission from Elsevier.

less macro-components except for starch (Filho et al., 2017; Valencia-Chamorro, 2003). Compared to quinoa leaves, quinoa seeds had less protein, iron, calcium, zinc, and phosphorus. The former had more oxalates and nitrates and less saponins than the latter (Villacrés et al., 2022).

2.4 Interpretation and comparability of the data on proximate composition of quinoa

The observations recorded above showed that the variations in the composition of quinoa grains from different studies varied considerably (Craine & Murphy, 2020; Encina-Zelada et al., 2017; Filho et al., 2017; Li et al., 2021; Nowak et al., 2016; Valencia-Chamorro, 2003). Different studies used quinoa samples from different batches or collected from different locations. The plant genetics, growing locations, environmental factors, and agronomic practices (e.g., intercrop and irrigation methods) for the quinoa growing significantly contributed to the variations of the results reported (Cole et al., 2020; Craine & Murphy, 2020; Lee & Sim, 2018; Matías et al., 2021; Tovar et al., 2022; Walters et al., 2016). For example, intercrop treatment (clover/medic or fescue/clover) affected the Mn concentration and protein content in quinoa (Walters et al., 2016). Three different quinoa varieties were grown in three different countries (Spain, Chile, and Peru) (Reguera et al., 2018). The composition of amino acids, proteins, phytate, and minerals was affected by the interactions between the varieties and the growing locations, whereas the contents of fibers or saponins were much less influenced by the interactions (Reguera et al., 2018). Matías et al. (2021) showed that heat stress decreased the contents of total lipids and carbohydrates and quinoa grain yields in the Mediterranean area, while increasing those of fiber and protein. The response to the heat stress was dependent on the quinoa variety (Matías et al., 2021). Tovar et al. (2022) showed that relatively short periods (11 days) of heat spells during anthesis significantly changed the contents of many elements in the seeds.

Apart from the abovementioned factors, differences in methodology used in different studies could also considerably affect the results. For example, Nowak et al. (2016) indicated that some methods (e.g., crude fiber measurement) used for the analysis of quinoa grains were sometimes not updated. In some cases, results were expressed on a different basis (e.g., dry vs wet) (Gómez et al., 2021; Vera et al., 2019). As a result of the differences in the methodology, direct comparisons of the results from studies using different analytical methods are hardly meaningful. There is a strong need to standardize the methods used by different researchers from different countries and laboratories. So that the effort of quinoa research community may be maximized. The comparative patterns in the proximate composition between quinoa and other food grains including cereals (e.g., as shown in Table 2.1) should be viewed critically. This is because of the wide genetic variations in the chemical composition of other food grains. For example, the protein contents of wheat varieties may range from 6% to 20% ("as-is" basis) (Caporaso et al., 2018). The lipid contents of different oat genotypes can vary from 3% to 11% (Campbell et al., 2021). Overall, for nutritional and food applications, both the quantity and the quality of the components in quinoa should be taken into consideration when in comparison with other food grains. The results on the composition of quinoa grains summarized in this chapter were mostly based on raw grains and flours. The composition of grain based food products may be significantly different taken processing effects into consideration (Deng et al., 2015; Mezzatesta et al., 2020; Nowak et al., 2016). For example, heating tends to oxidize unsaturated lipids and degrade vitamins in food grains. The results of this chapter should serve as a basis for the processing and formulations of quinoa-based food products.

2.5 Proteins and amino acids of quinoa

The proteins in quinoa are distributed through all the tissues in the grain and have the highest concentration in the embryo (Craine & Murphy, 2020; Dakhili et al., 2019). The protein content of quinoa grains can be very variable in different reports

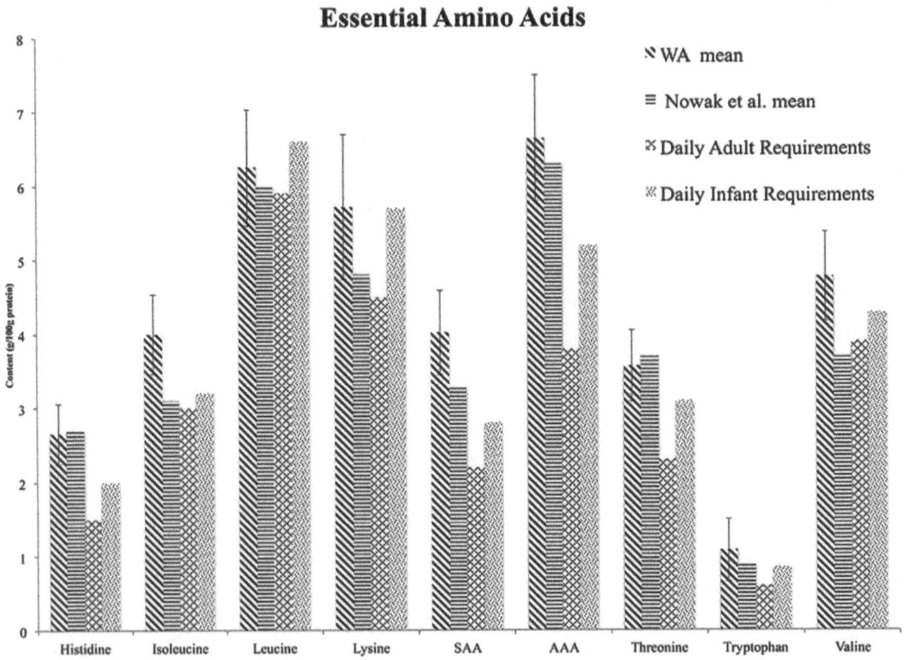

Figure 2.3 Comparison of essential amino acid mean values from the Washington grown samples (WA Mean) reported with error bars representing one standard deviation, mean values reported in the review by Nowak et al. (2016), and daily adult and infant essential amino acid requirements established by the World Health Organization. Differences among groups are not statistically significant. SAA, sulfur amino acids; AAA, aromatic amino acids (Craine & Murphy, 2020).
Source: Reprinted with permission under Creative Commons Attribution License.

ranging from 8% to 22% (Craine & Murphy, 2020). The amino acid composition of quinoa grains has been studied over many years (Craine & Murphy, 2020; Dakhili et al., 2019; Nowak et al., 2016; Repo-Carrasco et al., 2003) (Fig. 2.3). The nine essential amino acids were found in quinoa grains, though the amounts of some of the amino acids especially leucine may be limited (Fig. 2.3) (Craine & Murphy, 2020). Therefore, it is not suitable to call quinoa a "complete" source of protein for human nutrition. Quinoa protein was high in the lysine content, which is the limiting amino acid in cereal grains (Dakhili et al., 2019) (Table 2.2). It had more histidine than soy, barley, and wheat proteins. The tryptophan content of quinoa was comparable to that of cereals (Dakhili et al., 2019).

Nowak et al. (2016) analyzed the available data on the amino acid composition from the literature published up to the year 2014. They concluded that quinoa grain can provide enough amounts of essential amino acids to meet nutritional requirements for adults on a daily basis. Craine and Murphy (2020) systematically analyzed 100 quinoa genotypes for their amino acid composition. They concluded that

Table 2.2 Amino acid compositions of proteins of quinoa in comparison with wheat, rice, casein, and values recommended by FAO/WHO (Dakhili et al., 2019).

Amino acids (g/100 g protein)	Quinoa(1)	wheat(2)	Rice(3)	Casein(4)	FAO/ WHO Pattern(5)
Essential amino acids					
Histidine	2.9	2.3	2.1	2.7	1.6
Leucine	5.9	6.8	8.2	8.4	1.9
Isoleucine	3.6	3.6	4.1	4.9	1.3
Lysine	5.4	2.7	3.8	7.1	1.6
Methionine + cysteine	3.6	4.2	3.6	2.64	1.7
Phenylalanine + tyrosine	6.1	7.8	10.5	10.0	1.9
Threonine	3.0	2.9	3.8	3.7	0.9
Valine	4.2	4.4	6.1	6.0	1.8
Tryptophan	1.2	1.3	1.1	1.4	0.5
Non-essential amino acids					
Alanine	4.2	3.6	5.4	2.7	0.26
Glycine	4.9	4.2	4.25	1.6	0.20
Proline	5.5	10.2	4.9		0.61
Serine	4.0	4.6	4.8	4.6	0.53
Glutamic acid	13.2	31.7	17.2	19.0	1.75
Aspartic acid	8.0	5.1	8.6	6.3	0.88
Arginine	7.7	4.7	9.1	3.7	0.46

The superscripts (1), (2), (3), (4), and (5) denote related references as cited in the paper of Dakhili et al. (2019). Proline concentration in casein was 11% according to Hall et al. (2003)
Source: Reprinted with permission from Elsevier.

genotypes of quinoa significantly determined the nutritional quality of amino acids. A total of nine quinoa genotypes contained enough leucine and other essential amino acids to meet the nutritional requirements for all age groups including adults and infants. Not all the 100 genotypes analyzed in the study met the nutritional requirements for essential amino acids. In general, leucine was a limiting essential amino acid for quinoa grain (Craine & Murphy, 2020). Thus the breeding program should be focused on the improvement of the leucine content of quinoa seeds.

There are discrepancies in the amino acid composition reported from different studies (Dakhili et al., 2019; Nowak et al., 2016). Such discrepancies may be due to the different methods used for the analysis. This made the results from different studies difficult for direct comparison. Many studies only employed limited numbers of quinoa genotypes for the analysis. Such approaches also tend to lead to apparent differences in the results reported from different studies.

The protein quality of the quinoa grain has been analyzed (Lorusso et al., 2017; Nasir et al., 2015; Tavano et al., 2022). Variations in biological utilization and digestibility of quinoa protein were found among different genotypes (Nasir et al., 2015). Quinoa protein digestibility can be similar to that of casein (Tavano et al., 2022). Fermentation was found to increase the digestibility and nutritional indices (essential amino acid index, biological value, protein efficiency ratio, and nutritional index) of quinoa protein in the flour form (Lorusso et al., 2017).

2.6 Lipids of quinoa

Most of the lipids are found in the quinoa embryo (Prego et al., 1998). There is a significant variation in lipid composition of different quinoa genotypes (De Bock et al., 2021; Encina-Zelada et al., 2017; Filho et al., 2017; Vera et al., 2019). For example, De Bock et al. (2021) reported that the lipid contents of seven quinoa genotypes ranged from 2.7%–7.4%. The majority of the fatty acids was unsaturated with the most abundant one being linoleic acid (C18:2) (50%), which was followed by oleic acid (C18:1 c9) (24%) (Table 2.3). Sampaio et al. (2020) showed that a new quinoa variety developed for tropical environment had lipids consisting of 86% of unsaturated fatty acids. A total of 60% of these unsaturated fatty acids were polyunsaturated. The main saturated fatty acid in quinoa grain was palmitic acid (C16:0) (10%) (De Bock et al., 2021). Comparison in lipid composition was made between quinoa and other food grains including cereals, pseudocereals, and soy beans (Abugoch James, 2009; De Bock et al., 2021; Nowak et al., 2016). Compared to maize lipids, quinoa lipids had more linolenic acid and less linoleic acid with a higher ratio of unsaturated to saturated fatty acid. The composition of other fatty acids was similar between them (Filho et al., 2017). Compared to soy beans, quinoa had 10 times less saturated fatty acids. Quinoa had much less saturated fatty acids compared to rice (Nowak et al., 2016). De Bock et al. (2021) compared the fatty acid composition of quinoa with that of amaranth and buckwheat (Table 2.3). Quinoa and amaranth grains had similar amounts of linoleic acid and oleic acid and had more linoleic acid than buckwheat. Quinoa grain had less palmitic acid than

Table 2.3 Composition of fatty acids of quinoa in comparison with amaranth and buckwheat, *n* denotes the number of samples analyzed (De Bock et al., 2021).

Fatty acid (%)	Quinoa ($n = 7$)	Amaranth ($n = 8$)	Buckwheat ($n = 10$)
C14:0	0.15 − 0.22	0.18 − 0.23	0.12 − 0.15
C14:1	0.10 − 0.48	0.12 − 0.25	0.19 − 0.41
C16:0	9.3 − 10.7	18.8 − 19.4	13.4 − 14.9
C16:1 c	0.06 − 0.11	0.09 − 0.10	0.14 − 0.18
C18:0	0.55 − 0.76	3.16 − 3.67	1.71 − 1.98
C18:1 c9	20.2 − 27.0	22.4 − 25.4	32.7 − 35.2
C18:2 n-6	46.4 − 54.2	44.3 − 47.0	34.9 − 39.9
C18:3 n-3	3.71 − 8.10	0.91 − 1.06	2.20 − 2.26
C20:1	1.45 − 1.72	0.26 − 0.41	2.73 − 2.96
C22:1	0.00 − 0.05	1.21 − 1.96	0.16 − 0.21

Source: Reprinted with permission under Creative Commons Attribution License.

both amaranth and buckwheat grains (De Bock et al., 2021). Unfortunately, in this study, the specific species of both the amaranth and buckwheat were not given. Lipid-related components and properties are systematically described in Chapter 7.

2.7 Carbohydrates of quinoa

Carbohydrates of quinoa grain included sugars, oligosaccharides, starch and nonstarch polysaccharides (Abugoch James, 2009; De Bock et al., 2021; Filho et al., 2017; Gómez et al., 2021; Nowak et al., 2016). Starch is the major component of quinoa seeds. It is usually found in the perisperm of the quinoa seed (Fig. 2.2). It makes up ~50 − 70% of the dry matter of quinoa grain, depending on the genotypes (De Bock et al., 2021; Li & Zhu, 2018). Starch is composed of two types of molecules including linear amylose and branched amylopectin (Pérez & Bertoft, 2010). The amylose contents of quinoa ranged from 4% to 28%, depending on the genotypes and analytical methods (Li & Zhu, 2018). Resistant starch is a component of dietary fiber. The content of resistant starch in foods is significantly affected by processing. Lakshmi et al. (2021) showed that the resistant starch contents of flours of two quinoa varieties were similar (~0.16%). The values were significantly lower than that of cereal flours including rice, sorghum, and maize (Lakshmi et al., 2021).

Nonstarch polysaccharides in quinoa grains are major components of dietary fiber (Zhu, 2020). They consist of pectins, xyloglucans, and cellulose. Lignin (~1.7%) was also found in quinoa dietary fiber fraction, which was much lower in percentage than buckwheat (20%), mung bean (34%), pea (36%), and highland barley (15%) (Zhang et al., 2020). Dietary fiber contents of quinoa grains significantly varied among different genotypes (De Bock et al., 2021; Gómez et al., 2021; Nowak et al., 2016). For example, De Bock et al. (2021) showed that the range of dietary fiber contents of seven quinoa varieties was 7.2% to 15.3% (dry basis). Gómez et al. (2021) showed that the dietary fiber contents of six quinoa varieties varied from 12.6% to 22.7% (fresh weight). Quinoa grains have much more insoluble polysaccharides than soluble ones (Zhu, 2020). For example, Sobota et al. (2020) found that, for 25 different yellow-coated quinoa varieties, the ranges of soluble, insoluble, and total fiber contents were 3.7%–5.9%, 10%–14%, and 14%–20%, respectively. Nowak et al. (2016) surveyed the literature on quinoa dietary fiber contents and found that the results from different reports were difficult to compare. This is largely due to the different methods used by different researchers.

Simple sugars accounted for ~1% to 3% of fresh weight of quinoa grain, depending on the genotype (Gómez et al., 2021; Pellegrini et al., 2018; Pereira et al., 2019). The simple sugars of quinoa seeds included sucrose, glucose, arabinose, and fructose (Pereira et al., 2019). Variations in simple sugar composition were found in different quinoa genotypes. For example, in six quinoa varieties grown in Southern Europe, the concentration ranges of sucrose, glucose, and fructose were 1.29% − 1.81%, 0.31% − 0.99%, and 0.55% − 0.86% (fresh weight), respectively (Gómez et al., 2021).

2.8 Minerals, organic acids, vitamins, and carotenoids of quinoa

2.8.1 *Minerals*

Total ash contents of quinoa grains varied greatly among different genotypes (De Bock et al., 2021; Encina-Zelada et al., 2017; Gómez et al., 2021; Nowak et al., 2016; Vera et al., 2019). For example, Gómez et al. (2021) showed that the ash contents of six quinoa varieties grown in southern Europe varied from 3.0% to 3.8%. Encina-Zelada et al. (2017) analyzed 77 quinoa varieties and found that their ash contents ranged from 2.5% to 4.1%. Pellegrini et al. (2018) showed that ash contents of flours of six quinoa genotypes ranged from 1.7% to 2.6%. Nowak et al. (2016) surveyed the literature published up to the year of 2014 and showed that ash contents of quinoa grains ranged from 2.0% to 7.7%. Comparative analysis showed that ash contents of quinoa grains were similar to those of amaranth and were higher than those of buckwheat, maize, wheat, and rice grains (De Bock et al., 2021; Nowak et al., 2016).

The mineral composition of quinoa grains was analyzed (Abugoch James, 2009; De Bock et al., 2021; Gómez et al., 2021; Nowak et al., 2016; Prado et al., 2014). Great variations in the mineral composition of different quinoa genotypes were obtained. For example, De Bock et al. (2021) showed that, for seven quinoa varieties, the ranges of the concentrations of K, Mg, P, Ca, Fe, and Na were 6181–12453, 1222–2824, 2513–5738, 347–1041, 33.7–129.9, and 73.1–204.5 mg/kg (dry basis), respectively. Nowak et al. (2016) analyzed the data collected from the literature published up to the year of 2014 and found that wide ranges of concentrations of different mineral elements (Table 2.4). For example, the ranges of the concentrations of K, Mg, P, Ca, Fe, and Na were found to be

Table 2.4 Mineral composition (average and range) of raw quinoa seeds in comparison with wheat, corn and rice (mg/100 g dry matter) (Nowak et al., 2016).

Samples	Data source	Ca	Fe	Mg	P	K	Na	Zn	Cu
Quinoa, raw	Surveyed data[a]	87 28–149 ($n = 10$)	9.47 2.6–15 ($n = 10$)	362 207–502 ($n = 8$)	406 350–482 ($n = 9$)	907 656–1475 ($n = 7$)	20 11–31 ($n = 4$)	2.15 0.79–4 ($n = 9$)	7.84 5.7–9.5 ($n = 5$)
	Bolivia[b]	113 (102–119)	5.02 (3.1–6.7)	NA	251 (239–275)	NA	NA	NA	NA
	Peru, quinua[c]	63	8.47	NA	273	NA	NA	3.73	NA
	USDA, "quinoa, uncooked"[d]	54	5.27	227	527	649	6	3.57	0.68
Wheat, hard white	USDA[d]	35	5.04	103	393	478	2	3.68	0.40
Corn, yellow	USDA[d]	8	3.02	142	234	320	39	2.47	0.35
Rice, white, polished, raw	USDA[d]	22	1.36	NA	119	80	31	0.57	0.11

n: number of data points; NA: not available.
[a]Average of raw quinoa from the following publications: Alvarez-Jubete et al. (2009), Bruin (1964), Chauhan et al. (1992), Dini et al. (2005), Konishi et al. (2004), Ranhotra et al. (1993), White et al. (1955).
[b]Calculated as mean of six varieties of the Bolivian food composition table (Instituto Nacional de Laboratorios de Salud (INLASA), 2005).
[c]Adapted from the food entry "quinua" of the Peruvian food composition table (Ministerio de Salud (Perú) & Instituto Nacional de Salud, 2009).
[d]Adapted from the USDA nutrient database (US Department of Agriculture Agricultural Research Service, 2013).
(Please refer to the original publication for the references).
Source: Reprinted with permission from Elsevier.

660–1480, 210–500, 350–480, 28–150, 2.6–15, and 11–31 mg/100 g (dry basis), respectively (Nowak et al., 2016). Other mineral elements including Zn, Mn, and Zn were also found in quinoa in trace amounts (De Bock et al., 2021; Nowak et al., 2016). The mineral composition of quinoa seeds can be affected by growing locations and agronomic practice (Prado et al., 2014; Walters et al., 2016). For example, Tovar et al. (2022) found that relatively short periods of heat spells during anthesis changed the contents of many elements in quinoa seeds.

Comparative analysis showed that quinoa grain tended to have more K, Ca, and Mg than some cereal grains including wheat, maize, and rice (Nowak et al., 2016). Compared to other pseudocereals including amaranth and buckwheat grains, quinoa had more K and similar amounts of Fe. Quinoa grain had less Ca and Mg than amaranth (De Bock et al., 2021). It was noted that quinoa grain contained a significant amount of phosphorus (P) (De Bock et al., 2021; Gómez et al., 2021). P in plants can occur in the form of phytic acid as an antinutrient. Adding phytase can efficiently increase the mineral bioavailability in quinoa and other grain-based products (Iglesias-Puig et al., 2015). Castro-Alba, Lazarte, Perez-Rea, Carlsson et al. (2019) showed that fermentation of quinoa flour can increase the mineral in vitro bioaccessibility. The spontaneous fermentation or fermentation with added *Lactobacillus plantarum* affected the pH of the flour. The changed pH facilitated the endogenous phytase activity, degrading phytate in the quinoa flour and increasing the mineral bio-accessibility (Castro-Alba, Lazarte, Perez-Rea, Carlsson et al. (2019) and Castro-Alba, Lazarte, Perez-Rea, Sandberg et al., 2019). Similarly, Darwish et al. (2021) found that germination improved the iron bioavailability of quinoa seed sprouts with antianemic potential. P, Mg, K, and S were mostly found in the embryo of quinoa seeds, whereas Ca and K were concentrated in the pericarp (Konishi et al., 2004). Soil type and fertilizer application for quinoa growing can significantly affect the mineral composition.

Toxic species of As were found in the leaves of quinoa grown in a contaminated (e.g., copper mining) region of Chile (Pizarro et al., 2016). A total of 85% of As was present as As(V) and the rest were mostly As(III). In vitro simulated gastrointestinal digestion had little effect on the composition of the inorganic species of As. However, the dietary intake of the As species was significantly less than that of the safety guideline. No risk of eating the quinoa leaves containing As was expected (Pizarro et al., 2016). The presence of toxic species in quinoa seeds remains to be studied.

2.8.2 Phytic acid and organic acids

Phytic acid in food grains can affect mineral element absorption. Phytic acid content of quinoa is comparable to that of cereals. Duliński et al. (2017) analyzed the composition of myo-inositol phosphates of red, black, and white quinoa seeds. There was variation in the composition of myo-inositol phosphates among these quinoa genotypes. The total phytate contents of the three types of seeds in native form ranged from 13.8 to 16.5 μmol/g (dry matter). The ranges of the percentages of InsP6, InsP5, InsP4, InsP3, and InsP2–1 (s stands for sodium) were 65% – 69%, 2.4% – 4%,

0.8% – 1.1%, 2.2% – 3.7%, and 24% – 29%, respectively. The total contents of myo-inositol phosphates ranged from 29.4 to 32.2 μmol/g (dry matter) (Duliński et al., 2017). An effective way to reduce phytic acid content is to germinate quinoa, which allows the hydrolysis of phytic acid thus increasing mineral bioavailability (Ando et al., 2002; Duliński et al., 2017). Wang et al. (2020) showed that peeling significantly decreased the phytic acid content of quinoa flour. Duliński et al. (2017) revealed that fermentation decreased the phytate content from 14 to 5 μmol/g dry matter in red quinoa seeds. Cooking also reduced the phytate content (e.g., from 14 to 9 μmol/g dry matter). Despite the antinutritional properties of phytate, positive nutritional properties of inositol and inositol phosphates such as anticancer property have been reported (Weinberg et al., 2021). For example, lower inositol phosphates have special anticancer properties. Therefore a holistic view of the role of these antinutrient compounds should be adapted. Other antinutrients in quinoa include saponins, nitrates, trypsin inhibitors, and oxalates. They are mostly found in the outer layers of quinoa grain (Ahmed et al., 2021; Filho et al., 2017; Satheesh & Fanta, 2018). These antinutrients of quinoa are described in Chapter 8 of this book.

Sampaio et al. (2020) found four types of organic acids in quinoa seeds. They included oxalic acid (0.60 g/100 g fresh weight (fw)), quinic acid (3.6 g/100 g fw), malic acid (0.41 g/100 g fw), and fumaric acid (trace amount). Information on the composition of organic acids in quinoa and their functions is scarce.

2.8.3 Vitamins

Quinoa contains different types of vitamins (Granda et al., 2018; Motta et al., 2017; Nowak et al., 2016; Pereira et al., 2019; Sampaio et al., 2020; Vera et al., 2019). Nowak et al. (2016) showed that, in raw quinoa grain, there were thiamin, niacin, and riboflavin, vitamin E, and vitamin B6. They indicated that there was a lack of high-quality data for direct comparison (Nowak et al., 2016). Motta et al. (2017) analyzed the folate vitamers in quinoa. Raw quinoa seeds had higher total folate content (310 μg/100 g dry weight (dw)) than buckwheat (150 μg/100 g dw) and amaranth (230 μg/100 g dw) seeds. 5-Methyltetrahydrofolate (260 μg/100 g dw) was dominant followed by 10-formyltetrahydrofolate (53 μg/100 g dw) and folic acid (9 μg/100 g dw). Malting [steeped at 30°C for 48 h before germination (23°C, 48 h)], boiling (15 min, 100°C), and steaming (30 min) had little effect on the total folate content (Motta et al., 2017). Granda et al. (2018) systematically analyzed the composition of vitamins in nine quinoa varieties. Overall, quinoa grains had much more tocopherols (α-tocopherol and γ-tocopherol) and less riboflavin (0.24 – 0.46 mg/kg, dry basis) than barley (riboflavin: 0.84 – 1 mg/kg, dry basis) and wheat (riboflavin: 0.67 – 0.82 mg/kg, dry basis) grains. The ranges of concentrations of γ-tocopherol, α-tocopherol, and δ-tocopherol in the nine quinoa varieties were 25 – 87, 36 – 83, and 0.62 – 4.59 mg/kg (dry basis), respectively. β-Tocopherol was not detected in any of the quinoa genotypes. α-Tocotrienol and β-tocotrienol were not detected in any quinoa samples, but they were found in both wheat and barley samples. Overall, the activity of vitamin E of quinoa grains was much higher than that of wheat and barley. The composition of pyridoxine and thiamine in all grains were largely dependent on the genotype. For example, the ranges of concentrations of thiamine and pyridoxine of the nine quinoa varieties were 0.38 – 5.09 and 0.40 – 1.35 mg/kg (dry basis), respectively (Granda et al., 2018). Lakshmi et al. (2021) found that niacin concentrations in two quinoa varieties (230 and 330 μg/100 g) were much lower than those of maize, rice, and sorghum. The concentrations of pantothenic acid (980 and 1360 μg/100 g) and folates (98 and 147 μg/100 g) of the quinoa varieties were much higher than those of maize, rice, and sorghum (Lakshmi et al., 2021). A significant amount of vitamin C was found in quinoa seed sprouts, leaves, and microgreens (Darwish et al., 2021; Laus et al., 2017; Le et al., 2021; Pathan & Siddiqui, 2022). Le et al. (2021) reported that the vitamin C contents of 10 quinoa sprouts ranged from 36.6 to 69.5 mg/100 g FW.

2.8.4 Carotenoids

Different types of carotenoids were found in different varieties of quinoa grains (Lakshmi et al., 2021; Tang et al., 2015). Tang et al. (2015) systematically analyzed carotenoid composition in three quinoa grains with white, red, and black coatings. They found 15-*cis*-lutein, 13-*cis*-lutein, 13′-*cis*-lutein, 13-*cis*-zeaxanthin, 13′-*cis*-zeaxanthin, *trans*-lutein, *trans*-zeaxanthin, 9-*cis*-lutein, 90-*cis*-lutein, and 9-*cis*-zeaxanthin in all three quinoa grains. Quinoa grains had much more *trans*-carotenoids (8.0–11.9 μg/g) than *cis*-carotenoids (1.0–1.4 μg/g). The grains had much more lutein (8.3 – 12.4 μg/g) than zeaxanthin (~0.7 μg/g) (Tang et al., 2015). Multari et al. (2018) found zeaxanthin (0.22 mg/kg), lutein (1.47 mg/kg), lutein isomer A (0.09 mg/kg), lutein isomer B (0.21 mg/kg), neochrome A (0.24 mg/kg), and neochrome B (0.11 mg/kg) in quinoa seeds grown in Finland. In a comparison study, Lakshmi et al. (2021) showed that red and black quinoa had similar levels of carotenoids compared to red rice, red sorghum, and white maize, but had much less carotenoids than orange maize.

2.9 Conclusions

Quinoa grain is small-seeded. It has a large perisperm. An embryo layer surrounds the perisperm. A thin layer of endosperm surrounds the embryo. Starch granules are mostly located in the perisperm. Proteins in the form of globoids are mostly found in the embryo cells. Chemical analysis revealed significant variations in the composition of diverse nutrients. The variations in the composition are largely due to the genetic variations and as well as environmental effects. Nutritionally, quinoa grains compare favorably with common cereals in terms of human nutritional requirement. The quinoa protein is gluten free and is more balanced than cereal proteins in terms of the composition of essential amino acids. The quinoa lipids are more unsaturated than cereal lipids. Quinoa is a source of dietary fiber, minerals and vitamins, and carotenoids. Future studies should be focused on breeding new quinoa genotypes with wider ranges of concentrations of specific components.

References

Abderrahim, F., Huanatico, E., Segura, R., Arribas, S., Gonzalez, M. C., & Condezo-Hoyos, L. (2015). Physical features, phenolic compounds, betalains and total antioxidant capacity of coloured quinoa seeds (*Chenopodium quinoa* Willd.) from Peruvian Altiplano. *Food Chemistry, 183*, 83–90.

Abugoch James, L. E. (2009). Quinoa (*Chenopodium quinoa* Willd.): Composition, chemistry, nutritional, and functional properties. *Advances in Food and Nutrition Research, 58*, 1–31.

Ahmed, I. A. M., Juhaimi, F. A., & Özcan, M. M. (2021). Insights into the nutritional value and bioactive properties of quinoa (*Chenopodium quinoa*): past, present and future prospective. *International Journal of Food Science and Technology, 56*, 3726–3741.

Alvarez-Jubete, L., Arendt, E. K., & Gallagher, E. (2009). Nutritive value and chemical composition of pseudocereals as gluten-free ingredients. *International Journal of Food Sciences and Nutrition, 60*, 240–257.

Ando, H. ., Chen, Y., Tang, H., Shimizu, M., Watanabe, K., & Mitsunaga, T. (2002). Food components in fractions of quinoa seed. *Food Science and Technology Research, 8*, 80–84.

Bruin, A. D. (1964). Investigation of the food value of quinua and canihua seed. *Journal of Food Science, 29*, 872–876.

Campbell, M. T., Hu, H., Yeats, T. H., Caffe-Trem, M., Gutiérrez, L., Smith, K. P., et al. (2021). Translating insights from the seed metabolome into improved prediction for lipid-composition traits in oat (*Avena sativa* L.). *Genetics, 217*, iyaa043.

Caporaso, N., Whitworth, M. B., & Fisk, I. D. (2018). Protein content prediction in single wheat kernels using hyperspectral imaging. *Food Chemistry, 240*, 32–42.

Castro-Alba, V., Lazarte, C. E., Perez-Rea, D., Carlsson, N. G., Almgren, A., Bergenståhl, B., & Granfeldt, Y. (2019). Fermentation of pseudocereals quinoa, canihua, and amaranth to improve mineral accessibility through degradation of phytate. *Journal of the Science of Food and Agriculture, 99*, 5239–5248.

Castro-Alba, V., Lazarte, C. E., Perez-Rea, D., Sandberg, A. S., Carlsson, N. G., et al. (2019). Effect of fermentation and dry roasting on the nutritional quality and sensory attributes of quinoa. *Food Science & Nutrition, 7*, 3902–3911.

Chauhan, G. S., Eskin, N. A. M., & Tkachuk, R. (1992). Nutrients and antinutrients in quinoa seed. *Cereal Chemistry, 69*, 85–88.

Cole, D. L., Woolley, R. K., Tyler, A., Buck, R. L., & Hopkins, B. G. (2020). Mineral nutrient deficiencies in quinoa grown in hydroponics with single nutrient salt/acid/chelate sources. *Journal of Plant Nutrition, 43*, 1661–1673.

Craine, E. B., & Murphy, K. M. (2020). Seed composition and amino acid profiles for quinoa grown in Washington state. *Frontiers in Nutrition, 7*, 126.

Dakhili, S., Abdolalizadeh, L., Hosseini, S. M., Shojaee-Aliabadi, S., & Mirmoghtadaie, L. (2019). Quinoa protein: Composition, structure and functional properties. *Food Chemistry, 299*, 125161.

Darwish, A. M. G., Al- Jumayi, H. A. O., & Elhendy, H. A. (2021). Effect of germination on the nutritional profile of quinoa (*Cheopodium quinoa* Willd.) seeds and its anti-anemic potential in Sprague–Dawley male albino rats. *Cereal Chemistry, 98*, 315–327.

De Bock, P., Daelemans, L., Selis, L., Raes, K., Vermeir, P., Eeckhout, M., & Van Bockstaele, F. (2021). Comparison of the chemical and technological characteristics of wholemeal flours obtained from amaranth (*Amaranthus* sp.), quinoa (*Chenopodium quinoa*) and buckwheat (*Fagopyrum* sp.) seeds. *Foods, 10*, 651.

De Bock, P., Cnops, G., Muylle, H., Quataert, P., Eeckhout, M., & Van Bockstaele, F. (2022). Physicochemical characterization of thirteen quinoa (*Chenopodium quinoa* Willd.) varieties grown in North-West Europe—Part II. *Plants*, *11*, 265.

Deng, Y., Padilla-Zakour, O., Zhao, Y., & Tao, S. (2015). Influences of high hydrostatic pressure, microwave heating, and boiling on chemical compositions, antinutritional factors, fatty acids, *in vitro* protein digestibility, and microstructure of buckwheat. *Food and Bioprocess Technology*, *8*, 2235–2245.

Duliński, R., Starzyńska-Janiszewska, A., Byczyński, Ł., & Błaszczyk, U. (2017). Myo-inositol phosphates profile of buckwheatand quinoa seeds: Effects of hydrothermal processing and solid-state fermentation with *Rhizopus oligosporus*. *International Journal of Food Properties*, *20*, 2088–2095.

Encina-Zelada, C., Cadavez, V., Pereda, J., Gómez-Pando, L., & Salvá-Ruíz, B. (2017). Estimation of composition of quinoa (*Chenopodium quinoa* Willd.) grains by near-infrared transmission spectroscopy. *LWT-Food Science and Technology*, *79*, 126–134.

Filho, A. M. M., Pirozi, M. R., Borges, J. T. D. S., Sant'Ana, H. M. P., Chaves, J. B. P., & Coimbra, J. S. D. R. (2017). Quinoa: Nutritional, functional, and antinutritional aspects. *Critical Reviews in Food Science and Nutrition*, *57*, 1618–1630.

Ghumman, A., Mudgal, S., Singh, N., Ranjan, B., Kaur, A., & Rana, J. C. (2021). Physicochemical, functional and structural characteristics of grains, flour and protein isolates of Indian quinoa lines. *Food Research International*, *140*, 109982.

Gómez, M. J. R., Prieto, J. M., Sobrado, V. C., & Magro, P. C. (2021). Nutritional characterization of six quinoa (*Chenopodium quinoa* Willd) varieties cultivated in Southern Europe. *Journal of Food Composition and Analysis*, *99*, 103876.

Graf, B. L., Rojo, L. E., Delatorre-Herrera, J., Poulev, A., Calfio, C., & Raskin, I. (2016). Phytoecdysteroids and flavonoid glycosides among Chilean and commercial sources of *Chenopodium quinoa*: Variation and correlation to physico-chemical characteristics. *Journal of the Science of Food and Agriculture*, *96*, 633–643.

Granda, L., Rosero, A., Benešová, K., Pluháčková, H., Neuwirthová, J., & Cerkal, R. (2018). Content of selected vitamins and antioxidants in colored and nonpigmented varieties of quinoa, barley, and wheat grains. *Journal of Food Science*, *83*, 2439–2447.

Hall, W. L., Millward, D. J., Long, S. J., & Morgan, L. M. (2003). Casein and whey exert different effects on plasma amino acid profiles, gastrointestinal hormone secretion and appetite. *British Journal of Nutrition*, *89*, 239–248.

Iglesias-Puig, E., Monedero, V., & Haros, M. (2015). Bread with whole quinoa flour and bifidobacterial phytases increases dietary mineral intake and bioavailability. *LWT-Food Science and Technology*, *60*, 71–77.

Instituto Nacional de Laboratorios de Salud (INLASA). (2005). *Tabla Boliviana de composición de alimentos (Bolivian food composition table)*. La Paz: Agencia Española de Cooperación Internacional (AECI).

Konishi, Y., Hirano, S., Tsuboi, H., & Wada, M. (2004). Distribution of minerals in quinoa (*Chenopodium quinoa* Willd.) seeds. *Bioscience, Biotechnology, and Biochemistry*, *68*, 231–234.

Koziol, M. J. (1993). Quinoa: A potential new oil crop. In J. Janick, & J. E. Simon (Eds.), *New crops* (pp. 328–336). New York: Wiley.

Lakshmi, S., Goudar, G., Singh, M., Dhaliwal, H. S., Sharma, P., & Longvah, T. (2021). Variability in resistant starch, vitamins, carotenoids, phytochemicals and in-vitro antioxidant properties among diverse pigmented grains. *Journal of Food Measurement and Characterization*, *15*, 2774–2789.

Laus, M. N., Cataldi, M. P., Robbe, C., D'Ambrosio, T., & Amodio, M. L. (2017). Antioxidant capacity, phenolic and vitamin C contents of quinoa (*Chenopodium quinoa* Willd.) as affected by sprouting and storage conditions. *Italian Journal of Agronomy*, *12*, 816.

Le, L., Gong, X., An, Q., Xiang, D., Zou, L., & Peng, L. (2021). Quinoa sprouts as potential vegetable source: Nutrient composition and functional contents of different quinoa sprout varieties. *Food Chemistry*, *357*, 129752.

Lee, M. J., & Sim, K. H. (2018). Nutritional value and the kaempferol and quercetin contents of quinoa (*Chenopodium quinoa* Willd.) from different regions. *Korean Journal of Food Science and Technology*, *50*, 680–687.

Li, L., Lietz, G., & Seal, C. J. (2021). Phenolic, apparent antioxidant and nutritional composition of quinoa (*Chenopodium quinoa* Willd.) seeds. *International Journal of Food Science & Technology*, *56*, 3245–3254.

Li, G., & Zhu, F. (2018). Quinoa starch: Structure, properties, and applications. *Carbohydrate Polymers*, *181*, 851–861.

Lorusso, A., Verni, M., Montemurro, M., Coda, R., Gobbetti, M., & Rizzello, C. G. (2017). Use of fermented quinoa flour for pasta making and evaluation of the technological and nutritional features. *LWT-Food Science and Technology*, *78*, 215–221.

Martín, M. I. G., Moncada, G. W., Fischer, S., & Escuredo, O. (2014). Chemical characteristics and mineral composition of quinoa by near-infrared spectroscopy. *Journal of the Science of Food and Agriculture*, *94*, 876–881.

Matías, J., Rodríguez, M. J., Cruz, V., Calvo, P., & Reguera, M. (2021). Heat stress lowers yields, alters nutrient uptake and changes seed quality in quinoa grown under Mediterranean field conditions. *Journal of Agronomy and Crop Science*, *207*, 481–491.

Mezzatesta, P., Farah, S., Di Fabio, A., & Emilia, R. (2020). Variation of the nutritional composition of quinoa according to the processing used. *Proceedings*, *53*, 4.

Ministerio de Salud (Perú), & Instituto Nacional de Salud. (2009). *Tablas peruanas de composición de alimentos*. Lima: Ministerio de Salud, Instituto Nacional de Salud.

Motta, C., Delgado, I., Matos, A. S., Gonzales, G. B., Torres, D., Santos, M., Chandra-Hioef, M. V., et al. (2017). Folates in quinoa (*Chenopodium quinoa*), amaranth (*Amaranthus* sp.) and buckwheat (*Fagopyrum esculentum*): Influence of cooking and malting. *Journal of Food Composition and Analysis*, *64*, 181–187.

Multari, S., Marsol-Vall, A., Keskitalo, M., Yang, B., & Suomela, J. P. (2018). Effects of different drying temperatures on the content of phenolic compounds and carotenoids in quinoa seeds (*Chenopodium quinoa*) from Finland. *Journal of Food Composition and Analysis*, *72*, 75–82.

Nasir, M. A., Pasha, I., Butt, M. S., & Nawaz, H. (2015). Biochemical characterization of quinoa with special reference to its protein quality. *Pakistan Journal of Agricultural Sciences*, *52*, 731–737.

Ninfali, P., Panato, A., Bortolotti, F., Valentini, L., & Gobbi, P. (2020). Morphological analysis of the seeds of three pseudocereals by using light microscopy and ESEM-EDS. *European Journal of Histochemistry*, *64*, 3075.

Nowak, V., Du, J., & Charrondière, U. R. (2016). Assessment of the nutritional composition of quinoa (*Chenopodium quinoa* Willd.). *Food Chemistry*, *193*, 47–54.

Pathan, S., & Siddiqui, R. A. (2022). Nutritional composition and bioactive components in quinoa (*Chenopodium quinoa* Willd.) greens: A review. *Nutrients*, *14*, 558.

Pellegrini, M., Lucas-Gonzales, R., Ricci, A., Fontecha, J., Fernández-López, J., Pérez-Álvarez, J. A., & Viuda-Martos, M. (2018). Chemical, fatty acid, polyphenolic profile, techno-functional and antioxidant properties of flours obtained from quinoa (*Chenopodium quinoa* Willd) seeds. *Industrial Crops and Products*, *111*, 38–46.

Pereira, E., Encina-Zelada, C., Barros, L., Gonzales-Barron, U., Cadavez, V., & Ferreira, I. C. F. R. (2019). Chemical and nutritional characterization of *Chenopodium quinoa* Willd (quinoa) grains: a good alternative to nutritious food. *Food Chemistry*, *280*, 110–114.

Pérez, S., & Bertoft, E. (2010). The molecular structures of starch components and their contribution to the architecture of starch granules: A comprehensive review. *Starch/Stärke*, *62*, 389–420.

Pizarro, I., Gómez-Gómez, M., León, J., Román, D., & Palacios, A. M. (2016). Bioaccessibility and arsenic speciation in carrots, beets and quinoa from a contaminated area of Chile. *Science of the Total Environment*, *565*, 557–563.

Prado, F. E., Fernández-Turiel, J. L., Tsarouchi, M., Psaras, G. K., & González, J. A. (2014). Variation of seed mineral concentrations in seven quinoa cultivars grown in two agroecological sites. *Cereal Chemistry*, *91*, 453–459.

Präger, A., Munz, S., Nkebiwe, P. M., Mast, B., & Graeff-Hönninger, S. (2018). Yield and quality characteristics of different quinoa (*Chenopodium quinoa* Willd.) cultivars grown under field conditions in Southwestern Germany. *Agronomy*, *8*, 197.

Prego, I., Maldonado, S., & Otegui, M. (1998). Seed structure and localization of reserves in *Chenopodium quinoa*. *Annals of Botany*, *82*, 481–488.

Ranhotra, G., Gelroth, J., Glaser, B., Lorenz, K., & Johnson, D. (1993). Composition and protein nutritional quality of quinoa. *Cereal Chemistry*, *70*, 303–305.

Reguera, M., Conesa, C. M., Gil-Gómez, A., Haros, C. M., Pérez-Casas, M. Á., et al. (2018). The impact of different agroecological conditions on the nutritional composition of quinoa seeds. *Peer Journal*, *6*, e4442.

Ren, Y., Guo, K., Zhang, B., & Wei, C. (2020). Comparison of physicochemical properties of very small granule starches from endosperms of dicotyledon plants. *International Journal of Biological Macromolecules*, *154*, 818–825.

Repo-Carrasco, R., Espinoza, C., & Jacobsen, S. E. (2003). Nutritional value and use of the Andean crops quinoa (*Chenopodium quinoa*) and kañiwa (*Chenopodium pallidicaule*). *Food Reviews International*, *19*, 179–189.

Sampaio, S. L., Fernandes, Â., Pereira, C., Calhelha, R. C., Sokovic, M., Santos-Buelga, C., Barros, L., & Ferreira, I. C. F. R. (2020). Nutritional value, physicochemical characterization and bioactive properties of the Brazilian quinoa BRS Piabiru. *Food & Function*, *11*, 2969.

Satheesh, N., & Fanta, S. W. (2018). Review on structural, nutritional and anti-nutritional composition of teff (*Eragrostis tef*) in comparison with quinoa (*Chenopodium quinoa* Willd.). *Cogent Food & Agriculture*, *4*, 1546942.

Sobota, A., Świeca, M., Gęsiński, K., Wirkijowska, A., & Bochnak, J. (2020). Yellow-coated quinoa (*Chenopodium quinoa* Willd) – physicochemical, nutritional, and antioxidant properties. *Journal of the Science of Food and Agriculture*, *100*, 2035–2042.

Tang, Y., Li, X., Chen, P. X., Zhang, B., Hernandez, M., et al. (2015). Characterisation of fatty acid, carotenoid, tocopherol/tocotrienol compositions and antioxidant activities in seeds of three *Chenopodium quinoa* Willd. genotypes. *Food Chemistry*, *174*, 502–508.

Tavano, O. L., de Miguel Amistá., Ciello, G. D., Rodrigues, M. C. M., Nishida, A. M. B., Valadares, L. A., Siqueira, B. M., da Silva Gomes., Parolini, M. T., & da Silva Junior, S. I. (2022). Isolation and evaluation of quinoa (*Chenopodium quinoa* Willd.) protein fractions. A nutritional and bio-functional approach to the globulin fraction. *Current Research in Food Science*, *5*, 1028–1037.

Taylor, J. R. N., & Parker, M. L. (2002). Quinoa. In P. S. Belton, & J. R. N. Taylor (Eds.), *Pseudocereals and less common cereals* (pp. 93–122). Berlin: Springer.

Tovar, J. C., Berry, J. C., Quillatupa, C., Castillo, S. E., Acosta-Gamboa, L., Fahlgren, N., & Gehan, M. A. (2022). Heat stress changes mineral nutrient concentrations in *Chenopodium quinoa* seed. *Plant Direct*, *6*, e384.

U.S. Department of Agriculture, Agricultural Research Service. (2013). *USDA national nutrient database for standard reference*, release 26. Nutrient Data Laboratory, Beltsville Human Nutrition Research Center, Agriculture Research Service, United States Department of Agriculture.

Valencia-Chamorro, S. A. (2003). Quinoa. In L. Trugo, & P. M. Finglas (Eds.), *Encyclopedia of food science and nutrition* (pp. 4895–4902). Academic Press.

Varriano-Marston, E., & DeFrancisco, A. (1984). Ultrastructure of quinoa fruit (*Chenopodium quinoa* Willd). *Food Structure*, *3*, 165–173.

Vera, E. P., Alca, J. J., Saravia, G. R., Campioni, N. C., & Alpuy, I. J. (2019). Comparison of the lipid profile and tocopherol content of four Peruvian quinoa (*Chenopodium quinoa* Willd.) cultivars ('Amarilla de Maranganí', 'Blanca de Juli', INIA 415 'Roja Pasankalla', INIA 420 'Negra Collana') during germination. *Journal of Cereal Science*, *88*, 132–137.

Villacrés, E., Quelal, M., Galarza, S., Iza, D., & Silva, E. (2022). Nutritional value and bioactive compounds of leaves and grains from quinoa (*Chenopodium quinoa* Willd.). *Plants*, *11*, 213.

Walters, H., Carpenter-Boggs, L., Desta, K., Yan, L., Matanguihan, J., & Murphy, K. (2016). Effect of irrigation, intercrop, and cultivar on agronomic and nutritional characteristics of quinoa. *Agroecology and Sustainable Food Systems*, *40*, 783–803.

Wang, Y., Gong, X., Zhang, Y., Geng, D. H., Cao, L., Ruan, C., et al. (2020). Effect of peeling treatment on the physicochemical properties of quinoa flour. *Journal of Food Process Engineering*, *43*, e13387.

Weinberg, S. E., Sun, L. Y., Yang, A. L., Liao, J., & Yang, G. Y. (2021). Overview of inositol and inositol phosphates on chemoprevention of colitis-induced carcinogenesis. *Molecules*, *26*, 31.

White, P. L., Alvistur, E., Dias, C., Vinas, E., White, H. S., & Collazos, C. (1955). Nutritive values of crops, nutrient content and protein quality of quinua and cañihua, edible seed products of the Andes mountains. *Journal of Agricultural and Food Chemistry*, *3*, 531–534.

Zhang, D., Wang, L., Tan, B., & Zhang, W. (2020). Dietary fibre extracted from different types of whole grains and beans: a comparative study. *International Journal of Food Science and Technology*, *55*, 2188–2196.

Zhu, F. (2020). Dietary fiber polysaccharides of amaranth, buckwheat and quinoa grains: A review of chemical structure, biological functions and food uses. *Carbohydrate Polymers*, *248*, 116819.

Quinoa grain milling

3

3.1 Introduction

For application purposes, quinoa seeds may be wet or dry milled for processing (Haros & Wronkowska, 2017). The milling process converts the grains into their intermediate products such as flour, bran, protein, starch, and fiber fractions (Ballester-Sánchez et al., 2019; Ballester-Sánchez et al., 2020). The milling products have many different types of food and other applications. For example, a wet-milling coproduct of black quinoa seeds improved the quality and shelf-life of meat patties (Sayas-Barberá et al., 2021). Starch and protein isolated from quinoa were modified and used in the encapsulation of food ingredients for controlled release applications (Jiang et al., 2022; Zuo et al., 2022). The food applications of the quinoa milling products are described in Chapters 12 and 13.

The structure of quinoa grain is described in detail in Chapter 2. Different parts of quinoa grain have different compositions and concentrations of macro- and micronutrients. Starch granules are mainly found in the perisperm in the center of the grain and the embryo is rich in proteins and lipids (Ruiz et al., 2016). Saponins are mostly found at the outer layer of quinoa seeds. Saponins may be toxic when consumed in large amounts and contribute to a bitter taste to the grain and related foods (El Hazzam et al., 2020). Water alone only partially removes the saponins from the grains (Abugoch James, 2009). Slightly alkaline water may effectively remove saponins from quinoa seeds (El Hazzam et al., 2020). The different nutrients and fractions rich in specific component may be obtained by milling.

The milling process reduces the particle size and separates/enriches components (Thakur et al., 2019). The selected milling method for applications is mainly determined by the type of intermediate products targeted. Flour and bran may be produced by dry milling (roller milling), whereas purified products such as starch, protein, and fiber are normally obtained using wet milling. The steeping time, conditioning parameters, and the types of extraction solvents affect the yield, purity, and composition of the intermediate products as well as their physicochemical properties (Coțovanu et al., 2020; De Bock et al., 2021; Föste et al., 2015; Ghumman et al., 2021; Hemalatha et al., 2016; Mufari et al., 2018; Ray et al., 2021; Solaesa et al., 2020). This chapter describes the dry and wet milling processes and their impacts on the chemical composition of quinoa products. It also presents knowledge on the chemical composition of milled quinoa fractions in relation to their physicochemical properties.

Quinoa. DOI: https://doi.org/10.1016/B978-0-323-99909-0.00006-4

3.2 Quinoa milling processes

Quinoa fractions are obtained by two traditional milling processes: Wet milling and dry milling (Table 3.1). Wet and dry millings give different products with different properties. Selection of the processes used depends on the end uses of the products (Haros & Wronkowska, 2017). Dry and wet milling processes mainly differ by the amount of water, energy consumption, processing chemicals, and amplitude of the force applied during the process (Alonso-Miravalles & O'Mahony, 2018). The dry and wet milling processes and related products are schematically illustrated in Fig. 3.1.

3.2.1 *Dry milling*

The goal of dry milling is to separate different parts of quinoa grain (Fig. 3.1). Bran and flour fractions of different particle sizes are the end products of the dry milling process (Ballester-Sánchez et al., 2020). The process involves conditioning of the grains with small quantities of water. Dry milling involves seed grinding, sifting through sieves of different mesh sizes, separation, and regrinding. Quinoa grain has very small seeds (1.8–2.2 mm in diameter) (Bertero et al., 2004); they are generally milled as whole seeds. Steeping quinoa grains in water (tempering) may be included in the dry milling process (De Bock et al., 2021). The water absorbed by the seeds during tempering toughens the pericarp and germ, preventing their splintering and increasing the bran separation during milling. Tempering may be combined with heat to facilitate the separation of different parts of the grain and the milling process (Haros and Wronkowska, 2017). Dry milling also involves mechanical forces (Alonso-Miravalles & O'Mahony, 2018). The milling steps are repeated until the desirable products are produced. The main products of dry milling are bran and flour fractions. The fractions can be milled further to obtain differently sized products depending on the applied force and settings of roller systems. The components of the particles greatly affect the quality of the products and the particle sizes of the flour (Solaesa et al., 2020). The bran fractions of quinoa grains contain embryo and the seed coat and they are relatively high in the contents of dietary fiber, fat, and proteins (Alonso-Miravalles & O'Mahony, 2018).

The processing conditions such as tempering parameters and miller types can directly affect the yield of flour (De Bock et al., 2021; Ray et al., 2021). The tempering conditions can affect the protein yield of flour obtained by dry milling. Dry milling of unconditioned quinoa in an ultra-centrifugal mill yielded 71% of quinoa white flour and 28% of bran containing 24% of protein (Föste et al., 2015). Milling without tempering resulted in maximum yield of flour, but the resulting bran contained less proteins. If the moisture content was increased to 14% and the resting time was set to 20 h, the flour yield of dry milling was reduced to 64%, while the bran and bran protein yields increased to 35% and 28%, respectively (Föste et al., 2015). In a recent study, De Bock et al. (2021) analyzed the effect of tempering conditions (13% − 17% for moisture and 8 or 20 h for length of tempering) for roller milling of quinoa grains on the yield, particle size, and composition. Increasing tempering moisture increased the separation of bran from perisperm, while decreasing the yield. Increasing tempering length increased the amounts of

Table 3.1 Milling properties of quinoa grain.

Milling type	Miller type	Milling conditions	Targeted components/ fractions	Major findings	Targeted physicochemical property	Major findings	Reference
Dry	Abrasive grain mill	Quinoa grains were dehulled and pearled to obtain milled grain and bran fractions. These fractions were ground into fine powder using a coffee grinder	Phenolic compounds and their biological properties	Milling led to the loss of a third of the phenolics in the grains. Dehulled and milled fractions had higher in vitro antioxidant activities than other fractions			Hemalatha et al. (2016)
Dry	Coffee bean grinder	Quinoa grains were ground and sieved through a series of mesh size (>595 ~ >74 μm)	Chemical composition	Increasing particle size decreased the contents of protein, lipids, fibers, while increasing that of starch. Decreasing particle size	Rheological and water holding property	Increasing particle size decreased sediment volume and water holding capacity. Larger particles had higher complex	Ahmed et al. (2019)

(*Continued*)

Table 3.1 (Continued)

Milling type	Miller type	Milling conditions	Targeted components/fractions	Major findings	Targeted physicochemical property	Major findings	Reference
				increased the phenolics extractability and in vitro antioxidant activity		viscosity and rigidity. Increasing particle size increased setback and final viscosities during pasting	
Dry	Abrasive mill	Quinoa grains were milled for 5 s to obtain coarse grains and also fractions containing embryo and smaller particles. The fraction of coarse grains was used for dehulling. Eleven different	The composition of quinoa chemical components	Increasing dehulling degrees decreased the amounts of protein, dietary fiber, ash, and lipids and increased that of starch in the resulting milling fractions. The dehulling			Wu et al. (2020)

		dehulling degrees were obtained		rates should be no more than 8.6% to ensure that the nutritional components are significantly retained in the resulting products			
Dry	Abrasive milling	Quinoa grains were milled up to 70 min to obtain different fractions	FTIR (Fourier-transform infrared spectroscopy) spectra were obtained for the different fractions	Increasing milling time increased the ratio of 1545–1022/cm, while having less effect on that of 1653–1022/cm			Roa-Acosta et al. (2020)
Dry	Abrasive milling	Quinoa seeds were milled using abrasive mill for eight times with 1 min per time	The composition of quinoa chemical components	Increasing milling times increased the starch content of the resulting polished			D'Amico et al. (2019)

(Continued)

Table 3.1 (Continued)

Milling type	Miller type	Milling conditions	Targeted components/ fractions	Major findings	Targeted physicochemical property	Major findings	Reference
				grains, while decreasing that of ash, protein, and lipids. After eight times of milling, the polished grains lost over 60% of the original weight, while the contents of ash, protein, and lipids were <4% of the original values. The embryo was removed after four times of milling			

Dry	Disk mill	Quinoa seeds were milled using disk mill to obtain various degrees of milling	Composition of phenolics and saponins and in vitro antioxidant activity	Increasing the degree of milling to 27% decreased the contents of phenolics, flavonoids, and saponins 31%, 41%, and 42%, respectively. The antioxidant activities were decreased by ~40%. Degrees of milling of <16% should be used to retain the nutrients and the sensory properties of the resulting quinoa products			Han et al. (2019)

(Continued)

Table 3.1 (Continued)

Milling type	Miller type	Milling conditions	Targeted components/ fractions	Major findings	Targeted physicochemical property	Major findings	Reference
Dry	Laboratory grinder	Quinoa seeds were ground and sifted into fractions with different particle sizes (> 300 μm, between 180 and 300 μm, and <180 μm)	Chemical composition	Fraction with medium particle sizes had more proteins and ash and less carbo-hydrates	Water absorption, water retention, swelling, Falling number index	Fraction with medium particle sizes had higher water retention and absorption capacity. Increasing particle size decreased swelling capacity and falling number indexes and increased bulk density	Coţovanu et al. (2020)
Dry	Ball mill and hammer mill	Quinoa grains were milled into three fractions with different particle sizes (200, 500, and 1000 μm)	Yield and chemical composition of fractions	The yields of fractions with large, medium, and small particle sizes were 68%, 22%, and 11%.	Rheological and thermal	The fraction of small particle size had similar rheological properties compared to whole grain	Solaesa et al. (2020)

				The fraction of medium particle size had more lipids and protein and less starch compared to whole grain. The fraction of large particle size had more starch and less proteins		flour. The fraction of large size had much higher G' and G'' than that of other fractions and whole grain flour. Gelatinization temperatures were not much affected by particle sizes	
Dry	Roller mill	Quinoa grains were steeped under different moistures. Only 1 reduction passage was employed to obtain the germ and the flour fractions, and reduction rolls	Germ, bran, and perisperm fractions	The separation of germ by the process gave high recovery of 83% of protein, 81% of minerals and 80% of lipids. The bran fractions had two times the amount of fiber than the			Ray et al. (2021)

(*Continued*)

Table 3.1 (Continued)

Milling type	Miller type	Milling conditions	Targeted components/ fractions	Major findings	Targeted physicochemical property	Major findings	Reference
		were not used in the process		whole grain. The perisperm fractions had <1% of minerals			
Dry	Roller mill	Quinoa grains were steeped for different time lengths (8 or 20 h) and moisture contents (13%–17%) before milling and sieving to obtain flour and bran fractions	Flour and bran fractions	Decreasing moisture of tempering increased flour yield and decreased the separation of perisperm-bran. Larger grain size gave finer flour fraction with less protein and ash			De Bock et al. (2021)
Dry	Pearling mill	Quinoa grains of three varieties were pearled	Saponin and chemical composition	Pearling for 2 min significantly			Rafik et al. (2021)

		for different durations up to 8 min		reduced the saponin content of all the three varieties. For example, the saponin content of 1 variety was reduced from 0.49% to 0.09%. The chemical composition of protein, lipids, and moisture of the quinoa grains were not affected significantly by the pearling			
Dry	Pearling mill	Quinoa grains were pearled and then milled into flour. Pearling degrees of	Saponin and phenolic composition	A pearling degree of 30% was decrease the saponin content to be less than 110			Gómez-Caravaca et al. (2014)

(*Continued*)

Table 3.1 (Continued)

Milling type	Miller type	Milling conditions	Targeted components/ fractions	Major findings	Targeted physicochemical property	Major findings	Reference
		20% and 30% were obtained		mg/100 g. This increased the sensory quality of quinoa. The pearling decreased the bound and free phenolics of quinoa by 35% and 22%, respectively			
Dry	Roller mill	Quinoa grains were tempered to have 16% of moisture. Fractions of bran, break, short and break flours	Purity, yield, and properties of fiber fraction	Dry milling gave lower yield (9%), recovery (53%) and purity (59%) of the fiber fraction than wet milling (yield,			Ballester-Sánchez et al. (2020)

		were obtained (Fig. 3.1)		recovery and purity were 10%, 58% and 72%, respectively) Wet milling gave lower in vitro antioxidant activity and smaller particles than dry milling			
Wet	Homogenization	Quinoa grains were steeped in solution of sodium bisulfite and milled using a plate mill. This milling step removed the germ through flotation Then the de-germinated slurry fraction was homogenized	Purity, yield, and properties of fiber fraction	Wet milling gave the fiber fraction higher yield (10%), recovery (58%) and purity (72%) than drying milling (yield (9%), recovery (53%) and purity (59%)). Wet milling gave lower in vitro			Ballester-Sánchez et al. (2020)

(*Continued*)

Table 3.1 (Continued)

Milling type	Miller type	Milling conditions	Targeted components/ fractions	Major findings	Targeted physicochemical property	Major findings	Reference
		and sifted through mesh (sizes of 300 and 53 μm). The resulting filtrate slurry was centrifuged to obtain starch fraction		antioxidant activity and smaller particles of the fiber fraction than dry milling			
Wet	Plate mill and disperser of homogenization	Quinoa grains were steeped and milled using a plate mill. This milling step removed the germ through flotation Then the degerminated slurry fraction was homogenized and sifted	Starch	Different steeping conditions (time and temperature) greatly affected the starch yield and quality. The highest starch recovery rates were obtained at 30°C after			Ballester-Sánchez et al. (2019)

		through mesh (sizes of 300 and 53 μm). The resulting filtrate slurry was centrifuged to obtain starch fraction. The steeping conditions were optimized		6.5 h of steeping			
Wet	Roller mill and wet milling	Quinoa seeds were roller-milled using different opening between the rollers. The obtained fractions were wet milled through sieving, centrifugation and separation	Germ and starch fractions	The germ fraction had recovered 97% of proteins, 96% of lipids, and 93% of minerals			Mufari et al. (2018)
Dry and wet milling comb-ined	Roller mill and wet milling	Quinoa seeds were tempered before roller-milling to	Protein-rich fraction	The resulting bran fraction had 28% of protein content.			Föste et al. (2015)

(Continued)

Table 3.1 (Continued)

Milling type	Miller type	Milling conditions	Targeted components/ fractions	Major findings	Targeted physicochemical property	Major findings	Reference
		obtain white flour and bran fractions. The bran fraction was treated to obtain protein-rich fraction using isoelectric point method (Fig. 3.5)		Further extraction of protein from the bran gave 68% of protein yield which was higher than that extracted using whole grain flour			
Dry and wet milling comb-ined	Rotor mill, impact mill	Seeds were milled by rotor mill. The obtained flour was air-classified to obtain embryo-rich fraction. This fraction was further milled using impact mill	Protein-rich fraction	The resulting sample had 60% protein with a 62% recovery of protein from the seeds. This process of protein enrichment from quinoa seeds used less than			Ruiz et al., 2016

		(Fig. 3.4). The milled fraction was further treated with solution containing NaCl (0.5 M) to obtain protein-rich fraction		98% of water compared to the traditional wet milling process			
Dry and wet comb-ined	Lab scale mill	Quinoa grains were milled using a lab scale mill before air jet sieving (mesh sizes of 0.8–0.315 mm). Starch and protein fractions were isolated using wet milling. The quinoa flour was de-fatted and suspended in	Starch and protein-rich fractions	The dry milling process produced starch-rich fraction with up to 89% of starch (dry basis), and protein-rich fraction with 32% of protein	Physicochemical properties of starch and protein enriched fractions compared with those of isolated starch and protein from wet milling processes	The gelatinization and pasting properties of starch based fractions were different from those of purified starch due to the presence of protein residues. The protein-based fraction had higher solubility and	Opazo-Navarrete, Freire, et al. (2018) and Opazo-Navarrete, Schu-tyser et al. (2018)

(Continued)

Table 3.1 (Continued)

Milling type	Miller type	Milling conditions	Targeted components/ fractions	Major findings	Targeted physicochemical property	Major findings	Reference
		water before centrifugation to obtain the starch fraction. The defatted flour was also solubilized in water solution with a pH of 8 to obtain protein isolate				water retention The protein-rich fraction had higher digestibility than the purified protein isolate	

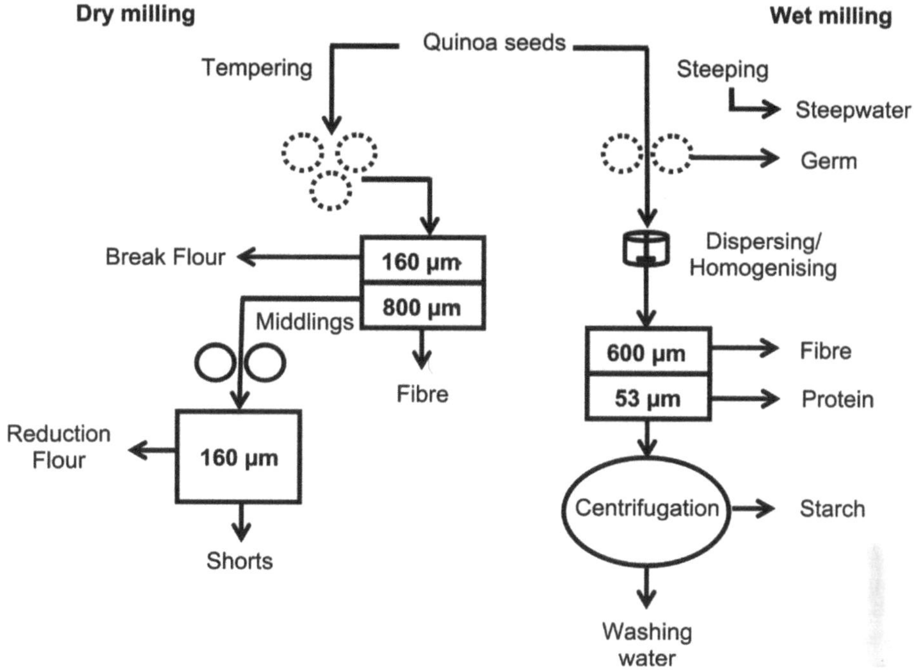

Figure 3.1 Flowchart of wet and dry quinoa milling processes and their products (Ballester-Sánchez et al., 2020).
Source: Reprinted with permission under Creative Commons Attribution License.

protein and ash in the final flour (De Bock et al., 2021). Similarly, Ballester-Sánchez et al. (2020) steeped quinoa in 16% of moisture for an unstated period of time. A flour yield of 52.8% was obtained using break rolls. Ray et al. (2021) analyzed the effect of tempering moisture (12%–20%) on the outcomes of quinoa fractionation using roller milling. They found that the conditioning moistures of 14%–16% gave good yield and quality of the fractions (Fig. 3.2). Under this condition, the roller milling process was modified to obtain germ fraction. The resulting germ fraction had 80% of fat, 83% of protein, and 81% of minerals of recovery. The different fractions obtained had significantly different microstructures (Fig. 3.3). By studying different milling and steeping conditions as variables, the optimal settings may be determined to maximize the yield of fractions with a good balance of purity of specific components. The applications of quinoa milled fractions are described in Chapter 12 of this book.

3.2.2 Wet milling

Wet milling separates quinoa seeds into its chemical components and fractions rich in specific components including starch, protein, fiber, and lipids (Ballester-Sánchez et al., 2019; Ballester-Sánchez et al., 2020; Fernández-López et al., 2020; Mufari et al., 2018; Solaesa et al., 2020). The wet milling process of quinoa grains

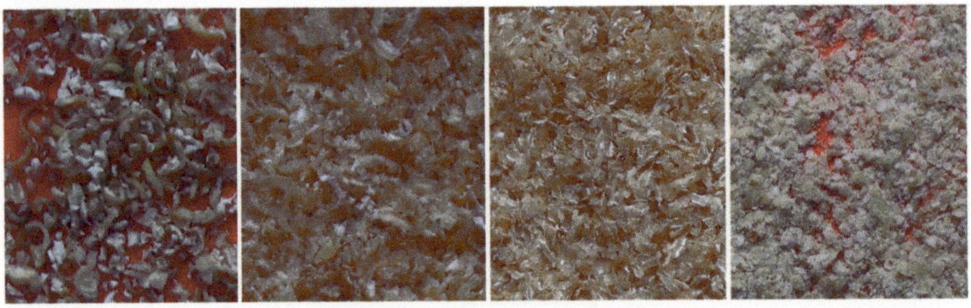

B1 B2 B3 C1

Figure 3.2 Photographs of intermediate milled fractions at different stages of roller milling. B1: Break one; B2: Break two; B3: Break three and C1: reduction one (Ray et al., 2021). *Source*: Reprinted with permission from Elsevier.

may be modified from the existing process of maize wet milling (Fig. 3.1). The maize milling process can be modified to cope with the very small size of starch granules and the presence of components that increase the viscosity of quinoa flour in water slurries (Haros & Wronkowska, 2017). In general, wet milling proceeds by the following steps: Quinoa seeds are washed and cleaned of foreign matter before steeping in solution to soften them by allowing absorption of a significant amount of water; solution containing SO_2 may also be used to facilitate the separation of the components (Ballester-Sánchez et al., 2019), and the seeds may be milled using a plate mill. The germ fraction is separated from the rest by density-difference method (flotation in water). This fraction may be washed with water to remove the starch residues. The de-germinated grains in the slurry are milled and sifted through a series of screens with decreasing mesh sizes to obtain fiber and protein fractions, respectively. The starch fraction may be obtained using centrifugation (Ballester-Sánchez et al., 2020). Starch and proteins may be separated by centrifugation as protein-rich fraction is lighter than starch fraction. The protein is then concentrated and dehydrated. The sediment (starch) layer is concentrated and washed. Finally, the yields and recoveries of the fractions are estimated. The yield and quality of the desired fraction from the wet milling determine the suitability of the milling conditions (Ballester-Sánchez et al., 2020; Haros & Wronkowska, 2017; Ruiz et al., 2016; Singh & Eckhoff, 1996).

Variables of the wet milling process such as the steeping time, temperature, and solvent types used have been modified to improve the yield, purity, and recovery of traditional quinoa wet milling (Guerreo-Ochoa et al., 2015; Haros & Wronkowska, 2017; Ballester-Sánchez et al., 2019). For example, Guerreo-Ochoa et al. (2015) used central composite design to optimize solvent/meal ratio, pH, temperature, and time for protein extraction from quinoa flour to obtain a yield of 62%. Ballester-Sánchez et al. (2019) optimized the steeping conditions for quinoa wet milling. They concluded that after steeping quinoa grains for 6.5 h at 30°C, the starch recovery rate was the highest at 86%. The properties and functions of isolated quinoa

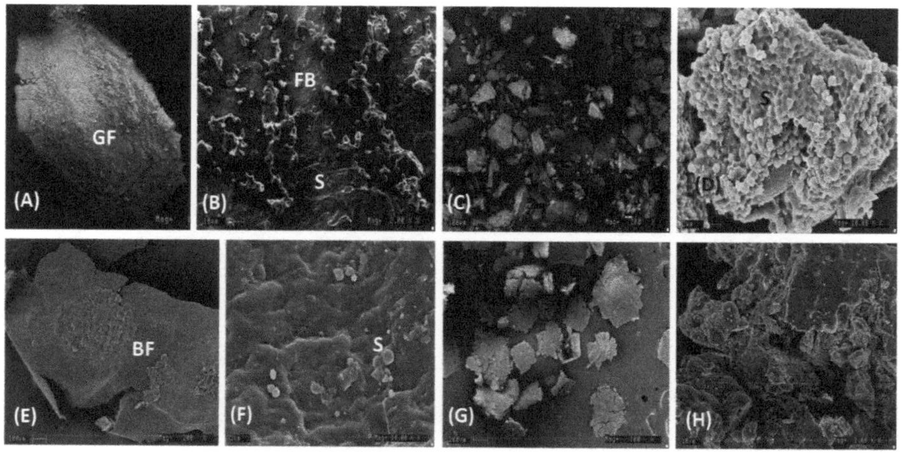

Figure 3.3 Scanning electron micrographs of milled fractions of quinoa seeds obtained using roller mill, A & B: Germ; C & D: Flour; E & F: Bran; G & H: Pollard (containing perisperm, bran and germ fractions), GF: Germ flake; FB: Fat bodies; S: Starches and BF: Bran flakes (Ray et al., 2021).

Source: Reprinted with permission from Elsevier.

components including starch, protein, dietary fiber, and lipids are described in the Chapters 4–7, respectively.

3.2.3 Alternatives to the conventional milling processes

Conventional wet milling consumes large amounts of water and energy (Ruiz et al., 2016). This process may be modified to reduce its negative impact on the environment. Ruiz et al. (2016) combined the dry and wet milling processes to obtain protein-rich fraction from quinoa seeds with reduced water and energy usages and to reduce the protein denaturation caused by wet milling (Fig. 3.4). In this hybridized process, quinoa seeds were initially milled by the normal dry milling process, and the milled and nonmilled fractions were fractionated by aqueous phase separation with NaCl solution (0.5 M). The fractionation was done for 3 h at room temperature. The slurry was then separated at room temperature under 350 kPa for 165 min. Compared to traditional wet milling process, the hybridized milling process gave lower protein purity, but a similar or higher protein yield. Although the hybrid milling did not improve the yield or purity of the products, it consumed less energy and water than the traditional wet milling process for ecofriendly applications (Ruiz et al., 2016).

3.2.4 Other factors affecting the quality of quinoa milling fractions

The factors affecting milling-fraction quality have been widely studied for common cereal grains such as maize, rice, and wheat but not so much for quinoa (Bao, 2019; Cappelli et al., 2020; Somavat et al., 2021). Genotypes and growing environmental conditions can influence the chemical composition and functional properties of quinoa seeds (López-Fernández et al., 2021; Thiam et al., 2021). Thus they may affect milling properties of quinoa grain (De Bock et al., 2021). This has been confirmed in other types of food grains. For example, growing temperature can influence the milling quality of rice

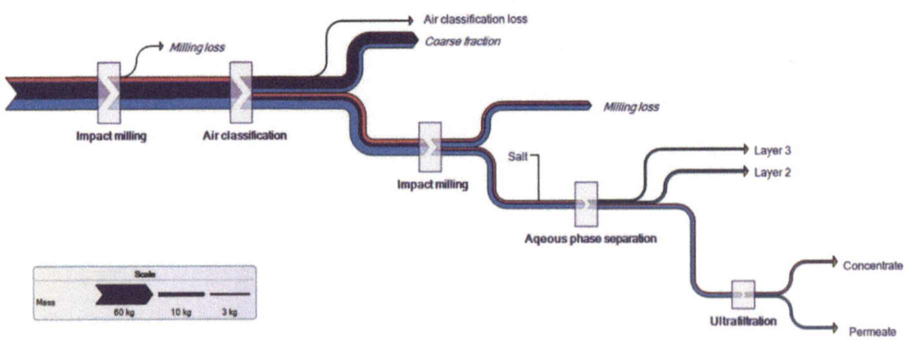

Figure 3.4 Diagram showing a hybrid dry and aqueous fractionation process for production of protein-rich fractions of quinoa grain (Ruiz et al., 2016).
Source: Reprinted with permission under Creative Commons Attribution License.

grains (Bao, 2019). The temperature was negatively correlated with some milling traits of rice fractions. Drier rice grains gave a lower yield than moisturized grains. Rice milling quality is also controlled by genes. The quality could be improved by breeding (Bao, 2019). In contrast, studies on the effects of genetics and growing environmental conditions on milling properties of quinoa grains are very limited. De Bock et al. (2021) analyzed the milling property of three different quinoa types. Larger grain size and lower hardness gave finer flour fraction with less protein and ash (De Bock et al., 2021). Factors influencing the milling quality of quinoa, such as the soil type during growth and quinoa seed size, color and density, should be investigated.

3.2.5 Milling devices

Milling devices have been neglected in quinoa studies. Roller mills, hammer mills, stone mills, ball mills, disk mills, jet mills, classifier mills, and plate mills are the common milling devices of different types of starchy food grains such as wheat and pulses (Thakur et al., 2019). These mills differ mainly by force/function types. As a result, the amount of heat generated during milling processes may be different. This difference in the heat and temperature may directly impact the composition (such as total amino acids and lipids) of the resulting milling products. For example, compared to hammer and roller milling, plate and stone milling of wheat kernels could generate higher temperatures, lowering the contents of total amino acids and some essential amino acids (e.g., lysine) (Prabhasankar & Rao, 2001). The heat generated from the milling induced Maillard reactions, reducing the arginine and lysine contents of the fractions. The high temperature generated during plate and stone milling also decreased the amount of unsaturated fatty acids such as linolenic acid in milled wheat samples (Prabhasankar & Rao, 2001). Liu et al. (2018) used different milling approaches (roller milling, ultrafine milling, wet milling, and stone milling) on Tartary buckwheat grains. The ultrafine milling significantly reduced the particle size, while increasing the brightness of flours. Stone and ultrafine milling gave higher amounts of fibers, proteins, phenolics, and minerals in the flours than the other two milling methods. Wet milling led to the partial transformation of rutin into quercetin (Liu et al., 2018). Therefore it is expected that the impact of milling device types on the milling quality of quinoa could be significant. Different studies tend to use different mills differing in geometry and designs for quinoa milling (Table 3.1). However, such impact remains to be studied in the future on quinoa grains in detail.

3.3 Effect of milling on chemical and physicochemical properties of quinoa milling fractions

3.3.1 Effect of milling on nutritional and chemical composition and particle size

3.3.1.1 Dry milling

Quinoa grains were subjected to dry milling to obtain different fractions with different particle sizes. These fractions tend to have different chemical compositions

(Table 3.1). The separation of germ by a modified dry milling process gave high recoveries of 83% of protein, 81% of minerals, and 80% of lipids. The bran fractions had two times the amount of fiber than the whole grain. The perisperm fractions had <1% of minerals (Ray et al., 2021). In general, increasing the degrees of milling/hulling decreased the contents of ash, proteins, lipids, and other micronutrients, while increasing that of starch (Ballester-Sánchez et al., 2020; D'Amico et al., 2019; De Bock et al., 2021; Gómez-Caravaca et al., 2014; Han et al., 2019; Roa-Acosta et al., 2020; Wang et al., 2020; Wu et al., 2020). For example, using a disk mill, Han et al. (2019) showed that increasing the degree of milling to 27% decreased the contents of phenolics, flavonoids and saponins by 31%, 41%, and 42%, respectively. The in vitro antioxidant activities were decreased by ~40%. Degrees of milling of <16% should be used to retain the nutrients and the sensory properties of the resulting quinoa products (Han et al., 2019). Similarly, using an abrasive mill, D'Amico et al. (2019) showed that increasing milling times increased the starch content of the resulting polished grains, while decreasing those of ash, protein, and lipids. After eight times of milling, the polished grains lost over 60% of the original weight, while the contents of ash, protein, and lipids became <4% of the original values. The embryo was removed after four times of milling (D'Amico et al., 2019). Rafik et al. (2021) reported that pearling for 2 min significantly reduced the saponin content of quinoa seeds and decreased the bitter taste of the quinoa products. Wang et al. (2020) studied the effect of peeling on chemical and physicochemical properties of quinoa flour. The peeling decreased the concentrations of phytic acid, ash, saponins, and insoluble dietary fiber. The processing quality and water-holding capacity of the flour after peeling increased as compared to those of whole flour and refined flour (Wang et al., 2020). Such changes in the chemical composition of the fractions during milling process could be related to the grain structure as described in Chapter 2. Germ fraction is rich in certain nutrients such as proteins and lipids, whereas the perisperm fraction is rich in starch.

Dry milling combined with sieving may yield different flour fractions with different particle sizes (Ahmed et al., 2019; Coțovanu et al., 2020; Solaesa et al., 2020). These fractions tend to have different chemical compositions (Table 3.1). For example, quinoa grains were milled into three fractions with different particle sizes (200, 500, and 1000 μm) (Solaesa et al., 2020). The yields of fractions with large, medium, and small particle sizes were 68%, 22%, and 11%, respectively. The fraction of medium particle size had more lipids and proteins and less starch compared to whole grain. The fraction of large particle size had fewer proteins and more starch (Solaesa et al., 2020). Therefore it may be possible to obtain different fractions with different compositions for diverse nutritional applications.

3.3.1.2 Wet milling

Quinoa grains were subjected to wet milling to obtain different fractions enriched with different chemical components. These fractions tend to have different chemical composition and purity (Table 3.1). For example, Ballester-Sánchez et al. (2019) used plate mill and disperser of homogenization for wet milling of quinoa grains.

The plate-milling step removed the germ through flotation. Then the degerminated slurry fraction was homogenized and sifted through mesh (sizes of 300 and 53 μm). The resulting filtrate slurry was centrifuged to obtain starch fraction. The steeping conditions were optimized for starch extraction. Different steeping conditions (time and temperature) greatly affected the starch yield and quality. The highest starch recovery rate was obtained at 30°C after 6.5 h of steeping (Ballester-Sánchez et al., 2019). Other fractions rich in specific components such as fiber and protein were also obtained using wet milling (Ballester-Sánchez et al., 2020; Föste et al., 2015) (Fig. 3.5).

3.3.2 Effect of milling on physicochemical and functional properties

The above sections showed that fractions with different compositions and particle sizes can be produced using a range of milling settings and conditions. As expected, these fractions tend to have different physicochemical and functional properties (Table 3.1). Ahmed et al. (2019) showed that increasing particle size of milled flour fractions decreased their sediment volume and water-holding capacity. The fraction with larger particles had higher complex viscosity and rigidity. Increasing particle size of the flour increased setback and final viscosities during pasting (Ahmed et al., 2019). Solaesa et al. (2020) showed that the fraction of small particle size had similar rheological properties compared to whole grain flour. The fraction of large size had much higher elastic modulus (G') and loss modulus (G'') than other fractions and whole grain flour. Gelatinization temperatures were not much affected by particle sizes (Solaesa et al., 2020). Therefore it may be possible to obtain different milling fractions with different physicochemical properties. Opazo-Navarrete, Freire et al. (2018), and Opazo-Navarrete, Schutyser et al. (2018) compared the physicochemical properties of starch- and protein-enriched fractions with those of isolated starch and protein from wet milling processes. The gelatinization and pasting properties of starch-based fractions were different from those of purified starch due to the presence of protein residues. The protein-based fraction had high solubility and water retention. The protein-rich fraction had higher digestibility than the purified protein isolate (Opazo-Navarrete, Freire, et al., 2018; Opazo-Navarrete, Schutyser et al., 2018). Therefore purified components could be obtained for specific applications using wet milling.

3.3.3 Comparison between wet and dry milling of quinoa grain

Ballester-Sánchez et al. (2020) compared the properties of fiber fractions obtained using dry milling and wet milling processes (Fig. 3.1). Dry milling gave lower yield (9%), recovery (53%), and purity (59%) of the fiber fraction than wet milling (yield, recovery, and purity were 10%, 58%, and 72%, respectively). Wet milling gave the fiber fraction lower in vitro antioxidant activity and smaller particles than

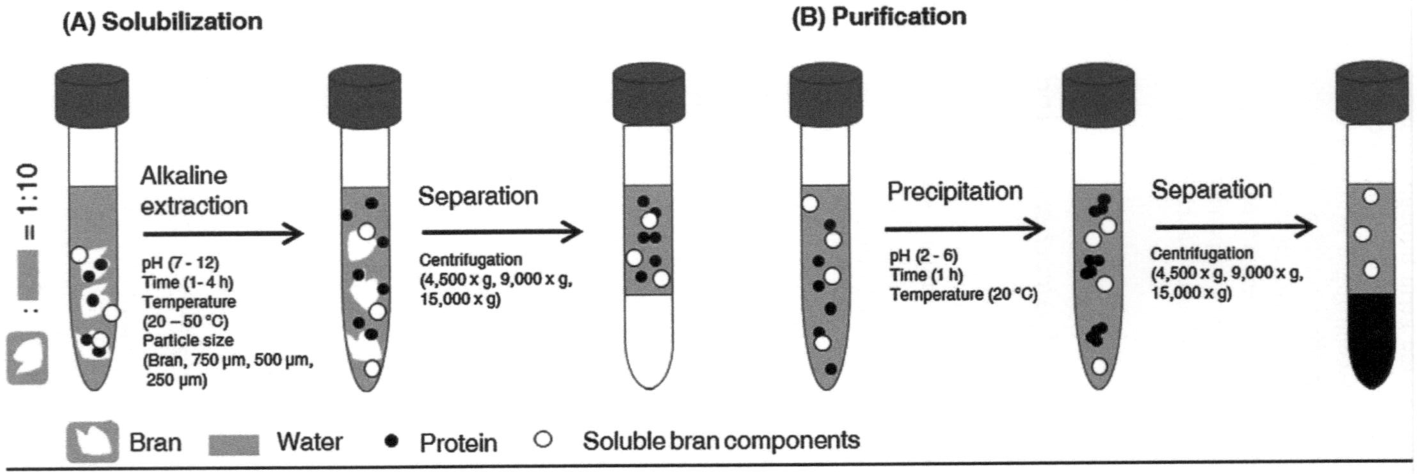

Figure 3.5 Solubilization (A) and purification (B) of proteins from quinoa bran (Föste et al., 2015).
Source: Reprinted with permission from Elsevier.

dry milling. In order to choose the suitable type of milling, the production cost and environmental impact should be analyzed in the context of specific applications.

3.4 Conclusions

Wet milling and dry milling processes of quinoa grains yield different intermediate products. The choice of milling method depends on the type of the intermediates demanded and specific applications. Dry milling produces grain/flour fractions or enriches components, whereas wet milling can separate/enrich the chemical components of quinoa grain in purer form. Alternative methods are required to reduce the water and energy consumptions of wet milling. The fractions produced from dry milling may have different particle sizes. These fractions tend to have different chemical compositions and physicochemical properties. Factors influencing the milling properties include tempering conditions, setting and types of mill, as well as the quinoa genetics. The conditions of milling process should be optimized to yield fractions with targeted composition and properties.

There are research gaps to be filled to better understand the milling properties of quinoa grains for specific applications. The effects of quinoa genetics and growing environments on milling process should be better studied. Environmental factors such as growth temperature, water absorption, and soil type during growth in relation to milling properties should be further explored. The effect of drying conditions and storage may be analyzed in relation to quinoa milling. Milling machines and their settings may be optimized to suit quinoa milling. Air classification and fine milling may be applied on quinoa grains. The interactions and entanglements between different components in the quinoa grains may be studied to better facilitate the milling and utilizations of quinoa. Both of the dry and wet milling processes should be modified to be more environmentally friendly.

References

Abugoch James, L. E. (2009). Quinoa (*Chenopodium quinoa* Willd.): Composition, chemistry, nutritional, and functional properties. *Advances in Food and Nutrition Research*, *58*, 1–31.

Ahmed, J., Thomas, L., & Arfat, Y. A. (2019). Functional, rheological, microstructural and antioxidant properties of quinoa flour in dispersions as influenced by particle size. *Food Research International*, *116*, 302–311.

Alonso-Miravalles, L., & O'Mahony, J. A. (2018). Composition, protein profile and rheological properties of pseudocereal-based protein-rich ingredients. *Foods*, *7*, 73.

Ballester-Sánchez, J., Fernández-Espinar, M. T., & Haros, C. M. (2020). Isolation of red quinoa fibre by wet and dry milling and application as a potential functional bakery ingredient. *Food Hydrocolloids*, *101*, 105513.

Ballester-Sánchez, J., Gil, J. V., Fernández-Espinar, M. T., & Haros, C. M. (2019). Quinoa wet-milling: Effect of steeping conditions on starch recovery and quality. *Food Hydrocolloids*, *89*, 837–843.

Bao, J. (2019). Rice milling quality. In J. S. Bao (Ed.), *Rice: Chemistry and technology* (4th ed., pp. 339–369). Elsevier.

Bertero, H. D., de la Vega, A. J., Correa, G., Jacobsen, S. E., & Mujica, A. (2004). Genotype and genotype-by-environment interaction effects for grain yield and grain size of quinoa (*Chenopodium quinoa* Willd.) as revealed by pattern analysis of international multi-environment trials. *Field Crops Research*, *89*, 299–318.

Cappelli, A., Oliva, N., & Cini, E. (2020). Stone milling vs roller milling: A systematic review of the effects on wheat flour quality, dough rheology, and bread characteristics. *Trends in Food Science & Technology*, *97*, 147–155.

Coţovanu, I., Batariuc, A., & Mironeasa, S. (2020). Characterization of quinoa seeds milling fractions and their effect on the rheological properties of wheat flour dough. *Applied Sciences*, *10*, 7225.

D'Amico, S., Jungkunz, S., Balasz, G., Foeste, M., Jekle, M., Tömösközi, S., & Schoenlechner, R. (2019). Abrasive milling of quinoa: Study on the distribution of selected nutrients and proteins within the quinoa seed kernel. *Journal of Cereal Science*, *86*, 132–138.

De Bock, P., Van Bockstaele, F., Raes, K., Vermeir, P., Van der Meeren, P., & Eeckhout, M. (2021). Impact of tempering process on yield and composition of quinoa flour. *LWT-Food Science and Technology*, *140*, 110808.

El Hazzam, K., Hafsa, J., Sobeh, M., Mhada, M., Taourirte, M., EL Kacimi, K., Yasri, A. (2020). An insight into saponins from quinoa (*Chenopodium quinoa* Willd): A review. *Molecules*, *25*, 1059

Fernández-López, J., Lucas-González, R., Roldán-Verdú, A., Viuda-Martos, M., Sayas-Barberá, E., Ballester-Sánchez, J., Haros, C. M., & Pérez-Álvarez, J. A. (2020). Effects of black quinoa wet-milling coproducts on the quality properties of Bologna-type sausages during cold storage. *Foods*, *9*, 274.

Föste, M., Elgeti, D., Brunner, A., Jekle, M., & Becker, T. (2015). Isolation of quinoa protein by milling fractionation and solvent extraction. *Food and Bioproducts Processing*, *96*, 20–26.

Ghumman, A., Mudgal, S., Singh, N., Ranjan, B., Kaur, A., & Rana, J. C. (2021). Physicochemical, functional and structural characteristics of grains, flour and protein isolates of Indian quinoa lines. *Food Research International*, *140*, 109982.

Gómez-Caravaca, A. M., Iafelice, G., Verardo, V., Marconi, E., & Caboni, M. F. (2014). Influence of pearling process on phenolic and saponin content in quinoa (*Chenopodium quinoa* Willd). *Food Chemistry*, *157*, 174–178.

Guerreo-Ochoa, M. R., Pedreschi, R., & Chirinos, R. (2015). Optimised methodology for the extraction of protein from quinoa (*Chenopodium quinoa* Willd.). *International Journal of Food Science and Technology*, *50*, 1815–1822.

Han, Y., Chi, J., Zhang, M., Zhang, R., Fan, S., Dong, L., Huang, F., & Liu, L. (2019). Changes in saponins, phenolics and antioxidant activity of quinoa (*Chenopodium quinoa* willd) during milling process. *LWT-Food Science and Technology*, *114*, 108381.

Haros, C. M., & Wronkowska, M. (2017). Pseudocereal dry and wet milling: Processes, products and applications. In C. M. Haros, & R. Schoenlechner (Eds.), *Pseudocereals: Chemistry and technology* (pp. 140–163). John Wiley & Sons.

Hemalatha, P., Bomzan, D. P., Rao, B. V. S., & Sreerama, Y. N. (2016). Distribution of phenolic antioxidants in whole and milled fractions of quinoa and their inhibitory effects on α-amylase and α-glucosidase activities. *Food Chemistry*, *199*, 330–338.

Jiang, F., Du, C., Zhao, N., Jiang, W., Yu, X., & Du, S. (2022). Preparation and characterization of quinoa starch nanoparticles as quercetin carriers. *Food Chemistry*, *369*, 130895.

Liu, F., He, C., Wang, L., & Wang, M. (2018). Effect of milling method on the chemical composition and antioxidant capacity of Tartary buckwheat flour. *International Journal of Food Science & Technology*, *53*, 2457–2464.

López-Fernández, M. P., Rodríguez, S. D., Favre, L. C., Busch, V. M., & Buera, M. D. P. (2021). Physicochemical, thermal and rheological properties of isolated Argentina quinoa starch. *LWT-Food Science and Technology*, *135*, 110113.

Mufari, J. R., Miranda-Villa, P. P., & Calandri, E. L. (2018). Quinoa germ and starch separation by wet milling, performance and characterization of the fractions. *LWT-Food Science and Technology*, *96*, 527–534.

Opazo-Navarrete, M., Freire, D. T., Boom, R. M., Janssen, A. E. M., & Schutyser, M. A. I. (2018). Dry fractionation of quinoa sweet varieties Atlas and Riobamba for sustainable production of protein and starch fractions. *Journal of Food Composition and Analysis*, *74*, 95–101.

Opazo-Navarrete, M., Schutyser, M. A. I., Boom, R. M., & Janssen, A. E. M. (2018). Effect of pre-treatment on in vitro gastric digestion of quinoa protein (*Chenopodium quinoa* Willd.) obtained by wet and dry fractionation. *International Journal of Food Sciences and Nutrition*, *69*, 1–11.

Prabhasankar, P., & Rao, P. H. (2001). Effect of different milling methods on chemical composition of whole wheat flour. *European Food Research & Technology*, *213*, 465–469.

Rafik, S., Rahmani, M., Rodriguez, J. P., Andam, S., Ezzariai, A., El Gharous, M., Karboune, S., et al. (2021). How does mechanical pearling affect quinoa nutrients and saponin contents? *Plants*, *10*, 1133.

Ray, A., Inamdar, A. A., Sakhare, S. D., & Srivastava, A. K. (2021). Development of physical process for quinoa fractionation and targeted separation of germ with physical, chemical and SEM studies. *LWT-Food Science and Technology*, *141*, 110957.

Roa-Acosta, D. F., Bravo-Gómez, J. E., García-Parra, M. A., Rodríguez-Herrera, R., & Solanilla-Duquea, J. F. (2020). Hyper-protein quinoa flour (*Chenopodium Quinoa* Wild): Monitoring and study of structural and rheological properties. *LWT-Food Science and Technology*, *121*, 108952.

Ruiz, G. A., Arts, A., Minor, M., & Schutyser, M. (2016). A hybrid dry and aqueous fractionation method to obtain protein-rich fractions from quinoa (*Chenopodium quinoa* Willd). *Food and Bioprocess Technology*, *9*, 1502–1510.

Sayas-Barberá, E., Valero-Asencio, M. M., Rodríguez-Vera, C. N., Fernández-López, J., Haros, C. M., Pérez-Álvarez, J. Á., & Viuda-Martos, M. (2021). Effect of different black quinoa fractions (seed, flour and wet-milling coproducts) upon quality of meat patties during freezing storage. *Foods*, *10*, 3080.

Singh, N., & Eckhoff, S. R. (1996). Wet milling of corn: A review of laboratory-scale and pilot plant-scale procedures. *Cereal Chemistry*, *73*, 659–667.

Solaesa, A. G., Villanueva, M., Vela, A. J., & Ronda, F. (2020). Protein and lipid enrichment of quinoa (cv. Titicaca) by dry fractionation. Techno-functional, thermal and rheological properties of milling fractions. *Food Hydrocolloids*, *105*, 105770.

Somavat, P., Liu, W., & Singh, V. (2021). Wet milling characteristics of corn mutants using modified processes and improving starch yields from high amylose corn. *Food and Bioproducts Processing*, *126*, 104–112.

Thakur, S., Scanlon, M. G., Tyler, R. T., Milani, A., & Paliwal, J. (2019). Pulse flour characteristics from a wheat flour miller's perspective: A comprehensive review. *Comprehensive Reviews in Food Science and Food Safety*, *18*, 775–797.

Thiam, E., Allaoui, A., & Benlhabib, O. (2021). Quinoa productivity and stability evaluation through varietal and environmental interaction. *Plants*, *10*, 714.

Wang, Y., Gong, X., Zhang, Y., Geng, D. H., Cao, L., Ruan, C., et al. (2020). Effect of peeling treatment on the physicochemical properties of quinoa flour. *Journal of Food Process Engineering*, *43*, e13387.

Wu, L. G., Wang, A., Shen, R., & Qu, L. (2020). The effect of slight milling on nutritional composition and morphology of quinoa (*Chenopodium*) grain. *International Journal of Food Engineering*, *16*, 20190371.

Zuo, Z., Zhang, X., Li, T., Zhou, J., Yang, Y., Bian, X., & Wang, L. (2022). High internal phase emulsions stabilized solely by sonicated quinoa protein isolate at various pH values and concentrations. *Food Chemistry*, *378*, 132011.

Quinoa starch

4

4.1 Introduction

Starch is a major ingredient in the food industry and a major component of our diet (Sjöö & Nilsson, 2018). Starch consists of amylopectin and amylose molecules. Amylopectin is highly branched and larger, whereas amylose is mostly linear and smaller (Pérez & Bertoft, 2010). They are polymers of glucose units. Amylose and amylopectin molecules are organized into different shapes of granules. The diameters of the granules from different plants and genotypes vary from ~1 to 100 μm (Pérez & Bertoft, 2010). Amylopectin has a building block structure. This type of structure facilitates the formation of crystals in the granules. Starch granules in the presence of heat and an excessive amount of water undergo a process termed gelatinization. The crystallinity is lost during the gelatinization. When the gelatinized starch is cooled, the starch molecules reassociate to develop a more ordered structure which can include a degree of crystallinity. This process is termed retrogradation (Sjöö & Nilsson, 2018). The gelatinization and retrogradation properties of starch are determined by starch composition and structure (Pérez & Bertoft, 2010; Sjöö & Nilsson, 2018).

The major component of quinoa seeds is starch. This component may amount to 70% of dried seed (Hussain et al., 2021). The sensory and textural properties of quinoa-based food products can be determined by the structural and physicochemical attributes of starch (Wang, Opassathavorn, et al., 2015). Compared with industrially important starches such as maize starch, quinoa starch is largely under-studied. On laboratory scale, quinoa starch in raw and modified forms has been used as an ingredient for food and other applications (Li and Zhu, 2018b). The chapter describes the extraction, composition, structure, and functional properties of quinoa starch. Research directions on better understanding this starch are suggested. This chapter is based on a previous publication of (Li and Zhu, 2018b) (Excerpts reprinted with permission from Elsevier) with recent updates.

4.2 Quinoa starch isolation

Most of the reports on starch isolation from quinoa seeds were on laboratory scale. Washed and cleaned quinoa seeds were soaked in solution/water. The solutions used may contain sodium hydroxide, sodium acetate, or sodium dodecyl sulfate (SDS). The use of these chemicals can facilitate the removal of nonstarch

Quinoa. DOI: https://doi.org/10.1016/B978-0-323-99909-0.00002-7

components in the seeds including proteins, saponins, lipids, polyphenols, and non-starch polysaccharides (Li et al., 2016; Valdez-Arana et al., 2020). The soaking can be done overnight at 4°C to avoid microbial issues (Li & Zhu, 2018a, 2018b, 2018c). The soaked seeds were homogenized in a high-speed blender (Jiang et al., 2020; López-Fernández et al., 2021). In some studies, quinoa seeds were subjected to dry milling to obtain flour, which was then soaked (Jan et al., 2017; Li, Wang, & Zhu, 2016; Mundigler, 1998; Valdez-Arana et al., 2020). Harsh milling conditions could damage the starch granules (Qian & Kuhn, 1999). The soaked flour slurry suspension was filtered through cloth/mesh to remove nonstarch materials. The slurry was centrifuged with a relatively high speed (e.g., 3500 × *g*) to recover the starch (Li et al., 2016; López-Fernández et al., 2021; Steffolani et al., 2013; Valdez-Arana et al., 2020). The starch cake was washed by resuspending in water and centrifuge. The brown layer on the surface of the starch cake obtained from the centrifugation was scraped off to improve the starch purity. The purity of the resulting quinoa starch was as high as 99% (Li & Zhu, 2018a,b,c).

4.3 Quinoa starch composition

Amylose ratio of quinoa starch was obtained using different methods in different studies (Table 4.1). The amylose contents obtained using iodine binding-based methods are seen to be higher than those obtained by concanavalin A binding-based method (Table 4.1). Such a discrepancy is due to the iodine-binding capacity of amylopectin component (Gibson, Solah, & McCleary, 1997). Linear polysaccharides such as amylose could not form any precipitate with Con A (Goldstein, Hollerman, & Merrick, 1965). On the other hand, endogenous lipids in quinoa grains may form inclusion complexes with amylose, decreasing the apparent amylose contents obtained using iodine binding-based methods (Srichuwong & Jane, 2007).

The amylose content obtained by ridding of the effect of amylopectin was much lower (7%) than that obtained based on whole starch with the effect of amylopectin component (25%) (Tang, Watanabe, & Mitsunaga, 2002). This indicated the existence of longer chains of quinoa amylopectin. Longer unit-chains of amylopectin can readily form inclusion complexes with iodine. This leads to an overestimation of amylose content in starch (Vilaplana, Hasjim, & Gilbert, 2012). The amylose contents obtained using size exclusion chromatography of whole starch (4%–11%) were seen to be less than those obtained using size exclusion chromatography of debranched whole starch (4%–27%) (Table 4.1). Such a difference also indicated the existence of longer chains of quinoa amylopectin (Li & Zhu, 2017b). Overall, it is seen that the quinoa amylose contents were lower compared with that of industrially important starches including potato and maize starches (Table 4.1).

There are minor components such as dietary fibers, lipids, proteins, and ash in isolated quinoa starch. The levels of these minor components were very low (Table 4.1). These minor components may be important in affecting the functional

Table 4.1 Chemical composition of quinoa starch.

No.	Method	Amylose content (%)^{A}	Protein (%)	Ash (%)	Lipid (%)	Fiber (%)	References
18	Iodine-SP	14.33–27.66					de Briceño and Briceño (1980)
1	Iodine-SP	11	0.11	0.04			Atwell et al. (1983)
1	Iodine-SP and SEC	20^{B}, 4^{C}					Praznik et al. (1999)
2	Iodine-SP	$19.8–20.6^{D}$, $20.6–21.0^{E}$					Wright et al. (2002)
1	Iodine-SP	25.4^{F}, 7.1^{G}					Tang et al. (2002)
1^{H}	Iodine-SP	11.1–23.9					Ando et al. (2002), Tari et al. (2003), Araujo-Farro et al. (2010)
8	Iodine-SP	0.3–12.1					Lindeboom, Chang, Falk, et al. (2005)
26	Iodine-SP	7.7–25.7					Li et al. (2016)
2	Iodine-SP	9.46–12.1	0.89–0.95	0.18–0.22	0.32–0.40	0.10–0.13	Jan et al. (2017)
4	Iodine-SP	9.4 – 10.9	0.97 – 1.95	0.07 – 0.29	0.26 – 0.91		Jiang et al. (2020)
6	Iodine-SP	9.8 – 12.3					López-Fernández et al. (2021)
1	Iodine-PT	9.28	0.91		0.11		Lorenz (1990)
1^{H}	Con A	8.25–12.2					Qian and Kuhn (1999), Linsberger-Martin Lukasch, and Berghofer (2012), Srichuwong et al. (2017)
3	Con A	8.22–9.30	1.09–4.13	1.46–1.64	1.94–2.56	0.19–0.24	Steffolani et al. (2013)
1	Con A	19.7					Nascimento et al. (2014)
9	Con A	6.06–8.44					Li and Zhu (2017b)
3	Con A	11.3 – 15.4					Valdez-Arana et al. (2020)
6	Con A	7.2 – 10.1					López-Fernández et al. (2021)
4	SEC, debranched	24.7–27.0					Inouchi et al. (1999)

(*Continued*)

Table 4.1 (Continued)

No.	Method	Amylose content (%)[A]	Protein (%)	Ash (%)	Lipid (%)	Fiber (%)	References
8	SEC, debranched	3.5–19.6	0.14–1.23				Lindeboom, Chang, Falk, et al. (2005)
16	SEC, debranched	3.5–19.5					Lindeboom, Chang, Tyler, and Chibbar (2005)
9	SEC, debranched	12.6–23.7					Li and Zhu (2017b)
9	SEC	7.49–10.88					Li and Zhu (2017b)
5	SEC, debranched	5.2–10.9					Watanabe et al. (2007)
1[H]			0.24–1.3	0.2–0.34	0.2–0.02	0.14–3	Wolf, Macmasters and Rist (1950), Mundigler (1998), Pagno et al. (2015)

No.: number of starch samples studied; Iodine-SP: colorimetric-iodine-based method; Iodine-PT: potentiometric titration-iodine based method; SEC: whole starch analyzed on size exclusion chromatography; SEC, debranched: debranched starch analyzed on size exclusion chromatography; Con A: concanavalin A precipitation based method; A: amylose content (%) is based on starch content; B: amylose content determined calorimetrically by whole starch solution; C: amylose content determined by SEC method; D: un-defatted starch sample; E: defatted starch sample; F: amylose content calculated by: 100% × BV(whole starch)/BV(amylose); G: amylose content calculated by: 100% × [BV(starch)−BV(amylopectin)]/[BV(amylose)−BV(amylopectin)]; H: Number of samples used in each report.
Source: Adapted from Li and Zhu (2018b) with permission from Elsevier.

properties of quinoa starch (Ma, Wen, et al., 2022; Ma, Zhu, et al., 2022; Zhu, 2018). For example, Ma, Zhu, et al. (2022) showed that the removal of proteins associated with quinoa starch granules reduced the breakdown, peak, and final viscosities during the pasting event and gelatinization temperatures, shear-thinning behavior, and pseudoplasticity, while increasing the amylose leaching, retrogradation, and gel strength. Using a similar approach, Ma, Wen, et al. (2022) found that the removal of granule-associated lipids of quinoa starch decreased the integrity of granular structure and gave a higher drop in viscosity during shearing. The lipid removal increased amylose leaching, retrogradation, and gel strength of quinoa starch.

4.4 Structure of quinoa starch

4.4.1 Shape and size of starch granules

Diameters of quinoa starch granules ranged from 0.4–2.0 μm (Table 4.2). The granules are smaller than those of most starches from other plants. Quinoa starch granules were mostly polygonal and irregular (Fig. 4.1). The variation in the shape and size of the starch granules is small (Jiang et al., 2020; Li & Zhu, 2017a; López-Fernández et al., 2021; Valdez-Arana et al., 2020). The outer layer of the starch granules was homogeneous and the granules have a hilum (Tang et al., 2002). The starch granules can aggregate due to the presence of proteins on the surface (Ruales & Nair, 1994). The aggregates may lead to wrong estimation of quinoa starch granule size distribution.

Figure 4.1 SEM photos of quinoa starch granules (Magnification: × 30,000) (Li & Zhu, 2017a). *Source*: Reprinted with permission from Elsevier.

Table 4.2 Morphology of quinoa starch granules.

No.	Method	Size distribution pattern	Granule size (μm)	Shape descriptions	Reference
1	LM		1.5–3.0		Wolf et al. (1950)
1	CC, SEM, LM	Bimodal^{A}, Unimodal^{B}	0.8–80^{A}, centered at 1.5^{B}	Aggregates of many starch granules were observed in crude preparation	Atwell et al. (1983)
4	SEM		0.4–2.0, 18–20^{C}	Spherical or oblong aggregates constituted of ~14,000 polygonal starch granules	Varriano-Marston and DeFrancisco (1984)
4	TEM	Bimodal	One centered at 0.5, another at 1.3	Some cells contained only the large granules while others contained small ones. Starch granules were surrounded by proteins and the bonding was weak	Varriano-Marston and DeFrancisco (1984)
1	SEM		0.4–2.0	Starch granules were found as single entities within the cells or as compound structures of spherical or oblong aggregates	Lorenz (1990)
1	SEM		0.6–2.0, 16–34^{C}	The matrix surrounding starch granules were digested by pepsin; aggregates were consisted of ~20,000 starch granules	Ruales and Nair (1994)
1	SEM		1–2	Polygonal	Qian and Kuhn (1999)
2	SEM		~1–2	Angular, polygonal	Wright et al. (2002)
1	SEM, LLD	Unimodal	~1	Only surface corrosion was observed after enzyme hydrolysis, suggesting a homogeneous structure of the outer layer	Tang et al. (2002)
1	TEM			Outer layer of starch granule was more homogeneous and denser than the inner part	Tang et al. (2002)

1	SEM, LLD	Unimodal	0.08–2.0, 20–30^{C}	Polygonal	Ando et al. (2002)
1	SEM		0.6–1.5	Spherical	Tari et al. (2003)
8	SEM, LLD	Unimodal	Centered at 1.5, some samples had a shoulder at 30	Irregular, polygonal	Lindeboom, Chang, Falk et al. (2005)
1	LM		0.41–0.92	Round	Linsberger-Martin et al. (2012)
3	SEM, LLD	Bimodal	1–3.5	Irregular, polygonal, and angular	Steffolani et al. (2013)
7	SEM		1–2	Polygonal; damage of starch was observed	Li and Zhu (2017c)
9	SEM, LLD	Unimodal	1.04–1.41^{D}	Irregular	Li and Zhu (2017a)
2	SEM		1.19–1.23	Irregular, angular, and polygonal	Jan et al. (2017)
1	SEM		0.3–1.7, 10–20^{C}	Single granule was polygonal; aggregates were spherical and oblong	Srichuwong et al. (2017)
3	SEM, LLD		1.1 – 1.5	Polygonal	Valdez-Arana et al. (2020)
4	SEM, LLD		0.46–5.56	polygonal and irregular	Jiang et al. (2020)
6	SEM		0.5 – 1.8	polygonal and irregular	López-Fernández et al. (2021)

No.: number of samples studied; LM: light microscopy; SEM: scanning electron microscopy; TEM: transmission electron microscopy; LLD: laser light diffraction based method for particle size analysis; CC: coulter counter; unimodal: unimodal size distribution; bimodal: bimodal size distribution; A: crude preparation; B: pure preparation; C: size of aggregates; D: number median diameter (D [n, 0.5]).

Source: Adapted from Li and Zhu (2018b) with permission from Elsevier.

4.4.2 Polymorphism

The polymorph type of quinoa starch is A and the degree of crystallinity ranged from 21.5%–43.0% (Jan et al., 2017; López-Fernández et al., 2021; Valdez-Arana et al., 2020; Watanabe et al., 2007; Xing et al., 2021). The percentage of crystallinity of starch may be much affected by the methods of estimation. The values obtained using the ratio of crystalline area to total areas tended to be lower than those obtained using peak fitting method. The degree of crystallinity of quinoa starch was less than that of garden orache, amaranth, and maize starches. It was close to that of potato starch but was more than that of starches from adzuki, barley, and kañiwa (Jiang et al., 2020; Qian & Kuhn, 1999; Steffolani et al., 2013; Tang et al., 2002). The variations in the crystallinity between various starches may be due to their different chemical composition and structure (Sjöö & Nilsson, 2018). Quinoa starch had a relatively high amount of fingerprint A-chains contributing to the formation of more imperfection in the crystallinity. Quinoa starch had a neglectable amount of amylose-lipid inclusion complexes (Li et al., 2016). Lamellar structure of starch granules from nine quinoa varieties was studied (Li, Hemar, & Zhu, 2021). The thickness values of amorphous and crystalline lamellae ranged from 2.3–2.8 and 5.9–6.2 nm, respectively. The repeating distance values of amorphous and crystalline lamellae varied from 8.4 to 8.9 nm (Li et al., 2021). The variations in the lamellar structure of quinoa starch granules were correlated with the molecular structure of the amylopectin component. More amylopectin short unit-chains contributed to the formation of longer crystalline lamellae, whereas longer internal unit-chains led to the formation of longer amorphous lamellae (Li et al., 2021). Such correlation patterns were readily explained using the two directional backbone model proposed by Bertoft (2013).

4.4.3 Chemical structure

4.4.3.1 Amylose

λ_{max} and blue values of purified quinoa amylose varied from 648–663 nm and 0.998–1.101 nm, respectively (Table 4.3). λ_{max} values of the amylose fraction from the size-exclusion chromatography of quinoa starch varied from 595 to 620 nm (Li & Zhu, 2017b; Ruales & Nair, 1994).

In comparison, quinoa amylose had lower blue value than adzuki and barley starches. The discrepancy may be due to the different amylose structure and content (Li & Zhu, 2017b; Tang et al., 2002). The number-based degree of polymerization and average chain length of quinoa amylose molecules were smaller compared to adzuki and barley amylose molecules. The number of chains per amylose molecule of quinoa amylose was higher than adzuki and barley amylose molecules (Watanabe et al., 2007). Quinoa amylose had relatively less long chain amylose fraction than short chain amylose fraction in comparison with normal maize amylose (Li & Zhu, 2017b).

Table 4.3 Chemical structure of quinoa starch, amylose, and amylopectin.

Sample type	No.	λ_{max} (nm)	BV	DP	CL	NC	Molecular mass	Analytical method	Reference
Whole starch	1				27			Debranched whole starch on GPC (Bio-Gel P-10)	Atwell et al. (1983)
Whole starch	1	AM: 595–604, AP:575						Whole starch on GPC (Sepharose CL-2B)	Ruales and Nair (1994)
Whole starch	4	587–596	0.32–0.41					Iodine absorption spectrum	Inouchi et al. (1999)
Whole starch	1			4,600–161,000, DP_w: 70,000			M_w: 1.13×10^7	Iodine absorption spectrum; whole starch on SEC system	Praznik et al. (1999)
Whole starch	1^A	602–609	0.287–0.305					Iodine absorption spectrum	Tang et al. (2002), Ando et al. (2002)
Whole starch	9	AM:601–621, AP:581–608						Whole starch on GPC (Sepharose CL-2B); debranched whole starch on GPC (Sepharose CL-6B)	Li and Zhu (2017b)
Amylopectin	4							Debranched AP analyzed by HPAEC-PAD	Inouchi et al. (1999)
Amylopectin	1	595		DP_n: 6,675	21	317	M_n: 1.1×10^6	Iodine absorption spectrum; debranched AP analyzed by SEC	Tang et al. (2002)

(*Continued*)

Table 4.3 (Continued)

Sample type	No.	λ_{max} (nm)	BV	DP	CL	NC	Molecular mass	Analytical method	Reference
Amylopectin	5	589–590	0.190 to 0.210	DP_n: 3,403–4,752	18–20	172–227	M_n: $5.5 \times 10^5 – 7.7 \times 10^5$	Iodine absorption spectrum; debranched AP analyzed by SEC	Watanabe et al. (2007)
Amylopectin	9				16.0–17.0			Debranched amylopectin analyzed by HPAEC-PAD	Li and Zhu (2017b)
Amylopectin	1							Isoamylase-treated starch analyzed on FACE	Srichuwong et al. (2017)
Amylose	1	663	1.014	DP_n: 921	73	11.6	M_n: 1.5×10^5	Iodine absorption spectrum; isolated amylose debranched and analyzed by SEC	Tang et al. (2002)
Amylose	5	648–650	0.998–1.101	DP_n: 822–1,054	98–119	6–9	M_n: $1.3 \times 10^5 – 1.7 \times 10^5$	Iodine absorption spectrum; isolated amylose debranched and analyzed by SEC	Watanabe et al. (2007)

No.: number of samples studied; λmax: the wavelength with the maximum absorbance; BV: blue value; DP: degree of polymerization; DPw: weight-average degree of polymerization; DPn: number-average degree of polymerization; CL: average chain length; NC: number of chains per molecule; Mw: weight-average molecular mass; Mn: number-average molecular mass; AM: amylose fraction; AP: amylopectin fraction; GPC: gel-permeation chromatography; SEC: size-exclusion chromatography; HPAEC-PAD: high-performance anion-exchange chromatography with pulsed amperometric detection; FACE: fluorophore-assisted capillary electrophoresis; A: number of samples applied for each literature.

Source: Adapted from Li and Zhu (2018b) with permission from Elsevier.

4.4.3.2 Amylopectin

Unit-chain length profiles of quinoa amylopectin from different varieties were studied (Li & Zhu, 2017b; Xing et al., 2021). A peak of short unit-chains of DP <12 and a protruding shoulder at DP 18–20 were obtained on the chromatograms (Li & Zhu, 2017b; Srichuwong et al., 2017; Xing et al., 2021) (Fig. 4.2). Germination of quinoa seeds tended to decrease the proportion of long unit-chains of amylopectin, while increasing that of short unit-chains (Xing et al., 2021). Quinoa amylopectin was seen to have relatively more short chains compared to the amylopectins from other plants (Bertoft, Piyachomkwan, Chatakanonda, & Sriroth, 2008; Inouchi et al., 1999; Li & Zhu, 2017b).

One study by Li and Zhu (2017b) showed that average chain length (CL) of nine quinoa amylopectins varied from 16–17 glucosyl residues. Another study by Xing et al. (2021) showed that the CL of three quinoa amylopectins ranged from 21 to 23 glucosyl residues. Such differences could be readily attributed to the different analytical methods used in the respective studies (Li & Zhu, 2017b; Xing et al., 2021). Comparison showed that the CL of quinoa amylopectin tended to be shorter than that of most other amylopectins except for oat amylopectin (Bertoft et al., 2008). Comparative analysis showed that chain lengths of both internal and external amylopectin unit-chains of quinoa were shorter than the lengths of those from A-type starches and most of C- and B-type starches (Bertoft et al., 2008; Li & Zhu, 2017b).

The amount of A-chains can be obtained using the definition by Peat, Whelan and Thomas (1952). Diversity in the percentages of A-chains of quinoa amylopectins from nine varieties (46.4%–50.1%) was obtained (Li & Zhu, 2017b). The amylopectins had more fingerprint A-chains than most other amylopectins as reported by Bertoft et al. (2008). Overall, quinoa amylopectin structure was seen to be special in

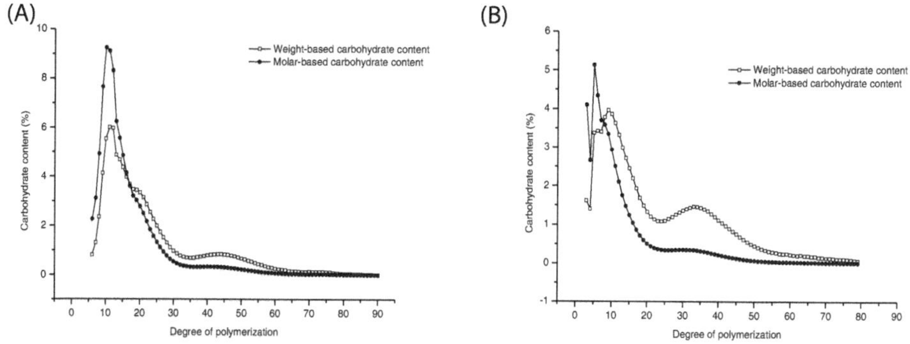

Figure 4.2 Weight- and molar-based unit-chain length distributions of typical quinoa amylopectin (A) and the φ, β-LDs (B). The carbohydrate content (%) is calculated based on the total weight of carbohydrate (weight-based carbohydrate content, empty dots) or total molar amount (molar-based carbohydrate content, solid dots) (Li & Zhu, 2017b).
Source: Reprinted with permission from Elsevier.

comparison with that of amylopectins from other sources. The building block composition of quinoa amylopectin remains to be studied. The percentages of super-long unit-chains of debranched quinoa amylopectins ranged from 13% to 19% (Tang et al., 2002; Watanabe et al., 2007). Super-long amylopectin chains have amylose-like properties, contributing to the functionalities of starch.

4.5 Physicochemical and functional characteristics of quinoa starch

4.5.1 Thermal property

Discrepancies in analytical methods and instrumental conditions render the results of different studies on the thermal property of quinoa starch unusable for direct comparison (Table 4.4). Diversity in gelatinization behaviors was obtained among starch samples of different quinoa genotypes (Jan et al., 2017; Jiang et al., 2020; Li et al., 2016; Lindeboom, Chang, Falk, et al., 2005; Xing et al., 2021; López-Fernández et al., 2021). For example, López-Fernández et al. (2021) found that the ranges of T_p and ΔH of six quinoa starches were 64.2°C–75.1°C and 9.8–12.2 J/g, respectively.

In comparison, quinoa starch had lower gelatinization temperatures than starches from other sources including garden orach, maize, adzuki, amaranth, kañiwa, millet, sorghum, and wheat. Quinoa starch had similar gelatinization temperatures to barley starch. Quinoa starch had more ΔH than wheat and barley starches, similar ΔH to garden orach, kañiwa, and sorghum starches, and less ΔH compared to maize, potato, amaranth, and adzuki starches (Jiang et al., 2020; Srichuwong et al., 2017; Steffolani et al., 2013; Tang et al., 2002; Wright et al., 2002). The discrepancies in the thermal properties between different starches could relate to their different structure and composition (Vamadevan & Bertoft, 2015; Zhu, 2018).

Gelatinization parameters are significantly correlated with amylopectin molecular structure (Srichuwong et al., 2017; Vamadevan & Bertoft, 2015; Zhu, 2018). A higher proportion of short unit-chains such as fingerprint A-chains combined with a lower proportion of long unit-chains of quinoa amylopectin could contribute to easy gelatinization of starch granules (Li & Zhu, 2017a). No endothermic phenomenon for amylose-lipid complexes was obtained for quinoa starch (Wright et al., 2002). In contrast, a comparative analysis showed that ΔH for melting the complexes of quinoa starch was less than that for melting the complexes of adzuki and barley starches (Tang et al., 2002). Thus quinoa starch has very low/neglectable amounts of amylose-lipid inclusion complexes.

4.5.2 Swelling and solubility

The degrees of leaching of soluble molecules and granular swelling during starch gelatinization were obtained (Jan et al., 2017; Jiang et al., 2020; Li et al., 2016; Valdez-Arana et al., 2020; López-Fernández et al., 2021). Variations in swelling

Table 4.4 Gelatinization properties of quinoa starch.

No.	Instrument	Starch concentration	Scanning rate (°C/min)	Scanning range (°C)	T_o (°C)	T_p (°C)	T_c (°C)	ΔH (J/g)[A]	References
1	DSC	32%–40% (w/w)	5		50.6–54.5	60.6–62.6	66.0–71.3	10.3–16.7 J/g	Atwell et al. (1983), Tang et al. (2002), Srichuwong et al. (2017)
1	DSC	21% (w/w)	10	30–120	59.9	64.5	71	1.66	Qian and Kuhn (1999)
4	DSC	33% (w/w)			52.2–57.4	54.2–61.9	67.6–68.5	7.3–10.5	Inouchi et al. (1999)
2	DSC	21% (w/w)	10	20–120	51.4–51.8	55.7–56.1	64.3–64.5	12.6	Wright et al. (2002)
1	DSC	20% (w/w)			54	62.2	71	11	Ando et al. (2002)
8	DSC	50% (w/v)	10	40–140	44.6–53.7	50.5–61.7		12.8–15.0	Lindeboom, Chang, Falk, et al. (2005)
3	DSC	25% (w/w)	10	25–120	54.25–55.72	61.66–63.01		8.45–9.00	Steffolani et al. (2013)
26	DSC	25% (w/w)	10	25–90	50.1–58.3	56.2–65.0	65.8–74.9	10.8–14.4	Li et al. (2016)
2	DSC	30% (w/w)	10	10–120	64.32–66.61	69.36–71.56	75.78–76.98	4.34–5.10	Jan et al. (2017)
4	DSC	25% (w/w)	10	10–100	57.9–61.8	63.8–67.4	71.9–77.2	7.8–11.8	Jiang et al. (2020)

(Continued)

Table 4.4 (Continued)

No.	Instrument	Starch concentration	Scanning rate (°C/min)	Scanning range (°C)	T_o (°C)	T_p (°C)	T_c (°C)	ΔH (J/g)^A	References
6	DSC	33% (w/w)	10	20–90	58.0–67.9	64.2–75.1	74.6–85.8	9.8–12.2	López-Fernández et al. (2021)
1	M				48		62		Wolf et al. (1950)
1	PM	1%	1		57		64		Atwell et al. (1983)
1	KHSM				61	64.5	68		Lorenz (1990)
1	RH	5% (w/w)	2		55				Praznik et al. (1999)

No.: number of samples studied; DSC: differential scanning calorimetry; M: microscope observation; PM: polarized microscope; KHSM: polarized microscope equipped with a Kofler hot stage; RH rheometer equipped with cone/plate; A: the unit cal/g is converted to J/g by a factor of 4.184; B: starch equivalent based on starch and moisture contents; T_o: onset temperature; T_p: peak temperature; T_c: conclusion temperature; ΔH: enthalpy change.

Source: Adapted from Li and Zhu (2018b) with permission from Elsevier.

and solubilization were obtained between quinoa starches from different genotypes (Li et al., 2016; Xing et al., 2021; López-Fernández et al., 2021). The order of swelling power (SP) values followed potato starch, quinoa starch, and maize starch (Jiang et al., 2020). The water solubility index (WSI) of quinoa starch was seen to be lower than that of potato starch and was similar to that of maize starch. The experimental conditions including centrifuge conditions, starch concentrations, stirring and separation methods were seen to be different among different studies (Jan et al., 2017; Jiang et al., 2020; Li et al., 2016; Valdez-Arana et al., 2020; López-Fernández et al., 2021). Consequently, it is impossible to compare the results of different reports.

Amylose was suggested to be able to restrict granular swelling of starch as indicated by the negative correlation between amylose contents and SP/WSI (Li et al., 2016; Lindeboom, Chang, Falk, et al., 2005). Quinoa amylopectin structure was related to its SP and WSI. For example, the amylose leaching (AML), WSI and SP at 95°C were signficantly correlated to the composition of internal B-chains of quinoa amylopectin (Li & Zhu, 2017a). Therefore both amylose and amylopectin are important for the determination of the swelling and solubility of starch.

4.5.3 Rheological properties

Great variation in pasting behaviors of quinoa starch samples of different genotypes was recorded in diverse studies (Table 4.5). The variation could be attributed to their different structures and compositions. The high peak viscosity (PV) was in agreement with its high SP of quinoa starch (Section 4.5.2) (Li et al., 2016). Long starch chains can inhibit the swelling of starch particles by better maintaining the granular structure (Srichuwong & Jane, 2007). Higher proportions of long amylopectin chains and amylose of quinoa starch contributed to its higher PV (Li & Zhu, 2017b). Some other studies reported the nagative correlation patterns between PV and amylose content and parabolic correlations between them (Lindeboom, Chang, Falk, et al., 2005). These different correlation patterns showed the complicated relationship between the two parameters. A range of factors including amylopectin unit-chain composition, granule volume, and the presence of minor nonstarch components may play roles in determining the pasting property of starch (Debet & Gidley, 2007; Srichuwong et al., 2017). In different comparative studies, quinoa starch had higher PV than starches from many other plants including sorghum, kañiwa, wheat, garden orach, buckwheat, maize, millet, and amaranth starch, but lower than potato starch (Jiang et al., 2020; Praznik et al., 1999; Qian & Kuhn, 1999; Srichuwong et al., 2017; Steffolani et al., 2013; Wright et al., 2002).

Dynamic oscillatory and flow analysis of quinoa starch have been done (Jan et al., 2017; Jiang et al., 2020; Li & Zhu, 2018a; 2018b; 2018c; Lindeboom, Chang, Falk, et al., 2005; Praznik et al., 1999). Diversity in these rheological properties of starch pastes of different quinoa varieties has been obtained (Jiang et al., 2020, Li & Zhu, 2018a, 2018b, 2018c). For example, the maximum of storage modulus (G')

Table 4.5 Pasting properties of quinoa starch.

No.	Instrument	Starch concentration	Viscosity unit	PT (°C)	PV	HPV	BD	CPV	SB	PKT (°C)	References
1	RVA	9.2% (w/w)	RVU	66.8	345.9	271.1	74.8	416.2	145.1		Qian and Kuhn (1999)
2	RVA	9.2% (w/w)	RVU	63.5–65.3	367.3–402.5	357.9–369.4	9.4–33.1	495.1–495.5	125.8–138.0		Wright et al. (2002)
8	RVA	~10% (w/w)	RVU	63.3–64.0	3.6–25.8	2.9–17.9	0.4–8.8	4.2–28.9	1.3–12.8		Lindeboom, Chang, Falk et al. (2005)
3	RVA	10.7% (w/w)	cP		3,322–4,306		439–1,161	3,725–4,109	676–963	62.7	Steffolani et al. (2013)
2	RVA	~10% (w/w)	RVU	69.45–72.85	207.33–386.4	149.16–307.8	58.17–78.58	223.67–405.75	74.5–97.9		Jan et al. (2017)
1	RVA	8% (w/w)	cP		2,860						Srichuwong et al. (2017)
3	RVA	10.7% (w/w)	cP	66.1–68.7	4498–5233		1035–2142	4442–5225	1351–1596		Valdez-Arana et al. (2020)
4	RVA		cP		2986–3551	2250–3205	313–733	2692–3705	442–730		Jiang et al. (2020)

1[A]	BV	5%–10%	BU	60–71	980–1960	790–1750		1200–2650		70	Atwell et al. (1983), Lorenz (1990), Ahamed, Singhal, Kulkarni, Pal, et al. (1996), Praznik et al. (1999)
26	RH	9.1% (w/w)	cP	51.3–65.8	3,756–5,688	1,920–3,588	900–2,604	3,804–5,436	1,164–2,136	75.5–94.1	Li et al. (2016)

No.: number of samples studied; RH: Rheometer equipped with a starch cell; RVA: Rapid Visco-analzer with RVU as the viscosity unit; BV: Brabender viscograph with BU as the viscosity unit; cP: centipoise, 1 cP equals to 1 mPa · s. PT: pasting temperature; PV: peak viscosity; PKT: peak temperature; HPV: hot paste viscosity; CPV: cold paste viscosity; BD: breakdown; SB: setback; A: number of samples studied by each study.

Source: Adapted from Li and Zhu (2018b) with permission from Elsevier.

of quinoa starches from nine varieties during the heating process of temperature sweep of dynamic oscillatory analysis ranged from 3997 to 7510 Pa, whereas that of loss modulus (G'') varied from 591 to 1019 Pa (Li & Zhu, 2018a, 2018b, 2018c). Among a range of different mathematical models, Mizrahi-Berk, Heinz-Casson, Herschel-Bulkley, and Robertson-Stiff models well described the flow curves of starch pastes of nine quinoa varieties (Li & Zhu, 2018a, 2018b, 2018c). The rheological stability of the starch pastes was found to be affected by amylose content and amylopectin unit-chain composition (Li & Zhu, 2018a, 2018b, 2018c; Lindeboom, Chang, Falk, et al., 2005). For example, the yield stresses of quinoa starch pastes were significantly affected by the internal chain profile of the amylopectin (Li & Zhu, 2018a, 2018b, 2018c). Quinoa starch pastes had better stability against shearing than maize, potato, millet, buckwheat, and amaranth starch pastes. It had similar stability to wheat starch (Jiang et al., 2020; Praznik et al., 1999). During the temperature sweep of dynamic oscillatory analysis, quinoa starch had less G' and G'' compared to maize starch (Li & Zhu, 2018a, 2018b, 2018c). Another study showed that quinoa starch had higher G' and G'' than potato starch (Jiang et al., 2020). The dynamic rheology of quinoa starch of both large and small deformation tests was found to be controlled by a range of parameters related to both the amylose content and amylopectin fine structure (Jiang et al., 2020, Li & Zhu, 2018a, 2018b, 2018c). Overall, designer starch from quinoa grain for desired rheological functions should be better researched focusing on function-structure relationships of the components.

4.5.4 Retrogradation property

Amylopectin retrogradation has mostly been studied using differential scanning calorimetry (Jiang et al., 2020; Srichuwong et al., 2017; Steffolani et al., 2013). The retrogradation can be greatly influenced by storage conditions and starch concentration. Quinoa starch had lower retrogradation rate than starches from different plants including kañiwa, wheat, sorghum, potato, maize, and millet, but had higher retrogradation rate than amaranth starch (Jiang et al., 2020; Srichuwong et al., 2017; Steffolani et al., 2013). Germination of quinoa seeds up to 48 h decreased the amylopectin retrogradation (Xing et al., 2021). This could be attributed to the reduced chain lengths of the starch molecules from the germination process (Xing et al., 2021). The temperatures needed to melt retrograded starch were in reverse correlations with the percentage of amylopectin unit-chains of DP 6–12. They were in positive correlations with that of DP 13–24 (Srichuwong et al., 2017). The percentage of quinoa starch retrogradation was positively correlated with the amylose content (López-Fernández et al., 2021). The external chains of quinoa amylopectin were relatively short. They were less able to form helices for retrogradation (Li & Zhu, 2017b). In most of the studies, the experimental conditions were different. This difference in the conditions makes the comparison of the results from different studies impossible (López-Fernández et al., 2021; Wang, Li, et al., 2015).

4.5.4.1 Gel textural property

Diversity in gel textural property of starches from different quinoa genotypes was obtained (Li et al., 2016; Valdez-Arana et al., 2020). For example, the hardness of starch gels of 26 quinoa varieties ranged from 13.3 to 33.6 g (Li et al., 2016). Quinoa starch gel had significantly less hardness than kañiwa (*Chenopodium pallidicaule*) starch gel (Steffolani et al., 2013). Effect of composition and structure of starch on gel texture was studied (Li et al., 2016; Steffolani et al., 2013). A higher amylose content was linked with harder quinoa starch gel. Amylose molecules reassociate for starch gelation. Super-long chains of quinoa amylopectin also had gelation function contributing to the gel hardness (Li & Zhu, 2017a; Tang et al., 2002; Watanabe et al., 2007).

4.5.4.2 Syneresis of starch gels

Syneresis of starch gel tends to occur during starch retrogradation and freeze-thaw cycles (Lindeboom, Chang, Falk, et al., 2005; Valdez-Arana et al., 2020). Diversity in the syneresis of starch gels was obtained among different quinoa starch samples (Lindeboom, Chang, Falk, et al., 2005; Valdez-Arana et al., 2020). For example, the syneresis of eight quinoa starch gels after freeze-thaw process varied from 0% to 35% (Lindeboom, Chang, Falk, et al., 2005). The syneresis values of three quinoa starch gels after storage at 4°C for 144 h ranged from 32% to 57% (Valdez-Arana et al., 2020). The syneresis of starch gels is affected by different parameters including starch composition and structure, processing methods, starch concentrations, and centrifugation conditions. Only the values of the same study with the same experimental conditions may be compared meaningfully (Zhu, 2015). The relationships between syneresis and starch structure remain to be established.

4.5.5 Starch digestibility

Variation in starch digestibility was obtained between different quinoa varieties (Inouchi et al., 1999; Li et al., 2016; Lorenz, 1990; Srichuwong et al., 2017; Tang et al., 2002). For example, after 1 h of incubation with porcine pancreatic α-amylase, the degrees of hydrolysis of starches from 26 quinoa varieties varied from 48% to 81% (Li et al., 2016). In another study, Inouchi et al. (1999) found that the hydrolysis extents for four quinoa starch varieties using glucoamylase of *Aspergillus* sp. K27 varied from 54% to 66% after 1 h. All the starch granules were completely degraded after 3 h of hydrolysis (Inouchi et al., 1999).

The results of different studies cannot be compared due to the differences in settings and conditions of the experiments including incubation time and temperature, pH, types of enzymes, and buffers used. For example, the digestibility of quinoa starch in food models can be significantly affected by the experimental conditions. Tagle-Freire et al. (2022) showed that starch digestibility of cooked quinoa grains after 3 h in static gastric digestion model was much lower (2.9%) than that in dynamic gastric digestion model (29%). Comparative studies showed that quinoa starch could be hydrolyzed by enzymes more easily than starches from many other sources including wheat, millet, barley, sorghum, maize, and adzuki (Inouchi et al., 1999; Srichuwong et al., 2017; Tang et al., 2002). The quinoa starch granules are small with high specific surface area, allowing efficient interactions between starch and enzyme (Li et al., 2016; Srichuwong et al., 2017). Factors affecting the in vitro starch digestibility of quinoa starch were analyzed (Li & Zhu, 2017a; Lorenz, 1990). A higher amylose content was related with a lower starch digestibility of quinoa starch (Lorenz, 1990). More short unit-chains of amaylopectin (DP 6−12 and A_{fp}-chains) contributed to increased enzyme susceptibiltiy of quinoa starch (Li & Zhu, 2017a). Unit-chains such as A_{fp}-chains can be too short, so that they cannot form double helical structural units in starch granules and can create structural imperfection (Vamadevan & Bertoft, 2015) (Fig. 4.2).

4.6 Conclusions

There is great diversity in composition, structure, and physicochemical properties of quinoa starch among different genotypes. The starch granules have A-type polymorph. They are small (~1−2 μm). Quinoa amylose has a degree of branching compared with amylose from other sources. The amylopectin has relatively less long unit-chains and more short unit-chains as compared with amylopectins from other sources. Quinoa starch has low gelatinization temperatures and retrogradation, and high SP and PV of pasting as well as high enzyme susceptibility.

Quinoa starch research is much limited when compared with that of industrially important starches such as maize starch. Diversity in starch composition and properties affected by genetics and growing conditions remains to be studied. The composition of clusters and building blocks of quinoa amylopectin is still not studied. There has been significant discrepancy in the experimental conditions and methodologies among different researchers. As a result, the results of different reports cannot be compared. Method standardization remains to be done to maximize our research efforts.

References

Ahamed, N. T., Singhal, R. S., Kulkarni, P. R., & Pal, R. (1996). Physicochemical and functional properties of *Chenopodium quinoa* starch. *Carbohydrate Polymers*, *31*, 99−103.

Ando, H., Chen, Y. C., Tang, H. J., Shimizu, M., Watanabe, K., & Mitsunaga, T. (2002). Food components in fractions of quinoa seed. *Food Science and Technology Research*, *8*, 80−84.

Araujo-Farro, P. C., Podadera, G., Sobral, P. J., & Menegalli, F. C. (2010). Development of films based on quinoa (*Chenopodium quinoa*, Willdenow) starch. *Carbohydrate Polymers*, *81*, 839−848.

Atwell, W. A., Patrick, B. M., Johnson, L. A., & Glass, R. W. (1983). Characterization of quinoa starch. *Cereal Chemistry*, *60*, 9−11.

Bertoft, E. (2013). On the building block and backbone concepts of amylopectin structure. *Cereal Chemistry*, *90*, 294−311.

Bertoft, E., Piyachomkwan, K., Chatakanonda, P., & Sriroth, K. (2008). Internal unit chain composition in amylopectins. *Carbohydrate Polymers*, *74*, 527–543.

de Briceño, Z. S., & Briceño, O. (1980). Evaluacion de la composición quimica (composición nutricional) de algunas entradas de quinua (*Chenopodium quinoa* Willd) del banco de germoplasma de la Universidad Nacional Técnica del Altiplano (Perú). *Anales Científicos*, *18*, 125–134.

Debet, M. R., & Gidley, M. J. (2007). Why do gelatinized starch granules not dissolve completely? Roles for amylose, protein, and lipid in granule "Ghost" integrity. *Journal of Agricultural and Food Chemistry*, *55*, 4752–4760.

Gibson, T. S., Solah, V. A., & McCleary, B. V. (1997). A procedure to measure amylose in cereal starches and flours with concanavalin A. *Journal of Cereal Science*, *25*, 111–119.

Goldstein, I., Hollerman, C., & Merrick, J. M. (1965). Protein-carbohydrate interaction I. The interaction of polysaccharides with concanavalin A. *Biochimica et Biophysica Acta (BBA)-General Subjects*, *97*, 68–76.

Hussain, M. I., Farooq, M., Syed, Q. A., Ishaq, A., Al-Ghamdi, A. A., & Hatamleh, A. A. (2021). Botany, nutritional value, phytochemical composition and biological activities of quinoa. *Plants*, *10*, 2258.

Inouchi, N., Nishi, K., Tanaka, S., Asai, M., Kawase, Y., Hata, Y., et al. (1999). Characterization of amaranth and quinoa starches. *Journal of Applied Glycoscience*, *46*, 233–240.

Jan, K. N., Panesar, P. S., Rana, J. C., & Singh, S. (2017). Structural, thermal and rheological properties of starches isolated from Indian quinoa varieties. *International Journal of Biological Macromolecules*, *102*, 315–322.

Jiang, F., Du, C., Guo, Y., Fu, J., Jiang, W., & Du, S. (2020). Physicochemical and structural properties of starches isolated from quinoa varieties. *Food Hydrocolloids*, *101*, 105515.

Li, G., & Zhu, F. (2017a). Amylopectin molecular structure in relation to physicochemical properties of quinoa starch. *Carbohydrate Polymers*, *164*, 396–402.

Li, G., & Zhu, F. (2017b). Molecular structure of quinoa starch. *Carbohydrate Polymers*, *158*, 124–132.

Li, G., & Zhu, F. (2017c). Physicochemical properties of quinoa flour as affected by starch interactions. *Food Chemistry*, *221*, 1560–1568.

Li, G., & Zhu, F. (2018a). Effect of high pressure on rheological and thermal properties of quinoa and maize starches. *Food Chemistry*, *241*, 380–386.

Li, G., & Zhu, F. (2018b). Quinoa starch: Structure, properties, and applications. *Carbohydrate Polymers*, *181*, 851–861.

Li, G., & Zhu, F. (2018c). Rheological properties in relation to molecular structure of quinoa starch. *International Journal of Biological Macromolecules*, *114*, 767–775.

Li, G., Hemar, Y., & Zhu, F. (2021). Relationships between supramolecular organization and amylopectin fine structure of quinoa starch. *Food Hydrocolloids*, *117*, 106685.

Li, G., Wang, S., & Zhu, F. (2016). Physicochemical properties of quinoa starch. *Carbohydrate Polymers*, *137*, 328–338.

Lindeboom, N., Chang, L. R., Tyler, R. T., & Chibbar, R. N. (2005). Granule-bound starch synthase I (GBSSI) in quinoa (*Chenopodium quinoa* Willd.) and its relationship to amylose content. *Cereal Chemistry*, *82*, 246–250.

Lindeboom, N., Chang, P. R., Falk, K. C., & Tyler, R. T. (2005). Characteristics of starch from eight quinoa lines. *Cereal Chemistry*, *82*, 216–222.

Linsberger-Martin, G., Lukasch, B., & Berghofer, E. (2012). Effects of high hydrostatic pressure on the RS content of amaranth, quinoa and wheat starch. *Starch/Stärke*, *64*, 157–165.

López-Fernández, M. P., Rodríguez, S. D., Favre, L. C., Busch, V. M., & Del Pilar Buera, M. (2021). Physicochemical, thermal and rheological properties of isolated Argentina quinoa starch. *LWT-Food Science and Technology*, *135*, 110113.

Lorenz, K. (1990). Quinoa (*Chenopodium quinoa*) starch—physico-chemical properties and functional characteristics. *Starch/Stärke*, *42*, 81—86.

Ma, M., Wen, Y., Zhang, C., Xu, Z., Li, H., Sui, Z., & Corke, H. (2022). Extraction and characterization of starch granule-associated surface and channel lipids from small-granule starches that affect physicochemical properties. *Food Hydrocolloids*, *126*, 107370.

Ma, M., Zhu, H., Liu, Z., Sui, Z., & Corke, H. (2022). Removal of starch granule-associated proteins alters the physicochemical properties of diverse small granule starches. *Food Hydrocolloids*, *124*, 107318.

Mundigler, N. (1998). Isolation and determination of starch from amaranth (*Amaranthus cruentus*) and quinoa (*Chenopodium quinoa*). *Starch/Stärke*, *50*, 67—69.

Nascimento, A. C., Mota, C., Coelho, I., Gueifão, S., Santos, M., Matos, A. S., et al. (2014). Characterisation of nutrient profile of quinoa (*Chenopodium quinoa*), amaranth (*Amaranthus caudatus*), and purple corn (*Zea mays* L.) consumed in the North of Argentina: Proximates, minerals and trace elements. *Food Chemistry*, *148*, 420—426.

Pagno, C. H., Costa, T. M., de Menezes, E. W., Benvenutti, E. V., Hertz, P. F., Matte, C. R., et al. (2015). Development of active biofilms of quinoa (*Chenopodium quinoa* W.) starch containing gold nanoparticles and evaluation of antimicrobial activity. *Food Chemistry*, *173*, 755—762.

Peat, S., Whelan, W., & Thomas, G. J. (1952). Evidence of multiple branching in waxy maize starch. *Journal of the Chemical Society*, 4536—4538.

Pérez, S., & Bertoft, E. (2010). The molecular structures of starch components and their contribution to the architecture of starch granules: A comprehensive review. *Starch/Stärke*, *62*, 389—420.

Praznik, W., Mundigler, N., Kogler, A., Pelzl, B., Huber, A., & Wollendorfer, M. (1999). Molecular background of technological properties of selected starches. *Starch/Stärke*, *51*, 197—211.

Qian, J., & Kuhn, M. (1999). Characterization of *Amaranthus cruentus* and *Chenopodium quinoa* starch. *Starch/Stärke*, *51*, 116—120.

Ruales, J., & Nair, B. M. (1994). Properties of starch and dietary fibre in raw and processed quinoa (*Chenopodium quinoa*, Willd) seeds. *Plant Foods for Human Nutrition*, *45*, 223—246.

Sjöö, M., & Nilsson, L. (Eds.), (2018). *Starch in food: Structure, function and applications* (Second ed.). Woodhead Publishing.

Srichuwong, S., Curti, D., Austin, S., King, R., Lamothe, L., & Gloria-Hernandez, H. (2017). Physicochemical properties and starch digestibility of whole grain sorghums, millet, quinoa and amaranth flours, as affected by starch and non-starch constituents. *Food Chemistry*, *233*, 1—10.

Srichuwong, S., & Jane, J. (2007). Physicochemical properties of starch affected by molecular composition and structures: A review. *Food Science and Biotechnology*, *16*, 663—674.

Steffolani, M. E., León, A. E., & Pérez, G. T. (2013). Study of the physicochemical and functional characterization of quinoa and kañiwa starches. *Starch/Stärke*, *65*, 976—983.

Tagle-Freire, D., Mennah-Govela, Y., & Bornhorst, G. M. (2022). Starch and protein hydrolysis in cooked quinoa (*Chenopodium quinoa* Willd.) during static and dynamic *in vitro* oral and gastric digestion. *Food & Function*, *13*, 920—932.

Tang, H., Watanabe, K., & Mitsunaga, T. (2002). Characterization of storage starches from quinoa, barley and adzuki seeds. *Carbohydrate Polymers*, *49*, 13—22.

Tari, T. A., Annapure, U. S., Singhal, R. S., & Kulkarni, P. R. (2003). Starch-based spherical aggregates: screening of small granule sized starches for entrapment of a model flavouring compound, vanillin. *Carbohydrate Polymers*, *53*, 45–51.

Valdez-Arana, J., Steffolani, M. E., Repo-Carrasco-Valencia, R., Pérez, G. T., & Condezo-Hoyos, L. (2020). Physicochemical and functional properties of isolated starch and their correlation with flour from the Andean Peruvian quinoa varieties. *International Journal of Biological Macromolecules*, *147*, 997–1007.

Vamadevan, V., & Bertoft, E. (2015). Structure-function relationships of starch components. *Starch/Stärke*, *67*, 55–68.

Varriano-Marston, E., & DeFrancisco, A. (1984). Ultrastructure of quinoa fruit (*Chenopodium quinoa* Willd). *Food Structure*, *3*, Article 9.

Vilaplana, F., Hasjim, J., & Gilbert, R. G. (2012). Amylose content in starches: Toward optimal definition and validating experimental methods. *Carbohydrate Polymers*, *88*, 103–111.

Wang, S., Li, C., Copeland, L., Niu, Q., & Wang, S. (2015). Starch retrogradation: a comprehensive review. *Comprehensive Reviews in Food Science and Food Safety*, *14*, 568–585.

Wang, S., Opassathavorn, A., & Zhu, F. (2015). Influence of quinoa flour on quality characteristics of cookie, bread, and Chinese steamed bread. *Journal of Texture Studies*, *46*, 281–292.

Watanabe, K., Peng, N. L., Tang, H. J., & Mitsunaga, T. (2007). Molecular structural characteristics of quinoa starch. *Food Science and Technology Research*, *13*, 73–76.

Wolf, M. J., Macmasters, M. M., & Rist, C. E. (1950). Some characteristics of the starches of three South American seeds used for food. *Cereal Chemistry*, *27*, 219–222.

Wright, K. H., Huber, K. C., Fairbanks, D. J., & Huber, C. S. (2002). Isolation and characterization of *Atriplex hortensis* and sweet *Chenopodium quinoa* starches. *Cereal Chemistry*, *79*, 715–719.

Xing, B., Teng, C., Sun, M., Zhang, Q., Zhou, B., Cui, H., Ren, G., Yang, X., & Qin, P. (2021). Effect of germination treatment on the structural and physicochemical properties of quinoa starch. *Food Hydrocolloids*, *115*, 106604.

Zhu, F. (2015). Isolation, composition, structure, properties, modifications, and uses of yam starch. *Comprehensive Reviews in Food Science and Food Safety*, *14*, 357–386.

Zhu, F. (2018). Relationships between amylopectin internal molecular structure and physicochemical properties of starch. *Trends in Food Science & Technology*, *78*, 234–242.

Quinoa dietary fiber and nonstarch polysaccharides

5

5.1 Introduction

The dietary fiber of quinoa consists of pectin, cellulose, xyloglucan, resistant starch, and lignin (Zhang, Wang et al., 2020; Zhu, 2020). The diverse claimed health effects of quinoa seeds such as antioxidant, immunomodulative, and anticancer properties could be partially attributed to the dietary fibers (Little et al., 2021; Tang & Tsao, 2017; Zhu, 2020). Quinoa has similar processing and cooking properties as cereals. The dietary fiber components of quinoa play a role in the cooking and processing of quinoa (Chen et al., 2021; Li & Zhu, 2017).

"Healthy" and "novel" food products such as bakery and meat-based products have been developed using quinoa-related products such as seeds and flours (Ballester-Sánchez et al., 2020; Fernández-López et al., 2020, 2021). The quality of the food products is determined by the properties of the chemical components including dietary fiber. Overall, the dietary fiber of quinoa is much less studied compared to other components including starch, protein, polyphenols, and saponins. Dietary fiber components include resistant starch, lignin, nondigestible oligosaccharides, and nonstarch polysaccharides (Phillips et al., 2019; Zhang, Wang et al., 2020). They are largely located in the bran layers of the grains except for resistant starch (Ninfali et al., 2020). Thin cell walls surrounding starch granules in quinoa perisperm contain small amounts of fiber, though this aspect has not been studied. Fiber-rich fractions of quinoa can be obtained using wet or dry milling process (Ballester-Sánchez et al., 2020). Less intake of dietary fiber contributes to higher risk of chronic diseases such as cardiovascular diseases, certain types of cancer, overweight, obesity, and diabetes (Gunness & Gidley, 2010; Khan et al., 2019; Nie et al., 2017; Wang et al., 2016; Zhang, Dong et al., 2020). Dietary fiber regulates a range of physiological functions and can enhance growth of gut microbiota (Gunness & Gidley, 2010; Muthusamy et al., 2021; Zeyneb et al., 2021; Zhang, Dong et al., 2020). Dietary fiber can be modified to create of wide range of functionalities to suit diverse applications (Gan et al., 2021). It plays significant functional roles in properties and quality of foods such as gluten-free breads (Djordjević et al., 2022). In different studies of quinoa dietary fiber, the methodologies used tend to be inconsistent with variations. This makes the results of the fiber functions difficult to compare.

A previous review summarized the composition, biological properties, and food uses of quinoa dietary fiber polysaccharide from the literature published up to 2020 (Zhu, 2020). Based on the research of Zhu (2020) (excerpts reprinted with permission from Elsevier), this chapter updates dietary fiber composition, structure,

Quinoa. DOI: https://doi.org/10.1016/B978-0-323-99909-0.00012-X

physicochemical properties, and biological functions of quinoa seeds. The applications of quinoa fiber and polysaccharide fractions are described in Chapters 12 and 13.

5.2 Dietary fiber composition of quinoa

Dietary fiber can be categorized into insoluble and soluble fiber fractions based on their water solubility. Overall, the insoluble fiber contents of quinoa are much higher than those of soluble fiber (Table 5.1). Variations in the fiber composition of quinoa seeds have been recorded (Miranda et al., 2012; Repo-Carrasco-Valencia & Serna, 2011; Sobota et al., 2020). For example, the ranges of total, insoluble, and soluble fiber contents were 14% – 20%, 10% – 14%, and 3.7% – 5.9%, respectively (Sobota et al., 2020). Factors other than quinoa genetics can affect the fiber composition. Growing locations affected the fiber composition of quinoa (Miranda et al., 2013). Extrusion processing decreased the total fiber content and increased the soluble fiber content (Repo-Carrasco-Valencia & Serna, 2011). This could be due to the chain scissions by mechanical forces during the extrusion. Cooking of quinoa seeds in water tended to decrease the fiber content by solubilizing the soluble fraction (Ruales & Nair, 1994). Resistant starch is a component of dietary fiber. There has been a lack of research on resistant starch content of quinoa products. Polyphenols are bound to insoluble fiber fraction of quinoa (Chen et al., 2021). In a comparative study, quinoa insoluble fiber fraction contained much fewer bound polyphenols (2.5 mg gallic acid equivalents/g, dry basis) than wheat insoluble fiber fraction (6.4 mg gallic acid equivalents/g, dry basis).

The fiber composition of quinoa was like that of amaranth grains. Proportionally, quinoa and amaranth had more soluble fiber than cereals. The fiber compositions of

Table 5.1 Dietary fiber composition of quinoa grains.

Number of genotypes	Total fiber fraction (%)	Insoluble fraction (%)	Soluble fraction (%)	References
4	14 – 16	12 – 14	1.4 – 1.6	Repo-Carrasco-Valencia and Serna (2011)
6	8 – 12			Miranda et al. (2012)
2[a]	12 – 15	10 – 12	0.4 – 3	Miranda et al. (2013)
1	10	7.8	2.2	Lamothe et al. (2015)
7	8.8 – 14			Nowak et al. (2016)
1	9.5	5.6	3.9	Srichuwong et al. (2017)
2	11 – 12	6 – 8	5 – 7	Kurek et al. (2018)
3	12	8 – 9	3 – 4	Jimenez et al. (2019)
25	14–20	10 – 14	3.7 – 5.9	Sobota et al. (2020)
1	12.1			Zhang et al. (2022)

[a]3 locations in Chile.

quinoa and amaranth were more similar to fruits than to cereals (Lamothe et al., 2015). Analytical methods used for quantification of dietary fibers significantly affect the results (Srichuwong et al., 2017). For example, the contents of total, soluble, and insoluble fibers of quinoa and amaranth seeds measured using AOAC 2011.25 were higher than those obtained using AOAC 991.43 (Srichuwong et al., 2017). This was due to the different recovery rates of resistant starch. Crude fiber measurement is incomplete, whereas the method of Prosky et al. (1985) is more complete to measure the fiber fractions (Nowak et al., 2016). Therefore the results from different reports should be critically compared taking the analytical methods into consideration.

5.3 Isolation of fiber fractions and polysaccharides of quinoa

The milling processes of quinoa grain are described in Chapter 3. Briefly, the first step to obtain fiber fractions and polysaccharides is to mill the seeds into flour (Ballester-Sánchez et al., 2020; Kurek et al., 2018; Lamothe et al., 2015; Muthusamy et al., 2021). Flour can be defatted to remove lipids using organic solvents. The other components including starch and proteins are removed using solvent-precipitation or enzyme hydrolysis (Kurek et al., 2018). Ultrasound/microwave processing can improve the extraction yields of fiber fractions (Hromádková & Ebringerová, 2003). For example, quinoa flour was suspended in phosphate buffer solution (pH 6.8, 20 mM) before ultrasound processing (150 W, 37 kHz, 7 min) and then adding amylase to remove starch (Kurek et al., 2018). The hydrolysis was done at 75°C for 30 min before adding subtilisin A for proteolysis for another 30 min (60°C, pH 7.5). The hydrolysates were centrifuged to obtain the supernatant containing the soluble fiber fraction, which was recovered using 96% ethanol. The fiber cake from the centrifugation step was the insoluble fraction. Enzyme (Termamyl SC and subtilisin A of *Bacillus licheniformis*) and ultrasound assistance gave the highest yield of fiber extraction (Kurek et al., 2018) (Fig. 5.1).

Fiber-rich fraction can also be obtained using dry milling (Ballester-Sánchez et al., 2020) (Fig. 5.2). The seeds were tempered to obtain a moisture content of 16% before passing through break rolls. Rotary sieves were used to obtain different fractions. The bran fraction was obtained to represent the fiber-rich fraction (Fig. 5.2). The two fractions from wet and dry millings had similar color. The fraction from wet milling had smaller size and higher dispersity than that from dry milling (Ballester-Sánchez et al., 2020).

The obtained crude fiber fractions may be subjected to further purification and separation to obtain polysaccharide fractions with a much higher purity (Fan et al., 2019). A combination of ion-exchange (e.g., anion-exchange) and gel permeation (e.g., Sepharose and Sephacryl gels) chromatography can be used for purification and fractionation of the polysaccharides (Fan et al., 2019; Yao et al., 2014; Teng et al., 2021). The ion-exchange chromatography (e.g., DEAE columns) separates acidic polysaccharides from neutral fractions. Gel permeation chromatography

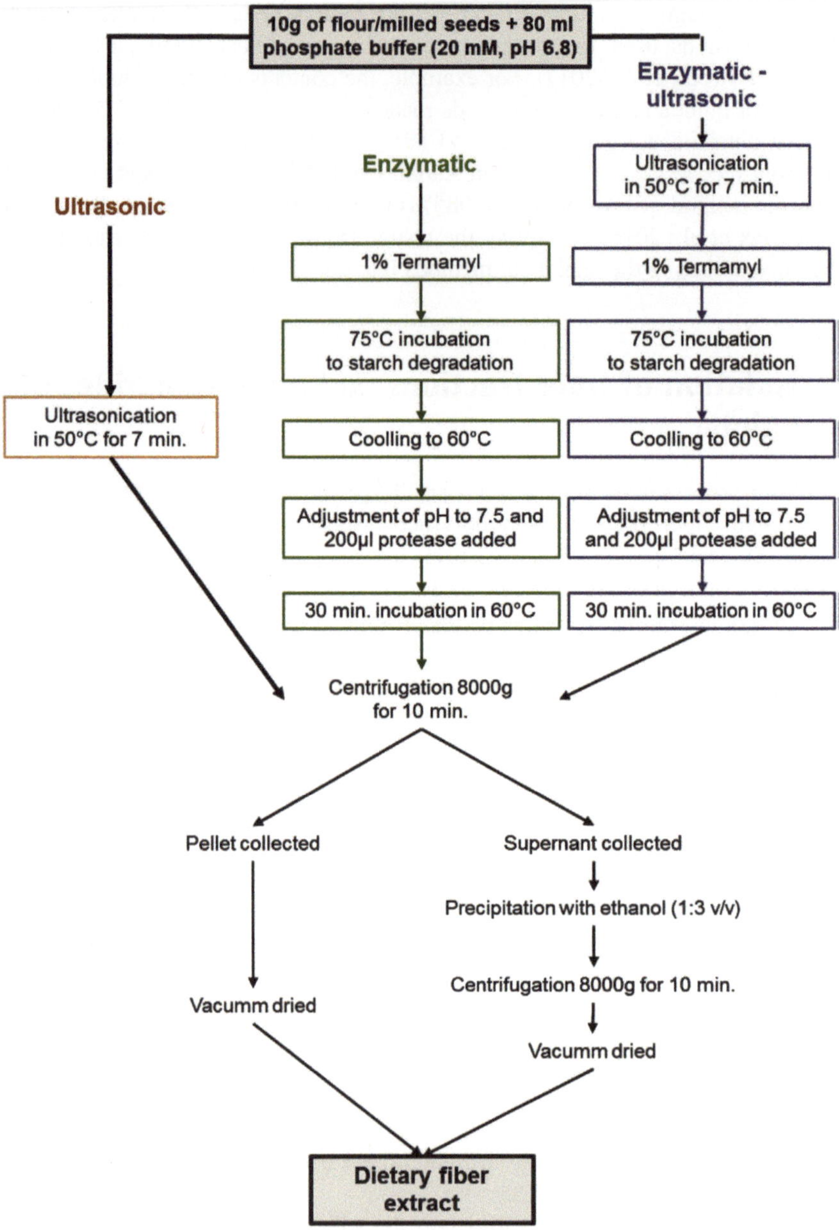

Figure 5.1 Procedures for enzyme/ultrasound assisted extraction of fiber fractions from quinoa seeds (Kurek et al., 2018).
Source: Reprinted with permission from Elsevier.

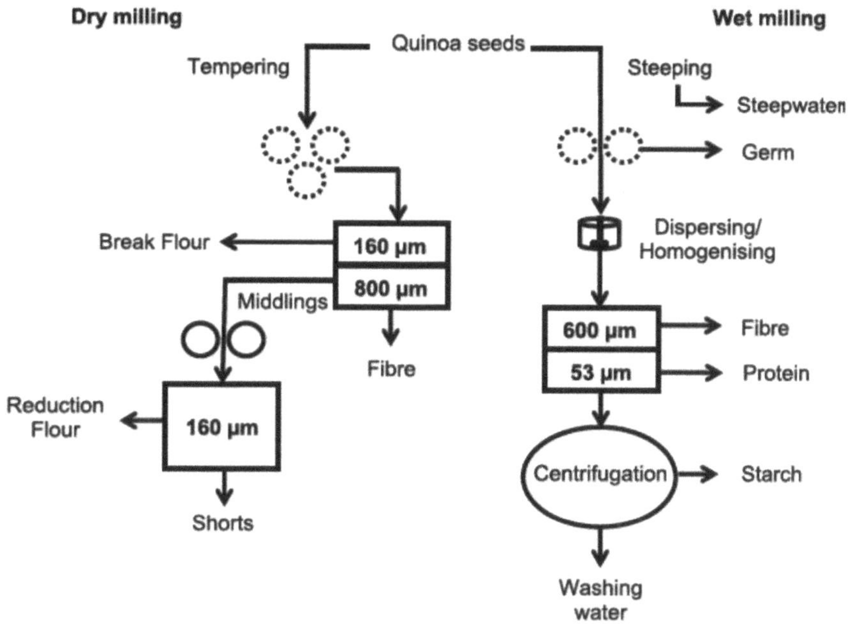

Figure 5.2 Dry and wet milling of quinoa seeds (Ballester-Sánchez et al., 2020).
Source: Reprinted with permission from Elsevier.

separates different fractions based on molecular "size" (Nie et al., 2017). Different fractions of polysaccharides obtained tend to have different structures, properties, and uses as described in the sections below. The instrumental and experimental conditions of extractions should be carefully selected for different grains due to the differences in the structure and composition of polysaccharides.

5.3.1 Structure of fiber polysaccharides of quinoa

5.3.1.1 Chemical structure

The fiber components of quinoa seeds (1 genotype) were pectins (4.7%), hemicellulose (41%), lignins (1.7%), and cellulose (52%) (Zhang, Wang et al., 2020). The use of "hemicellulose" should be discouraged to avoid confusion. The insoluble fiber fraction of quinoa consisted of homogalacturonans, which were interspersed with rhamnogalacturonan I units having galactans and arabinans as sidechains (55%), xyloglucans (30%) mostly with sidechains of di- and trisaccharides and cellulose. The soluble fiber fraction mostly consisted of arabinans and homogalacturonans (Lamothe et al., 2015). Arabinans as sidechain of quinoa pectin had a sidechain of β-(1→3)-linked arabinobiose (Wefers et al., 2014). Alkaline extracts contained rhamnogalacturonan type I (branched at O-3) with arabinans and galactans as sidechains (Cordeiro et al., 2012).

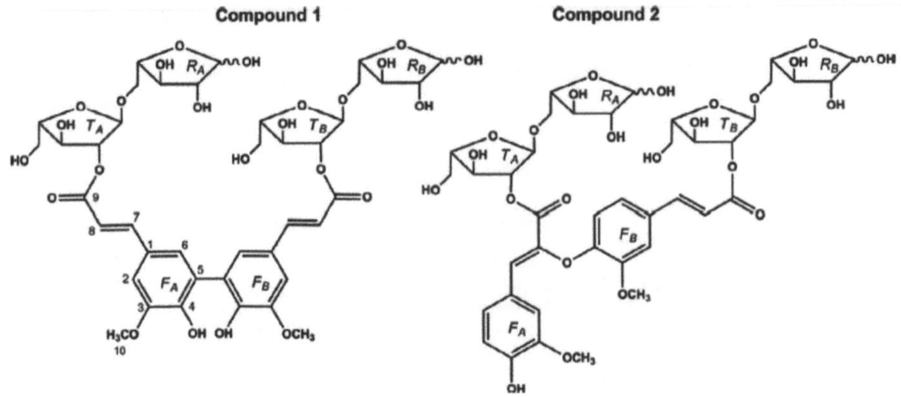

Figure 5.3 Chemical structure of di-feruloyated oligosaccharides released from quinoa fiber polysaccharides (Wefers, Gmeiner et al., 2015).
Source: Reprinted with permission from Elsevier.

Diferulic acid oligosaccharides were found in the pectin fraction of quinoa seeds after hydrolysis (Wefers, Gmeiner et al., 2015) (Fig. 5.3). 8-O-4- and 5−5-diferulic acid attached to O-2-position of the nonreducing ends of 2 (1→5)-arabinobioses (Wefers, Gmeiner et al., 2015). Diferulic acid was intermolecularly and/or intramolecularly involved with the arabinan sidechains of the pectin (Wefers, Gmeiner et al., 2015). Pectin sidechains (arabinans) were linked to ferulic acid as well as *p*-coumaric acid. The oligosaccharides O-(2-O-trans-coumaroyl-α-L-arabinofuranosyl)-(1→5)-L-arabinofuranose and O-(2-O-*trans*-feruloyl-α-L-arabinofuranosyl)-(1→5)-L-arabinofuranose were found in quinoa cell wall hydrolysates (Wefers et al., 2015b).

Teng et al. (2021) obtained a polysaccharide fraction from alkaline extract of quinoa seeds. The polysaccharide fraction of glucosyl units was purified using column chromatography and the molecular weight was determined to be 2.0×10^3 Da. The main sugar linkage was →4)-α-D-Glcp-(1→glycoside linkage). α-D-Glcp-(1→(the terminal group) was bonded by O-6 to the main chain (Teng et al., 2021). It was not certain if this polysaccharide was degraded from quinoa starch during the extraction process. Cao et al. (2022) obtained a hot water-soluble polysaccharide fraction from quinoa seeds. The fraction was further treated with α-amylase and proteinase for purification. This polysaccharide fraction was composed of (1→3, 4), (1→6), (1→3), and (1→4) linked galactopyranosyl units with glucopyranosyl and arabinopyranosyl residues (Cao et al., 2022).

Overall, the nonstarch polysaccharides of quinoa were structurally similar to those of other pseudocereals including amaranth and buckwheat (Wefers & Bunzel, 2015). The composition of nonstarch polysaccharides of quinoa was more similar to fruits and vegetables than the cereals (Lamothe et al., 2015). Arabinoxylans and β-glucans are the two major fiber components in cereals (Cui & Wang, 2009). For example, wheat and rye are rich in arabinoxylans, whereas oats and barley contain high levels of β-glucans (Cui & Wang, 2009).

5.3.1.2 *Physical structure*

A few studies on the physical structure of dietary fiber of quinoa seeds were done. Most of them were focused on the physical structure of dietary fiber fractions instead of purified polysaccharides (Chen et al., 2021; Kurek et al., 2018; Ren & Liu, 2020; Zhang, Wang et al., 2020). The physical structure of dietary fiber fractions containing nonstarch polysaccharides isolated from grains was affected by purification and separation methods (Ren & Liu, 2020). The separation process (resin adsorption, anion-exchange, and gel filtration) led to the formation of crystals and triple helical structure, smaller and more uniformed particles of isolated fiber fraction of quinoa (Ren & Liu, 2020). Wide angle X-ray scattering analysis of isolated fiber fractions of quinoa showed a major peak of the diffractogram at 21°C (Zhang, Wang et al., 2020). However, the nature of this peak in relation to fiber components is unknown. A comparative study showed that quinoa soluble fiber had more compacted structure than wheat soluble fiber (Chen et al., 2021). Volume moment mean diameter (D(4, 3)) of fiber fractions from quinoa flour ranged from 180 to 230 μm, and the sizes were largely dependent on isolation method (e.g., with or without enzyme/ultrasound assistance) (Kurek et al., 2018). Another study showed completely different particle sizes of purified quinoa polysaccharides to be 0.5 − 1 μm (Ren & Liu, 2020). Therefore morphology of polysaccharides and fiber fractions are significantly affected by the separation and purification procedures of preparation. Overall, the physical structure of polysaccharides and fiber fractions remains to be correlated with the functional and biological properties for specific applications.

5.4 Physicochemical properties of dietary fiber of quinoa

The physicochemical properties of dietary fiber fractions of quinoa grains have been studied (Table 5.2).

5.4.1 *Thermal and rheological properties*

The thermal stability of quinoa dietary fibers extracted from different genotypes and using different extraction methods was studied (Kurek et al., 2018; Liu et al., 2020) (Fig. 5.4). The analysis was done without the addition of water into the fiber samples, though the quality of the data reported is in question. The fiber samples produced using ultrasound-assisted process had the highest enthalpy change (ΔH). Samples obtained using enzymes tended to have higher temperatures of endothermic peaks than those obtained using ultrasound. The thermal properties of dietary fiber fractions of quinoa were like those of the fractions of amaranth and millets, though the results were inconclusive (Kurek et al., 2018). Soluble dietary fibers of red seeds had higher thermal stability than those of white and black seeds (Liu et al., 2020). The onset and conclusion temperatures of endothermic peak of quinoa fiber fractions were around 160°C − 170°C and 170°C−180°C, whereas the enthalpy changes were 130−180 J/g (Kurek et al., 2018). It is not clear if the

Table 5.2 Physicochemical properties of dietary fiber fractions of quinoa grains.

Property	Samples	Methods	Major findings	Reference
Thermal stability	Dietary fibers extracted from quinoa flour and whole seeds. The extraction process was assisted by ultrasound and/or enzymes	Differential scanning calorimetry, 20 to 200°C at 10°C/min	The fiber samples produced using ultrasound-assisted process had the highest ΔH. Samples obtained using enzymes tended to have higher temperatures of endothermic peaks than those obtained using ultrasound. The thermal properties of dietary fiber fractions of quinoa were similar to those of the fractions of amaranth and millets, though the results were inconclusive	Kurek et al. (2018)
Thermal stability	Soluble dietary fibers of white, red, and black seeds	Thermal gravimetric analyzer, 30°C–600°C, 10°C/min	Soluble dietary fibers of red seeds had higher thermal stability than those of white and black seeds (Fig. 5.4)	Liu et al. (2020)
Rheological properties	Soluble dietary fibers of white, red and black seeds	Flow and oscillatory dynamic properties	Viscosity of gels of fibers from red seeds was lower than that of the fiber gels from white and black seeds during the shear-thinning process. During the oscillatory process, increasing angular frequency rendered more gel-like property to the fibers of red seeds than to those of black and white seeds	Liu et al. (2020)
Water-holding capacity	Dietary fiber fractions extracted from flour or whole seeds using ultrasound and/or enzymes	Fiber (0.5 g) and water (20 mL) were mixed at room temperature for 24 h before centrifugation at 4000 g for 10 min	Quinoa fiber fractions had higher water-holding capacity than amaranth and millet fiber fractions. The effect of ultrasound and/or enzyme treatments on the water-holding capacities was not clear	Kurek et al. (2018)
Water-holding capacity	Soluble dietary fibers of white, red and black seeds	Soluble fiber (1 g) and water (40 mL) were mixed	Soluble dietary fibers of red seeds had higher water-binding capacity than those of white and black seeds	Liu et al. (2020)

Water-holding capacity	Soluble and insoluble dietary fibers	Soluble fiber (1 g) and water (30 mL) were mixed at room temperature for 18 h before centrifugation at 3,000 g for 20 min	Quinoa insoluble fiber had higher water-holding capacity (3.7 g/g) than wheat insoluble fiber (3.1 g/g)	Chen et al. (2021)
Oil-holding capacity	Dietary fibers extracted from quinoa flour and whole seeds. The extraction process was assisted by ultrasound and/or enzymes	Soluble fiber (0.5 g) and sunflower seed oil (10 mL) were mixed at room temperature for 1 h before centrifugation at 4000 g for 10 min	The comparative analysis in the oil-holding capacities between the fiber fractions of quinoa, amaranth and millets was inconclusive. Combined ultrasound and enzyme treatments gave the fibers higher oil-holding capacities than individual treatments	Kurek et al. (2018)
Oil-holding capacity	Soluble and insoluble dietary fibers	Soluble fiber (0.2 g) and sunflower seed oil (30 mL) were mixed at room temperature for 18 h before centrifugation at 2000 g for 20 min	Insoluble fiber had much higher oil-holding capacity (1.22 g/g) than soluble fiber (0.55 g/g). Quinoa soluble fiber had higher oil-holding capacity (0.55 g/g) than wheat soluble fiber (0.35 g/g). Quinoa insoluble fiber had lower oil-holding capacity (1.22 g/g) than wheat insoluble fiber (1.50 g/g)	Chen et al. (2021)
Glucose adsorption capacity	Soluble dietary fibers of white, red and black seeds	Soluble fiber (50 mg) was mixed with 5 mL of glucose solution (80 mM)	Soluble dietary fibers of red seeds had higher glucose binding capacity than those of white and black seeds	Liu et al. (2020)
Cholesterol adsorption capacity	Soluble and insoluble dietary fibers	Diluted fresh egg yolk was made into emulsions before mixing with fibers. The cholesterol concentrations of the emulsions before and after the fiber-mixing were measured	Soluble dietary fibers absorbed much more cholesterol (18 mg/g) than insoluble fibers (9 mg/g). Quinoa fiber fractions had less cholesterol adsorption capacities than wheat fiber fractions	Chen et al. (2021)
Bile acid binding capacity	Soluble dietary fibers of white, red, and black seeds	Under simulated intestinal condition	Soluble dietary fibers of red seeds had higher bile acid binding than those of white and black seeds	Liu et al. (2020)

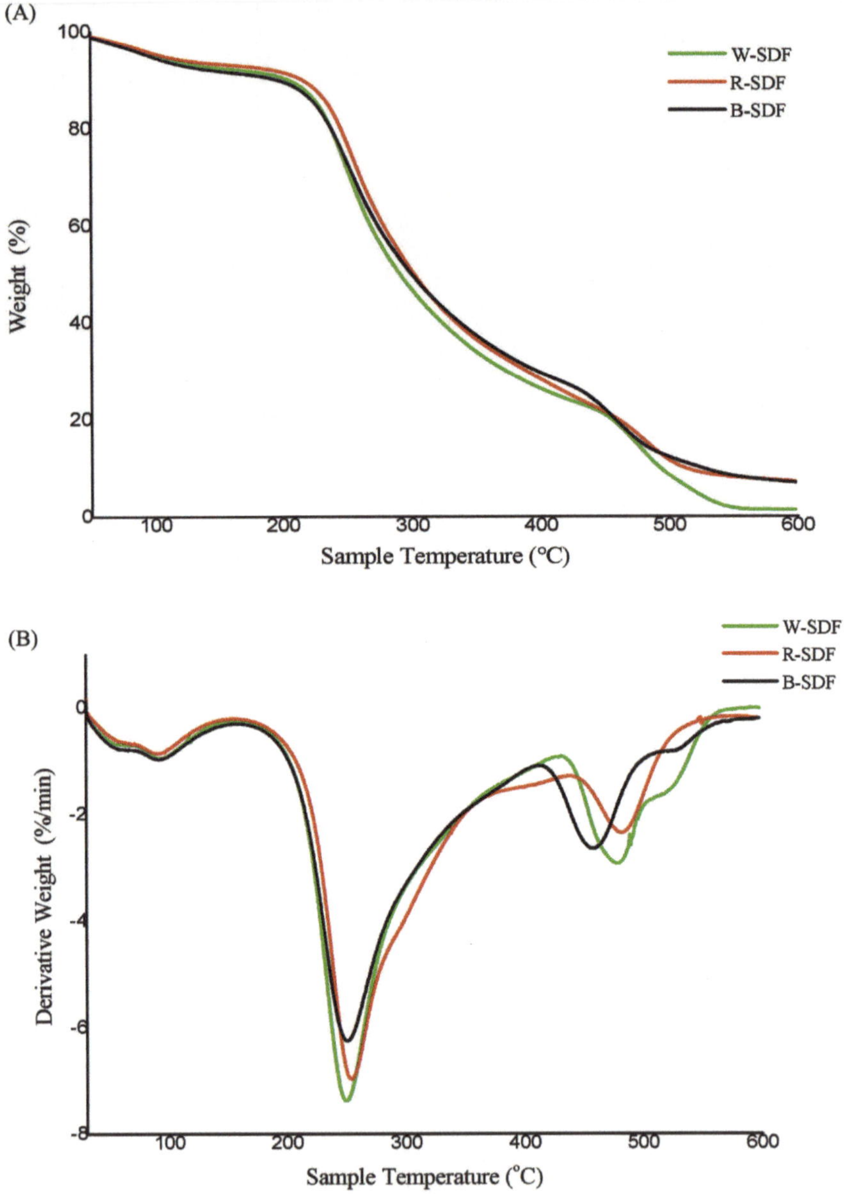

Figure 5.4 Thermal gravimetric analyzer (A) and differential thermogravimetric analysis (B) diagrams for thermodynamic properties of soluble dietary fiber (SDF) from the brans of different colored quinoa seeds. W-SDF, R-SDF, and B-SDF represent SDF from white, red, and black quinoa brans, respectively (Liu et al., 2020).
Source: Reprinted with permission from Elsevier.

thermal stability of the fiber fractions was related to the quality attributes of quinoa-based products. Overall, more studies on the thermal properties of quinoa fiber fractions should be done to confirm the reported data. Liu et al. (2020) analyzed the flow and oscillatory dynamic properties of soluble dietary fiber fractions of white, red, and black quinoa seeds. The viscosity of gels of fibers from red seeds was lower than that of the fiber gels from white and black seeds during the shear-thinning process. During the oscillatory process, increasing angular frequency rendered more gel-like properties to the fibers of red seeds than to those of black and white seeds (Liu et al., 2020).

5.4.2 Water- and oil-holding capacities

The dietary fiber fractions of quinoa were studied for their water- and oil-holding capacities (Chen et al., 2021; Kurek et al., 2018; Liu et al., 2020). Kurek et al. (2018) showed that quinoa fiber fractions had higher water-holding capacity than amaranth and millet fiber fractions. The effect of ultrasound and/or enzyme treatments on the water-holding capacities was not clear. Liu et al. (2020) showed that soluble dietary fibers of red seeds had higher water-binding capacity than those of white and black seeds. Chen et al. (2021) found that quinoa insoluble fiber had higher water-holding capacity (3.7 g/g) than wheat insoluble fiber (3.1 g/g).

Quinoa fiber fractions showed oil-holding capacities. Combined ultrasound and enzyme treatments gave the fibers higher oil-holding capacities than individual treatments (Kurek et al., 2018). The comparative analysis in the oil-holding capacities between the fiber fractions of quinoa, amaranth, and millets was inconclusive. Quinoa insoluble fiber had much higher oil-holding capacity (1.22 g/g) than soluble fiber (0.55 g/g). Quinoa soluble fiber had higher oil-holding capacity (0.55 g/g) than wheat soluble fiber (0.35 g/g). Quinoa insoluble fiber had lower oil-holding capacity (1.22 g/g) than wheat insoluble fiber (1.50 g/g) (Chen et al., 2021).

These studies used different experimental settings. For example, the centrifugation conditions between different studies tended to be different. For the analysis of water-holding capacity, in the study of Kurek et al. (2018), fiber (0.5 g) and water (20 mL) were mixed at room temperature for 24 h before centrifugation at 4000 g for 10 min. In the study of Chen et al. (2021), soluble fiber (1 g) and water (30 mL) were mixed at room temperature for 18 h before centrifugation at 3000 g for 20 min. The differences in the experimental conditions of different reports made the direct comparison of the data of these reports impossible. Such differences in methods hinder our progress towards understand of quinoa.

5.4.3 Glucose, cholesterol, and bile acid adsorption capacity

The adsorption capacities of different nutrients including glucose, cholesterol, and bile acid by quinoa dietary fiber fractions were studied in vitro (Chen et al., 2021; Liu et al., 2020). Liu et al. (2020) found that soluble dietary fibers of red quinoa seeds had higher glucose-binding capacity than those of white and black quinoa seeds. Chen et al. (2021) showed that soluble dietary fibers of quinoa seeds absorbed much more cholesterol (18 mg/g) than insoluble fibers (9 mg/g). Quinoa fiber fractions had less cholesterol adsorption capacities than wheat fiber fractions. Liu et al. (2020) indicated that soluble dietary fibers of red quinoa seeds had higher bile acid-binding activity than those of white and black quinoa seeds. It was not clear on which components of the dietary fiber fractions played a major role in the binding to these nutrients. These binding properties of quinoa fiber fractions contributed to some of their biological functions as described in section 5.5 below.

5.5 Biological properties of fiber polysaccharides and fractions of quinoa

Dietary fiber fractions and polysaccharides of quinoa showed a range of biological functions. They included antioxidant activity, inhibition on food digestion, bile acid-binding, anticancer property, gastro-protection, and immunomodulation (Table 5.3). Most of the studies were done in vitro. Only few clinical studies were done employing humans.

5.5.1 Antioxidant activity

In vitro antioxidant activities of quinoa fiber polysaccharides and fractions were measured using ABTS (2,2'-azino-bis(3-ethylbenzthiazoline-6-sulfonic acid)), DPPH (2,2-diphenyl-1-picrylhydrazyl), and ORAC (oxygen radical absorbance capacity) based chemical assays. Fractions of fiber and isolated polysaccharides of quinoa had a range of antioxidant activities (Ballester-Sánchez et al., 2020; Chen et al., 2021; Hu et al., 2017; Kurek et al., 2018; Teng et al., 2021; Yao et al., 2014). Fiber fraction obtained from wet milling of quinoa had lower in vitro antioxidant activity than that from dry milling (Ballester-Sánchez et al., 2020). Two fractions from water extraction and another two from alkaline extraction purified showed a range of antioxidant activities (Yao et al., 2014). Antioxidant activity of a purified water-soluble polysaccharide fraction was lower than that of vitamin C (Hu et al., 2017). Quinoa soluble fiber had higher FRAP-based and lower DPPH-based antioxidant activity than wheat soluble fiber (Chen et al., 2021). The hydroxyl groups of the sugar units in the polysaccharides contributed to the radical scavenging capacity. The antioxidant activity may also be due to the bound phenolics (Chen et al., 2021; Wang et al., 2016). However, in most of the studies, the effect of phenolics was not differentiated from that of the polysaccharides. Another major concern is if these in vitro results have any relevance to in vivo and human studies. This is a major challenge in the area of food and nutritional sciences in general, which has been hardly explored so far.

5.5.2 Inhibition on food digestion

Polysaccharide and fiber fractions of quinoa suppressed enzymatic hydrolysis of food macronutrients including protein and starch in vitro (Chen et al., 2021; Liu et al., 2020;

Table 5.3 Biological activities of fiber polysaccharides and fractions of quinoa.

Bio-functions	Assays	Fiber form	Major findings	Reference
Antioxidant	In vitro, DPPH, ORAC	Two fractions from water extraction and two fractions from alkaline extraction purified using DEAE Sepharose and Sephacryl columns	The four fractions showed a range of antioxidant activities	Yao et al. (2014)
Antioxidant	In vitro, DPPH, ABTS	A purified water-soluble fraction of 8852 Da	The fraction had strong antioxidant activity that was even higher than some reported polysaccharides, though the activities were lower than vitamin C	Hu et al. (2017)
Antioxidant	In vitro, DPPH, FRAP	Obtained from wet and dry milling	Fiber fraction obtained from wet milling had lower antioxidant activity than that from dry milling	Ballester-Sánchez et al. (2020)
Antioxidant	In vitro, DPPH, ABTS, superoxide radicals, hydroxyl, and ferric reducing	A polysaccharide fraction (2.0×10^3 Da) purified from alkaline extract using column chromatography	The fraction showed antioxidant activities according to the chemical assays	Teng et al. (2021)
Antioxidant	DPPH, hydroxyl radical scavenging, FRAP	Soluble and insoluble dietary fiber fractions of quinoa and wheat	Quinoa soluble fiber had higher FRAP based and lower DPPH based antioxidant activity than wheat soluble fiber	Chen et al. (2021)
Inhibition on proteolysis	In vitro	Obtained from wet milling	Fiber fraction significantly inhibited the pepsin hydrolysis of quinoa protein isolates. The fiber and starch isolates synergistically inhibited the protein digestion	Opazo-Navarre-te et al. (2019)
Glycemic control	In vitro	Obtained from wet milling	The fibers showed glucose adsorption and α-amylase inhibitory activity	Zhang, Wang et al. (2020)

(Continued)

Table 5.3 (Continued)

Bio-functions	Assays	Fiber form	Major findings	Reference
α-Glucosidase inhibition	In vitro	Soluble and insoluble dietary fiber fractions of quinoa and wheat	Quinoa insoluble fiber had lower α-glucosidase inhibitory activity than wheat insoluble fiber. Quinoa soluble fiber had higher α-glucosidase inhibitory activity than wheat soluble fiber	Chen et al. (2021)
Antidiabetic and glucagon-like peptide-1 release	In vitro	Polyphenol-polysaccharide conjugates obtained from sprouted quinoa based yoghurt	Human endothelial NCI-H716 cells were treated by the polyphenol-polysaccharide conjugates, polyphenol or polysaccharide extracts obtained from sprouted quinoa based yoghurt. Glucagon-like peptide-1 release from the NCI-H716 cells was dependent on the germination conditions of quinoa and the dose of the polyphenol-polysaccharide conjugates. The treatment of NCI-H716 cells with the conjugates increased the expression of proglucagon mRNA. Sprouted quinoa based yogurt can regulate energy and glucose homeostasis	Obaroakpo et al. (2020)
Cholesterol adsorption	In vitro	Soluble and insoluble dietary fiber fractions of quinoa and wheat	Quinoa insoluble and soluble fiber had lower cholesterol adsorption capacity than wheat insoluble and soluble fibers	Chen et al. (2021)
Anticancer	In vitro	A purified water-soluble fraction of 8852 Da	The fraction had cytotoxicity against human liver cancer (SMMC-7721) and breast cancer (MCF-7) cells, while having no effect on normal cells	Hu et al. (2017)

Gastro-protection	In vivo	Cell wall polysaccharides were extracted using water and alkaline (10% KOH) solution and ultrafiltration	Wistar rats were fed with ethanol to induce acute gastric lesions. Arabinans and arabinan-rich pectins were orally fed to the rats and protected the gastric mucosa by forming a mucus barrier	Cordeiro et al. (2012)
Immunomodulatory	In vitro	Two fractions from water extraction and two fractions from alkaline extraction purified using DEAE Sepharose and Sephacryl columns	A water-soluble fraction stimulated the production of NO, IL-6 and TNF-α in RAW 264.7 macrophages. Water-soluble fractions had higher activity than alkaline fractions	Yao et al. (2014)
Immunomodulatory	In vitro	A purified water-soluble fraction of 8852 Da	The fraction promoted the growth of RAW264.7 macrophage, while inhibiting the production of nitric oxide on inflammatory macrophage in a dose-dependent manner	Hu et al. (2017)
Immunomodulatory	In vivo	Polysaccharide fraction of 34 kDa from aqueous extraction and was purified using DEAE and gel permeation	ICR mice were treated with cyclophosphamide to induce immune-suppression. The polysaccharide treatment increased the serum content of IL-6, IFN-γ, IgM, IFN-α, and lysozyme, improved the phagocytic activities of macrophages and relieved allergy in mice	Fan et al. (2019)
Immunomodulatory	In vitro	RAW 264.7 cells were treated with a quinoa nonstarch polysaccharide fraction before and after protein removal	After protein removal, a water-soluble polysaccharide fraction of quinoa stimulated the RAW 264.7 cells by increasing the production of inducible nitric oxide synthase, nitric oxide, tumor necrosis factor-alpha (TNF-α), and interleukins (IL)-1β, IL-6, IL-10. The MAPKs and NF-κB signaling pathways were involved in the stimulation of the RAW 264.7 cells by the polysaccharide	Cao et al. (2022)

(*Continued*)

Table 5.3 (Continued)

Bio-functions	Assays	Fiber form	Major findings	Reference
Gut microbiota regulation	In vitro	Human fecal microbiota was used to ferment raw, cooked quinoa or quinoa polysaccharide extract after in vitro human digestion	Anaerobic fermentation of the samples for 24 h increased contents of short-chain fatty acids and decreased the pH, while increasing the amounts of *Bacteroides* and *Prevotella* species. In particular, quinoa polysaccharide fraction increased the amounts of *Collinsella* and *Bifidobacterium* species with prebiotic function	Zeyneb et al. (2021)

ABTS, 2,2'-azino-bis(3-ethylbenzthiazoline-6-sulfonic acid); DPPH, 2,2-diphenyl-1-picrylhydrazyl; ORAC, oxygen radical absorbance capacity; FRAP, ferric reducing antioxidant power.

Opazo-Navarrete et al., 2019; Wang et al., 2016; Zhang, Wang et al., 2020). A quinoa fiber fraction from wet milling largely suppressed pepsin hydrolysis of protein isolates. Adding starch synergistically decreased the protein hydrolysis (Opazo-Navarrete et al., 2019). Polysaccharide and fiber fractions had glycemic control properties by inhibiting α-glucosidase/α-amylase activity and decreasing glucose adsorption in vitro (Zhang, Wang et al., 2020). Quinoa insoluble fiber had lower α-glucosidase inhibitory activity than wheat insoluble fiber. Quinoa soluble fiber had higher α-glucosidase inhibitory activity than wheat soluble fiber (Chen et al., 2021). Glucose adsorption was correlated with pectin and cellulose components of quinoa fiber fraction (Liu et al., 2020; Zhang, Wang et al., 2020). The pectins and other nonstarch polysaccharides formed gel layers surrounding protein and starch, reducing the interactions with hydrolytic enzymes. The polysaccharides can bind with the hydrolytic enzymes directly, reducing their activities and interactions. The polysaccharides with high water adsorption decreased the mobility of the substrates and enzymes in solutions and increased the viscosity of the systems. These factors contributed to the decreased digestion of food components and reduced glycemic response (Opazo-Navarrete et al., 2019; Zhang, Wang et al., 2020).

5.5.3 Antidiabetic activity

Polyphenol-polysaccharide conjugates obtained from sprouted quinoa-based yogurt were tested in vitro for antidiabetic activity (Obaroakpo et al., 2020). Human endothelial NCI-H716 cells were treated by the polyphenol-polysaccharide conjugates, polyphenol or polysaccharide extracts obtained from sprouted quinoa-based yogurt. Glucagon-like peptide-1 release from the NCI-H716 cells was dependent on the germination conditions of quinoa and the dose of the polyphenol-polysaccharide conjugates. The treatment of NCI-H716 cells with the conjugates increased the expression of proglucagon mRNA. Sprouted quinoa-based yogurt could regulate energy and glucose homeostasis (Obaroakpo et al., 2020). Little et al. (2021) reviewed the literature on potential antiobesity and antidiabetic properties of quinoa. They concluded that quinoa dietary fiber is a component contributing to its antiobesity and antidiabetic capacity (Little et al., 2021).

5.5.4 Cholesterol and bile acid adsorption

The nutrient adsorption capacities of quinoa dietary fiber fractions were described in section 5.4.3 above. Briefly, soluble and insoluble dietary fiber fractions of quinoa and wheat were tested for cholesterol adsorption in vitro (Chen et al., 2021). Quinoa insoluble and soluble fiber had lower cholesterol adsorption capacity than wheat insoluble and soluble fiber (Chen et al., 2021). Soluble dietary fibers of red quinoa seeds had higher bile acid binding than those of white and black quinoa seeds (Liu et al., 2020). Fermentation products of the fiber polysaccharides in human gut can lead to decreased hepatic cholesterol production (Gunness & Gidley, 2010). The bile acid-binding capacity of the fiber fractions contributed to the hyperlipidemic and cholesterol-lowering properties of quinoa. This could also lead to reduced occurrences of cardiovascular diseases in humans (Gunness & Gidley, 2010). However, these deductive arguments on quinoa fiber polysaccharides remain to be studied using in vivo and clinical models.

5.5.5 Anticancer property

The in vitro anticancer property of fiber polysaccharides from quinoa was confirmed (Hu et al., 2017). A purified water-soluble fraction of 8852 Da had cytotoxicity against human liver cancer (SMMC-7721) and breast cancer (MCF-7) cells, while having no effect on normal cells (Hu et al., 2017). Previous studies showed that many polysaccharides and fibers from different sources had anticancer effects (Khan et al., 2019). Different mechanisms are involved, such as arresting cell cycle, DNA damage, immunotherapy and nitric oxide production, and mitochondrial membrane disruption (Khan et al., 2019). A major advantage of using dietary fiber as anticancer agents is related to much less adverse side effects. The anticancer properties of quinoa fiber should be further studied using in vivo and clinical models. The differential effects of different types of the polysaccharides such as pectins and xyloglucans from quinoa remain to be studied.

5.5.6 Gastro-protection

The in vivo gastro-protective effect of quinoa polysaccharides was studied using Wistar rats with acute gastic lesions induced by ethanol (Cordeiro et al., 2012). Arabinans and arabinan-rich pectins were orally fed to the rats. The pectic polysaccharides with gelation capacity formed a mucus barrier, preventing the penetration of the necrotizing agent in to the rats (Cordeiro et al., 2012).

5.5.7 Immunomodulation

Fiber polysaccharide fractions of quinoa showed immunomodulatory activity in vitro and in vivo (Cao et al., 2022; Fan et al., 2019; Hu et al., 2017; Yao et al., 2014). In vivo, mice were treated with cyclophosphamide to induce immune suppression before treatment with the water-soluble polysaccharide fractions of quinoa (Fan et al., 2019). The polysaccharide treatment increased the serum contents of IL-6, IFN-γ, IgM, IFN-α, and lysozyme, the thymus and spleen indices, the leukocyte count of the blood, and improved the phagocytic activities of macrophages and relieved allergy in the mice (Fan et al., 2019). However, the polysaccharides used were not pure and contained other components including proteins and polyphenols. The latter may also be immunomodulatory (Mileo et al., 2019). RAW 264.7 cells were treated with a quinoa nonstarch polysaccharide fraction before and after protein removal (Cao et al., 2022). After protein removal, a water-soluble polysaccharide fraction of quinoa stimulated the RAW 264.7 cells by increasing the production of inducible nitric oxide synthase, nitric oxide, tumor necrosis factor-alpha (TNF-α), and interleukins (IL)-1β, IL-6, and IL-10. The MAPKs and NF-κB signaling pathways were involved in the stimulation of the RAW 264.7 cells by the polysaccharide (Cao et al., 2022).

5.5.8 Fermentative properties and gut microbiota

Dietary fibers from various sources positively affect gut microbiota and immune systems of humans (Tang et al., 2019). Fiber polysaccharides promote the growth of bacteria that produce short chain fatty acid (SCFA) in gut. They improve immunity, while decreasing stimulated excessive immune responses through regulating gut microbiota (Tang et al., 2019). The bound phenolic acids should also contribute to the regulation of gut microbiota (Zhang, Dong et al., 2020). Human fecal microbiota was used to ferment raw, cooked quinoa or quinoa polysaccharide extract after in vitro human digestion (Zeyneb et al., 2021). Anaerobic fermentation of the samples for 24 h increased contents of SFCA and decreased the pH, while increasing the amounts of *Bacteroides* and *Prevotella* species. In particular, quinoa polysaccharide fraction increased the amounts of *Collinsella* and *Ifidobacterium* species with prebiotic function (Zeyneb et al., 2021). Agarwal et al. (2022) studied the in vivo effects of quercetin and quinoa soluble fiber on the brush border membrane and gut microbiome using red junglefowl (*Gallus gallus*). The paneth and goblet cell numbers increased from the fiber and quercetin intake. The fiber and quercetin intake had no effect on the gene expression in the liver and duodenum, while increasing the *Lactiplantibacillus plantarum* populations (Agarwal et al., 2022). Indeed, previous studies have shown the synergistic interactions between fiber polysaccharides and polyphenols and their effects on gut health (Siemińska-Kuczer et al., 2022). The role of quinoa fiber on gut microbiota and related health benefits remains to be better studied using in vivo and clinical models.

5.6 Discussion

Polysaccharides and fiber fractions from different sources have many health benefits such as antioxidation, gastro-protection, inhibition on food digestion, immunomodulation, bile acid-binding, and anticancer properties (Cao et al., 2022; Chen et al., 2021; Gunness & Gidley, 2010; Khan et al., 2019; Nie et al., 2017; Wang et al., 2016; Zhang, Wang et al., 2020). These health effects were also noted for fiber polysaccharides and fractions of quinoa as described in this chapter. Many plant foods including cereals, fruits, and vegetables contain fibers. It is important to compare the health effects of polysaccharides and fibers of quinoa with those of other sources. This aspect has been little studied. It is hypothesized that quinoa polysaccharides and fibers can play an important roles in human wellness.

Gidley and Yakubov (2019) proposed a new categorization of dietary fibers to replace the commonly used divisions based on the solubility. Dietary fibers may be categorized based on their local density and particle/molecule size, which can more accurately reflect their nutritional functionality. Such concept remains to be applied to include quinoa dietary fibers.

Phenolics such as coumaric and ferulic acids can be bound in plant cell walls, which may positively affect the nutritional properties of fiber fractions (Siemińska-Kuczer et al., 2022; Zhang, Dong et al., 2020). The biological activities of phenolics and those of polysaccharides and fibers are significantly similar in many aspects (Mileo et al., 2019; Zheng et al., 2020). For example, both have antioxidant, immunomodulatory, and anticancer activities (Mileo et al., 2019). Interactions of fiber components and phenolics can significantly affect functional and nutritional properties of related foods (Obaroakpo et al., 2020; Siemińska-Kuczer et al., 2022; Zhu, 2018). Further studies should be done on the differences and interactions of nonfiber and fiber components to obtain their structure-function relationships. The methods used for measuring the biological functions of quinoa fiber and polysaccharide fractions are not complete. Most of the studies used in vitro models, while some studies employed animal models. The usefulness of in vitro and animal models in relation to human applications is questionable in so many cases (Brockman, 2015). This kind of mentality to do research continues to plague our scientific discourse.

5.7 Conclusions

The bran of quinoa grains is rich in dietary fiber. The fiber polysaccharides and fractions were obtained using wet or dry milling, and solvent/aqueous and chromatographic extractions. Quinoa had much more insoluble polysaccharides and fibers than soluble ones. The fiber polysaccharides included pectins, xyloglucans, and cellulose. A large portion of these polysaccharides are cross-linked with coumaric and ferulic acids. The quinoa polysaccharides are more like those of vegetables and fruits than to those of cereals. Claimed biological activities of quinoa fiber fractions and polysaccharides include antioxidation, inhibition on food digestion, bile acid binding and cholesterol lowering, anticancer, gastro-protection, and immunomodulation. It can be concluded that quinoa fibers and polysaccharides have great potential to improve the functional and nutritional properties of food formulations including quinoa grain-based products.

In most of the studies, the fiber and polysaccharide samples used were mixtures containing other components such as phenolics. Interactions of fiber components such as polysaccharides with phenolics can affect functional and nutritional properties of foods. Purified fiber components such as polysaccharides should be used to analyze the component–property relationships. Rheological properties such as dynamic rheology and flow and gelation of quinoa fiber fractions and components should be analyzed for food applications. Most of the reported biological functions of quinoa fiber fractions/components were tested in vitro and in vivo, and clinical studies remain to be done. The effect of geographic and growing conditions on the properties of fiber polysaccharides of quinoa should be analyzed.

References

Agarwal, N., Kolba, N., Khen, N., Even, C., Turjeman, S., Koren, O., & Tako, E. (2022). Quinoa soluble fiber and quercetin alter the composition of the gut microbiome and improve brush border membrane morphology *in vivo* (*Gallus gallus*). *Nutrients*, *14*, 448.

Ballester-Sánchez, J., Fernández-Espinar, M. T., & Haros, C. M. (2020). Isolation of red quinoa fibre by wet and dry milling and application as a potential functional bakery ingredient. *Food Hydrocolloids*, *101*, 105513.

Brockman, J. (2015). *This idea must die: Scientific theories that are blocking progress*. New York: Harper Perennial.

Cao, R. A., Ma, N., Subramanian, P., Talapphet, N., Zhang, J., Wang, C. Y., & You, S. G. (2022). Structural elucidation and immunostimulatory activities of quinoa non-starch polysaccharide before and after deproteinization. *Journal of Polymers and the Environment*, *30*, 2291–2303.

Chen, H., Xiong, M., Bai, T., Chen, D., Zhang, Q., Lin, D., et al. (2021). Comparative study on the structure, physicochemical, and functional properties of dietary fiber extracts from quinoa and wheat. *LWT-Food Science and Technology*, *149*, 111816.

Cordeiro, L. M. C., de Fátima Reinhardt, V., Baggio, C. H., de Paula Werner, M. F., Burci, L. M., Sassaki, G. L., & Iacomini, M. (2012). Arabinan and arabinan-rich pectic polysaccharides from quinoa (*Chenopodium quinoa*) seeds: Structure and gastroprotective activity. *Food Chemistry*, *130*, 937–944.

Cui, S. W., & Wang, Q. (2009). Cell wall polysaccharides in cereals: Chemical structures and functional properties. *Structural Chemistry*, *20*, 291–297.

Djordjević, M., Djordjević, M., Šoronja-Simović, D., Nikolić, I., & Šereš, Z. (2022). Delving into the role of dietary fiber in gluten-free bread formulations: Integrating fundamental rheological, technological, sensory, and nutritional aspects. *Polysaccharides*, *3*, 59–82.

Fan, S., Li, J., & Bai, B. (2019). Purification, structural elucidation and in vivo immunity-enhancing activity of polysaccharides from quinoa (*Chenopodium quinoa* Willd.) seeds. *Bioscience, Biotechnology, and Biochemistry*, *83*, 2334–2344.

Fernández-López, J., Lucas-González, R., Viuda-Martos, M., Sayas-Barberá, E., Ballester-Sánchez, J., Haros, C. M., et al. (2020). Chemical and technological properties of bologna-type sausages with added black quinoa wet-milling coproducts as binder replacer. *Food Chemistry*, *310*, 125936.

Fernández-López, J., Viuda-Martos, M., & Pérez-Alvarez, J. A. (2021). Quinoa and chia products as ingredients for healthier processed meat products: Technological strategies for their application and effects on the final product. *Current Opinion in Food Science*, *40*, 26–32.

Gan, J., Xie, L., Peng, G., Xie, J., Chen, Y., & Yu, Q. (2021). Systematic review on modification methods of dietary fiber. *Food Hydrocolloids*, *119*, 106872.

Gidley, M. J., & Yakubov, G. E. (2019). Functional categorisation of dietary fibre in foods: Beyond 'soluble' vs 'insoluble'. *Trends in Food Science & Technology*, *86*, 563–568.

Gunness, P., & Gidley, M. J. (2010). Mechanisms underlying the cholesterol-lowering properties of soluble dietary fibre polysaccharides. *Food & Function*, *1*, 149–155.

Hromádková, Z., & Ebringerová, A. (2003). Ultrasonic extraction of plant materials—investigation of hemicellulose release from buckwheat hulls. *Ultrasonics Sonochemistry*, *10*, 127–133.

Hu, Y., Zhang, J., Zou, L., Fu, C., Li, P., & Zhao, G. (2017). Chemical characterization, antioxidant, immune-regulating and anticancer activities of a novel bioactive polysaccharide from *Chenopodium quinoa* seeds. *International Journal of Biological Macromolecules, 99*, 622–629.

Jimenez, M. D., Lobo, M., & Sammán, N. (2019). Influence of germination of quinoa (*Chenopodium quinoa*) and amaranth (*Amaranthus*) grains on nutritional and techno-functional properties of their flours. *Journal of Food Composition and Analysis, 84*, 103290.

Khan, T., Date, A., Chawda, H., & Patel, K. (2019). Polysaccharides as potential anticancer agents—A review of their progress. *Carbohydrate Polymers, 210*, 412–428.

Kurek, M. A., Karp, S., Wyrwisz, J., & Niu, Y. (2018). Physicochemical properties of dietary fibers extracted from gluten-free sources: Quinoa (*Chenopodium quinoa*), amaranth (*Amaranthus caudatus*) and millet (*Panicum miliaceum*). *Food Hydrocolloids, 85*, 321–330.

Lamothe, L. M., Srichuwong, S., Reuhs, B. L., & Hamaker, B. R. (2015). Quinoa (*Chenopodium quinoa* W.) and amaranth (*Amaranthus caudatus* L.) provide dietary fibres high in pectic substances and xyloglucans. *Food Chemistry, 167*, 490–496.

Li, G., & Zhu, F. (2017). Physicochemical properties of quinoa flour as affected by starch interactions. *Food Chemistry, 221*, 1560–1568.

Little, A., Murphy, K., & Solverson, P. (2021). Quinoa's potential to enhance dietary management of obesity and type-2 diabetes: A review of the current evidence. *Diabetology, 2*, 77–94.

Liu, J., Wang, Z., Wang, Z., Hao, Y., Wang, Y., Yang, Z., Li, W., & Wang, J. (2020). Physicochemical and functional properties of soluble dietary fiber from different colored quinoa varieties (*Chenopodium quinoa* Willd). *Journal of Cereal Science, 95*, 103045.

Mileo, A. M., Nisticò, P., & Miccadei, S. (2019). Polyphenols: Immunomodulatory and therapeutic implication in colorectal cancer. *Frontiers in Immunology, 10*, 729.

Miranda, M., Vega-Gálvez, A., Martínez, E. A., López, J., Marín, R., Aranda, M., et al. (2013). Influence of contrasting environments on seed composition of two quinoa genotypes: Nutritional and functional properties. *Chilean Journal of Agricultural Research, 73*, 108–116.

Miranda, M., Vega-Gálvez, A., Martinez, E., López, J., Rodríguez, M. J., Henríquez, K., et al. (2012). Genetic diversity and comparison of physicochemical and nutritional characteristics of six quinoa (*Chenopodium quinoa* Willd.) genotypes cultivated in Chile. *Ciência e Tecnologia de Alimentos, 32*, 835–843.

Muthusamy, S., Udayakumar, G. P., & Narala, V. R. (2021). Recent advances in the extraction and characterization of seed polysaccharides, and their bioactivities: A review. *Bioactive Carbohydrates and Dietary Fibre, 26*, 100276.

Nie, S., Cui, S. W., & Xie, M. (2017). *Bioactive polysaccharides*. Academic Press.

Ninfali, P., Panato, A., Bortolotti, F., Valentini, L., & Gobbi, P. (2020). Morphological analysis of the seeds of three pseudocereals by using light microscopy and ESEM-EDS. *European Journal of Histochemistry, 64*, 3075.

Nowak, V., Du, J., & Charrondière, U. R. (2016). Assessment of the nutritional composition of quinoa (*Chenopodium quinoa* Willd.). *Food Chemistry, 193*, 47–54.

Obaroakpo, J. U., Liu, L., Zhang, S., Lu, J., Liu, L., Pang, X., & Lv, J. (2020). In vitro modulation of glucagon-like peptide release by DPP-IV inhibitory polyphenol-polysaccharide conjugates of sprouted quinoa yoghurt. *Food Chemistry, 324*, 126857.

Opazo-Navarrete, M., Freire, D. T., Boom, R. M., & Janssen, A. E. M. (2019). The influence of starch and fibre on *in vitro* protein digestibility of dry fractionated quinoa seed (Riobamba variety). *Food Biophysics, 14*, 49–59.

Phillips, K. M., Haytowitz, D. B., & Pehrsson, P. R. (2019). Implications of two different methods for analyzing total dietary fiber in foods for food composition databases. *Journal of Food Composition and Analysis*, *84*, 103253.

Prosky, L., Asp, N. G., Furda, I., Devries, J. W., Schweizer, T. F., & Harland, B. F. (1985). Determination of total dietary fiber in foods and food products: Collaborative study. *Journal of Association of Official Analytical Chemists*, *68*, 677–679.

Ren, Y., & Liu, S. (2020). Effects of separation and purification on structural characteristics of polysaccharide from quinoa (*Chenopodium quinoa* willd). *Biochemical and Biophysical Research Communications*, *522*, 286–291.

Repo-Carrasco-Valencia, R. A. M., & Serna, L. A. (2011). Quinoa (*Chenopodium quinoa*, Willd.) as a source of dietary fiber and other functional components. *Ciência e Tecnologia de Alimentos*, *31*, 225–230.

Ruales, J., & Nair, B. M. (1994). Properties of starch and dietary fibre in raw and processed quinoa (*Chenopodium quinoa*, Willd) seeds. *Plant Foods for Human Nutrition*, *45*, 223–246.

Siemińska-Kuczer, A., Szymańska-Chargot, M., & Zdunek, A. (2022). Recent advances in interactions between polyphenols and plant cell wall polysaccharides as studied using an adsorption technique. *Food Chemistry*, *373*, 131487.

Sobota, A., Świeca, M., Gęsiński, K., Wirkijowska, A., & Bochnak, J. (2020). Yellow-coated quinoa (*Chenopodium quinoa* Willd) – physicochemical, nutritional, and antioxidant properties. *Journal of the Science of Food and Agriculture*, *100*, 2035–2042.

Srichuwong, S., Curti, D., Austin, S., King, R., Lamothe, L., & Gloria-Hernandez, H. (2017). Physicochemical properties and starch digestibility of whole grain sorghums, millet, quinoa and amaranth flours, as affected by starch and non-starch constituents. *Food Chemistry*, *233*, 1–10.

Tang, C., Ding, R., Sun, J., Liu, J., Kan, J., & Jin, C. (2019). The impacts of natural polysaccharides on intestinal microbiota and immune responses – A review. *Food & Function*, *10*, 2290–2312.

Tang, Y., & Tsao, R. (2017). Phytochemicals in quinoa and amaranth grains and their antioxidant, anti-inflammatory, and potential health beneficial effects: A review. *Molecular Nutrition & Food Research*, *61*, 1600767.

Teng, C., Qin, P., Shi, Z., Zhang, W., Yang, X., Yao, Y., & Ren, G. (2021). Structural characterization and antioxidant activity of alkali-extracted polysaccharides from quinoa. *Food Hydrocolloids*, *113*, 106392.

Wang, X. T., Zhu, Z. Y., Zhao, L., Sun, H. Q., Meng, M., Zhang, J. Y., et al. (2016). Structural characterization and inhibition on α-D-glucosidase activity of non-starch polysaccharides from *Fagopyrum tartaricum*. *Carbohydrate Polymers*, *153*, 679–685.

Wefers, D., & Bunzel, M. (2015). Characterization of dietary fiber polysaccharides from dehulled common buckwheat (*Fagopyrum esculentum*) seeds. *Cereal Chemistry*, *92*, 598–603.

Wefers, D., Gmeiner, B. M., Tyl, C. E., & Bunzel, M. (2015). Characterization of diferuloylated pectic polysaccharides from quinoa (*Chenopodium quinoa* WILLD.). *Phytochemistry*, *116*, 320–328.

Wefers, D., Tyl, C. E., & Bunzel, M. (2014). Novel arabinan and galactan oligosaccharides from dicotyledonous plants. *Frontiers in Chemistry*, *2*, 100.

Wefers, D., Tyl, C. E., & Bunzel, M. (2015b). Pectic arabinans in quinoa seeds (*Chenopodium quinoa* Willd.) are acylated with *p*-coumaric acid. *Cereal Chemistry*, *92*, 401–404.

Yao, Y., Shi, Z., & Ren, G. (2014). Antioxidant and immunoregulatory activity of polysaccharides from quinoa (*Chenopodium quinoa* Willd.). *International Journal of Molecular Sciences*, *15*, 19307–19318.

Zeyneb, H., Pei, H., Cao, X., Wang, Y., Win, Y., & Gong, L. (2021). In vitro study of the effect of quinoa and quinoa polysaccharides on human gut microbiota. *Food Science and Nutrition*, *9*, 5735–5745.

Zhang, D., Wang, L., Tan, B., & Zhang, W. (2020). Dietary fibre extracted from different types of whole grains and beans: A comparative study. *International Journal of Food Science and Technology*, *55*, 2188–2196.

Zhang, S., Hu, J., Sun, Y., Ji, H., Liu, F., Peng, X., Zhong, Y., Geng, F., & Nie, S. (2022). In vitro digestion of eight types of wholegrains and their dietary recommendations for different populations. *Food Chemistry*, *370*, 131069.

Zhang, X., Dong, L., Jia, X., Liu, L., Chi, J., Huang, F., Ma, Q., Zhang, M., & Zhang, R. (2020b). Bound phenolics ensure the antihyperglycemic effect of rice bran dietary fiber in db/db mice via activating the insulin signaling pathway in skeletal muscle and altering gut microbiota. *Journal of Agricultural and Food Chemistry*, *68*, 4387–4398.

Zheng, Y., Tian, J., Yang, W., Chen, S., Liu, D., Fang, H., Zhang, H., & Ye, X. (2020). Inhibition mechanism of ferulic acid against α-amylase and α-glucosidase. *Food Chemistry*, *317*, 126346.

Zhu, F. (2018). Interactions between cell wall polysaccharides and polyphenols. *Critical Reviews in Food Science and Nutrition*, *58*, 1808–1831.

Zhu, F. (2020). Dietary fiber polysaccharides of amaranth, buckwheat and quinoa grains: A review of chemical structure, biological functions and food uses. *Carbohydrate Polymers*, *248*, 116819.

Quinoa protein

6

6.1 Introduction

There has been increasing interest to explore plant-based proteins for food applications (Alrosan et al., 2022; Yuliarti et al., 2021). Protein is a major component of quinoa grains. Quinoa protein is considered gluten free and has the potential to be used in different food and nutritional applications (Alrosan et al., 2022; Gupta et al., 2021; Janssen et al., 2017; López-Alarcón et al., 2019). Although some quinoa genotypes contain celiac-toxic epitopes, quinoa consumption of 50 g (over 6 weeks) was considered safe (Nowak et al., 2016; Zevallos et al., 2014). The amino acid composition of quinoa proteins is more balanced for essential amino acids as compared with that of major cereals such as wheat and rice (Craine & Murphy, 2020). Apart from providing nutrients for human consumption, quinoa protein can be extracted for different food applications. The functional and physicochemical properties (e.g., solubility, gelling, emulsifying, and foaming capacity) of quinoa protein isolates and concentrates have significant relevance for different formulations and applications (López-Castejón et al., 2020; Patole et al., 2022; Shi et al., 2020; Solaesa et al., 2020; Yang et al., 2022).

The protein isolate can be hydrolyzed using different proteinases to obtain hydrolysates and peptides. These hydrolysates and peptides have a range of biological functions (Cao et al., 2021; Dakhili et al., 2019; Daliri et al., 2021; Guo et al., 2020a, 2020b, 2021; Joye, 2019; Mudgil et al., 2019; Salami et al., 2022; Shi et al., 2019; Vilcacundo et al., 2018). They have the potential to be used as food ingredients in the formulations of functional foods and nutraceuticals.

To expand the range of functional properties, different types of physical, chemical, and enzymatic modifications and processes have been applied on quinoa protein and grains (Cao et al., 2021; Dakhili et al., 2019; Daliri et al., 2021; de Vondel et al., 2021; Galante et al., 2020; He et al., 2021; Mir et al., 2019; Shen et al., 2021; Suárez-Estrella et al., 2020). Factors affecting the outcomes of these modifications include processing conditions, quinoa genotypes, composition of processing materials, and experimental conditions (e.g., pH and temperature). These factors should be carefully compared between different studies to understand the mechanisms behind the effects of modification processes.

The protein content of quinoa seeds and amino acid composition of quinoa protein are reviewed in Chapter 2. In the current chapter, the composition and structure of Osborne-type protein fractionation, functional properties, processing, and biological properties of quinoa seed protein are described by focusing on recent literature. The gaps in quinoa protein research are discussed. The food and other applications of quinoa protein and related products are described in Chapter 13.

Quinoa. DOI: https://doi.org/10.1016/B978-0-323-99909-0.00008-8

6.2 Composition, fractionation, and structure of quinoa proteins

Osborne (1907) categorized wheat protein into albumins (water extractable), prolamins (aqueous alcohol extractable), globulins (aqueous salt solution extractable), and glutelins (the remaining part) based on their solubility in different types of solvents. The contents and types of proteins affect functional properties of proteins and food systems. Osborne-type protein composition of different quinoa genotypes was analyzed by different researchers (de Vondel et al., 2020; Janssen et al., 2017; López et al., 2018). Janssen et al. (2017) and López et al. (2018) systematically reviewed the protein composition of quinoa in comparison with other types of cereals and pseudocereals. The major Osborne fractions of pseudocereals including quinoa are globulins and albumins. In contrast, most cereals including wheat had more prolamins and glutelins (Janssen et al., 2017). Variations in the protein composition were obtained. The ranges of percentages of albumins, globulins, prolamins, and glutelins of four quinoa genotypes were 41%–50%, 7%–11%, 4%–6%, and 14%–16%, respectively (de Vondel et al., 2020). Some other studies showed that globulins were the dominant protein type in quinoa protein isolate, followed by albumins (Janssen et al., 2017; López et al., 2018). Such differences could be due to the difference in the genotypes and also the variations in the methods and experimental conditions for protein extraction (Janssen et al., 2017). For example, different temperatures, length, types, and concentrations of salts in solutions were used in different studies (Janssen et al., 2017). The water-soluble albumins have lower molecular weight than globulins (López et al., 2018). High performance size-exclusion liquid chromatography and mass spectrometry analysis showed that water extracts of protein contained vicilin-like and legumin-like proteins and cytoplasmic and chloroplastic enzymes. The 2 S albumin was the major storage type (Janssen et al., 2017). The salt extracts mostly had legumin-like proteins. The 11 S globulin of quinoa is termed chenopodin of 320 kDa (Janssen et al., 2017). There was variation in the stability of the disulfide bond between the basic and acidic subunits of the 11 S globulin. Proteins from dilute base extracts had similar structure compared to those of dilute salt and water extracts (de Vondel et al., 2020). Overall, the Osborne-type protein composition of quinoa was more similar to that of pulses and other pseudocereals than to cereals (Janssen et al., 2017).

Lunasin is a bioactive peptide of 43 amino acids found in different cereals and pseudocereals including quinoa (Ren et al., 2017). The concentration of lunasin in 15 quinoa samples grown in China varied from 1.01×10^{-3} to 4.89×10^{-3} g/kg dry seed (Ren et al., 2017). The occurrence of lunasin in quinoa grown in the other part of the world remains to be studied.

6.3 Quinoa protein isolates

Quinoa protein isolates were mostly extracted using the isoelectric method. The isolates had protein contents of 70%–73% and starch contents of 21%–23% (Wang et al., 2020). The most abundant amino acid of the protein isolates was

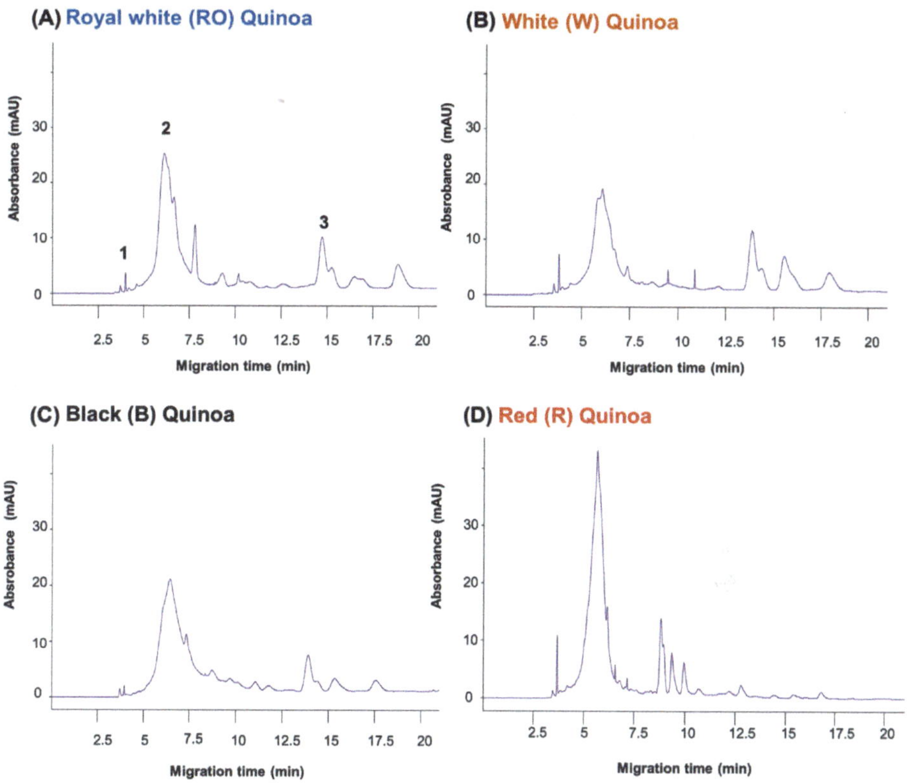

Figure 6.1 Electropherograms obtained at 214 nm after protein extraction for (A) royal white (RO), (B) white (W), (C) black (B), and (D) red (R) quinoa samples. The peaks labeled as 1, 2, and 3 in (A) were considered for the repeatability studies (Galindo-Luján et al., 2021).
Source: Reprinted with permission from Elsevier.

glutamic acid. The molecular weights of the isolates ranged from 10 k to 50 kDa. β-Sheet was the major secondary structure (31%−37%) in the protein isolates. Quinoa protein isolate had 16% of α-helices, 22% of β-rotations, 31% of β-sheets, and 31% of random coils (Vera et al., 2019). Diversity in the composition and secondary structure among protein isolates from six quinoa genotypes was obtained (Wang et al., 2020). Soluble fractions of quinoa protein from NaOH solubilization process at pH 10 were characterized using capillary electrophoresis-UV absorption (Galindo-Luján et al., 2021). The chemometric approach was used to categorize different quinoa varieties based on the protein composition (Fig. 6.1).

6.4 Functional properties of quinoa proteins and peptides

The most studied functional properties of quinoa proteins and hydrolysates included solubility, gelling capacity, water- and oil-binding property, thermal property, emulsifying and foaming capacity, and digestibility as outlined in the sections below.

6.4.1 Solubility

Protein solubility relates to the hydrophobic-hydrophilic balance of the molecules and the thermodynamics of solvent. The protein solubility of quinoa protein isolate was low in the range of 3–5 (Steffolani et al., 2016). The lowest solubility of quinoa protein isolate was obtained at pH 4.5 (Elsohaimy et al., 2015). Increasing pH from 5 to 7 drastically increased the solubility. Further increase in pH increased the solubility to less extent or had little effect on it, depending on the genotypes. The genotype of quinoa affected the protein solubility. For example, at pH 11, the solubility of protein isolates from six genotypes ranged from 55% to 90% (Steffolani et al., 2016). Ghumman et al. (2021) found that the range of solubility of protein isolates from six quinoa genotypes was 30%–53%. Elsohaimy et al. (2015) observed that the highest solubility of quinoa protein isolate was obtained at pH 10. Shi et al. (2020) noted the lowest solubility of quinoa protein isolate was at pH 5, whereas the pH for obtaining the highest solubility was dependent on the genotype. The pH of the solution for protein extraction had impact on the solubility of protein isolate (Abugoch et al., 2008). The solubility of the proteins extracted at pH 11 decreased by nearly 50% compared to that extracted at pH 9. This suggested that some of the proteins were denaturalized and aggregated in the extreme pH environment (Abugoch et al., 2008). The solubility of protein of milled fractions was affected by the temperature. For example, the solubility of a milled fraction rich in proteins was maximized (17.5%) at 60°C, whereas at 95°C, the protein solubility was reduced to 8%. Both albumins and globulins were denatured at higher temperature. This partially contributed to the decreased solubility of the fraction at 95°C (Solaesa et al., 2020).

6.4.2 Gelling property

The gelation properties of quinoa protein isolate affected by different food ingredients were studied (Kaspchak et al., 2017; Patole et al., 2022). Kaspchak et al. (2017) studied the gelation property of quinoa protein isolate (mostly 11 S globulin) at pH 3.5 and 7.0 in the presence of $MgCl_2$ and $CaCl_2$. The gels formed at pH 7.0 were looser and less viscoelastic as compared with those obtained at pH 3.5. The addition of divalent salts increased the firmness of the gels formed at pH 3.5 but led to the formation of a weaker gel at pH 7. The differences of the protein isolate in solubility, structural flexibility, and surface charges at pH 7 and 3.5 contributed to the different gelling property (Kaspchak et al., 2017). Patole et al. (2022) analyzed the effect of adding different types of polysaccharides (guar gum, xanthan gum, and locust bean gum) at different concentrations (0.05%–0.2%, weight basis) on structural and rheological properties of quinoa protein isolate gels formed upon heating (30 min, 85°C). The addition of the polysaccharides increased the gel firmness, gelation rate, and breaking stress in a polysaccharide concentration-dependent manner. The extents of the effect followed the order of xanthan gum, locust bean gum, and guar gum. The gum addition increased the compactness of the gels (Patole et al., 2022). Indeed, many polysaccharides are known to have excellent gelling capacity (Shi et al., 2022). The interactions between proteins and polysaccharides have great potential to be used for food formulations.

6.4.3 Emulsifying property

Emulsification relies on the amphiphilic nature of proteins; specifically, the hydrophobic and hydrophilic parts react with the oil and water at interfaces, respectively. Reduction of surface tension is important for emulsifying capacity and stability (Ravera et al., 2021). The interfacial properties of quinoa protein concentrations were affected by the pH (9 or 11) of extraction solution as revealed in a comparative study (López-Castejón et al., 2020). The quinoa protein extracted using pH 11 formed more sustained films on the interface and led to the formation of smaller oil droplets of O/W emulsions. In comparison, the protein extracted using pH 9 gave more compact interfacial structure and lower tension. Increasing protein concentration increased the emulsion stability and decreased oil droplet size (López-Castejón et al., 2020). For example, Elsohaimy et al. (2015) found that increasing concentrations of quinoa protein isolate from 0.1% to 3% increased the emulsion ability indexes from 1.2 to 3.1 m^2/g. Shi et al. (2020) found that the stability of quinoa protein isolate-based emulsions was pH dependent. The emulsifying properties of quinoa protein isolate were much affected by genotypes. For example, Ghumman et al. (2021) found that the ranges of emulsifying stability and emulsion activity indexes of protein isolates from six quinoa genotypes were 97–150 min and 19–26 m^2/g, respectively. More random coils and β-structure of the protein isolates contributed to their higher emulsifying capacity (Ghumman et al., 2021). Variations in the composition of polypeptides with the sizes of 33 and 21–24 kDa were recorded in the protein isolates from the six genotypes. The results of the two studies of Elsohaimy et al. (2015) and Ghumman et al. (2021) were too different to be compared meaningfully. These results were sensitive to the experimental methods and calculations. Both the studies measured the emulsion activity indexes, following the method of Pearce and Kinsella (1978), in which the quinoa protein was simply added to a water and peanut oil mixture that was then shaken by hand. The oil content of the emulsions was calculated against 1 mL of aliquot (Pearce & Kinsella 1978). Elsohaimy et al. (2015) used sunflower oil instead of peanut oil and their test samples were based on different volumes. They also calculated the emulsifying capacity and emulsifying stability indexes using different equations. These deviations led to different results of the emulsion activity indexes, creating confusions in the literature and for ther readers. The emulsifying stability indexes of quinoa protein exceeded those of wheat and soybean and were similar to that of bovine serum albumin (Elsohaimy et al., 2015).

6.4.4 Foaming property

The foaming capacity measures the ability of a protein to form a continuous cohesive film around air bubbles in foams. Meanwhile, foaming stability determines the ability of protein to remain surface-active in continuous phase (e.g., water) over a period of time (Solaesa et al., 2020). The foaming capacity of proteins was dependent on pH and quinoa genotypes (Ghumman et al., 2021; Shi et al., 2020; Steffolani et al., 2016). For example, Ghumman et al. (2021) found that the ranges of foaming capacity and stability of protein isolates from six quinoa genotypes were 51%–60% and 39%–50%, respectively. Steffolani et al. (2016) showed that protein isolates from six quinoa genotypes had different foaming property. At pH 7, the foaming capacity ranged from 58% to 197%, whereas the foaming stability varied from 39% to 100%. At pH 3, the protein isolates from the quinoa genotype of "Chucapaca" had foaming capacity and stability of 79% and 46%, respectively, whereas, at pH 11, the capacity and stability were 167% and 76%, respectively. Therefore the interactions between pH and genotypes determined the foaming property of the protein isolate. Foaming capacity increased with increasing concentration of quinoa protein isolate (Elsohaimy et al., 2015). Comparative analysis showed that quinoa protein had a lower foaming capacity and similar foaming stability compared with egg albumin (Elsohaimy et al., 2015). Foaming capacity depends on the interfacial properties, moisture retaining capacity, and hydrophobicity of protein. A protein containing more hydrophobic moieties is easily spread at the air–water interface to prevent aggregation. The differences in the foaming properties of protein isolates from quinoa varieties could be attributed to their different protein compositions and structures that were affected by the crop genetics and processing methods.

6.4.5 Water and oil-binding property

Water-binding capacity of quinoa protein isolate was dependent on pH and genotypes (Elsohaimy et al., 2015; Shi et al., 2020; Steffolani et al., 2016). For example, Steffolani et al. (2016) found that protein isolates from six quinoa genotypes had different water-binding capacity at different pH. At pH 5, water-binding capacity of the six samples varied from 0.81 to 3.35 g/g, whereas at pH 11, the capacity ranged from 3.9 to 11 g/g. The oil-binding capacity of the six samples varied from 2.81 to 4.42 g/g. Shi et al. (2020) found that protein isolates from two quinoa genotypes had different water-binding capacity (4.8 and 2.9 g/g) and similar oil-binding capacity (8.6 g/g). Such differences could be attributed to the differences in the protein composition and structure.

6.4.6 Thermal property

The thermal property of quinoa protein isolate depended on the genotypes (Steffolani et al., 2016). For example, Steffolani et al. (2016) used differential scanning calorimetry to study the thermal property of protein isolates from six quinoa genotypes. The onset temperature (T_o) of denaturation ranged from 88°C to 94°C, whereas enthalpy change (ΔH) of denaturation varied from 2.0 to 7.8 J/g. The differences in the thermal properties indicated that the proteins had differences in structure (e.g., the extent of folding) and the amount of hydrogen bond (Steffolani et al., 2016).

6.4.7 Digestibility

Protein digestibility is one of the most important functions of the quinoa protein and isolates (Ghumman et al., 2021). There was diversity in protein digestibility among different quinoa genotypes (Ghumman et al., 2021; Shi et al., 2020). For example, Shi et al. (2020) showed that in vitro digestibility values of protein isolates from two quinoa genotypes were different by 3%. Ghumman et al. (2021) showed that in vitro digestibility of protein isolates from six quinoa genotypes varied from 71% and 78%. Higher protein digestibility was correlated with fewer random coils (Ghumman et al., 2021). The presence of starch and fiber significantly reduced the in vitro protein digestibility of quinoa flour (Opazo-Navarrete et al., 2019). Synergistic interactions between starch and fiber in reducing the protein digestibility were recorded. Heating the quinoa flour system reduced the protein digestion rates due to the protein–protein interactions and aggregation (Opazo-Navarrete et al., 2019). Native proteins may be more digestible than denatured proteins. Denaturation led to the formation of protein aggregates, which reduced their accessibility to hydrolytic enzymes (Opazo-Navarrete et al., 2018). The studies on protein digestibility were done in vitro under different experimental models for comparative purposes. Tagle-Freire et al. (2022) showed that protein digestibility of cooked quinoa grains after 3 h in static gastric digestion model was higher (4.9%) than that in dynamic gastric digestion model (3.9%). This again confirmed the importance of food matrix and structure on the protein digestibility.

Amino acid composition and structural properties of quinoa proteins influence the digestibility (López et al., 2018). Albumins contain more sheets and have a more aperiodic structure than globulins; they also contain fewer cysteine amino acids. Therefore they are less digestible than globulins. High contents of sulfur-containing amino acids are negatively correlated with digestibility because proteins enriched in these amino acids are less accessible to proteases (Ghumman et al., 2021). Quinoa protein was more digestible than wheat protein (Elsohaimy et al., 2015). This could be due to the differences in protein composition and structure.

6.5 Biological properties of quinoa proteins and peptides

Different biological functions of quinoa protein concentration/isolate and its hydrolysates were measured in vivo and in vitro (Table 6.1). Quinoa protein isolate was subjected to different types of hydrolytic enzymes including trypsin, bromelain, chymotrypsin, protease, pancreatin, pepsin, papain, and ficin or under simulated gastrointestinal conditions (Daliri et al., 2021; Guo, Richel et al., 2020; Vilcacundo, Barrio et al., 2017; Vilcacundo, Martínez-Villalueng et al., 2017). The hydrolysates obtained showed a range of biological activities including antioxidant, antihaemolytic, antibacterial, antihypertensive, antidiabetic, adipogenesis inhibition, anti-inflammatory, and anticancer properties (Daliri et al., 2021; Guo et al., 2021; Guo,

Table 6.1 Biological properties of quinoa proteins and peptides.

Biological function	Protein/food form	In vitro/ in vivo	Experimental information	Major results	Reference
Antioxidant	Lunasin isolated from quinoa	In vitro	ABTS, DPPH, ORAC	Lunasin showed strong antioxidant activity measured by ABTS and ORAC assays but not by DPPH assay	Ren et al. (2017)
Antioxidant	Protein concentrate (up to 65% of protein content) digested under simulated gastrointestinal conditions	In vivo	Zebrafish larva model	Gastric digest by pepsin obtained at pH 1.2 had the strongest capacity in lipid peroxidation. Digest obtained at full gastrointestinal conditions had inhibition activity against lipid peroxidation comparable to BHT	Vilcacundo et al. (2017)
Antioxidant	Bromelain, chymotrypsin, protease hydrolyzed up to 6 h	In vitro	DPPH, ABTS	Hydrolysate obtained using chymotrypsin up to 4 h had the highest antioxidant activity	Mudgil et al. (2019)
Antioxidant	Pancreatin hydrolysate	In vitro	DPPH assay	The hydrolysate had a higher degree of antioxidant activity compared to the original protein	Daliri et al. (2021)
Antihaemolytic	Bromelain, chymotrypsin, protease hydrolyzed up to 6 h	In vitro	Red blood cells	Hydrolysates obtained using bromelain and protease (6 h) had the highest antihaemolytic capacity. The antihaemolytic activity of quinoa protein hydrolysate was comparable to that of amaranth protein hydrolysate	Mudgil et al. (2019)
Antibacterial	Bromelain, chymotrypsin, protease hydrolyzed up to 6 h	In vitro	*Salmonella typhimurium*, *Enterobacter aerogenes*, *Escherichia coli*	Hydrolysates obtained using bromelain (4 h), protease (4 h), and protease (2 h) had the highest antibacterial property. Quinoa protein hydrolysate had higher antibacterial capacity than amaranth protein hydrolysate	Mudgil et al. (2019)

Antihypertension	Pepsin-pancreatin hydrolysate	In vivo and in vitro	Hypertensive rats were fed with protein hydrolysate up to 400 mg/kg body weight	The hydrolysate feeding decreased the levels of systolic and diastolic blood pressure in the rats. The hydrolysate had in vitro angiotensin converting enzyme (ACE) inhibitory activity. Three peptides including FHPFPR, NIFRPF and NWFPLPR were identified in the hydrolysate as potent candidates for high blood pressure control	Guo, Hao, et al. (2020)
Antihypertension	Papain, bromelain and ficin hydrolysates	In vitro; in silico	Angiotensin-I-converting enzyme (ACE) inhibition assay	A total of four peptides (MCG, HPF, NMF, and MAF) were identified as candidates with antihypertensive property and angiotensin-I-converting enzyme (ACE) inhibition capacity	Guo, Richel et al. (2020)
Antihypertension	Protein extracted from isoelectric precipitation (pH 4.5)	In vivo	Hypertensive rats were fed with protein for 5 weeks	A range of bioactive peptides after gastrointestinal digestion were obtained. Protein eating decreased blood pressure and increased microbial diversity and abundance of certain microbial groups. The composition of gut microbiota of the rats became more similar to that of normal rats without hypertension	Guo et al. (2021)
Adipogenesis inhibition	Protein isolate was hydrolyzed using papain, pancreatin and pepsin up to 2 h	In vitro	3T3-L1 cells were treated with the hydrolysates	Hydrolysis increased emulsifying capacity of the protein. The hydrolysate showed the activity of adipogenesis inhibition with the product from pepsin hydrolysis (120 min) showing the highest activity. The pathway of peroxisome proliferator-activated receptor γ (PPARγ) was involved in the inhibition. A total of five peptides were identified to have strong antithrombotic and ACE inhibiting capacity	Shi et al. (2019)

(*Continued*)

Table 6.1 (Continued)

Biological function	Protein/food form	In vitro/ in vivo	Experimental information	Major results	Reference
Antidiabetic	Protein isolate was hydrolyzed using papain and papain-similar enzyme of microbial origin	In vitro	Dipeptidyl peptidase IV (DPP-IV) inhibitory activity	The hydrolysates had higher DPP-IV inhibitory activity as well as antioxidant activity than the protein isolate	Nongonierma et al. (2015)
Antidiabetic	Hydrolysate from simulated gastrointestinal digestion	In vitro	α-Glucosidase, α-amylase and dipeptidyl peptidase IV (DPP-IV) inhibitory activity	Duodenal digestion released peptides with the highest inhibitory activities against the enzymes. IQAEGGL, TDKDYPK, and GEHGSDGNV were identified as good candidates for antidiabetic capacity	Vilcacundo et al. (2017)
Antidiabetic	Papain, bromelain and ficin hydrolysates	In vitro, in silico	Dipeptidyl peptidase-IV (DPP-IV) inhibitory activity	A total of four peptides (MCG, HPF, NMF, and MAF) were identified as candidates with antidiabetic property and DPP-IV inhibition capacity	Guo, Richel et al. (2020)
Antidiabetic	Hydrolysates of proteins in germinated (1 day) and raw quinoa seeds after simulated gastrointestinal digestion	In vitro	A homologue of human α-glucosidase	The hydrolysates of the proteins in the germinated seeds after the digestion had the highest inhibitory activity against the homolog of human α-glucosidase as competitive-type inhibitors	Salami et al. (2022)
Anticancer	Hydrolysate from simulated gastrointestinal digestion	In vitro	Oxygen radical absorbance capacity; inhibition on colon cancer cell lines (Caco-2, HT-29, and HCT-116)	Digests from intestinal phase had the highest antioxidant and anticancer capacity. The fraction of < 5 kDa had the highest antioxidant activity, whereas that of $>$ 5 kDa had the highest anticancer capacity. A total of 17 bioactive peptides were identified as potential candidates with potent anticancer property	Vilcacundo et al. (2018)

Anticancer	Peptides in the hydrolysates from simulated gastrointestinal digestion using trypsin and pancreatin	In vitro	Anticancer properties of the peptides against Caco-2 cells of colon cancer	Three peptides of NWFPLPR, FHPFPR and HYNPYFPG showed significant anticancer property. Molecular simulation showed that the peptides were docked into the active "pocket" of histone deacetylase 1. The peptides decreased the expressions of histone deacetylase 1, Bcl-2, NFκB, IL-8 and IL-6, while increasing that of caspase3	Fan et al. (2022)
Anti-inflammatory	Lunasin isolated from quinoa	In vitro	RAW264.7 macrophages treated by lipopolysaccharide	Lunasin (0.40 g/L) showed significant anti-inflammatory activity and significantly inhibited the production of tumor necrosis factor-α (by 40%), interleukin-6 (by 34%), and nitric oxide (NO) (by 45%) (in the RAW264.7 macrophages)	Ren et al. (2017)

Hao, et al., 2020; Guo, Richel et al., 2020; Mudgil et al., 2019; Nongonierma et al., 2015; Ren et al., 2017; Salami et al., 2022; Shi et al., 2019; Vilcacundo et al., 2018; Vilcacundo, Barrio et al., 2017; Vilcacundo, Martínez-Villalueng et al., 2017).

6.5.1 Antioxidant activity

The antioxidant activities of different quinoa protein hydrolysates were studied in vitro and in vivo (Table 6.1) (Daliri et al., 2021; Mudgil et al., 2019; Vilcacundo, Barrio et al., 2017; Vilcacundo, Martínez-Villalueng et al., 2017). In vitro chemical assays based on different free radicals including 1,1-diphenyl-2-picryl hydrazine (DPPH) and (2,2′-azinobis-(3-ethylbenzthiazolin-6-sulfonic acid)) (ABTS) were used to obtain antioxidant activities of quinoa protein hydrolysates. The antioxidant activity of the hydrolysates tended to be higher than that of the original protein samples (Daliri et al., 2021). The types of enzymes and hydrolytic conditions affected the antioxidant activity of the resulting hydrolysates. For example, the hydrolysate obtained from the hydrolysis of quinoa protein isolate using chymotrypsin up to 4 h had the highest antioxidant activity compared to that obtained using other enzymes and under different time lengths of hydrolysis (Mudgil et al., 2019). The antioxidant activity of hydrolysate obtained under simulated gastrointestinal conditions was confirmed using in vivo zebrafish larva model (Vilcacundo, Barrio et al., 2017; Vilcacundo, Martínez-Villalueng et al., 2017). Gastric digest by pepsin obtained at pH 1.2 had the strongest capacity in lipid peroxidation. Digest obtained at full gastrointestinal conditions had inhibition activity against lipid peroxidation comparable to BHT (Vilcacundo, Barrio et al., 2017; Vilcacundo, Martínez-Villalueng et al., 2017). Lunasin isolated from quinoa showed strong antioxidant activity as measured by ABTS and ORAC assay but not by DPPH assay (Ren et al., 2017). The underlying reason for such a discrepancy is obscure.

The in vitro antioxidant activities of food products obtained using the chemical assays have been in question of their relevance to food nutrition (Granato et al., 2018). It appears that almost any food components and products show any degrees of antioxidant activity. However, these in vitro assays for antioxidant activity are easy to perform and can be done in almost of any chemistry laboratory. Such a fact greatly contributes to the wide applications of these chemical assays. Efforts should be conducted to curb the uses of these methods. Debate on if these methods belong to the category of pseudoscience should be done.

6.5.2 Antihaemolytic activity

Quinoa protein hydrolysates obtained using bromelain, chymotrypsin, or protease were tested for their in vitro antihaemolytic activity using red blood cells (Mudgil et al., 2019). Hydrolysates obtained using bromelain and protease (6 h) had the highest antihemolytic capacity. The antihaemolytic activity of quinoa protein hydrolysate was comparable to that of amaranth protein hydrolysate (Mudgil et al., 2019).

Oxidation of erythrocyte membrane through free radical damages may lead to hemolysis. The antioxidant and free radical scavenging activities of the quinoa protein hydrolysates may contribute to the prevention of hemolysis (Mudgil et al., 2019).

6.5.3 Antibacterial activity

In vitro antibacterial activity of quinoa protein hydrolysates obtained from using bromelain, chymotrypsin, and protease was tested against *Salmonella typhimurium*, *Enterobacter aerogenes*, and *Escherichia coli* (Mudgil et al., 2019). The antibacterial activity of the hydrolysates depended on the type of enzymes used and the hydrolysis length of time. The hydrolysates obtained using bromelain (4 h), protease (4 h), and protease (2 h) had the highest antibacterial property. Quinoa protein hydrolysate had higher antibacterial capacity than amaranth protein hydrolysate. The mechanisms of antibacterial peptides may be related to biofilm degradation, RNA/DNA/protein targeting, and stringent response inhibition (Raheem & Straus, 2019). However, the antibacterial mechanisms of quinoa protein hydrolysates have not been studied.

6.5.4 Antihypertension property

The antihypertension property of quinoa protein isolates and different hydrolysates of the protein isolates was measured in vitro, in silico, and in vivo (Guo et al., 2021; Guo, Hao, et al., 2020; Guo, Richel et al., 2020). The hydrolysates obtained using papain, bromelain and ficin showed in vitro angiotensin-I-converting enzyme (ACE) inhibition activity. In vivo, feeding both the hydrolysates and quinoa protein isolates reduced the systolic and diastolic blood pressure levels of hypertensive rats (Guo et al., 2021; Guo, Hao, et al., 2020). Different types of peptides including FHPFPR, NIFRPF, NWFPLPR, MCG, HPF, NMF, and MAF were identified as potential candidates with antihypertensive capacity. Mechanistically, the feeding of the quinoa protein hydrolysates increased microbial diversity and abundance of certain microbial groups in the rats. The composition of gut microbiota of the rats became more similar to that of normal rats without hypertension (Guo et al., 2021).

6.5.5 Adipogenesis inhibition

The adipogenesis inhibition property of quinoa protein hydrolysates obtained using papain, pancreatin, and pepsin was measured in vitro (Shi et al., 2019). 3T3-L1 cells were treated with the hydrolysates. The hydrolysate obtained from using pepsin (120 min) had the highest adipogenesis inhibition activity. Mechanically, the pathway of peroxisome proliferator-activated receptor γ (PPARγ) was involved in the inhibition. A total of five peptides were identified as having strong antithrombotic and ACE inhibiting capacity (Shi et al., 2019).

6.5.6 Antidiabetic capacity

The antidiabetic capacity of quinoa protein hydrolysates was studied in vitro (Guo, Richel et al., 2020; Nongonierma et al., 2015; Salami et al., 2022; Vilcacundo, Barrio et al. 2017; Vilcacundo, Martínez-Villalueng et al., 2017; Zamudio & Campos, 2022). The inhibitory capacity on the activities of α-glucosidase, α-amylase, and dipeptidyl peptidase IV (DPP-IV) of these hydrolysates was measured. Nongonierma et al. (2015) showed that hydrolysis of quinoa protein isolate by papain and papain-similar enzyme of microbial origin increased the dipeptidyl peptidase IV (DPP-IV) inhibitory activity. Vilcacundo, Barrio et al. (2017) and Vilcacundo, Martínez-Villalueng et al. (2017) showed that simulated duodenal digestion released peptides with the highest inhibitory activities against the enzymes compared with other phases of digestion. Different peptides in these hydrolysates were identified as potential candidates with antidiabetic property. They included MCG, HPF, NMF MAF, IQAEGGL, TDKDYPK, and GEHGSDGNV (Guo, Richel et al., 2020; Vilcacundo, Barrio et al. 2017; Vilcacundo, Martínez-Villalueng et al., 2017). Hydrolysates of proteins in germinated (1 day) and raw quinoa seeds after simulated gastrointestinal digestion were tested as inhibitors on a homolog of human α-glucosidase in vitro (Salami et al., 2022). The hydrolysates of the proteins in the germinated quinoa seeds after the digestion had the highest inhibitory activity against the homologue of human α-glucosidase as competitive-type inhibitors (Salami et al., 2022). There are many different types of peptides from different sources. Different mechanisms of antidiabetic properties have been recorded for different peptides (Yan et al., 2019; Zamudio & Campos, 2022). Comparative studies should be done on the feasibility of using quinoa peptides and other commonly used and mostly studied peptides for antidiabetic applications. Online databases such as the BIOPEP-UWM Database may be used to explore the potential peptides generated from quinoa protein for targeted nutritional and medicinal properties (Minkiewicz et al., 2019; Orona-Tamayo et al., 2018; Zamudio & Campos, 2022).

6.5.7 Anticancer capacity

Quinoa protein hydrolysate from simulated gastrointestinal digestion was measured for its in vitro anticancer capacity (Fan et al., 2022; Vilcacundo et al., 2018). The colon cancer cell lines (Caco-2, HT-29, and HCT-116) were used as model systems. The digests from the intestinal phase had the highest anticancer capacity. The fraction of >5 kDa had the highest anticancer capacity. A total of 17 bioactive peptides were identified as potential candidates with potent anticancer properties (Vilcacundo et al., 2018). In the study of Fan et al. (2022), anticancer properties of quinoa peptides against Caco-2 cells of colon cancer were studied. Three peptides of NWFPLPR, FHPFPR, and HYNPYFPG showed significant anticancer properties. Molecular simulation showed that these peptides were docked into the active "pocket" of histone deacetylase 1. On the molecular level, the peptides decreased the expressions of histone deacetylase 1, Bcl-2, NFκB, IL-8, and IL-6, while increasing that of caspase3 (Fan et al., 2022). There are many different types of peptides with anticancer potential from different sources. There are different therapeutic agents and anticancer drugs that are commercially used (Sharma et al., 2021). Comparative studies remain to be done to reveal if quinoa peptides have any practical applications for anticancer potential.

6.5.8 Antiinflammatory

Lunasin isolated from quinoa showed in vitro antiinflammatory activity and significantly inhibited the production of tumor necrosis factor-α, interleukin-6, and nitric oxide (NO) in the RAW264.7 macrophages treated by lipopolysaccharide (Ren et al., 2017).

6.5.9 Safety of quinoa protein, hydrolysates, and peptides

Consumption of quinoa seeds may give intestinal disorders to a certain group of people. This has been attributed to the presence of bioactive components in the seeds such as saponins (Landi et al., 2021). Ragucci et al. (2021) and Landi et al. (2021) reported the existence of quinoin, which is a type 1 ribosome-inactivating protein in quinoa seeds (Fig. 6.2). This protein was found to be toxic to keratinocytes and fibroblast. Heating of the quinoa seeds should be done to inactivate quinoin before consumption (Landi et al., 2021).

Quinoa protein hydrolysates and peptides have shown a range of different biological functions (Table 6.1). To be able to apply these hydrolysates and peptides for food and nutraceutical formulations, their safety profiles should be studied. For example, while the hydrolysate of quinoa protein isolate had in vitro anticancer capacity (Vilcacundo et al., 2018), they should not have any toxicity towards normal cells of human tissues. However, the toxicity data of quinoa protein hydrolysates and peptides is largely missing. Such a lack of toxicity data seriously hinders the development of quinoa for functional food production. Apart from traditional in vitro and in vivo methods using cell lines and animal models, in silico methods may be used to complement the other methods to obtain a whole picture (Iwaniak et al., 2019). Eventually, the data generated using these methods should be translated into clinical results related to humans. Unfortunately, the majority of the published research has never had a chance to be subjected to any test of clinical trials.

6.6 Effect of modifications and processing on physicochemical, functional, and biological properties of quinoa protein and hydrolysates

Different physical and chemical processing methods affect functional, structural, and biological properties of quinoa protein and hydrolysates (Table 6.2).

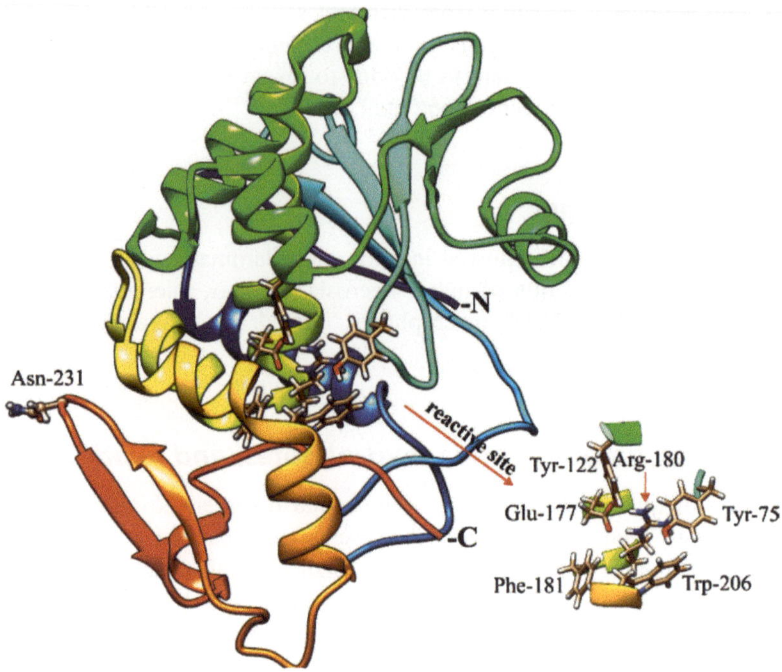

Figure 6.2 Three-dimensional molecular model obtained for quinoin. In this model, reported as a backbone ribbon, the secondary structure topology is shown. N- and C- indicate N- and C-terminus amino acid residues of the protein, respectively. In the insert, six amino acid residues are present in the catalytic site, common to all ribosome inactivating proteins (Ragucci et al., 2021).
Source: Reprinted with permission under Creative Commons Attribution License.

6.6.1 Physical treatments

6.6.1.1 Fractionation

Fractionation methods and conditions may affect the properties of isolated proteins. For example, Navarro-Lisboa et al. (2017) used ultrafiltration to separate quinoa proteins into fractions using ceramic membranes at pH 7.0 and 9.5. The pH values were above the isoelectric point (pH $= \sim 5$) of quinoa protein. The protein recovery was higher at pH 9.5 than at pH 7.0. The pH significantly affected the permeate flux, structure, and amino acid composition of the protein concentrate fractions obtained. At pH 7, α-helix structure was obtained for the concentrate. The ultrafiltration process modified the protein structure by decreasing the ΔH of protein denaturation. Overall, pH was a critical factor affecting the separation of protein fractions (Navarro-Lisboa et al., 2017).

Table 6.2 Processing and modifications of quinoa proteins and peptides.

Processing/ Applications	Protein form	Formulations/processing	Major findings	References
Ultrasound	Protein isolate	Alcalase hydrolysis of quinoa protein isolate under ultrasound. Ultrasound of up to 600 W and 158.79 W/cm^2, 20 kHz, 20 min	Ultrasound at 400 W led to increased hydrolysis and unfolding of the protein isolate and enhanced in vitro antioxidant activity of the resulting hydrolysate. Further increase of the power to 600 W led to protein aggregation and thus reduced enzyme hydrolysis	Li, Da et al. (2018)
Ultrasound	Protein isolate	20 kHz, 700 W; amplitude (20%), up to 30 min, at 20°C, sweep cycles (1 – 10 s) and ratio of on-time/off-time were 16.7% – 83.3%	Peptides of >190 kDa were produced from ultrasound processing (Vera et al., 2019). Ultrasound process induced the transformation of β-folds to α-helices and increased the denaturation temperature, *Z*-average, and solubility of the protein isolate. *Z*-potential was not affected	Vera et al. (2019)
Ultrasound	Protein isolate	20 kHz, 500 W, 25% amplitude, up to 35 min, protein concentration (8%)	Ultrasound treatment led to protein aggregation, larger particles, higher turbidity, and increased gelation. The treatment increased oil and water-binding and emulsifying capacity. The molecular composition of the protein isolate was not affected	Mir et al. (2019)
Ultrasound	Oxidized protein isolate	300 W, up to 40 min, on (3 s) and off (5 s)	Ultrasound processing up to 30 min increased water solubility, *Z*-potential, and uniformity of particle size, while decreasing the particle size. The processing increased the surface hydrophobicity and the amounts of β-turns and α-helices. Excessive ultrasound processing of 40 min led to aggregation of the proteins	Cao et al. (2021)

(Continued)

Table 6.2 (Continued)

Processing/ Applications	Protein form	Formulations/processing	Major findings	References
High-pressure homogenization	Protein isolate	High-pressure homogenization at 10, 30, and 50 MPa (each for 5 passes)	High-pressure homogenization at 50 MPa decreased the particle size from ~8.2 to 0.5 μm by dispersing large aggregates of quinoa protein, while increasing emulsifying and foaming capability and water solubility. The treatment increased the firmness of the gels and its microstructural compactness. The treatment had little effect on the secondary structure of the protein	Luo et al. (2022)
Heating	Protein isolate (globulins)	Heat denaturation at 100°C for 15 min (different pH of 6.5–10.5)	Heating at pH of 8.5 and 10.5 (not 6.5) disrupted disulfide bonds of the protein. Heating at 8.5 and 6.5 induced protein aggregation which became soluble upon reduction. The secondary structure of the protein was not affected by the heating, while the protein became unfolded Heating at pH 8.5 increased surface hydrophobicity, and reduced solubility and acid-induced gelation. Heating at pH 10.5 improved the solubility, and acid-induced gelation a high storage modulus (G')	Mäkinen et al., (2015, 2016)
Heating	Protein isolate	Protein was extracted at different pH (8–11). Protein suspensions were heated at 60°C–120°C (30 min)	Increasing pH of extraction conditions increased the protein denaturation and yield, while decreasing the purity, solubility, and the amounts of larger molecules. The thermal treatment denatured the protein and induced aggregation. Decreasing pH of extraction condition led to more aggregation and firmer texture of heat-set gels	Ruiz, Opazo-Navarrete et al. (2016) and Ruiz, Xiao et al. (2016)

			Increasing denaturation temperature and or extraction pH decreased in vitro gastric protein digestibility by inducing more aggregation. Digestibility of protein isolate was higher than that of protein in flour	
Heating	Protein isolate	Heat denaturation and freezing-lyophilization	Freezing-lyophilization process increased the emulsifying and foaming ability, and water adsorption of the protein. Heat-induced denaturation increased the water/oil retention and particle sizes of the protein. The two modifications increased the porosity of the isolate	Cerdán-Leal et al. (2020)
Heating	Albumins (water extracted) and globulins (dilute salt solution extracted)	Heating the protein fractions at different pH (5–7), temperatures (45°C–100°C), and time (1–15 min) with different protein concentrations (25–100 mg/mL)	Protein solubility was not affected by the protein concentrations. The extents of protein aggregation were affected by heating length, pH, and temperature, whereas the types of aggregation were affected by pH and temperature. Aggregation of noncovalent interactions was dominant when the pH was near the isoelectric point and the temperature was $<70°C$. Aggregation of covalent interactions was increased to 25% when heated at pH of 7 and 100°C for 15 min. The globulins 7 S and 11 S (not 2 S) were polymerized mostly through disulfide bonds	de Vondel et al. (2021)

(*Continued*)

Table 6.2 (Continued)

Processing/ Applications	Protein form	Formulations/processing	Major findings	References
Heating	Protein isolate	Heating protein suspension (8%) (80°C–100°C, 15–30 min)	Heating led to the loss of protein secondary structure (β-sheet and α-helix) and the amount of SH groups, formation of soluble aggregates, decreased particle size and crystallinity, increased turbidity and solubility. Increasing temperature and heating length led to decreased binding of water and oils and emulsifying capacity. Foaming capacity and stability may be optimized under suitable heating temperature (80°C) and time (30 min)	Mir et al. (2021)
Heating	Steaming, microwave heating, baking and boiling	Quinoa protein isolate was treated by different heating methods	Boiling dissociated the protein and disrupted disulfide bond. Steaming and baking led to the formation of large molecules which were not resolved by SDS-PAGE. Boiling and microwave heating decreased the amount of α-helices and increased that of β-sheets. Steaming and baking reduced the surface hydrophobicity and the amount of free sulfhydryl group. The solubility, water-holding, and emulsifying capacity of native protein isolate, and those treated by microwaving and boiling were higher than those treated by steaming and baking. These differences could be related to the molecular aggregation induced by steaming and baking	Wang et al. (2021)

Drying	Protein isolate	Freeze drying, vacuum drying, and spray drying	Spray drying gave the highest water absorption capacity. Freeze-drying gave the highest capacity for oil binding and emulsification and the highest G' (elastic) and G'' (viscous) modulus. The lowest ΔH and the highest denaturation temperature were found in the freeze-dried sample due to the presence of less β-sheets and α-helices and more random coils compared to other samples	Shen et al. (2021)
Supercritical fluid	Protein hydrolysate	Quinoa protein was treated with supercritical CO_2 and ethanol for extraction of lipids, phenolics, and other components before hydrolysis	The treatment increased the hydrolysis and antioxidant activities of the resulting products compared to that with petroleum ether extraction	Olivera-Montenegro et al. (2021)
Oxidation	Protein isolate	2,2'-Azobis (2-amidinopropane) dihydrochloride was used to oxidize protein isolate	Oxidation reaction led to carbonylation of sulfhydryl groups and protein aggregation. The oxidation decreased turbidity, solubility, emulsifying capacity of the protein	Cao et al. (2021)
Sprouting	Protein in seeds	Quinoa seeds were sprouted for 72 h	Sprouting for at least 48 h led to proteolysis and increased foaming capacity of the protein	Suárez-Estrella et al. (2020)
Sprouting and enzyme hydrolysis	Protein isolate and hydrolysates from 3 types of quinoa grains (red, white and black)	Quinoa seeds were germinated at 28°C for up to 72 h. Then, the protein isolates were obtained before hydrolysis under simulated gastric and duodenal digestion	Overall, sprouting and digestion led to the formation of antioxidant hydrolysates. The antioxidant activity was dependent on the genotypes, digestion type, and sprouting length. The antioxidant activity was confirmed using in vivo zebrafish embryo model	Piñuel et al. (2019)

(*Continued*)

Table 6.2 (Continued)

Processing/ Applications	Protein form	Formulations/processing	Major findings	References
Enzyme hydrolysis	Protein isolate	Protein was hydrolyzed using alcalase and pepsin	The highest antioxidant activity and degree of hydrolysis of the resulting hydrolysates could be obtained by optimizing temperatures and length of the hydrolysis and enzyme-to-substrate ratio	Mahdavi-Yekta et al. (2019)
Enzyme hydrolysis	Protein isolate	Protein was hydrolyzed using peptidase of *Aspergillus niger* at 40°C up to 3 h	A hydrolysis degree of 17% was obtained. The hydrolysis had no effect on in vitro antioxidant activity and decreased surface hydrophobicity. Acid-induced gels from hydrolyzed proteins had lower water retention and softer texture than those from untreated protein	Galante et al. (2020)
Enzyme hydrolysis	Protein isolate	Protein was hydrolyzed by pancreatin	A degree of hydrolysis of 19% was obtained. Aromatic and hydroxylic amino acid residues were found. The hydrolysis increased emulsifying, solubility, and foaming capacities of the protein	Daliri et al. (2021)

6.6.1.2 Ultrasound

Ultrasound (up to 600 W and 158.79 W/cm^2, 20 kHz, 20 min) was used to facilitate enzyme (alcalase) hydrolysis of quinoa protein isolate (Li, Da et al., 2018). The ultrasound at 400 W led to increased hydrolysis and unfolding of the protein isolate and enhanced in vitro antioxidant activity of the resulting hydrolysate. Further increase of the power to 600 W reduced the enzyme hydrolysis as the ultrasound treatment-induced protein aggregation. The aggregation made the protein less accessible to the hydrolytic enzymes (Li, Da et al., 2018).

Ultrasound was used to treat quinoa protein isolate and to oxidize protein (Cao et al., 2021; Mir et al., 2019; Vera et al., 2019). Vera et al. (2019) showed that ultrasound processing led to the formation of peptides of $>$190 kDa. Ultrasound process induced the transformation of β-folds to α-helices, while increasing the denaturation temperature, Z-average (hydrodynamic diameter), and solubility of the quinoa protein isolate. Z-potential (zeta potential, the electrical potential at the interface) was not affected. In contrast, Mir et al. (2019) showed that ultrasound treatment led to protein aggregation, formation of larger particles, higher turbidity, and increased gelation. The treatment increased oil- and water-binding and emulsifying capacity of the resulting protein isolate. The molecular composition of the protein isolate was not affected (Mir et al., 2019). The differences in the ultrasound effects on the protein functional properties could be due to the different experimental conditions such as ultrasound settings (i.e., power level, energy density, amplitude, temperature, processing time, shape of processing unit, and mode of action) and protein concentrations used in different studies. Indeed, Cao et al. (2021) showed that ultrasound processing up to 30 min increased water solubility, Z-potential, and uniformity of particle size of oxidized quinoa protein isolate, while decreasing the particle size. The processing increased the surface hydrophobicity and the amounts of β-turns and α-helices of the protein. Excessive ultrasound processing of 40 min led to aggregation of the proteins. Overall, ultrasound processing conditions should be selectively chosen to obtain targeted structural and functional properties of protein isolate.

6.6.1.3 High-pressure homogenization

Quinoa protein isolate was treated by high-pressure homogenization at 10, 30, and 50 MPa (each for five passes). High-pressure homogenization at 50 MPa had the most significant effects on the isolate and its gels. The treatment decreased the particle size from ~8.2 to 0.5 μm by dispersing large aggregates of quinoa protein, while increasing the emulsifying and foaming capability and water solubility. The treatment increased the firmness of the gels and its microstructural compactness (Fig. 6.3). The treatment had little effect on the secondary structure of the quinoa protein (Luo et al., 2022).

6.6.1.4 Heating and drying

Different heating and drying methods with different conditions were applied on different quinoa protein isolates and fractions (Cerdán-Leal et al., 2020;

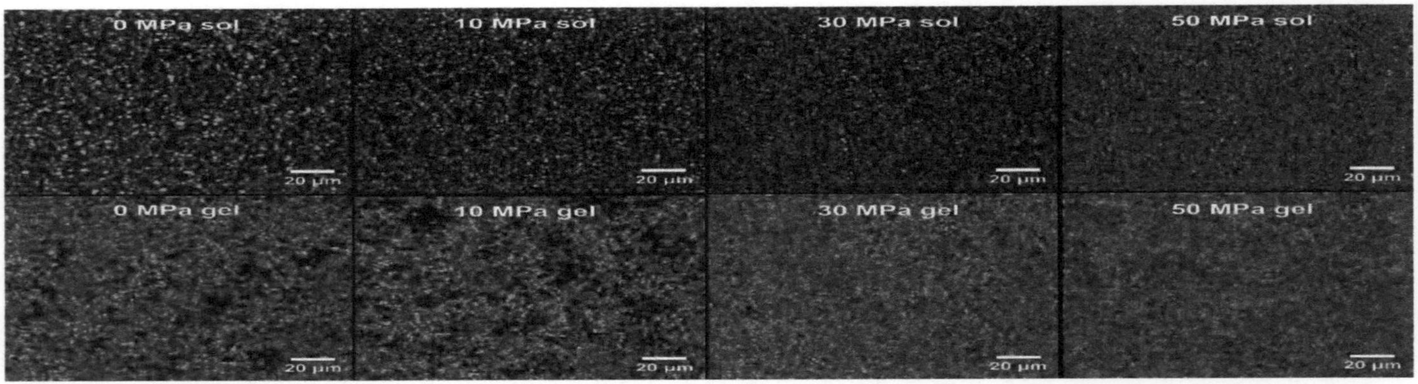

Figure 6.3 Confocal laser scanning microscopy images of various untreated and high-pressure homogenization treated suspensions (sol) of quinoa protein isolate and their corresponding gels. The gels were formed by heating at 85°C for 30 min (Luo et al., 2022).
Source: Reprinted with permission from Elsevier.

de Vondel et al., 2021, 2022; Mäkinen et al., 2015, 2016; Mir et al., 2021; Ruiz, Opazo-Navarrete et al., 2016; Ruiz, Xiao et al., 2016; Shen et al., 2021; Wang et al., 2021).

Cerdán-Leal et al. (2020) showed that heat induced quinoa protein denaturation and increased the water/oil retention and particle sizes of the protein. Quinoa protein aggregation behaviors are largely affected by heating temperature, length, and pH (de Vondel et al., 2021). Mäkinen et al. (2015, 2016) showed that heating at pH of 8.5 and 10.5 (not 6.5) disrupted disulfide bonds of quinoa protein isolate. Heating at pH 8.5 and 6.5 induced protein aggregation that became soluble upon reduction. The secondary structure of the protein was not affected by the heating, while the protein became unfolded. Heating at pH 8.5 increased surface hydrophobicity and reduced solubility and acid-induced gelation. Heating at pH 10.5 improved the solubility and acid-induced gelation with a high storage modulus (G') for the formed gels (Mäkinen et al., 2015, 2016). Aggregates bound with disulfide bonds were formed with heating at pH 8.5. Upon acidification, repulsive forces were reduced among the aggregates to form a coarse gel with a lower storage modulus (G'). The protein when heated at pH 10.5 had high solubility due to the high surface charge. Upon acidification, the soluble proteins crosslinked to form a gel with compacted structure and high G' (Mäkinen et al., 2015). Ruiz, Opazo-Navarrete et al. (2016) and Ruiz, Xiao et al. (2016) showed that thermal treatment denatured the protein and induced aggregation. Decreasing pH of protein extraction condition led to more aggregation and firmer heat-set gels. Increasing denaturation temperature and/or extraction pH decreased in vitro gastric protein digestibility by inducing more aggregation (Ruiz, Opazo-Navarrete et al., 2016; Ruiz, Xiao et al., 2016). de Vondel et al. (2021) showed that the extents of protein aggregation were affected by heating length, pH, and temperature, whereas the types of aggregation of different protein fractions including albumins and globulins were affected by pH and temperature. Aggregation of noncovalent interactions was dominant when the pH was near the isoelectric point and the temperature was $<70°C$. Aggregation of covalent interactions was increased to 25% when heated at pH of 7 and 100°C for 15 min. The globulins 7 S and 11 S (not 2 S) were polymerized mostly through disulfide bonds (de Vondel et al., 2021).

Wang et al. (2021) compared the effects of different heating methods on protein structure and properties. Boiling dissociated the protein and disrupted disulfide bonds. Steaming and baking led to the formation of large molecules that were not resolved by SDS-PAGE. Boiling and microwave heating decreased the amount of α-helices and increased that of β-sheets. Steaming and baking reduced the surface hydrophobicity and the amount of free sulfhydryl group. The solubility, water-holding, and emulsifying capacity of native protein isolate, and those treated by microwaving and boiling were higher than those treated by steaming and baking. These differences could be related to the different behaviors in their molecular aggregation induced by different heating methods (Wang et al., 2021).

The presence of other components can significantly affect the protein properties during the heating process (de Vondel et al., 2022; Yang et al., 2022). For example, Yang et al. (2022) showed that the presence of NaCl and $CaCl_2$ at different concentrations significantly affected the microstructural and rheological properties of the gels of quinoa protein isolate upon heating. Increasing NaCl concentration up to 200 mM increased the complex moduli (G^*) from 70 up to 1300 Pa, while decreasing the gelation temperature from 73°C to 40°C. Adding $CaCl_2$ decreased the particle size, while increasing the tendency of phase separation (Yang et al., 2022). de Vondel et al. (2022) analyzed the effect of heating on the aggregation and denaturation of protein in whole meal and in grains of quinoa. Inside the intact grains, the percentage of proteins that became aggregated through covalent bond formation upon boiling was up to 37%, whereas the embryonic cell wall enveloped proteins were not affected by the boiling treatment. A portion of 2 S albumins were leached out of the seeds. During heating, the monomers of 11 S globulin were degraded into basic and acidic subunits that were further crosslinked by disulfide bond. The denaturation and aggregation of proteins were more in whole seeds than in whole flour, indicating the effect of food matrix (de Vondel et al., 2022). Overall, it may be concluded that food matrix, heating temperature, method and length, and pH of the solution and methods used for protein extraction are the factors affecting the outcomes on structure and functional properties of quinoa protein isolate.

Quinoa protein isolate was dried using different drying methods (Cerdán-Leal et al., 2020; Shen et al., 2021). Freezing–lyophilization process increased the emulsifying and foaming ability, water adsorption of quinoa protein isolate, and the porosity of the isolate (Cerdán-Leal et al., 2020). Shen et al. (2021) used different drying processes to treat quinoa protein isolate. Spray drying gave the highest water absorption capacity. Freeze-drying gave the highest capacity for oil binding and emulsification and the highest G' (elastic) and G'' (viscous) modulus. The lowest ΔH and the highest denaturation temperature were found in the freeze-dried samples due to the presence of less β-sheets and α-helices and more random coils in these samples than in other samples (Shen et al., 2021). The implications of these results are that results from different studies on functional properties of quinoa protein isolate may be difficult for direct comparison if different drying methods are used. However, many researchers tend to forget this aspect of the experiment when analyzing the properties of protein isolate. Unfortunately, this lack of standardized procedures for experiments continues to plague our discourse in food-related scientific research.

6.6.2 Chemical and enzymatic treatments

6.6.2.1 Oxidation

Protein oxidation may lead to decreased functional performance for food applications (He et al., 2021). Quinoa protein isolate was subjected to oxidation using 2,2′-azobis (2-amidinopropane)dihydrochloride. The oxidation reaction led to carbonylation of sulfhydryl groups and protein aggregation (Cao et al., 2021). The oxidation decreased turbidity, solubility, and emulsifying capacity of the protein. Similar results were recorded for soy protein subjected to oxidation (He et al., 2021).

Processing of the oxidized protein with other methods such as ultrasound under suitable settings could improve the functional properties (Cao et al., 2021).

6.6.2.2 Sprouting

Quinoa seeds were subjected to sprouting of up to 72 h (Piñuel et al., 2019; Suárez-Estrella et al., 2020). Sprouting for at least 48 h led to proteolysis, increasing the foaming capacity of the protein (Suárez-Estrella et al., 2020). Piñuel et al. (2019) showed that sprouting led to the formation of antioxidant hydrolysates. The in vitro antioxidant activity was dependent on the quinoa genotypes and sprouting length. Seed sprouting in combination with enzyme hydrolysis of quinoa protein isolates produced hydrolysates with improved in vitro and in vivo (zebrafish embryo model) antioxidant activity (Piñuel et al., 2019). Therefore sprouting can be a feasible method to produce antioxidant protein hydrolysates. Indeed, sprouted food grains tend to contain more nutrients than raw food grains (Benincasa et al., 2019).

6.6.2.3 Enzyme hydrolysis

Quinoa protein isolate was hydrolyzed using different enzymes to produce peptides (Chew, Toh, & Ismail, 2019; Daliri et al., 2021; Galante et al., 2020; Mahdavi-Yekta et al., 2019). Improved antioxidant activity and degree of hydrolysis of the resulting hydrolysates could be obtained by optimizing temperatures and length of the hydrolysis and enzyme-to-substrate ratio (Mahdavi-Yekta et al., 2019). Galante et al. (2020) obtained a hydrolysis degree of 17% using peptidase of *Aspergillus niger*. The hydrolysis had no effect on in vitro antioxidant activity, while decreasing the surface hydrophobicity. Acid-induced gels from hydrolyzed proteins had lower water retention and softer texture than those from untreated protein (Galante et al., 2020). Daliri et al. (2021) showed that hydrolysates of quinoa protein isolate obtained using pancreatin had increased emulsifying, solubility, and foaming capacity compared to the native protein isolate. Alrosan et al. (2022) reviewed the literature and showed that fermentation of quinoa protein may improve the functional, nutritional, and digestive properties. The fermentation conditions were critical for the outcomes. Therefore it appeared that the biological and functional properties of the resulting hydrolysates were dependent on hydrolysis conditions such as the enzyme type, enzyme-to-substrate ratio, and hydrolysis length.

6.6.3 Interactions of quinoa protein and peptides with other components

There are other components in quinoa grains including phenolics as described in Chapters 2 and 8. These compounds may interact with proteins and peptides in quinoa food systems during processing. For example, Pérez-Gregorio et al. (2020) reviewed the interactions between polyphenols and peptides. The interactions between them can be reversible or irreversible through covalent or noncovalent bonds. In the food industry, it is important to manage the interactions of proteins and peptides with other components in quinoa-based food products for quality control. Some technologies such as chromatography, electrospray mass spectrometry, and nuclear magnetic resonance spectroscopy can be involved to analyze the polyphenol-peptide complexes (Pérez-Gregorio et al., 2020). In many of the reported studies summarized in this chapter, quinoa protein isolate used was not pure. Other components such as phenolics, lipids, and starch may be present (Wang et al., 2020), contributing to the physicochemical and biological properties of the protein isolates and related hydrolysates. Alrosan et al. (2022) reviewed the literature and showed that quinoa protein complexes with other proteins such as animal-based proteins may improve the functional and nutritional properties. This aspect of quinoa protein chemistry remains to be better studied.

6.7 Conclusions

Quinoa protein is better balanced in the composition of essential amino acids than cereal proteins. Quinoa protein has more albumins and globulins than the proteins of most cereals including wheat and is higher in lysine content than cereals. The amino acid composition and structure influence the solubility of quinoa protein. The solubility influences the foaming/emulsifying capacity and stability of the protein, and itself depends on the charge on the protein. The secondary structures of quinoa albumins and globulins influence the digestibility. Quinoa protein isolate was processed to obtain a range of functional and biological properties using chemical, physical, and/or enzymatic methods. External factors of processing including pH, temperature, process parameters, and composition of materials influenced the functional properties of the resulting protein isolates and peptides. The quinoa protein hydrolysates showed a range of biological functions including antioxidant, antihaemolytic, antibacterial, antihypertension, antidiabetic, antiobesity, and anticancer capacity.

The research gap to bridge the practical applications of quinoa protein for food applications is too large. This is largely due to the fact that there is a lack of comparison studies using currently used industrial and commercial standards. It is not known if most of the results summarized in this chapter have any potential for practical applications due to the nature of the methodologies used. Perhaps, most of the published data on quinoa protein processing and biological and functional properties are purely of any academic curiosity. Gut microbiota as affected by quinoa protein, hydrolysates, and peptides remains to be studied. Antinutritional factors should also be considered for the nutritional applications of quinoa proteins and peptides.

References

Abugoch, L. E., Romero, N., Tapia, C. A., Silva, J., & Rivera, M. (2008). Study of some physicochemical and functional properties of quinoa (*Chenopodium quinoa* Willd) protein isolates. *Journal of Agricultural and Food Chemistry*, *56*, 4745–4750.

Alrosan, M., Tan, T. C., Easa, A. M., Gammoh, S., & Alu'datt, M. H. (2022). Recent updates on lentil and quinoa protein-based dairy protein alternatives: Nutrition, technologies, and challenges. *Food Chemistry*, *383*, 132386.

Benincasa, P., Falcinelli, B., Lutts, S., Stagnari, F., & Galieni, A. (2019). Sprouted grains: A comprehensive review. *Nutrients*, *11*, 421.

Cao, H., Sun, R., Shi, J., Li, M., Guan, X., Liu, J., Huang, K., & Zhang, Y. (2021). Effect of ultrasonic on the structure and quality characteristics of quinoa protein oxidation aggregates. *Ultrasonics-Sonochemistry*, *77*, 105685.

Cerdán-Leal, M. A., López-Alarcón, C. A., Ortiz-Basurto, R. I., Luna-Solano, G., & Jiménez-Fernández, M. (2020). Influence of heat denaturation and freezing-lyophilization on physicochemical and functional properties of quinoa protein isolate. *Cereal Chemistry*, *97*, 373–381.

Chew, L. Y., Toh, G. T., & Ismail, A. (2019). Application of proteases for the production of bioactive peptides. In M. Kuddus (Ed.), *Enzymes in food biotechnology* (pp. 247–261). Academic Press.

Craine, E. B., & Murphy, K. M. (2020). Seed composition and amino acid profiles for quinoa grown in Washington state. *Frontiers in Nutrition*, *7*, 126.

Dakhili, S., Abdolalizadeh, L., Hosseini, S. M., Shojaee-Aliabadi, S., & Mirmoghtadaie, L. (2019). Quinoa protein: Composition, structure and functional properties. *Food Chemistry*, *299*, 125161.

Daliri, H., Ahmadi, R., Pezeshki, A., Hamishehkar, H., Mohammadi, M., Beyrami, H., Heshmati, M. K., & Ghorbani, M. (2021). Quinoa bioactive protein hydrolysate produced by pancreatin enzyme functional and antioxidant properties. *LWT-Food Science and Technology*, *150*, 111853.

de Vondel, J. V., Lambrecht, M. A., & Delcour, J. A. (2020). Osborne extractability and chromatographic separation of protein from quinoa (*Chenopodium quinoa* Willd.) wholemeal. *LWT-Food Science and Technology*, *126*, 109321.

de Vondel, J. V., Lambrecht, M. A., & Delcour, J. A. (2022). Heat-induced denaturation and aggregation of protein in quinoa (*Chenopodium quinoa* Willd.) seeds and whole meal. *Food Chemistry*, *372*, 131330.

de Vondel, J. V., Lambrecht, M. A., Housmans, J. A. J., Rousseau, F., Schymkowitz, J., & Delcour, J. A. (2021). Impact of hydrothermal treatment on denaturation and aggregation of water-extractable quinoa (*Chenopodium quinoa* Willd.) protein. *Food Hydrocolloids*, *115*, 106611.

Elsohaimy, S. A., Refaay, T. M., & Zaytoun, M. A. M. (2015). Physicochemical and functional properties of quinoa protein isolate. *Annals of Agricultural Science*, *60*, 297–305.

Fan, X., Guo, H., Teng, C., Zhang, B., Blecker, C., & Ren, G. (2022). Anti-colon cancer activity of novel peptides isolated from *in vitro* digestion of quinoa protein in Caco-2 cells. *Foods*, *11*, 194.

Galante, M., De Flaviis, R., Boeris, V., & Spelzini, D. (2020). Effects of the enzymatic hydrolysis treatment on functional and antioxidant properties of quinoa protein acid-induced gels. *LWT-Food Science and Technology*, *118*, 108845.

Galindo-Luján, R., Pont, L., Sanz-Nebot, V., & Benavente, F. (2021). Classification of quinoa varieties based on protein fingerprinting by capillary electrophoresis with ultraviolet absorption diode array detection and advanced chemometrics. *Food Chemistry*, *341*, 128207.

Ghumman, A., Mudgal, S., Singh, N., Ranjan, B., Kaur, A., & Rana, J. C. (2021). Physicochemical, functional and structural characteristics of grains, flour and protein isolates of Indian quinoa lines. *Food Research International*, *140*, 109982.

Granato, D., Shahidi, F., Wrolstad, R., Kilmartin, P., Melton, L. D., et al. (2018). Antioxidant activity, total phenolics and flavonoids contents: Should we ban *in vitro* screening methods? *Food Chemistry*, *264*, 471–475.

Guo, H., Hao, Y., Fan, X., Richel, A., Everaert, N., Yang, X., & Ren, G. (2021). Administration with quinoa protein reduces the blood pressure in spontaneously hypertensive rats and modifies the fecal microbiota. *Nutrients*, *13*, 2446.

Guo, H., Hao, Y., Richel, A., Everaert, N., Chen, Y., Liu, M., Yang, X., & Ren, G. (2020). Antihypertensive effect of quinoa protein under simulated gastrointestinal digestion and peptide characterization. *Journal of the Science of Food and Agriculture*, *100*, 5569–5576.

Guo, H., Richel, A., Hao, Y., Fan, X., Everaert, N., Yang, X., & Ren, G. (2020). Novel dipeptidyl peptidase-IV and angiotensin-I-converting enzyme inhibitory peptides released from quinoa protein by *in silico* proteolysis. *Food Science and Nutrition*, *8*, 1415–1422.

Gupta, A., Sharma, S., & Surasani, V. K. R. (2021). Quinoa protein isolate supplemented pasta: Nutritional, physical, textural and morphological characterization. *LWT-Food Science and Technology*, *135*, 110045.

He, M., Wu, C., Li, L., Zheng, L., Tian, T., Jiang, L., Li, Y., et al. (2021). Effects of cavitation jet treatment on the structure and emulsification properties of oxidized soy protein isolate. *Foods*, *10*, 2.

Iwaniak, A., Darewicz, M., Mogut, D., & Minkiewicz, P. (2019). Elucidation of the role of in silico methodologies in approaches to studying bioactive peptides derived from foods. *Journal of Functional Foods*, *61*, 103486.

Janssen, F., Pauly, A., Rombouts, I., Jansens, K. J. A., Deleu, L. J., & Delcour, J. A. (2017). Proteins of amaranth (*Amaranthus* spp.), buckwheat (*Fagopyrum* spp.), and quinoa (*Chenopodium* spp.): A food science and technology perspective. *Comprehensive Reviews in Food Science and Food Safety*, *16*, 39–58.

Joye, I. (2019). Protein digestibility of cereal products. *Foods*, *8*, 199.

Kaspchak, E., de Oliveira, M. A. S., Simas, F. F., Franco, C. R. C., Silveira, J. L. M., Mafra, M. R., & Igarashi-Mafra, L. (2017). Determination of heat-set gelation capacity of a quinoa protein isolate (*Chenopodium quinoa*) by dynamic oscillatory rheological analysis. *Food Chemistry*, *232*, 263–271.

Landi, N., Ruocco, M. R., Ragucci, S., Aliotta, F., Nasso, R., Pedone, P. V., & Di Maro, A. (2021). Quinoa as source of type 1 ribosome inactivating proteins: A novel knowledge for a revision of its consumption. *Food Chemistry*, *342*, 128337.

Li, X., Da, S., Li, C., Xue, F., & Zang, T. (2018). Effects of high-intensity ultrasound pretreatment with different levels of power output on the antioxidant properties of alcalase hydrolyzates from quinoa (*Chenopodium quinoa* Willd.) protein isolate. *Cereal Chemistry*, *95*, 518–526.

López, D. N., Galante, M., Robson, M., Boeris, V., & Spelzini, D. (2018). Amaranth, quinoa and chia protein isolates: Physicochemical and structural properties. *International Journal of Biological Macromolecules*, *109*, 152–159.

López-Alarcón, C. A., Cerdán-Leal, M. A., Beristain, C. I., Pascual-Pineda, L. A., Azuara, E., & Jiménez-Fernández, M. (2019). The potential use of modified quinoa protein isolates in cupcakes: Physicochemical properties, structure and stability of cupcakes. *Food & Function*, *10*, 4432–4439.

López-Castejón, M. L., Bengoechea, C., Díaz-Franco, J., & Carrera, C. (2020). Interfacial and emulsifying properties of quinoa protein concentrates. *Food Biophysics*, *15*, 122–132.

Luo, L., Cheng, L., Zhang, R., & Yang, Z. (2022). Impact of high-pressure homogenization on physico-chemical, structural, and rheological properties of quinoa protein isolates. *Food Structure*, *32*, 100265.

Mahdavi-Yekta, M., Nouri, L., & Aziz, M. H. (2019). The effects of hydrolysis condition on antioxidant activity of protein hydrolyzate from quinoa. *Food Science and Nutrition*, *7*, 930–936.

Mäkinen, O. E., Zannini, E., & Arendt, E. K. (2015). Modifying the cold gelation properties of quinoa protein isolate: Influence of heat-denaturation pH in the alkaline range. *Plant Foods for Human Nutrition*, *70*, 250–256.

Mäkinen, O. E., Zannini, E., Koehler, P., & Arendt, E. K. (2016). Heat-denaturation and aggregation of quinoa (*Chenopodium quinoa*) globulins as affected by the pH value. *Food Chemistry*, *196*, 17–24.

Minkiewicz, P., Iwaniak, A., & Darewicz, M. (2019). BIOPEP-UWM Database of bioactive peptides: Current opportunities. *International Journal of Molecular Sciences*, *20*, 5978.

Mir, N. A., Riar, C. S., & Singh, S. (2019). Structural modification of quinoa seed protein isolates (QPIs) by variable time sonification for improving its physicochemical and functional characteristics. *Ultrasonics-Sonochemistry*, *58*, 104700.

Mir, N. A., Riar, C. S., & Singh, S. (2021). Improvement in the functional properties of quinoa (*Chenopodium quinoa*) protein isolates after the application of controlled heat-treatment: Effect on structural properties. *Food Structure*, *28*, 100189.

Mudgil, P., Omar, L. S., Kamal, H., Kilari, B. P., & Maqsood, S. (2019). Multi-functional bioactive properties of intact and enzymatically hydrolysed quinoa and amaranth proteins. *LWT-Food Science and Technology*, *110*, 207–213.

Navarro-Lisboa, R., Herrera, C., Zúñiga, R. N., Enrione, J., Guzmánd, F., Matiacevich, S., & Astudillo-Castroe, C. (2017). Quinoa proteins (*Chenopodium quinoa* Willd.) fractionated by ultrafiltration using ceramic membranes: The role of pH on physicochemical and conformational properties. *Food and Bioproducts Processing*, *102*, 20–30.

Nongonierma, A. B., Le Maux, S., Dubrulle, C., Barre, C., & FitzGerald, R. J. (2015). Quinoa (*Chenopodium quinoa* Willd.) protein hydrolysates with *in vitro* dipeptidyl peptidase IV (DPP-IV) inhibitory and antioxidant properties. *Journal of Cereal Science*, *65*, 112–118.

Nowak, V., Du, J., & Charrondière, U. R. (2016). Assessment of the nutritional composition of quinoa (*Chenopodium quinoa* Willd.). *Food Chemistry*, *193*, 47–54.

Olivera-Montenegro, L., Best, I., & Gil-Saldarriaga, A. (2021). Effect of pretreatment by supercritical fluids on antioxidant activity of protein hydrolyzate from quinoa (*Chenopodium quinoa* Willd.). *Food Science and Nutrition*, *9*, 574–582.

Opazo-Navarrete, M., Freire, D. T., Boom, R. M., & Janssen, A. E. M. (2019). The influence of starch and fibre on *in vitro* protein digestibility of dry fractionated quinoa seed (Riobamba variety). *Food Biophysics*, *14*, 49–59.

Opazo-Navarrete, M., Schutyser, M. A. I., Boom, R. M., & Janssen, A. E. M. (2018). Effect of pre-treatment on *in vitro* gastric digestion of quinoa protein (*Chenopodium quinoa* Willd.) obtained by wet and dry fractionation. *International Journal of Food Sciences and Nutrition*, *69*, 1–11.

Orona-Tamayo, D., Valverde, M. E., & Paredes-López, O. (2018). Bioactive peptides from selected Latin American food crops – A nutraceutical and molecular approach. *Critical Reviews in Food Science and Nutrition*, *59*, 1949–1975.

Osborne, T. B. (1907). *The proteins of the wheat kernel*. Carnegie Institution of Washington.

Patole, S., Cheng, L., & Yang, Z. (2022). Impact of incorporations of various polysaccharides on rheological and microstructural characteristics of heat-induced quinoa protein isolate gels. *Food Biophysics*. Available from https://doi.org/10.1007/s11483-022-09720-3.

Pearce, K. N., & Kinsella, J. E. (1978). Emulsifying properties of proteins: Evaluation of a turbidimetric technique. *Journal of Agricultural and Food Chemistry*, *26*, 716–723.

Pérez-Gregorio, R., Soares, S., Mateus, N., & Freitas, V. (2020). Bioactive peptides and dietary polyphenols: Two sides of the same coin. *Molecules*, *25*, 3443.

Piñuel, L., Boeri, P., Zubillaga, F., Barrio, D. A., Torreta, J., Cruz, A., et al. (2019). Production of white, red and black quinoa (*Chenopodium quinoa* Willd Var. Real) protein isolates and its hydrolysates in germinated and non-germinated quinoa samples and antioxidant activity evaluation. *Plants*, *8*, 257.

Ragucci, S., Bulgari, D., Landi, N., Russo, R., Clemente, A., Valletta, M., Chambery, A., Gobbi, E., Faoro, F., & Di Maro, A. (2021). The structural characterization and anti-pathogenic activities of quinoin, a type 1 ribosome-inactivating protein from quinoa seeds. *International Journal of Molecular Sciences*, *22*, 8964.

Raheem, N., & Straus, S. K. (2019). Mechanisms of action for antimicrobial peptides with antibacterial and antibiofilm functions. *Frontiers in Microbiology*, *10*, 2866.

Ravera, F., Dziza, K., Santini, E., Cristofolini, L., & Liggieri, L. (2021). Emulsification and emulsion stability: The role of the interfacial properties. *Advances in Colloid and Interface Science*, *288*, 102344.

Ren, G., Zhu, Y., Shi, Z., & Li, J. (2017). Detection of lunasin in quinoa (*Chenopodium quinoa* Willd.) and the *in vitro* evaluation of its antioxidant and anti-inflammatory activities. *Journal of the Science of Food and Agriculture*, *97*, 4110–4116.

Ruiz, G. A., Opazo-Navarrete, M., Meurs, M., Minor, M., Sala, G., van Boekel, M., Stieger, M., et al. (2016). Denaturation and in vitro gastric digestion of heat-treated quinoa protein isolates obtained at various extraction pH. *Food Biophysics*, *11*, 184–197.

Ruiz, G. A., Xiao, W., van Boekel, M., Minor, M., & Stieger, M. (2016). Effect of extraction pH on heat-induced aggregation, gelation and microstructure of protein isolate from quinoa (*Chenopodium quinoa* Willd). *Food Chemistry*, *209*, 203–210.

Salami, M., Motahar, S. F. S., Ariaeenejad, S., Mamaghani, A. S. A., Kavousi, K., Moosavi-Movahedi, A. A., & Salekdeh, G. H. (2022). The novel homologue of the human α-glucosidase inhibited by the non-germinated and germinated quinoa protein hydrolysates after *in vitro* gastrointestinal digestion. *Journal of Food Biochemistry*, *46*, e14030.

Sharma, P., Kaur, H., Kehinde, B. A., Chhikara, N., Sharma, D., & Panghal, A. (2021). Food-derived anticancer peptides: A review. *International Journal of Peptide Research and Therapeutics*, *27*, 55–70.

Shen, Y., Tang, X., & Li, Y. (2021). Drying methods affect physicochemical and functional properties of quinoa protein isolate. *Food Chemistry*, *339*, 127823.

Shi, D., Fidelis, M., Ren, Y., Stone, A. K., Ai, Y., & Nickerson, M. T. (2020). The functional attributes of Peruvian (Kankolla and Blanca juli blend) and Northern quinoa (NQ94PT) flours and protein isolates, and their protein quality. *Food Research International*, *128*, 108799.

Shi, X. D., Huang, J. J., Wu, J. L., Cai, X. X., Tian, Y. Q., Rao, P. F., Huang, J. L., & Wang, S. Y. (2022). Fabrication, interaction mechanism, functional properties, and applications of fish gelatin-polysaccharide composites: A review. *Food Hydrocolloids*, *122*, 107106.

Shi, Z., Hao, Y., Teng, C., Yao, Y., & Ren, G. (2019). Functional properties and adipogenesis inhibitory activity of protein hydrolysates from quinoa (*Chenopodium quinoa* Willd.). *Food Science and Nutrition*, *7*, 2103–2112.

Solaesa, A. G., Villanueva, M., Vela, A. J., & Ronda, F. (2020). Protein and lipid enrichment of quinoa (cv. Titicaca) by dry fractionation. Techno-functional, thermal and rheological properties of milling fractions. *Food Hydrocolloids*, *105*, 105770.

Steffolani, M. E., Villacorta, P., Morales-Soriano, E. R., Repo-Carrasco, R., León, A. E., & Pérez, G. T. (2016). Physicochemical and functional characterization of protein isolated from different quinoa varieties (*Chenopodium quinoa* Willd.). *Cereal Chemistry, 93*, 275–281.

Suárez-Estrella, D., Bresciani, A., Iametti, S., Marengo, M., Pagani, M. A., & Marti, A. (2020). Effect of sprouting on proteins and starch in quinoa (*Chenopodium quinoa* Willd.). *Plant Foods for Human Nutrition, 75*, 635–641.

Tagle-Freire, D., Mennah-Govela, Y., & Bornhorst, G. M. (2022). Starch and protein hydrolysis in cooked quinoa (*Chenopodium quinoa* Willd.) during static and dynamic *in vitro* oral and gastric digestion. *Food & Function, 13*, 920–932.

Vera, A., Valenzuela, M. A., Yazdani-Pedram, M., Tapia, C., & Abugoch, L. (2019). Conformational and physicochemical properties of quinoa proteins affected by different conditions of high-intensity ultrasound treatments. *Ultrasonics-Sonochemistry, 51*, 186–196.

Vilcacundo, R., Barrio, D., Carpio, C., García-Ruiz, A., Rúales, J., Hernández-Ledesma, B., & Carrillo, W. (2017). Digestibility of quinoa (*Chenopodium quinoa* Willd.) protein concentrate and its potential to inhibit lipid peroxidation in the zebrafish larvae model. *Plant Foods for Human Nutrition, 72*, 294–300.

Vilcacundo, R., Martínez-Villalueng, C., & Hernández-Ledesma, B. (2017). Release of dipeptidyl peptidase IV, α-amylase and α-glucosidase inhibitory peptides from quinoa (*Chenopodium quinoa* Willd.) during *in vitro* simulated gastrointestinal digestion. *Journal of Functional Foods, 35*, 531–539.

Vilcacundo, R., Miralles, B., Carrillo, W., & Hernández-Ledesma, B. (2018). *In vitro* chemopreventive properties of peptides released from quinoa (*Chenopodium quinoa* Willd.) protein under simulated gastrointestinal digestion. *Food Research International, 105*, 403–411.

Wang, L., Dong, J. L., Zhu, Y. Y., Shen, R. L., Wu, L. G., & Zhang, K. Y. (2021). Effects of microwave heating, steaming, boiling and baking on the structure and functional properties of quinoa (*Chenopodium quinoa* Willd.) protein isolates. *International Journal of Food Science and Technology, 56*, 709–720.

Wang, X., Zhao, R., & Yuan, W. (2020). Composition and secondary structure of proteins isolated from six different quinoa varieties from China. *Journal of Cereal Science, 95*, 103036.

Yan, J., Zhao, J., Yang, R., & Zhao, W. (2019). Bioactive peptides with antidiabetic properties: A review. *International Journal of Food Science & Technology, 54*, 1909–1919.

Yang, Z., de Campo, L., Gilbert, E. P., Knott, R., Cheng, L., Storer, B., Lin, X., et al. (2022). Effect of NaCl and $CaCl_2$ concentration on the rheological and structural characteristics of thermally-induced quinoa protein gels. *Food Hydrocolloids, 124*, 107350.

Yuliarti, O., Kovis, T. J. K., & Yi, N. J. (2021). Structuring the meat analogue by using plant-based derived composites. *Journal of Food Engineering, 288*, 110138.

Zamudio, F. V., & Campos, M. R. S. (2022). Amaranth, quinoa and chia bioactive peptides: A comprehensive review on three ancient grains and their potential role in management and prevention of type 2 diabetes. *Critical Reviews in Food Science and Nutrition, 62*, 2707–2721.

Zevallos, V. F., Herencia, L. I., Chang, F., Donnelly, S., Ellis, H. J., & Ciclitira, P. J. (2014). Gastrointestinal effects of eating quinoa (*Chenopodium quinoa* Willd.) in celiac patients. *The American Journal of Gastroenterology, 109*, 270–278.

Quinoa lipids

7

7.1 Introduction

Lipids are a major category of human foods. Like cereals and other pseudocereals, wholegrain quinoa contains a significant amount of lipids. The quinoa seeds may contain lipids of up to over 7% of the total seed weight (fresh basis) (Nowak et al., 2016). Most of the lipids are found in the embryo of the seeds (Prego et al., 1998). The lipid components may have an impact on the nutritional and physicochemical properties of quinoa flour and related food products (Rolandelli et al., 2022; Solaesa et al., 2020). Quinoa lipids include neutral and polar lipids as well as free fatty acids. There are different unsaponifiable compounds of lipid fraction such as tocopherols, phytosterols, carotenoids, and squalenes (Fanali et al., 2015; Graf et al., 2016). The percentage of unsaturated fatty lipids in quinoa seed is high (e.g., 80%) (Curti et al., 2020; Peiretti et al., 2013). They contribute to the beneficial effects of quinoa grains and related food products on human health. The germ of a quinoa seed is rich in lipids and can be used for oil extraction and functional food formulations (Mufari et al., 2020). In this chapter, the composition of neutral and polar lipids, triglycerides, fatty acids, and unsaponifiable lipid fractions of quinoa seeds are reviewed. The effect of different processing operations on the lipid profile and the role of lipids in affecting the properties of quinoa flour are described. Research directions to improve the utilization of quinoa lipids are discussed.

7.2 Extraction of quinoa lipids

Nowak et al. (2016) reviewed literature published up to the year 2014 and showed that the total lipid contents of quinoa seeds ranged from 4.0% to 7.6% (fresh weight). Recent studies showed that there was a significant variation in lipid contents of different quinoa genotypes (De Bock et al., 2021; Encina-Zelada et al., 2017; Filho et al., 2017; Vera et al., 2019). For example, De Bock et al. (2021) reported that the lipid contents of seven quinoa genotypes ranged from 2.7% to 7.4%. Such a high quantity warrants detailed studies. Therefore lipids were extracted from quinoa seeds for analysis using different organic solvents (Table 7.1). For example, Wood et al. (1993) used chloroform-methanol mixture (2:1) to extract lipids from quinoa flour. Ruales and Nair (1993) used petroleum ether to extract lipids. Mufari et al. (2020) used hexane as the solvent for lipid extraction from quinoa seeds, whereas Curti et al. (2020) employed

Quinoa. DOI: https://doi.org/10.1016/B978-0-323-99909-0.00014-3

Table 7.1 Analysis of lipids in quinoa seeds.

Analysis	Methods for lipid content or composition	Method for fatty acid composition	Major findings	References
Lipid content and fatty acid composition	Chloroform-methanol (2:1) for total lipid extraction; hexane was used to extract neutral lipids	Gas chromatography with a flame ionization detector. Boron trifluoride in methanol was used for transmethylation	Total lipid content was 5.3%. A total of 89% of fatty acids were unsaturated. The percentages of linoleic, oleic and linolenic acids were 52%, 23% and 8.1%, respectively. Palmitic acid was the major saturated fatty acid	Wood et al. (1993)
Lipid content and fatty acid composition	Dichloroethane–chloroform-acetone-acetic acid for neutral lipids, acetone-water-acetic acid for galactolipids, chloroform-methanol-water-acetic acid for polar lipids	Gas chromatography with a flame ionization detector. Lipids were treated with 10% hydrochloric acid in methanol	Total lipid content was 7.6%. Composition of neutral and polar lipids in quinoa seeds was analyzed. Free fatty acids accounted for 19% and 15% of total lipids. Triglycerides and diglycerides accounted for 50% and 20% of neutral lipids, respectively. Lysophosphatidyl ethanolamine (45%) and phosphatidyl choline (12%) were the major polar lipids in quinoa. The bran fraction had more neutral lipids (76%) and free fatty acid (11%) and less polar lipids (13%) than flour (neutral lipids (70%), free	Przybylski et al. (1994)

			fatty acid (9%) and polar lipids (21%). Linoleic, oleic, and palmitic acids were the major fatty acids in quinoa seeds	
Lipid content and fatty acid composition	Petroleum ether for extraction	Sulfuric acid in methanol for lipid hydrolysis, gas chromatography	Total lipid content was 9.7% (dry basis). The percentages of oleic, linoleic and linolenic acids were 25%, 52%, and 4%, respectively	Ruales and Nair (1993)
Composition of triacylglycerols and fatty acids	Petroleum ether for extraction	American Oil Chemists' Society (AOCS) method (Ce 2–66), gas chromatography with a flame ionization detector for fatty acid analysis	Linoleic acid (53%), oleic acid (26%), and palmitic acid (11%) were predominant in quinoa lipids. The most abundant triacylglycerol was OLL (~20%) consisting of 2 linoleoyl and 1 oleoyl residues	Jahaniaval et al. (2000)
Lipid content and fatty acid composition	Methyl *tert*-butyl ether/tetrahydrofuran (1:1, v/v) (Soxtec extraction)	Methylation with methanol. gas chromatography-flame ionization detector and mass spectrometry were used	The oil yields of 3 quinoa varieties were 6.6% – 7.2%. A total of 90% of the fatty acids were unsaturated. The ratios of ω-6 to ω-3 were 6. The ranges of percentages of linoleic acid, oleic acid, palmitic acid, and α-linolenic acid were 47%–50%, 25%–29%, 9.6%–10% and 7.3%–8.5%, respectively	Tang et al. (2015)

(*Continued*)

Table 7.1 (Continued)

Analysis	Methods for lipid content or composition	Method for fatty acid composition	Major findings	References
Composition of triacylglycerols, fatty acids and unsaponifiable lipidic compounds	Non-aqueous reversed-phase high-performance liquid chromatography with mass spectrometry detection was used for analysis of triacylglycerols	Methylation with sodium methylate in methanol, gas chromatography-flame ionization detector	The major triacylglycerols included OLL, OOL, and OLP. The OLL was the most abundant (18%). Oleic and linoleic acids accounted for approximately 28% and 48% of the total fatty acids. A total of 66 compounds were tentatively identified in the unsaponifiable fraction	Fanali et al. (2015)
Lipid content and fatty acid composition	The Association of Official Analytical Chemists (AOAC) official method 930.09.7, chloroform-methanol (2:1, v/v) mixture was used as solvent	10% of methanolic boron trifluoride was added to the chloroform phase to obtain fatty acid methyl esters. Gas chromatograph with a flame ionization detector for fatty acid analysis	The lipid contents of different varieties sowed at different times ranged from 4.7% to 7.2%. Their ratios of unsaturated fatty acids to saturated fatty acids ranged from 5.9 to 8.5. The ranges of percentages of linoleic (C18:2 w-6), oleic (C18:1 w-9), palmitic (C16:0), and α-linolenic (C18:3 w-3) acids were 28% − 44%, 8.6% − 18%, 5.2% − 8.5%, and 2.8% − 5.1%, respectively. Lipid contents and fatty acid composition of quinoa grains	Curti et al. (2020)

			were affected by genetics and sowing dates (via solar radiation and temperature)	
Fatty acid composition and oil extraction	AOAC Standard Method 920.39, *n*-hexane as the solvent	Gas chromatography–mass spectrophotometry	Lipid yields by Soxhlet extraction from whole quinoa flour and germ flour were 5.9% and 27%, respectively	Mufari et al. (2020)
Fatty acid composition and lipid oxidization	Soxhlet extraction with *n*-hexane as the solvent	Gas chromatography–mass spectrophotometry (ISO 5509 method)	The lipid content of quinoa seeds (5.4%) was lower than that of amaranth seeds (7.3%). The unsaturated fatty acids tended to be at the sn-2 of triacylglycerols, whereas the saturated fatty acids were mostly located at the external position of triacylglycerols. Quinoa oils had 11% of saturated fatty acids, 31% of monounsaturated fatty acids and 58% of polyunsaturated fatty acids. Quinoa oil had less saturated fatty acids and more polyunsaturated fatty acids compared to amaranth oil	Wirkowska-Wojdyła et al. (2022)

chloroform-methanol (2:1, v/v). Soxhlet extraction gave higher yields of oils from quinoa than cold extraction (Mufari et al., 2020). Lipid yields by Soxhlet extraction from whole quinoa flour and germ flour were 5.9% and 27%, respectively (Mufari et al., 2020). Wirkowska-Wojdyła et al. (2022) showed that the lipid yield of quinoa seeds (5.4%) (1 genotype) was lower than that of amaranth seeds (7.3%) (1 genotype).

Wejnerowska and Ciaciuch (2018) used supercritical carbon dioxide coupled with cosolvents to extract lipid components from quinoa seeds in the form of flour. Process optimization was done to improve the lipid extraction. A recovery of 89% under optimized process conditions of 40°C, 25 MPa, and 80 min was achieved. Decreasing particle size of the flour and adding 20% methanol/ethanol (1:1, v/v) as cosolvents increased the extraction efficiency without much affecting the composition of fatty acids (Wejnerowska & Ciaciuch, 2018).

7.3 Analysis of lipid content and fatty acid composition in quinoa grain

Most studies on quinoa lipid content analysis used organic solvent for extraction (Curti et al., 2020; Ruales & Nair, 1993; Wirkowska-Wojdyła et al., 2022). Different studies used different experimental conditions and lipid extraction settings (Table 7.1). For example, Curti et al. (2020) used the AOAC (Association of Official Analytical Collaboration) official method 930.09.7 where chloroform-methanol (2:1, v/v) mixture was used as solvent for the lipid extraction from quinoa seeds. Tang et al. (2015) used methyl tert-butyl ether/tetrahydrofuran (1:1, v/v) to extract lipids from quinoa seeds. Also, different methods were used by various reports to analyze the fatty acid composition of quinoa lipids. For example, Ruales and Nair (1993) used sulfuric acid in methanol for lipid hydrolysis and analysis of fatty acid composition on gas chromatography (GC). Fanali et al. (2015) did methylation of the fatty acids using sodium methylate in methanol for GC analysis. Such differences in the use of different solvents and methylation methods may contribute to the differences in the results from different reports (Bahrami et al., 2013; Liu et al., 2018). As a result, data from different studies cannot be compared directly. Future research should be focused on using standardized methods for the analysis of quinoa lipids.

7.4 Composition of different types of lipids including neutral and polar lipids in quinoa

Lipids in cereals and pseudocereals can be categorized as neutral lipids and polar lipids or bound and free lipids (Przybylski et al., 1994; Tong & Bao, 2019). Przybylski et al. (1994) analyzed the lipid composition of quinoa seeds and fractions. The percentages of free fatty acids in seeds and their hulls were 19% and 15%, respectively. The percentages of neutral and polar lipids in wholegrain seeds were 56% and 25%, respectively (Przybylski et al., 1994). Park and Mori (2004) also found that quinoa seeds contained 50% of neutral lipids. For the neutral lipids of quinoa seeds, triglycerides and diglycerides accounted for >50% and 20%, respectively. There were more 1,2-diglycerides (13%) than 1,3-diglycerides (7%). Waxy components accounted for 3% of the neutral lipids. For the polar lipids, the percentages of lysophosphatidyl ethanolamine, phosphatidyl ethanolamine, and phosphatidylcholine (three types of phospholipids) were 45%, 19%, and 12%, respectively (Przybylski et al., 1994). Phospholipids have a hydrophilic head of a negatively charged phosphate group covalently bound to the sn (stereospecifically numbered)-3 carbon on a glycerol molecule. The hydrophobic tails consisted of two neutral fatty acids chains esterified to the glycerol molecule at carbon sn-1 and sn-2 (Lichtenstein, 2013; Wang et al., 2021). Phosophilipids are integral components of cellular and plasma membranes and may have beneficial health effects on humans (Kullenberg et al., 2012). Similar to quinoa, other food grains and cereals such as rice and wheat contain phosphatidyl ethanolamine and phosphophrylcholine (Liu et al., 2013; Yoshida et al., 2011). Phospholipids may be beneficial in breadmaking. Wheat contained more phospholipids (mostly phosphophrylcholine and lysophosphatidylcholine) than rice (Gonzalez-Thuillier et al., 2015). The comparative studies between quinoa and cereals and other pseudocereals remain to be done in the composition of neutral and polar lipids. Genetic variations in the composition of neutral and polar lipids may be expected, though more quinoa samples need to be analyzed.

7.5 Triacylglycerol composition of quinoa lipids

The major triacylglycerides of quinoa lipids were OLL (21%), OOL (13%), LLL (19%), PLL (11%), POL (9.5%), and mixtures of OOP + PSL (4%), OOO + MSO (2%), and POP (2%) (S, stearic acid; P, palmitic acid; O, oleic acid; L, linoleic acid; Ln, linolenic acid; Po, palmitoleic acid; M, myristic acid). Quinoa lipids had more LLL than amaranth and buckwheat lipids (Jahaniaval et al., 2000). Similarly, another study showed that the major triacylglycerides in quinoa seeds included OLL, OOL, and OLP. The OLL was the most abundant (18%) (Fanali et al., 2015). In comparison, the composition of POL, PPL, OLL, OOP, and OOO of quinoa lipids was different from that of amaranth lipids. Amaranth species contained predominantly PLL, POL, and OLL. Rice bran oil had the highest abundance of OOL and second being POL, whereas soybean had the highest abundance of LLL and second being OLL (Jahaniaval et al., 2000).

7.6 Fatty acid composition of quinoa lipids

Fatty acids are hydrophobic, fat-soluble organic molecules consisting of a straight chain of even or odd numbered carbon atoms with one side of the chain consisting of a methyl group and the other, a carboxyl group (Orsavova et al., 2015; Zaloga, 2021). Fatty acids contain single bond and/or double bonds. Saturated fatty acids consist of carbon chains of single bonds. Unsaturated fatty acids have one double bond classifying it as monounsaturated and two or more double bonds as polyunsaturated (Orsavova et al., 2015). The placement of the double bonds in the carbon chain classifies it as either ω-3 or ω-6 fatty acid. Unsaturated fatty acids can have *cis* or *trans* configuration (Orsavova et al., 2015).

Przybylski et al. (1994) showed that linoleic, oleic, and palmitic acids were the major fatty acids in quinoa seeds. Jahaniaval et al. (2000) also showed that linoleic acid (53%), oleic acid (26%), and palmitic acid (11%) were predominant in quinoa lipids. The fatty acid composition of soy oil was similar to that of quinoa oil. Quinoa had relatively more linolenic acid and less palmitic acid, and thus higher levels of unsaturated fatty acids than amaranth (Jahaniaval et al., 2000). A recent study showed that, in quinoa grain, the majority of the fatty acids was unsaturated with the most abundant one being linoleic acid (C18:2) (50%), which was followed by oleic acid (C18:1 c9) (24%) (Table 7.2). The major saturated fatty acid was palmitic acid (C16:0) (10%) (De Bock et al., 2021). Wirkowska-Wojdyła et al. (2022) showed that the unsaturated fatty acids tended to be at the sn-2 of triacylglycerols, whereas the saturated fatty acids were mostly located at the external position of triacylglycerols. Quinoa oils had 11% of saturated fatty acids, 31% of monounsaturated fatty acids, and 58% of polyunsaturated fatty acids. Quinoa oil had less saturated fatty acids and more polyunsaturated fatty acids compared to amaranth oil (Wirkowska-Wojdyła et al., 2022). Quinoa oils extracted using supercritical CO_2 with or without co-solvents and under different extraction conditions had similar fatty acid composition (Wejnerowska & Ciaciuch, 2018).

Genetic variation in fatty acid composition was obtained in seed oils of 28 quinoa varieties (Chen et al., 2019). For example, the percentages of palmitic acid methyl esters of these 28 samples ranged from 3.6% to 8.4%, whereas those of

Table 7.2 Composition of fatty acids of quinoa in comparison with amaranth and buckwheat (De Bock et al., 2021).

Fatty acid (%)	Quinoa (n = 7)	Amaranth (n = 8)	Buckwheat (n = 10)
C14:0	0.15 – 0.22	0.18 – 0.23	0.12 – 0.15
C14:1	0.10 – 0.48	0.12 – 0.25	0.19 – 0.41
C16:0	9.3 – 10.7	18.8 – 19.4	13.4 – 14.9
C16:1 c	0.06 – 0.11	0.09 – 0.10	0.14 – 0.18
C18: 0	0.55 – 0.76	3.16 – 3.67	1.71 – 1.98
C18:1 c9	20.2 – 27.0	22.4 – 25.4	32.7 – 35.2
C18:2 n−6	46.4 – 54.2	44.3 – 47.0	34.9 – 39.9
C18:3 n−3	3.71 – 8.10	0.91 – 1.06	2.20 – 2.26
C20:1	1.45 – 1.72	0.26 – 0.41	2.73 – 2.96
C22:1	0.00 – 0.05	1.21 – 1.96	0.16 – 0.21

n, Number of samples.
Source: Reprinted with permission under Creative Commons Attribution License.

polyunsaturated fatty acids varied from 37% to 63% (Chen et al., 2019). Genetic means may be used to breed new quinoa varieties with increased amounts of unsaturated fatty acids for nutritional applications.

Comparison in lipid composition was made between quinoa and other food grains including cereals, other pseudocereals, and soy bean (Abugoch James, 2009; De Bock et al., 2021; Nowak et al., 2016; Ruales & Nair, 1993; Wood et al., 1993). The fatty acid composition of quinoa oil was similar to that of soybean oil (Wood et al., 1993). The ratio of unsaturated fatty acids to saturated fatty acids in quinoa was higher than that of maize oil (Ruales & Nair, 1993). Compared to maize lipids, quinoa lipids had more linolenic acid and less linoleic acid with a higher ratio of unsaturated to saturated fatty acid. The composition of other fatty acids was similar between them (Filho et al., 2017). Compared to soy bean, quinoa had 10 times less saturated fatty acids. Quinoa had much less saturated fatty acids compared to rice (Nowak et al., 2016). De Bock et al. (2021) compared the fatty acid composition of quinoa with that of amaranth and buckwheat grains (Table 7.2). Quinoa and amaranth grains had similar amounts of linoleic acid and oleic acid and had more linoleic acid than buckwheat. Quinoa grain had less palmitic acid than both amaranth and buckwheat grains (De Bock et al., 2021). Unfortunately, in this study, the specific species of both the amaranth and buckwheat were not given.

7.7 Unsaponifiable lipidic fraction of quinoa lipids

Unsaponifiable components of lipid fractions of quinoa seeds were analyzed using comprehensive 2D GC (GC × GC) with dual mass spectrometer-flame ionization detector (MS/FID) detection (Fanali et al., 2015). Sixty-six unsaponifiable compounds were tentatively identified. Hydrocarbons of the C12−C31 range were found with squalene being the most abundant in this category. Among the 4-desmethylsterols, β-sitosterol (27.2%), Δ7-stigmastenol (51.3%), and Δ7-avenasterol (8.7%) were the most abundant. Among the tocopherols, γ-tocopherol was the most abundant. α-Tocopherol, β-tocopherol, and δ-tocopherol were also found (Fanali et al., 2015; Mufari et al., 2020). Squalene, which is an unsaturated hydrocarbon compound, can be used in pharmaceuticals and skin care products (Ahamed et al., 1998). Squalene was tentatively identified in quinoa (3.4%) (Jahaniaval et al., 2000). Tocopherols and phytosterols have antioxidant properties against lipid oxidation (Filho et al., 2017; Jahaniaval et al., 2000). The compositional data of the unsaponifiable components in quinoa seeds is described in Section 7.11 below.

7.8 Melting properties of quinoa lipids

The melting point and behaviors of extracted quinoa lipids were measured using differential scanning calorimetry (DSC) (Wirkowska-Wojdyła et al., 2022) (Fig. 7.1). Three endothermic peaks of quinoa lipid melting were obtained with their peak

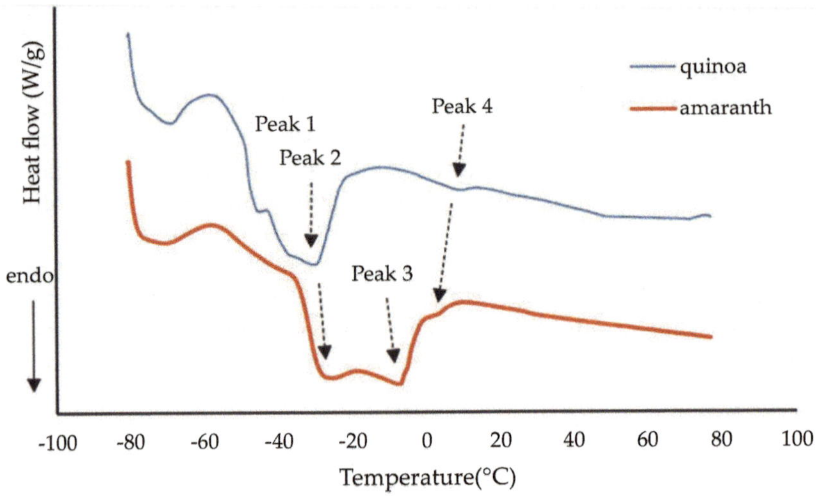

Figure 7.1 Melting curves of quinoa (top curve) and amaranth (bottom curve) oils measured by differential scanning calorimetry (DSC) (Wirkowska-Wojdyła et al., 2022).
Source: Reprinted with permission under Creative Commons Attribution License.

temperatures at −45°C, −20°C, and 9°C. In contrast, the peak temperatures of melting amaranth lipids were −26°C, −7°C, and 4°C. The occurrence of the melting peaks at low temperatures (peak 1 and peak 2) was attributed to the high percentages of unsaturated fatty acids in the quinoa lipids (Wirkowska-Wojdyła et al., 2022). The melting behaviors of quinoa lipids are largely similar to those of common vegetable oils.

7.9 Effect of processing on quinoa lipid composition

7.9.1 Effect of thermal processing

Babiker et al. (2022) found changes in the fatty acid composition of the lipids in quinoa seeds upon roasting up to 15 min. However, the changes were random according to the data, which was of low quality due to low standard of research and peer review. More studies are needed to better study the thermal processing effect on lipids.

7.9.2 Effect of aging and oxidation

Lipid oxidation in foods produces undesirable compounds such as free radicals and off-flavors, which contribute to the rancidity and reduced shelf life of the products (Jacobsen et al., 2019). The oxidative stability of quinoa lipids was studied using accelerated aging process (Ng, 2003; Ng et al., 2007). Quinoa flour was stored under a range of temperatures of up to 55°C for 30 days. The measured compounds included hexanal, free fatty acids, and conjugated diene hydroperoxides. The interactions between storage temperature and time had little effect on conjugated diene hydroperoxides. The two parameters of storage affected the three parameters greatly. The unsaturated fatty acids were not easily oxidized in the tests. Overall, quinoa lipids were stable in the tested conditions (Ng, 2003). The oxidative stability of quinoa lipids may be attributed to the presence of antioxidant components such as tocopherols in quinoa. Indeed, Mufari et al. (2020) showed that quinoa germ oil contained a significant amount of tocopherols (1350 mg/kg oil). Mufari et al. (2020) tested the oxidative stability of oils from quinoa wholegrain and germ flours using accelerated aging process at 60°C for 12 days. Different physicochemical and compositional parameters of the oils were measured. Increasing storage length increased acid values, the amounts of conjugated dienes and trienes, while decreasing the in vitro radical scavenging capacity of the oils. The germ oil had higher stability during the accelerated aging process than the wholegrain oil. This again could be attributed to the presence of natural antioxidants of quinoa (e.g., tocopherols) (Mufari et al., 2020). Similar comparative pattern in oxidative stability was obtained between crude and refined rice bran oils (Ali et al., 2019). Wirkowska-Wojdyła et al. (2022) compared the oxidative stability of oils extracted from quinoa and amaranth (*Amaranthus cruentus*) seeds using pressure DSC at a temperature range of 100°C–140°C (Wirkowska-Wojdyła et al., 2022). Quinoa oils were less stable against oxidation than amaranth oils. Other types of oxidation process such as cooking and roasting have impacted the lipid profiles of other grains such as wheat (Zou et al., 2018). These processes should also be tested on quinoa oils.

7.9.3 Effect of germination

Germination at 30°C for 72 h affected the composition of the types of lipids in quinoa seeds (Park & Mori, 2004). Germination for 72 h increased the amount of neutral lipids from 50% to 61%, while decreasing that of phospholipids from 27% to 8.7%. The germination process increased the amount of oleic acid and decreased that of linoleic acid (Park & Mori, 2004). Vera et al. (2019) analyzed the changes in the lipid profiles of the seeds of four quinoa varieties during germination (25°C, up to 72 h). The lipid content increased or had no change during the germination process, depending on the quinoa variety (Vera et al., 2019). Similarly, Aguilar et al. (2019) showed that malting (100% of relative humidity, 48 h, 25°C, 45%–55% moisture content of the grains) had little effect or decreased the lipid content of the quinoa, depending on the quinoa variety. Germination increased the concentration of polyunsaturated fatty acids and the ω-6/ω-3 ratios, regardless of the specific varieties (Vera et al., 2019). Overall, the effect of germination on lipid profiles of quinoa seeds depends on the germination conditions and the specific genotypes used. Germination involves a range of metabolic processes and enzymatic reactions, contributing to the altered lipid profiles (Ruan et al., 2021).

7.9.4 Effect of electron beam irradiation

Luo et al. (2021) studied the effect of electron beam irradiation (up to 8 kGy) on fatty acid composition of quinoa seeds. During the seed storage, the irradiation inhibited the increase of the free fatty acid content and the lipase activity. The fatty acid composition was not significantly affected by the irradiation of up to 8 kGy (Luo et al., 2021). Higher doses of irradiation, though remaining to be justified for the purpose of applications, may be used to induce significant changes in the fatty acid composition.

7.10 Effect of lipids on physicochemical properties of quinoa flour

The presence of lipids may affect the physicochemical properties of quinoa flour (Rolandelli et al., 2022; Solaesa et al., 2020). Rolandelli et al. (2022) compared the hygroscopicity and glass transition of quinoa flour and maize flour through water-solid interactions. They indicated that the lower glass transition temperature and hygroscopicity of quinoa flour than maize flour were due to the higher lipid contents of quinoa (Rolandelli et al., 2022). It may be expected that some nonlipid components could also play a role in these properties. Solaesa et al. (2020) obtained different flour fractions from dry milling process. The physicochemical properties of one fraction richer in lipids (11%) was compared with those of other fractions with less lipids. More lipids and less carbohydrates contributed to lower density of the lipid-rich fraction. More lipids led to lower swelling and gelation capacity and lower peak viscosity during the pasting event and formation of weaker gels. Similar results were obtained in wholegrain quinoa flour regarding the role of lipids in affecting the physicochemical properties (Li & Zhu, 2017). Starch can form V-type inclusion complexes with lipids. The complexes can inhibit starch swelling and decrease starch digestibility. Lipids can also form ternary complexes with protein and starch, which can play a role in the functional properties of flour systems (Wang et al., 2020). However, this aspect of quinoa lipids has not been studied.

7.11 Quinoa seed oil

Tang et al. (2015) extracted oils from the seeds of three types of quinoa varieties (black, red, and white). The oil yields of three quinoa varieties were 6.6% − 7.2%. A total of 90% of the fatty acids in the oils were unsaturated. The ratios of ω-6 to ω-3 were around 6. The ranges of percentages of linoleic acid, oleic acid, palmitic acid, and α-linolenic acid were 47%−50%, 25%−29%, 9.6%−10%, and 7.3%−8.5%, respectively. The oils contained tocopherols (mostly γ-tocopherol with smaller amounts of α- and β-tocopherols) of 38−60 μg/g. *Trans*-lutein (~85%) and zeaxanthin were found in the seed oils. Overall, the oil of black quinoa seeds contained more tocopherols and carotenoids than that of red and white quinoa seeds. The in vitro antioxidant activities of the quinoa seed oils were significantly contributed by the carotenoids, tocopherols, and polyunsaturated fatty acids (Tang et al., 2015). Quinoa oil also contained squalene (Toderich et al., 2020). Squalene is a potential antioxidant with cardiovascular protective function (Micera et al., 2020). Different types of salinity stress on the quinoa growing affected the squalene contents to different extents. The applications of soils containing NaCl, Na_2SO_4 or NaCl and Na_2SO_4 decreased the squalene content from 3.5 (control) to 3.1, 2.9 or 2.7 g/100 g seed mass, respectively (Toderich et al., 2020).

Oil yield by Soxhlet extraction from quinoa germ flour was 27% (Mufari et al., 2020). The oil from the germ had more tocopherols (1354 mg/kg oil) than that from the wholegrain flour (735 mg/kg oil). The oil of wholegrain had higher oxidative stability than that of the germ. The cold-extracted oils from wholegrain flour and germ flour had similar physicochemical properties including refractive index, density, iodine value, saponification value, and radical scavenging capacity. They also had similar fatty acid composition. On the other hand, the germ oil had lower acid and peroxide values than the whole grain oil. This indicated that the former had higher initial oxidative stability than the latter (Mufari et al., 2020). The quinoa germ oil contained a significant amount of tocopherols (1350 mg/kg oil) including α-tocopherol (362 mg/kg oil), β-tocopherol (21 mg/kg oil), γ-tocopherol (953 mg/kg oil), and δ-tocopherol (18 mg/kg oil). The total tocopherol content of the germ oil was higher than that of the cold-extracted quinoa flour oil (734 mg/kg oil) (Mufari et al., 2020). Przygoda and Wejnerowska (2015) developed supercritical fluid-based extraction to obtain quinoa oils rich in tocopherols. Response surface methodology was used to optimize the extraction conditions of the supercritical fluid extraction to obtain quinoa seed oils with a tocopherol concentration of 340 mg/100 g of oil. This concentration was four times of that obtained using the traditional Soxhlet extraction (Przygoda & Wejnerowska, 2015).

Chen et al. (2019) analyzed the composition and properties of seed oils of 28 quinoa varieties grown in the Washington State, USA. The ranges of the contents of total phenolics and carotenoids in the oils were 14−63 mg/g and 106−246 μg/g, respectively, whereas those of squalene, campesterol, stigmasterol, and β-sitosterol were 27−68, 0.4−2.0, 8.7−20, and 20−47 mg/g, respectively. The presence of these components in the quinoa seed oils contributed to their antioxidant activities (Chen et al., 2019). These 28 quinoa varieties may be categorized into two groups based on the compositional data of the seed oils. One group had more phenolics, carotenoids, squalene and phytosterols, and the other had more long-chain fatty acids and linolenic acid (Chen et al., 2019). The composition of polyphenols in quinoa oils should be analyzed for potential nutritional and food applications.

To better utilize quinoa seed oil for food formulations, the water dispersibility and stability of bioactive compounds of the oil may be improved. da Silva Lira et al. (2020) used whey protein isolate and gelatin-based system for the nanoencapsulation of quinoa oils. The emulsions with optimized formulations were freeze dried to make powder (Fig. 7.2). This encapsulation increased the water-dispersibility of quinoa oil in an emulsion system, while improving the stability of the bioactive compounds (e.g., tocopherols) present in the oil.

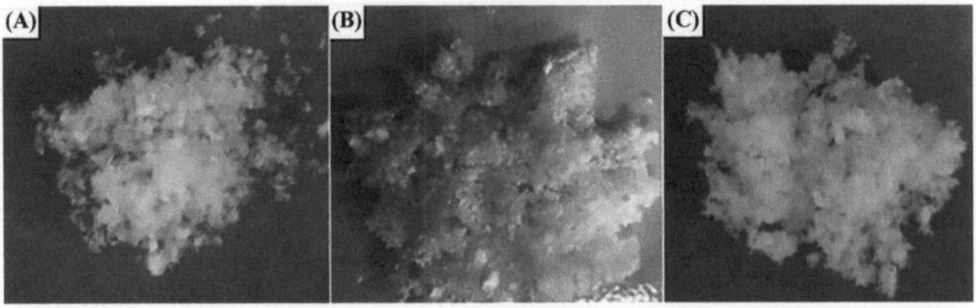

Figure 7.2 Powder encapsulating quinoa oil obtained from O/W emulsification. (A) Quinoa oil and porcine gelatin in aqueous phase 2; (B) Quinoa oil, whey protein isolate, and porcine gelatin in aqueous phase 2; (C) Quinoa oil, whey protein isolate in aqueous phase 1, and porcine gelatin in aqueous phase 2 (da Silva Lira, 2020).
Source: Reprinted with permission under Creative Commons Attribution License.

Overall, it is questionable whether quinoa grain can be developed as a source of oil for extraction. There are other important oilseeds that have high crop yields and much higher percentage of lipids for extraction, desired fatty acid composition, and a range of bioactive compounds. In comparison, quinoa has relatively too low lipid content. Unless new quinoa genotypes are discovered/engineered with much higher percentage of oils, the current quinoa genotypes are not up to the standard of being a commercially viable oil source. Published research suggesting that quinoa may be a source of oils could be rather misleading.

7.12 Oils of quinoa leaves

Quinoa leaves may be used as vegetables. The lipophilic compounds of the oils extracted from quinoa leaves were studied (Tang et al., 2014). The powdered leaves were subjected to Soxtec extraction with methyl *tert*-butyl ether/tetrahydrofuran (1:1, v/v). The oil yields of the leaves of six quinoa genotypes including one of *Chenopodium formasanum* ranged from 2.1% to 3.6%, which were higher than those of the levels of amaranth leaves (14 genotypes) (2.1%–3.6%). The quinoa leaf oils contained essential fatty acids, with linolenic acid (35%–44%), linoleic acid (12%–16%), and palmitic acid (11%–16%) being dominant. The ω3/ω6 ratios ranged from 2.3 to 3.9, which were similar to those of amaranth leaf oils, and were considered nutritionally favorable for human health (Domingues-Avila & Gonzalez-Aguilar, 2019). The quinoa leaf oils contained a range of lipophilic compounds mainly including α- and β-carotenes, violaxanthin, lutein, and α- and β-tocopherols. Overall, the nutritional quality of quinoa and amaranth leaf oils was similar (Tang et al., 2014). The analysis of quinoa leaf oil showed that the leaves can be used as healthy vegetables.

7.13 Conclusions

Quinoa seeds especially the germ contains a significant amount of lipids rich in essential fatty acids and unsaturated fatty acids with linoleic acid and oleic acid being dominant. The major triacylglycerides of quinoa lipids are OLL, OOL, and LLL. The polar lipids of quinoa are mostly phospholipids with lysophosphatidyl ethanolamine, phosphatidyl ethanolamine, and phosphatidylcholine being the major ones. Unsaponifiable lipidic components of quinoa seeds mainly include squalene, β-sitosterol, Δ7-stigmastenol, Δ7-avenasterol, *trans*-lutein (~85%) and zeaxanthin, and tocopherols (mostly γ-tocopherol). Lipids play important roles in the stability and physicochemical properties of quinoa flour. Quinoa germ oil has high nutritional quality and is rich in bioactive components. It may be expected that the oil could have a range of biological functions such as mitigation of diabetes, cardiovascular, and cancer risks and improving overall health by inhibiting proinflammatory pathways and free radical induced oxidation.

Genetic diversity in lipid profiles of quinoa genotypes should be systematically analyzed. Quinoa genotypes with increased lipid content may be developed using genetic means. Such quinoa genotypes would have a high potential for oil extraction and utilization. The rheological, cooking, and sensory properties of quinoa oils should be studied. The genetic basis of lipid production in quinoa grains should be analyzed. The role of lipids in quinoa quality including cooking, textural, and sensory (flavor and aroma) properties of quinoa-based food products should be studied. The effect of different processing operations such as heating, storage, roasting, and food formulations should be explored. The relationships between lipid content and quinoa grain hardness may be studied for milling applications and textural analysis.

References

Abugoch James, L. E. (2009). Quinoa (*Chenopodium quinoa* Willd.): Composition, chemistry, nutritional, and functional properties. *Advances in Food and Nutrition Research*, *58*, 1–31.

Aguilar, J., Miano, A. C., Obregon, J., Soriano-Colchado, J., & Barraza-Jáuregui, G. (2019). Malting process as an alternative to obtain high nutritional quality quinoa flour. *Journal of Cereal Science*, *90*, 102858.

Ahamed, N. T., Singhal, R. S., Kulkarni, P. R., & Pal, M. (1998). A lesser-known grain, *Chenopodium quinoa*: Review of the chemical composition of its edible parts. *Food and Nutrition Bulletin*, *19*, 61–70.

Ali, A., Islam, M. A., Othman, N. H., Noor, A. M., Hossen, J., & Ibrahim, M. (2019). Effect of heating on compositional characteristics and oxidative stability of crude and refined rice bran oil. *Journal of Oleo Science*, *68*, 1085–1097.

Babiker, E. E., Uslu, N., Ghafoor, K., AL-Juhaimi, F., Özcan, M. M., & Ahmed, I. A. M. (2022). Variations in bioactive properties, fatty acid compositions, and phenolic compounds of quinoa grain and oils roasted in a pan. *Journal of Food Processing and Preservation*, *46*, e16161.

Bahrami, N., Yonekura, L., Linforth, R., da Silva, M. C., Hill, S., Penson, S., Chope, G., & Fisk, I. D. (2013). Comparison of ambient solvent extraction methods for the analysis of fatty acids in non-starch lipids of flour and starch. *Journal of the Science of Food and Agriculture*, *94*, 415–423.

Chen, Y. S., Aluwi, N. A., Saunders, S. R., Ganjyal, G. M., & Medina-Meza, I. G. (2019). Metabolic fingerprinting unveils quinoa oil as a source of bioactive phytochemicals. *Food Chemistry*, *286*, 592–599.

Curti, R. N., Sanahuja, M. D. C., Vidueiros, S. M., Curti, C. A., Pallaro, A. N., & Bertero, H. D. (2020). Oil quality in sea-level quinoa as determined by cultivar-specific responses to temperature and radiation conditions. *Journal of the Science of Food and Agriculture*, *100*, 1358–1361.

da Silva Lira, K. H. D., Passos, T. S., Ramalho, H. M. M., da Silva Ribeiro Rodrigues, K. D., de Andrade Vieira, É., de Magalhães Cordeiro, A. M. T., Maciel, B. L. L., et al. (2020). Whey protein isolate-gelatin nanoparticles enable the water-dispersibility and potentialize the antioxidant activity of quinoa oil (*Chenopodium quinoa*). *PLOS One*, *15*, e0240889.

De Bock, P., Daelemans, L., Selis, L., Raes, K., Vermeir, P., Eeckhout, M., & Van Bockstaele, F. (2021). Comparison of the chemical and technological characteristics of wholemeal flours obtained from amaranth (*Amaranthus* sp.), quinoa (*Chenopodium quinoa*) and buckwheat (*Fagopyrum* sp.) seeds. *Foods*, *10*, 651.

Domingues-Avila, J., & Gonzalez-Aguilar, G. (2019). Lipids. In E. Yahia (Ed.), *Postharvest physiology and biochemistry of fruits and vegetables* (pp. 273–292). Woodhead Publishing.

Encina-Zelada, C., Cadavez, V., Pereda, J., Gómez-Pando, L., Salvá-Ruíz, B., Teixeira, J. A., Ibañez, M., Liland, K. H., & Gonzales-Barron, U. (2017). Estimation of composition of quinoa (*Chenopodium quinoa* Willd.) grains by Near-Infrared Transmission spectroscopy. *LWT-Food Science and Technology*, *79*, 126–134.

Fanali, C., Beccaria, M., Salivo, S., Tranchida, P., Tripodo, G., Farnetti, S., et al. (2015). Non-polar lipids characterization of quinoa (*Chenopodium quinoa*) seed by comprehensive two-dimensional gas chromatography with flame ionization/mass spectrometry detection and non-aqueous reversed-phase liquid chromatography with atmospheric pressure chemical ionization mass spectrometry detection. *Journal of Separation Science*, *38*, 3151–3160.

Filho, A. M. M., Pirozi, M. R., Borges, J. T. D. S., Sant'Ana, H. M. P., Chaves, J. B. P., & Coimbra, J. S. D. R. (2017). Quinoa: Nutritional, functional, and antinutritional aspects. *Critical Reviews in Food Science and Nutrition*, *57*, 1618–1630.

Gonzalez-Thuillier, I., Salt, L., Chope, G., Penson, S., Skeggs, P., Tosi, P., Powers, S. J., et al. (2015). Distribution of lipids in the grain of wheat (cv. Hereward) determined by lipidomic analysis of milling and pearling fractions. *Journal of Agriculture and Food Chemistry*, *63*, 10705–10716.

Graf, B. L., Rojo, L. E., Delatorre-Herrera, J., Poulev, A., Calfio, C., & Raskin, I. (2016). Phytoecdysteroids and flavonoid glycosides among Chilean and commercial sources of *Chenopodium quinoa*: Variation and correlation to physico-chemical characteristics. *Journal of the Science of Food and Agriculture*, *96*, 633–643.

Jacobsen, C., Paiva-Martins, F., Schwarz, K., & Bochkov, V. (2019). Lipid oxidation and antioxidants in food and nutrition. *European Journal of Lipid Science and Technology*, *121*, 1900298.

Jahaniaval, F., Kakuda, Y., & Marcone, M. (2000). Fatty acid and triacylglycerol compositions of seed oils of five *Amaranthus* accessions and their comparison to other oils. *Journal of the American Oil Chemists, Society*, *77*, 847–852.

Kullenberg, D., Taylor, L. A., Schneider, M., & Massing, U. (2012). Health effects of dietary phospholipids. *Lipids in Health and Disease*, *11*, Article number: 3.

Li, G., & Zhu, F. (2017). Physicochemical properties of quinoa flour as affected by starch interactions. *Food Chemistry*, *221*, 1560–1568.

Lichtenstein, A. H. (2013). Fats and oils. In B. Caballero (Ed.), *Encyclopedia of human nutrition* (pp. 201–208). Academic Press.

Liu, L., Waters, D. L. E., Rose, T. J., Bao, J., & King, G. J. (2013). Phospholipids in rice: Significance in grain quality and health benefits: A review. *Food Chemistry*, *139*, 1133–1145.

Liu, Z., Ezernieks, V., Rochfort, S., & Cocks, B. (2018). Comparison of methylation methods for fatty acid analysis of milk fat. *Food Chemistry*, *261*, 210–215.

Luo, X., Du, Z., Yang, K., Wang, J., Zhou, J., Liu, J., et al. (2021). Effect of electron beam irradiation on phytochemical composition, lipase activity and fatty acid of quinoa. *Journal of Cereal Science*, *98*, 103161.

Micera, M., Botto, A., Geddo, F., Antoniotti, S., Bertea, C. M., Levi, R., Gallo, M. P., & Querio, G. (2020). Squalene: More than a step toward sterols. *Antioxidants*, *9*, 688.

Mufari, J. R., Gorostegui, H. A., Miranda-Villa, P. P., Bergesse, A. E., & Calandri, E. L. (2020). Oxidative stability and characterization of quinoa oil extracted from wholemeal and germ flours. *Journal of the American Oil Chemists' Society*, *97*, 57–66.

Ng, S. C. (2003). *Effects of accelerated aging on lipid oxidation in quinoa (Chenopodium quinoa)* (M.Sc. thesis). University of Wisconsin-Stout.

Ng, S. C., Anderson, A., Coker, J., & Ondrus, M. (2007). Characterization of lipid oxidation products in quinoa (*Chenopodium quinoa*). *Food Chemistry*, *101*, 185–192.

Nowak, V., Du, J., & Charrondiere, R. (2016). Assessment of the nutritional composition of quinoa (*Chenopodium quinoa* Willd.). *Food Chemistry*, *193*, 47–54.

Orsavova, J., Misurcova, L., Ambrozova, J. V., Vicha, R., & Mlcek, J. (2015). Fatty acids composition of vegetable oils and its contribution to dietary energy intake and dependence of cardiovascular mortality on dietary intake of fatty acids. *International Journal of Molecular Sciences*, *16*, 12871–12890.

Park, S. H., & Mori, N. (2004). Changes of bound lipids and composition of fatty acids in germination of quinoa seeds. *Food Science and Technology Research*, *10*, 303–306.

Peiretti, P. G., Gai, F., & Tassone, S. (2013). Fatty acid profile and nutritive value of quinoa (*Chenopodium quinoa* Willd.) seeds and plants at different growth stages. *Animal Feed Science and Technology*, *183*, 56–61.

Prego, I., Maldonado, S., & Otegui, M. (1998). Seed structure and localization of reserves in *Chenopodium quinoa*. *Annals of Botany*, *82*, 481–488.

Przybylski, R., Chauhan, G. S., & Eskin, N. A. M. (1994). Characterization of quinoa (*Chenopodium quinoa*) lipids. *Food Chemistry*, *51*, 187–192.

Przygoda, K., & Wejnerowska, G. (2015). Extraction of tocopherol-enriched oils from quinoa seeds by supercritical fluid extraction. *Industrial Crops and Products*, *63*, 41–47.

Rolandelli, G., Farroni, A. E., & del Pilar Buera, M. (2022). Analysis of molecular mobility in corn and quinoa flours through 1H NMR and its relationship with water distribution, glass transition and enthalpy relaxation. *Food Chemistry*, *373*, 131422.

Ruales, J., & Nair, B. (1993). Content of fat, vitamins and minerals in quinoa (*Chenopodium quinoa*, Willd) seeds. *Food Chemistry*, *48*, 131–136.

Ruan, Y., Cai, Z., Deng, Y., Pan, D., Zhou, C., Cao, J., et al. (2021). An untargeted metabolomic insight into the high-pressure stress effect on the germination of wholegrain *Oryza sativa* L. *Food Research International*, *140*, 109984.

Solaesa, A. G., Villanueva, M., Vela, A. J., & Ronda, F. (2020). Protein and lipid enrichment of quinoa (cv. Titicaca) by dry fractionation. Techno-functional, thermal and rheological properties of milling fractions. *Food Hydrocolloids*, *105*, 105770.

Tang, Y., Li, X., Chen, P. X., Zhang, B., Hernandez, M., Zhang, H., et al. (2014). Lipids, tocopherols, and carotenoids in leaves of amaranth and quinoa cultivars and a new approach to overall evaluation of nutritional quality traits. *Journal of Agricultural and Food Chemistry*, *62*, 12610–12619.

Tang, Y., Li, X., Chen, P. X., Zhang, B., Hernandez, M., Zhang, H., et al. (2015). Characterisation of fatty acid, carotenoid, tocopherol/tocotrienol compositions and antioxidant activities in seeds of three *Chenopodium quinoa* Willd. genotypes. *Food Chemistry*, *174*, 502–508.

Toderich, K. N., Mamadrahimov, A. A., Khaitov, B. B., Karimov, A. A., Soliev, A. A., Nanduri, K. R., & Shuyskaya, E. V. (2020). Differential impact of salinity stress on seeds minerals, storage proteins, fatty acids, and squalene composition of new quinoa genotype, grown in hyper-arid desert environments. *Frontiers in Plant Science*, *11*, 607102.

Tong, C., & Bao, J. (2019). Rice lipids and rice bran oil. In J. Bao (Ed.), *Rice: Chemistry and technology* (4th ed., pp. 131–168). Woodhead Publishing and AACC International Press.

Vera, E. P., Alca, J. J., Saravia, G. R., Campioni, N. C., & Alpuy, I. J. (2019). Comparison of the lipid profile and tocopherol content of four Peruvian quinoa (*Chenopodium quinoa* Willd.) cultivars ('Amarilla de Maranganí', 'Blanca de Juli', INIA 415 'Roja Pasankalla', INIA 420 'Negra Collana') during germination. *Journal of Cereal Science*, *88*, 132–137.

Wang, J., Tang, J., Ruan, S., Lv, R., Zhou, J., Tian, J., et al. (2021). A comprehensive review of cereal germ and its lipids: Chemical composition, multi-objective process and functional application. *Food Chemistry*, *362*, 130066.

Wang, S., Chao, C., Cai, J., Niu, B., Copeland, L., & Wang, S. (2020). Starch–lipid and starch–lipid–protein complexes: A comprehensive review. *Comprehensive Reviews in Food Science and Food Safety*, *19*, 1056–1079.

Wejnerowska, G., & Ciaciuch, A. (2018). Optimisation of oil extraction from quinoa seeds with supercritical carbon dioxide with co-solvents. *Czech Journal of Food Sciences*, *36*, 81–87.

Wirkowska-Wojdyła, M., Ostrowska-Ligęza, E., Górska, A., & Bryś, J. (2022). Application of chromatographic and thermal methods to study fatty acids composition and positional distribution, oxidation kinetic parameters and melting profile as important factors characterizing amaranth and quinoa oils. *Applied Sciences*, *12*, 2166.

Wood, S. G., Lawson, L. D., Fairbanks, D. J., Robison, L. R., & Andersen, W. R. (1993). Seed lipid content and fatty acid composition of three quinoa cultivars. *Journal of Food Composition and Analysis*, *6*, 41–44.

Yoshida, H., Tanigawa, T., Kuriyama, I., Yoshida, N., Tomiyama, Y., & Mizushina, Y. (2011). Variation in fatty acid distribution of different acyl lipids in rice (*Oryza sativa* L.) brans. *Nutrients*, *3*, 505–514.

Zaloga, G. P. (2021). Narrative review of n-3 polyunsaturated fatty acid supplementation upon immune functions, resolution molecules and lipid peroxidation. *Nutrients*, *13*, 662.

Zou, Y., Gao, Y., He, H., & Yang, T. (2018). Effect of roasting on physico-chemical properties, antioxidant capacity, and oxidative stability of wheat germ oil. *LWT-Food Science and Technology*, *90*, 246–253.

Polyphenols, saponins, and other bioactive components of quinoa

8

8.1 Introduction

Quinoa contains a range of different types of bioactive components including polyphenols, saponins, phytoecdysteroids, betalains, and alkylresorcinols (Ahmed et al., 2021; Chen et al., 2022; El Hazzam et al., 2020; Melini & Melini, 2021c; Nowak et al., 2016; Ross et al., 2017). These bioactive compounds have a range of well-established physiological and biological functions such as anticancer and antioxidant activity. For example, polyphenols are found in different plants. The intake of polyphenol-rich foods has been associated with reduction in risk factors for diverse chronic diseases such as cardiovascular diseases, cancers, cognitive disorders, and diabetes (Di Lorenzo et al., 2021). Saponins exist in more than 100 plant families. Saponins have structural diversity and diverse biological activities (e.g., anticancer) and low toxicity (El Hazzam et al., 2020; Lin, Han, et al., 2019; Lin, Ke, et al., 2019). On the other hand, saponins contribute to the bitter taste of quinoa seeds with decreased sensory quality (Präger et al., 2018). Some parts of quinoa plants such as roots can be rich in saponins. In view of the unique resource and price advantages of quinoa saponins, the exploitation and utilization of underutilized parts such as roots and leaves can reduce the cost of research and development, while improving the economic benefits of quinoa. A range of biological functions such as antidiabetes and antioxidant activities were also found for the minor components of quinoa including phytoecdysteroids and betalains (Foucault et al., 2014; Graf et al., 2014; Sadowska-Bartosz & Bartosz, 2021). The components contribute to the role of quinoa as a type of "super grain." Therefore it is critical to understand the composition of these bioactive components and analyze their biological activities.

Consumption and utilization of quinoa grains are associated with different processing operations including home cooking such as germination, roasting, and washing (Gu et al., 2021; Rocchetti et al., 2019; Sharma et al., 2022; Suárez-Estrella et al., 2021). It is essential to understand the role of these processing operations on the chemical composition and biological activities of the bioactive components including polyphenols and saponins. This chapter summarizes the composition and biological activities of polyphenols, saponins, phytoecdysteroids, betalains, and alkylresorcinols in different parts of quinoa plants focusing on the seeds. The effects of different types of processing operations including thermal and non-thermal methods on the composition and biological activities are also described.

Quinoa. DOI: https://doi.org/10.1016/B978-0-323-99909-0.00015-5

8.2 Composition of polyphenols and saponins in quinoa

8.2.1 Extraction of phenolics from quinoa

Free and bound (conjugated) phenolics exist in quinoa seeds. Different solvents and solutions were used for the extractions of different fractions (Table 8.1). Most of the studies used organic solvents in aqueous solutions for the extractions of free phenolics in quinoa seeds. Alkaline or acidic solutions were used to break the ether and ester bonds of bound phenolics attached to the cell wall polysaccharides (Abderrahim et al., 2015; Antognoni et al., 2021; Han, Chi, Zhang, Zhang, Fan, Dong, et al., 2019; Han, Chi, Zhang, Zhang, Fan, Huang, et al., 2019; Tang et al., 2015). For example, Tang et al. (2016) used acidified aqueous methanol for free phenolic fraction extraction, whereas conjugated phenolics were extracted using NaOH and HCl solutions. Conjugated phenolics are a significant portion of the total phenolics in quinoa seeds (Antognoni et al., 2021; Tang et al., 2016). Unfortunately, some of the studies did not use different means to differentiate the free and bound phenolics from each other in quinoa seeds (Graf et al., 2016; Stikić et al., 2020). Such a lack of a complete picture of the phenolic composition decreases the value of the reports significantly. The role of conjugated phenolics of food grains in human nutrition can be significant (Rocchetti et al., 2022; Zhu, 2018). Future studies should be thorough in the analysis of both of the free and bound phenolics. Different alternative methods have been used to extract bioactive compounds from quinoa seeds. For example, Mufari et al. (2021) used ethanol aqueous solutions of subcritical status to extract bioactive components from malted quinoa seeds. Compared to the traditionally used method, the application of subcritical solutions increased the extract yield and antioxidant activity (Mufari et al., 2021). Melini and Melini (2021b) showed that ultrasound-assisted extraction of phenolics from black seeded quinoa may be optimized using response surface methodology to improve the yield. It should be kept in mind that these different methods used for phenolics extraction may lead to different results of the composition in quinoa seeds. Such aspect remains to better studied.

8.2.2 Phenolic acids and derivatives in quinoa

The type of phenolic acids is diverse and different studies reported a range of different phenolic acids in quinoa (Table 8.1). For example, in one study of Pellegrini et al. (2017), the identified phenolic acids in quinoa seeds included gallic acid, *p*-coumaric acid, *o*-coumaric acid, syringic acid, ferulic acid, vanillic acid, 4-hydroxybenzoic acid, rosmarinic acid, and chlorogenic acid. 4-Hydroxybenzoic acid (97–129 μg/g, fresh weight) and syringic acid (83–132 μg/g, fresh weight) were the most abundant (Pellegrini et al., 2017). Liu et al. (2020) showed that protocatechuic acid was the major phenolic acid in black and red quinoa seeds. The types of free and bound phenolics of quinoa seeds tend to be similar with a degree of difference (Tang et al., 2015, 2016) (Fig. 8.1). Major bound phenolics included ferulic acid, 2,4-dihydroxy benzoic acid, 8,5′-diferulic acid, rosmarinic acid, and

Table 8.1 Composition of polyphenols, betalains, and saponins in quinoa.

Samples	Solvent extraction	Identified phenolic acids and glycosides	Identified flavonoids	Other minor compounds including betalains and saponins	Composition and concentrations	References
Fifteen colored quinoa genotypes	Acidic methanol and acetone aqueous solutions for free phenolics extraction; methanol/H_2SO_4 solution for bound phenolics extraction			Betacyanin and betaxanthin	The amounts of free and bound phenolics ranged from 1.23 to 3.41 and 1.28 to 4.52 mg gallic acid equivalents/g, respectively. The amounts of free and bound flavonols were 0.30–1.13 and 0.16–1.47 mg quercetin equivalents/g. The amounts of betacyanins and betaxanthins ranged from 0.18 to 5.23 and from 0 to 1.63 mg/100 g, respectively	Abderrahim et al. (2015)
Three quinoa genotypes (white, red, and black)	Acidified aqueous methanol for free phenolic fraction; conjugated phenolics were extracted using NaOH and HCl solutions; betalains were extracted using methanol (50%) containing HCl at a pH of 3	Free form: 3,4-Dihydroxybenzoic acid, *p*-coumaric acid 4-glucoside, *p*-hydroxybenzoic acid, vanillic acid 4-glucoside, 2,5-dihydroxybenzoic acid, caffeic acid, vanillic acid, *p*-coumaric acid, ferulic acid, ferulic acid 4-glucoside, isoferulic acid	Free form: Epigallocatechin, epicatechin, kaempferol 3,7-dirhamnoside, kaempferol 3-galactoside, quercetin-3-rutinoside, kaempferol 3-glucoside, quercetin 3-arabinoside, quercetin, biochanin A	Free form: Vanillin, acacetin/questin/apigenin-7-methyl ether Betanin and isobetanin were found in black and red quinoa seeds	Vanillic acid (39–70 mg/kg) was the one of the most abundant phenolic acids in the free form, whereas quercetin-3-rutinoside (57–71 mg/kg) and quercetin 3-arabinoside (25– 66 mg/kg) were the most abundant flavonoids In alkaline hydrolyzed samples, ferulic acid (208–257 mg/kg) and vanillic acid (207–235 mg/kg) were the most abundant phenolic acids Overall, a total of 23 phenolic compounds were found. Kaempferol and quercetin and their glycosides as well as ferulic acid and vanillic acid and related derivatives were the major phenolics	Tang et al. (2015)

(Continued)

Table 8.1 (Continued)

Samples	Solvent extraction	Identified phenolic acids and glycosides	Identified flavonoids	Other minor compounds including betalains and saponins	Composition and concentrations	References
Seventeen samples	Ethanol extracts		Flavonoid glycosides, quercetin 3-O-(2,6-di-α-L-rhamno-pyranosyl)-β-D-galactopyranoside as one of the most abundant		The range of the contents of flavonoid glycosides was 190–800 μg/g. Varieties of highland regions had more flavonoid glycosides than those of lowland	Graf et al. (2016)
Twenty eight quinoa genotypes	The extraction of saponins and sapogenins flowchart (Fig. 8.3)			Sapogenins included phytolaccagenic acid, hederagenin, oleanolic acid, and serjanic acid	Total saponin contents varied from 3.8 to 27 mg/g. Phytolaccagenic acid and hederagenin accounted for ~70% and 30% of the total sapogenins	Medina-Meza et al. (2016)
One genotype (*Chenopodium berlandieri*): native and sprouted under selenium stress	80% methanol			A total of 12 saponins were identified. Sapogenins in native grain mostly included oleanolic and phytolaccagenic acids. Sprouting decreased the saponin content significantly, as reflected in over 90% of decrease in phytolaccagenic acid content	Selenium stress increased the content of the most abundant saponin in quinoa seeds during germination	Lazo-Vélez et al. (2016)
Red, white, and black quinoa seeds	Acid, alkaline, and enzymatic hydrolysis to obtain bound phenolics	Major bound phenolics included ferulic acid, 2,4-dihydroxy benzoic acid, 8,5′-diferulic acid, rosmarinic acid, sinapic acid	Biochanin A, quercetin, kaempferol, catechin		Black quinoa seeds had the highest amounts of bound phenolics, followed by red and white seeds	Tang et al. (2016

Six quinoa genotypes	Methanol extract	Gallic acid, *p*-coumaric acid, *o*-coumaric acid, syringic acid, ferulic acid, vanillic acid, 4-hydroxybenzoic acid, rosmarinic acid, and chlorogenic acid	Rutin, quercetin, isoquercetin, kaempferol, neohesperidin, and hesperidin		4-Hydroxybenzoic acid (97–129 μg/g, fresh weight) and syringic acid (83–132 μg/g, fresh weight) were the most abundant phenolic acids. Neohesperidin (39–103 μg/g, fresh weight), isoquercetin (44–54 μg/g, fresh weight) and kaempferol (23–62 μg/g, fresh weight) were the most abundant flavonoids	Pellegrini et al. (2017)
Twenty nine Peruvian genotypes	Sodium acetate solution containing sodium ascorbate for extraction			Betalains: amaranthin, iso-amaranthin, betanin, iso-betanin, dopaxanthin, dopamine-betaxanthin, proline-betaxanthin (Fig. 8.5) Saponins: 3-O-β-D-glucuronopyranosyl oleanolic acid 28-O-β-D-glucopyranosyl ester, 3-O-α-Larabinopyranosyl phytolaccagenic acid 28-O-β-D-glucopyranosyl ester, 3-O-β-D-xylopyranosyl-(1→3)-β-D-glucuronopyranosyl oleanolic acid 28-O-β-D-glucopyranosyl ester, 3-O-β-D-xylopyranosyl-(1→3)-β-D-glucuronopyranosyl hederagenin 28-O-β-D-glucopyranosyl ester, 3-O-β-D-glucopyranosyl-(1→3)-α-L-arabinopyranosyl serjanic acid 28-O-β-D-glucopyranosyl ester	Amaranthin (0–149 mg/kg), iso-amaranthin (0–146 mg/kg), betanin (0–9.7 mg/kg), iso-betanin (0–7.7 mg/kg), dopaxanthin (0–85.3 mg/kg), dopamine-betaxanthin (0–11 mg/kg), proline-betaxanthin (0–7.5 mg/kg)	Escribano et al. (2017)

(*Continued*)

Table 8.1 (Continued)

Samples	Solvent extraction	Identified phenolic acids and glycosides	Identified flavonoids	Other minor compounds including betalains and saponins	Composition and concentrations	References
Seven colored genotypes	Bound and free phenolics were extracted using different methods. The free phenolics were extracted using aqueous acetone (80%). The bound phenolics were extracted using NaOH solution and then HCl solution before ethyl acetate based extraction. Methanol was used to extract saponins at 80°C for 1 h	Gallic acid and ferulic acid are the major phenolic acids	Quercetin, rutin		The ranges of total gallic acid and ferulic acid contents were 168–313 and 147–274 μg/g. Total saponin contents varied from 7.5 to 12.1 oleanolic acid equivalents/g DW	Han, Chi, Zhang, Zhang, Fan, Dong, et al. (2019); Han, Chi, Zhang, Zhang, Fan, Huang, et al. (2019)
Three quinoa genotypes	Bound and free phenolics of white, red, and black quinoa seeds	Protocatechuic acid was the major phenolic acid in black and red quinoa seeds			Red and black quinoa seeds had more bound phenolics than white ones. Black quinoa seeds had more phenolic acids than red and white ones, whereas red ones had more phenolics and flavonoids than the other two. The ranges of total phenolic and total flavonoid contents were 514–1410 mg gallic acid equivalent /100 g and 178–408 mg rutin equivalent /100 g	Liu et al. (2020)
Two quinoa genotypes	Extractable phenolics	Ferulic acid, 5-O-caffeoylquinic acid, gentisic acid, *p*-coumaric acid, ellagic acid	Quercetin, isorhamnetin, quercetin-3-O-galactoside, isorhamnetin-3-O-rutinoside, rutin, naringin, aesculin, phlorizin, eriodictyol	Pterostilbene, coniferyl aldehyde	Ferulic acid (56 mg/kg dry weight) was the most abundant phenolic acid in one variety. Rutin was the most abundant flavonoid (30 and 34 mg/kg dry weight)	Stikić et al. (2020)

[illegible] genotypes grown in Italy, Chile, and Denmark	[illegible] aqueous methanol for free phenolic fraction; conjugated phenolics were extracted using NaOH and HCl solutions	In free and conjugated forms, vanillic acid, *p*-coumaric acid and ferulic acid were the major ones	In the free form, epicatechin, quercetin-3-O-glucoside, quercetin-3-O-rutinoside, kaempferol-3-O-rutinoside, kaempferol-3-O-glucoside, genistein, daidzein. In conjugated form, in alkaline hydrolyzed extract, rutin, and kaempferol-3-O-glucoside were the main ones		The growing locations influenced the composition of certain phenolic compounds	Antognoni et al. (2021)
One quinoa genotype	Ethanol was the best solvent for extraction compared to the mixtures of ethanol and water			7-Isopentenyloxycoumarin, auraptene, umbelliprenin, boropinic acid, 4, 4′-geranyloxyferulic acid		Antognoni et al. (2021)
Seeds, flowers, leaves, stems, and roots	96% ethanol				Total phenolic contents (TPC) ranged from 6.0 to 43.5 mg of gallic acid equivalents per g plant material, and total flavonoid contents (TFC) from 1.3 to 12.3 mg of quercetin equivalents per g of plant material. Freeze-drying gave higher TPC and TFC than air drying	Buitrago et al. (2019)
Quinoa seed coats	80% methanol			The major aglycones of saponins were hederagenin (the most abundant), phytolaccagenic acid (the second abundant) and oleanolic acid. A total of 24 saponins were identified with different sugar moieties	The composition of the saponins was affected by the quinoa genotype	Ruiza et al. (2017)

(*Continued*)

Table 8.1 (Continued)

Samples	Solvent extraction	Identified phenolic acids and glycosides	Identified flavonoids	Other minor compounds including betalains and saponins	Composition and concentrations	References
Quinoa bran of black, white, and red quinoa	n-Hexane and methanol were used for saponins extraction			Phytolaccageni, hederagenin, and oleanolic acid were released from the hydrolysis of saponins extract. Variations in the composition of the sapogenins were found among the three quinoa varieties. The contents of phytolaccagenin, hederagenin, and oleanolic acid varied from 27 to 88, 8 to 35, and 13 to 55 g/100 g extract, respectively	Black and white quinoa had higher yields of saponins than red one	Chen et al (2022)
Quinoa stalk	Pressurized hot water extraction (10 MPa). The extraction conditions were optimized using response surface method			The highest yield of saponins was 13 mg diosgenin equivalents per g of raw materials		Gil-Ramirez et al. (2018)

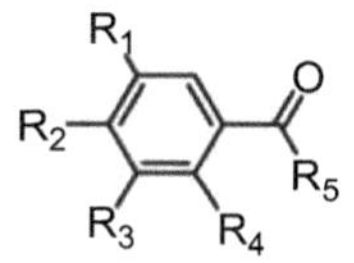

p-Hydroxybenzoic acid, R_1=R_2=R_3=R_4=H, R_5=OH;
3,4-Dihydroxybenzoic acid, R_1=R_4=H, R_2=R_3=R_5=OH;
2,5-Dihydroxybenzoic acid, R_1=R_4=R_5=OH, R_2=R_3=H;
2,4-Dihydroxybenzoic acid, R_1=R_3=R_5=OH, R_2=R_4=H;
Vanillic acid, R_1=R_4=H, R_2=R_5=OH,R_3=OCH_3;
Vanillic acid 4-glucoside, R_1=R_4=H, R_5=OH,R_3=OCH_3,R_2=O-glucose;
Vanillic, R_1=R_4=R_5=H, R_2=OH,R_3=OCH_3;

Benzoic acids

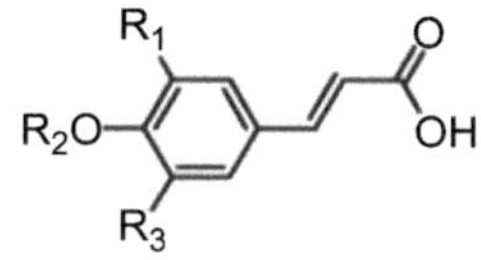

p-Coumaric acid, R_1=R_2=R_3=H;
p-Coumaric acid 4-glucoside, R_1=R_3=H, R_2=glucose;
Ferulic acid, R_1=R_2=H, R_3=OCH_3;
Isoferulic acid, R_1=H, R_2=CH_3, R_3=OH;
Ferulic acid 4-glucoside, R_1=H, R_3=OCH_3, R_2=glucose;
Caffeic acid, R_1=R_2=H, R_3=OH;

Cinnamic acids

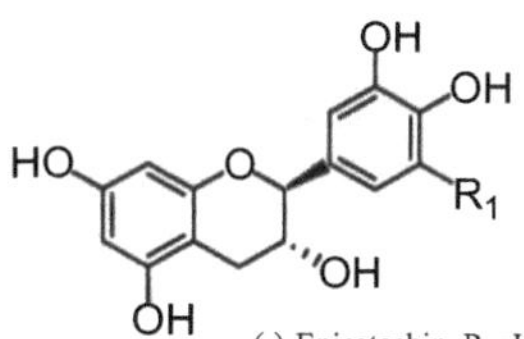

(-)-Epicatechin, R_1=H;
(-)-Epigallocatechin, R_1=OH

Monomeric and Dimeric Favanols

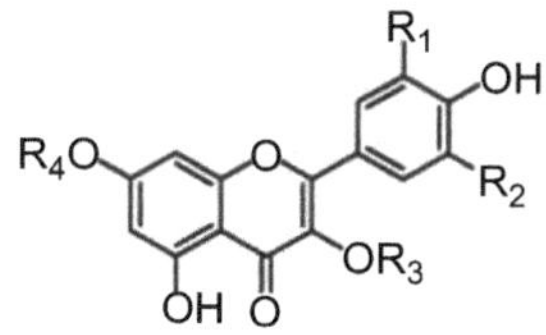

Kaempferol, R_1=R_2=R_3=R_4=H;
Kaempferol 3-glucoside, R_1=R_2=R_4=H, R_3=glucose;
Kaempferol 3-galactoside, R_1=R_2=R_4=H, R_3=galactose;
Kaempferol 3,7-dirhamnoside, R_1=R_2=H, R_3=R_4=rhamnose;
Quercetin, R_1=R_3=R_4=H, R_2=OH;
Quercetin 3-rutinoside, R_1=R_4=H, R_2=OH, R_3=rutinose;
Quercetin 3-arabinoside, R_1=R_4=H, R_2=OH, R_3=arabinose;

Flavonols

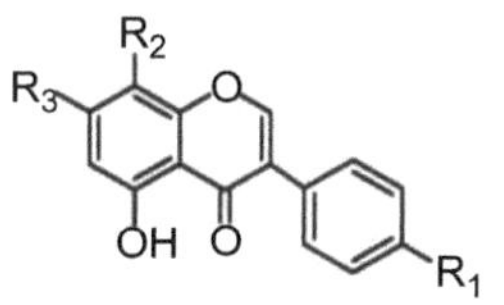

Biochanin A, R_1=OCH_3, R_2=H, R_3=OH;
Puerarin, R_1=OH, R_2=glucose, R_3=OH;

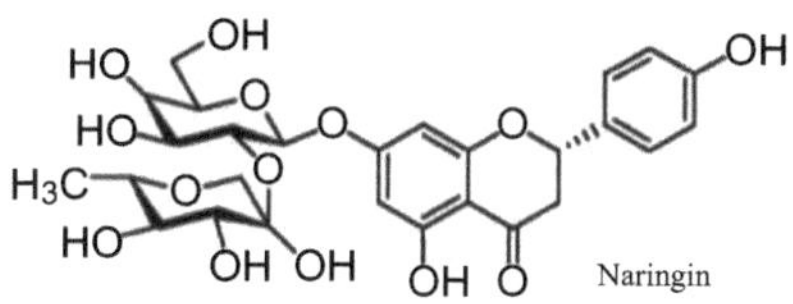

Naringin

Other flavonoids

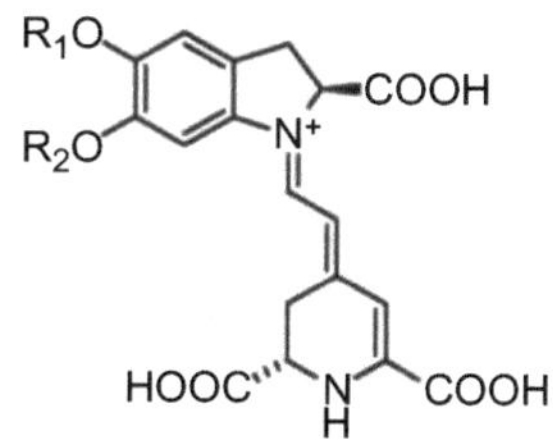

Betanin, R_1=glucose, *S*, R_2=H;
Isobetanin, R_1=glucose, *R*, R_2=H;

Betacyanins

Figure 8.1 Chemical structures of phenolic compounds and betacyanins identified in quinoa seeds (Tang et al., 2015).
Source: Reprinted with permission from Elsevier.

sinapic acid, whereas free phenolics included 3,4-dihydroxybenzoic acid, *p*-coumaric acid 4-glucoside, *p*-hydroxybenzoic acid, vanillic acid 4-glucoside, 2,5-dihydroxybenzoic acid, caffeic acid, vanillic acid, *p*-coumaric acid, ferulic acid, ferulic acid 4-glucoside, and isoferulic acid (Tang et al., 2015, 2016). Zhang, Bai, et al. (2022) showed that extracts of bound phenolics mainly contained ferulic acid and its derivatives and those of free phenolics mainly had hydroxybenzoic acid and its derivatives.

Genetic diversity in phenolic acid composition was obtained among different quinoa varieties. For example, among six quinoa genotypes, the concentrations of syringic acid ranged from 83 to 132 μg/g (fresh weight) (Pellegrini et al., 2017). The contents of total phenolic acids of two quinoa varieties were 73 and 24 mg/kg dry weight (DW) (Stikić et al., 2020). 3,4-Dihydroxybenzoic acid in the free form was found in red (30 mg/kg) and black (47 mg/kg) quinoa seeds but not in white ones (Tang et al., 2015). The differences in the composition of phenolic acids in quinoa seeds between different studies may be attributed to the differences in the quinoa genotypes and analytical methods. For example, different types of solvents used could lead to different extraction results of phenolics (Mokrani & Madani, 2016).

8.2.3 Flavonoids and derivatives in quinoa

Villacrés et al. (2022) found that the content of total flavonoids in one quinoa variety was 9.5 mg quercetin/100 g (dry weight). There are different types of flavonoids and derivatives in quinoa seeds. For example, Tang et al. (2015) found epigallocatechin, epicatechin, kaempferol 3,7-dirhamnoside, kaempferol 3-galactoside, quercetin-3-rutinoside, kaempferol 3-glucoside, quercetin 3-arabinoside, quercetin, and biochanin A in quinoa seeds (Fig. 8.1). The main flavonoids were kaempferol and quercetin (Tang et al., 2016). In a different study, the identified flavonoids in quinoa seeds included rutin, quercetin, isoquercetin, kaempferol, neohesperidin, and hesperidin (Pellegrini et al., 2017). In another report, kaempferol and quercetin were only found in certain quinoa varieties (Lee & Sim, 2018). Quercetin 3-*O*-(2,6-di-α-L-rhamno-pyranosyl)-β-D-galactopyranoside was one of the most abundant flavonoids as revealed by Graf et al. (2016). Some quinoa varieties had more kaempferol but not quercetin than the others. The growing locations can significantly affect the flavonoid composition (Lee & Sim, 2018). Sampaio et al. (2020) showed that a new quinoa variety developed for tropical region had kaempferol and quercetin glycosides as the major phenolics. Antognoni et al. (2021) showed that flavonoids in the free form contained epicatechin, quercetin-3-*O*-glucoside, quercetin-3-*O*-rutinoside, kaempferol-3-*O*-rutinoside, kaempferol-3-*O*-glucoside, genistein, and daidzein. The conjugated flavonoids included rutin and kaempferol-3-*O*-glucoside as the main ones found in the alkaline hydrolyzed extract (Antognoni et al., 2021).

Genetic variations in quinoa flavonoid composition and concentrations were obtained (Lee & Sim, 2018; Stikić et al., 2020). For example, the contents of naringin of two quinoa varieties were 0.07 and 0.12 mg/kg DW (Stikić et al., 2020). Black quinoa seeds had the highest amounts of bound phenolics, followed by red and white seeds (Tang et al., 2016). *Chenopodium formosanum* (locally known as red quinoa) is indigenous to Taiwan. It is rich in rutin (Lin, Han, et al., 2019; Lin, Ke, et al., 2019). The concentrations of phenolics in seeds, flowers, leaves, stems, and roots of quinoa tended to be different and were affected by growing time (Buitrago et al., 2019). The contents of phenolics and flavonoids in quinoa seeds were affected by growing locations (Lee & Sim, 2018). For example, quinoa seeds from Korea had less phenolics and flavonoids than those from Peru, though genetics could be a influencing factor. Liu et al. (2022) analyzed the synthesis of flavonoids in quinoa varieties with four different colors using metabolomics and transcriptomics. A total of 90 flavonoid-related metabolites were found in the quinoa seeds, and 18 of them were key players in the flavonoid biosynthesis. A total of 25 flavonoid-related genes were important in the biosynthesis of flavonoids (Liu et al., 2022). Overall, the types and composition of flavonoids in quinoa depend on the genetics, plant parts, growing environment, and also the methods of their extraction, identification and quantification.

8.2.4 Saponins in quinoa

Saponins exist in various parts of quinoa including flowers, leaves, stalks, and seeds (Gil-Ramirez et al., 2018; Tang & Tsao, 2017). The presence of saponins in quinoa seeds may be feasibly detected using a foaming test (Fig. 8.2) (Stanschewski et al., 2021). Saponins consist of aglycones with sugar moieties that can be based on glucose, glucuronic acid, galactose, galacturonic acid, xylose, rhamnose, and arabinose (Ahmed et al., 2021). They can be divided into monosaccharide-, disaccharide-, and trisaccharide-based saponins, depending on the number of glycosides linked chains. Saponins contribute to the bitter taste of quinoa by decreasing the palatability (Suárez-Estrella et al., 2018). While the majority of quinoa varieties contain a significant level of saponins, sweet quinoa cultivars with little saponins have been developed (Präger et al., 2018). For example, the content of saponins ranged from 0.2 to 0.4 g/kg in sweet quinoa seeds and was 1.13 g/kg in bitter quinoa seeds (Palombini et al., 2013). Villacrés et al. (2022) found that the content of total saponins in one quinoa variety was 199 mg/100 g (dry weight). Saponins were found in

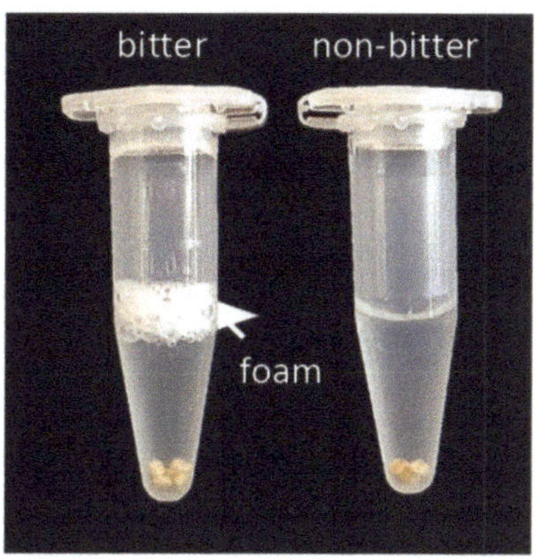

Figure 8.2 Foaming test for presence or absence of saponins in quinoa seeds, saponins create foams (Stanschewski et al., 2021).
Source: Reprinted with permission under under Creative Commons Attribution License.

different parts of quinoa plant including stems, leaves, seeds, roots, and sprouts (Lim et al., 2019). The roots had the highest level of total saponins (13%), which was followed by quinoa bran. These figures of saponins contents should be evaluated with caution since the extraction methods and conditions (e.g., the types of solvents used) can greatly affect the yields of saponins (Gianna et al., 2012). These compounds can be important for the physiology of quinoa. For example, dipping quinoa seeds in saponin solution increased the salt tolerance of quinoa (Yang et al., 2018). Saponins can be used in different applications. Quinoa extracts rich in saponins can be used for emulsion fabrication in cosmetic, pharmaceutical, or food products (Góral et al., 2021). Different methods have been developed to quantify saponins in quinoa seeds (Yousif et al., 2021). For example, a method based on water droplet surface tension was developed to measure saponin system with a range of concentrations (0.05–0.15 mg/mL). For structural and compositional analysis, sapogenins may be extracted using hydrolysis and solvent extraction as described in Fig. 8.3 (Medina-Meza et al., 2016). Saponins are not only found in quinoa seeds but also in other parts of quinoa such as stalk (Gil-Ramirez et al., 2018). Ultrasound-assisted extraction of betacyanins, betaxanthins, and saponins from the hulls of colored quinoa seeds has been developed (Laqui-Vilca et al., 2018).

The types of saponins found in quinoa are structurally diverse (Dong et al., 2020) (Fig. 8.4). A total of 14 different saponins of the bidesmosidic family were identified in quinoa seeds (Segura et al., 2020). Ruiza et al. (2017) showed that the major aglycones of saponins (sapogenins) were hederagenin (the most abundant), phytolaccagenic acid (the second abundant) and oleanolic acid. A total of 24 saponins with different sugar moieties were identified (Ruiza et al., 2017). Chen et al. (2022) found variations in the composition of sapogenins released from the hydrolysis of saponins among three quinoa varieties with three different colors (red, white, and black). Black and white quinoa seeds had similar concentrations and composition of sapogenins, which were different from those of red quinoa (Chen et al., 2022). Escribano et al. (2017) found the saponins in quinoa seeds to be 3-O-β-D-glucuronopyranosyl oleanolic acid 28-O-β-D-glucopyranosyl ester, 3-O-α-Larabinopyranosyl phytolaccagenic acid 28-O-β-D-glucopyranosyl ester, 3-O-β-D-xylopyranosyl-(1→3)-β-D-glucuronopyranosyl oleanolic acid 28-O-β-D-glucopyranosyl ester, 3-O-β-D-xylopyranosyl-(1→3)-β-D-glucuronopyranosyl hederagenin 28-O-β-D-glucopyranosyl ester, and 3-O-β-D-glucopyranosyl-(1→3)-a-L-arabinopyranosyl serjanic acid 28-O-β-D-glucopyranosyl ester. An earlier study of Woldemichael and Wink (2001) identified 16 saponins in quinoa seeds, and the major ones were 3-O-α-L- arabinopyranosyl hederagenin 28-O-β-D-glucopyranosyl ester, 3-O-β-D-glucuronopyranosyl oleanolic acid 28-O-β-D-glucopyranosyl ester, 3-O-β-D-glucuronopyranosyl-(1–3)-α-L-arabinopyranosyl phytolaccagenic acid 28-O-β-D-glucopyranosyl ester, 3-O-α-L-arabinopyranosyl phytolaccagenic acid 28-O-β-D-glucopyranosyl ester, 3-O-β-D-glucuronopyranosyl-(1–3)-α-L-arabinopyranosyl hederagenin 28-O-β-D-glucopyranosyl ester, and a novel saponin 3-O-β-D-glucuronopyranosyl-(1–3)-α-L-arabinopyranosyl phytolacca-genic acid.

There is great genetic variation in the concentrations of saponins. For example, among 28 quinoa genotypes, the total saponin contents varied from 3.8 to 27 mg/g.

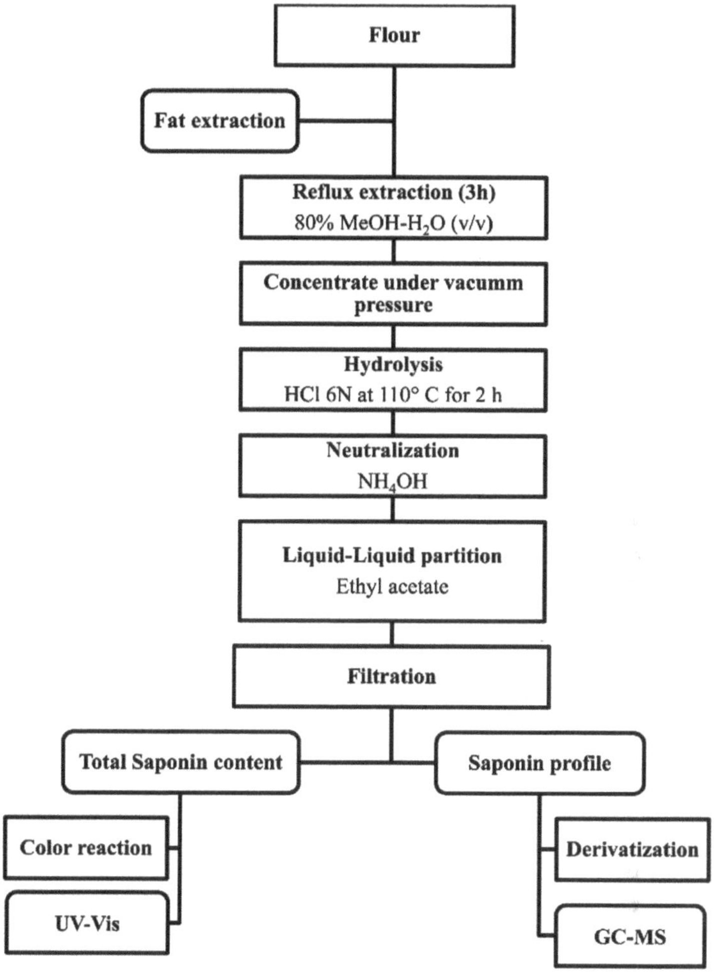

Figure 8.3 Flowchart of the extraction protocol used for the quantification of sapogenins (Medina-Meza et al., 2016).
Source: Reprinted with permission from American Chemical Society.

Phytolaccagenic acid and hederagenin accounted for ~70% and 30% of the total sapogenins (Medina-Meza et al., 2016). Among seven colored quinoa genotypes, the total saponin contents ranged from 7.5 to 12.1 oleanolic acid equivalents/g (dry weight) (Han, Chi, Zhang, Zhang, Fan, Dong, et al., 2019; Han, Chi, Zhang, Zhang, Fan, Huang, et al., 2019). The composition of the saponins was affected by the quinoa genotype (Escribano et al., 2017; Han, Chi, Zhang, Zhang, Fan, Dong, et al., 2019; Han, Chi, Zhang, Zhang, Fan, Huang, et al., 2019; Medina-Meza et al., 2016; Ruiza et al., 2017). Metabolomic analysis showed that 25 white quinoa genotypes could be categorized into two groups based on the composition of phenolics and saponins

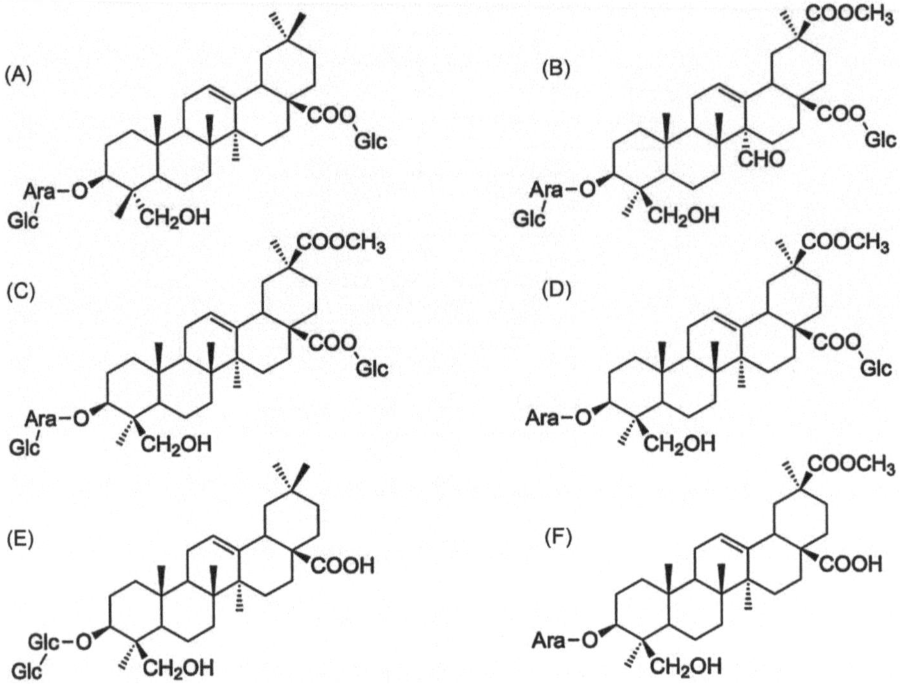

Figure 8.4 Six saponins (monomer structure) from quinoa husk: (A) 3-O-β-D-glucopyranosyl-(1→3)-α-L-arabinopyranosyl hederagenin-28-O-β-D-glucopyranosyl ester; (B) 3-O-β-D-glucopyranosyl-(1→3)-α-L-arabinopyranosyl-phytolaccagenic acid-27-oxo-28-O-β-D-glucopyranosyl ester; (C) 3-O-β-D-glucopyranosyl-(1→3)-α-L-arabinopyranosyl phytolaccagenic acid 28-O-β- D-glucopyranosyl ester; (D) 3-O-α-L-arabinopyranosyl phytolaccagenic acid 28-O-β-D-glucopyranosyl ester; (E) 3-O-β-D-glucopyranosyl-(1→4)-β-D-glucopyranosyl-28-O-hederagenin; (F) 3-O-α-L-arabinopyranosyl-phytolaccagenic acid (Dong et al., 2020).
Source: Reprinted with permission from Elsevier.

(Cabanillas et al., 2021). One group had a low content of phenolics and a high content of saponins, and the other one had the opposite compositional pattern. The quinoa growing conditions also can affect the saponin content. For example, selenium stress increased the content of the most abundant saponin in quinoa seeds during germination (Lazo-Vélez et al., 2016). Sprouting decreased the saponin content significantly, as reflected in over 90% of decrease in the phytolaccagenic acid content (Lazo-Vélez et al., 2016).

8.2.5 Phytoecdysteroids, betalains, and other minor components in quinoa

Apart from phenolics and saponins, quinoa contains a range of other bioactive components including phytoecdysteroids, phytosterols, betalains, and derivatives (Abderrahim et al., 2015; Antognoni et al., 2021; Escribano et al., 2017; Graf et al., 2016; Stikić et al., 2020; Tang et al., 2015).

Phytoecdysteroids are mostly concentrated in the embryo of quinoa seeds (Graf et al., 2016). Among 17 quinoa genotypes, 20-hydroxyecdysone is the most abundant phytoecdysteroid. The contents of phytoecdysteroids ranged from 138 to 570 μg/g. The content was positively related to the lipid content of quinoa seeds (Graf et al., 2016). There is chemical diversity in the structure of betalains in quinoa (Fig. 8.5) (Escribano et al., 2017). Amaranthin, iso-amaranthin, betanin, iso-betanin, dopaxanthin, dopamine-betaxanthin, and proline-betaxanthin were identified in quinoa (Fig. 8.5). Betacyanin and betaxanthin were found in colored quinoa genotypes (Abderrahim et al., 2015). Betanin and isobetanin were found in black and red quinoa seeds (Tang et al., 2015). Yellow-orange quinoa grains contained betaxanthins, which were responsible for the color of the grain (Escribano et al., 2017). Genetic diversity in betalain composition was obtained. For example, the concentrations of amaranthin (0−149 mg/kg), iso-amaranthin (0−146 mg/kg), betanin (0−9.7 mg/kg), iso-betanin (0−7.7 mg/kg), dopaxanthin (0−85.3 mg/kg), dopamine-betaxanthin (0−11 mg/kg), and proline-betaxanthin (0−7.5 mg/kg) were found to be varied among different genotypes (Escribano et al., 2017). Apart from phytoecdysteroids and betalains, other minor components were also found in quinoa. For example, quinoa (also spinach and goji) contained 7-isopentenyloxycoumarin, auraptene, umbelliprenin, boropinic acid, and 4, 4′-geranyloxyferulic acid (Antognoni et al., 2021). Letelier et al. (2011) discovered thiol compounds in the extract of quinoa seed coats, though the exact chemical structure of these compounds remains to be studied. These thiol compounds together with polyphenols contributed to the antioxidant potential of the extracts (Letelier et al., 2011). It is expected that more types of minor components and their biological activities may be revealed in quinoa grains in the future. The genetic mechanisms for the synthesis of bioactive components such as betalains, saponins, and phenolics in quinoa seeds remain to be better studied (Fiallos-Jurado et al., 2016; Wang, Cao, et al., 2020). New quinoa varieties with increased amounts of specific bioactive compounds should be developed using genetic means.

8.3 Biological activities of polyphenols, saponins, and other bioactive compounds in quinoa

The biological activities of quinoa extracts rich in bioactive components including phenolics and saponins were analyzed (Table 8.2). Most of the studies used in vitro model, whereas only a limited number of studies employed in vivo and human clinical models.

8.3.1 Phenolics-related biological activities

8.3.1.1 Antioxidant activity

Different quinoa extracts containing phenolics were found to have in vitro antioxidant activities (Abderrahim et al., 2015; Antognoni et al., 2021; Liu et al., 2020; Paśko et al., 2019; Stikić et al., 2020; Tang et al., 2015; Zhang, Bai, et al., 2022). Great genetic diversity in in vitro antioxidant activities of quinoa grains was

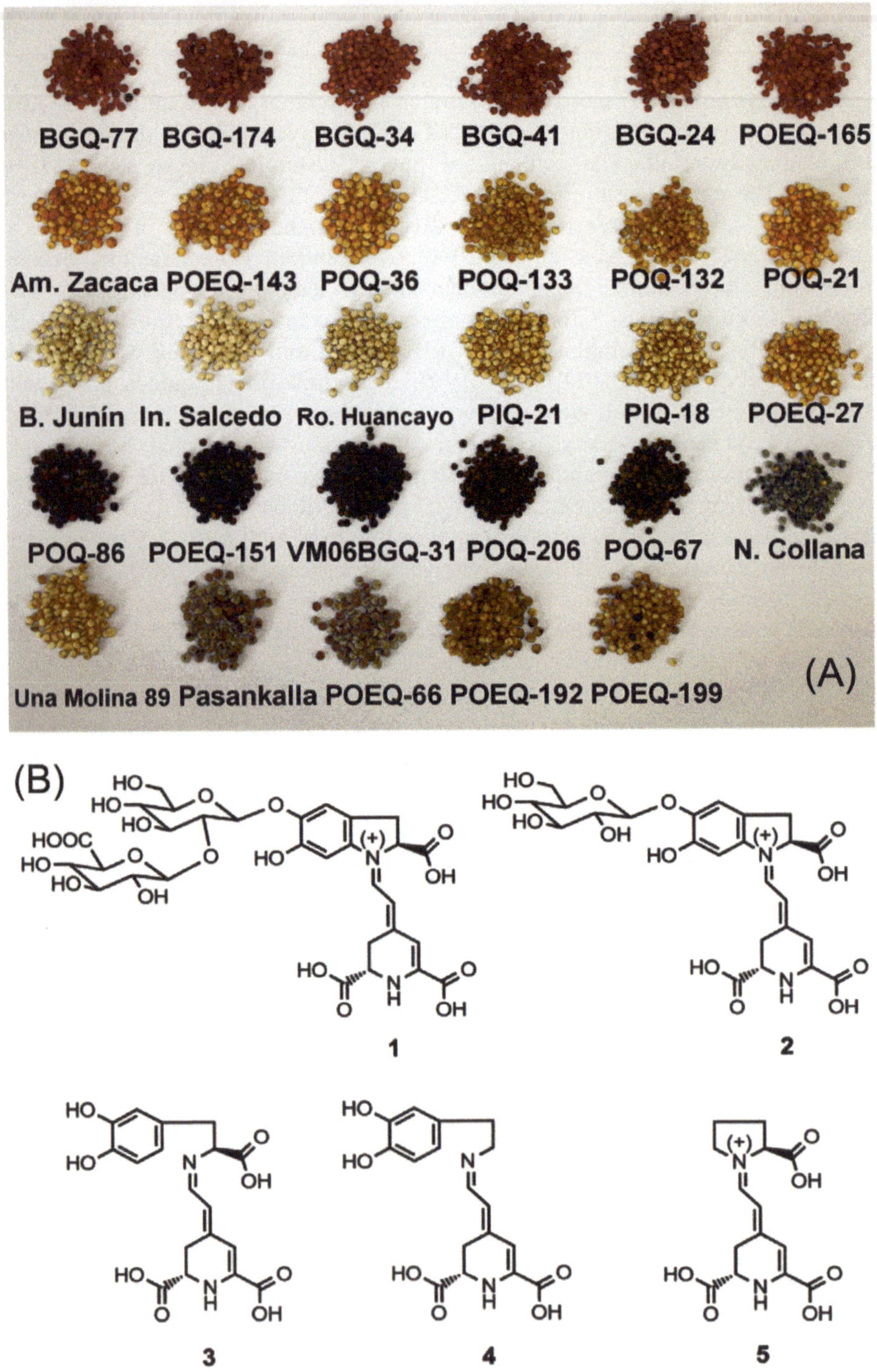

Figure 8.5 (A) Photos of different varieties of *Chenopodium quinoa* grains. (B) Betalain structures identified in the colored quinoa grains: Amaranthin (1), betanin (2), dopaxanthin (3), miraxanthin V (4), and indicaxanthin (5) (Escribano et al., 2017).
Source: Reprinted with permission under Creative Commons Attribution License.

Table 8.2 Biological properties of quinoa extracts.

Biological property	Mode	Method	Sample form	Major findings	References
Phenolics related					
Antioxidant	In vitro	QUENCHER-CUPRAC (cupric ion reducing antioxidant capacity)	Extracts of free and bound phenolics	Antioxidant activity was positively correlated with the content of betalains, and free and bound phenolics of the seeds	Abderrahim et al. (2015)
Antioxidant	In vitro	ORAC (oxygen radical absorbance capacity), FRAP (ferric reducing antioxidant power), DPPH (2,2 diphenyl-1-picrylhydrazyl)	Free phenolics, extracts obtained using alkaline or acid solutions	Darker seeds had higher antioxidant activity	Tang et al. (2015)
Antioxidant	In vitro	FRAP, ABTS (2,2′-azino-bis-3-ethylbenzthi) and CUPRAC	Methanol extract	The antioxidant activity of quinoa extract was lower than that of buckwheat and higher than that of amaranth	Paśko et al. (2019)
Antioxidant	In vitro	FRAP, ABTS	Methanol extract and extract of bound phenolics	Extractable phenolics showed much higher antioxidant than the bound ones	Stikić et al. (2020)
Antioxidant	In vitro	ABTS, DPPH	Bound and free phenolics of white, red, and black quinoa seeds	Red quinoa samples had higher antioxidant activities than black and white ones	Liu et al. (2020)
Antioxidant	In vitro	ABTS, ORAC, FRAP	Extracts of free and bound phenolics	Antioxidant activity was positively correlated with the content of quercetin derivatives	Antognoni et al. (2021)
Antioxidant	In vitro	DPPH, ABTS	Extracts of free and bound phenolics of red quinoa seeds	Extracts of bound phenolics (mainly containing ferulic acid and its derivatives) had higher antioxidant activities than those of free phenolics (mainly containing hydroxybenzoic acid and its derivatives)	Zhang, Bai et al. (2022)
Antiobesity	In vitro	Lipase inhibition	Bound phenolics	Bound phenolics from black, red, and white quinoa seeds showed a range of lipase inhibition activities	Tang et al. (2016)

(Continued)

Table 8.2 (Continued)

Biological property	Mode	Method	Sample form	Major findings	References
Antiobesity	In vitro	Lipase inhibition	Ethanol extract	Quinoa extract had lower lipase inhibition activity than fenugreek extract. In vitro intestinal digestion decreased the lipase inhibition activity	Herrera et al. (2019)
Antiobesity	In vitro	α-Glucosidase	Bound phenolics	Bound phenolics from black, red, and white quinoa seeds showed a range of α-glucosidase inhibition activities	Tang et al. (2016)
Antihypoglycemia	In vitro	α-Amylase	Ethanol extract	Quinoa extract had higher α-amylase inhibition activity than fenugreek extract. In vitro intestinal digestion had little effect on the α-amylase inhibition activity	Herrera et al. (2019)
Antihypoglycemia	In vitro	α-Glucosidase	Extracts of free and bound phenolics of red quinoa seeds	Extracts of bound phenolics had higher inhibitory activity against α-glucosidase hydrolysis than those of free phenolics	Zhang, Bai et al. (2022)
Anticancer	In vitro	Human cervical carcinoma cell lines (C4-I, HTB-35, and HTB-34), and normal human fibroblasts	Methanol extract	The extract had moderate inhibitory activity against the cancer cells, while having little effect on normal human fibroblasts. The anticancer effect of quinoa extract was less than that of buckwheat and more than that of amaranth	Paśko et al. (2019)
Anticancer	In vitro	Human colorectal cancer cell line HCT-116, MTT assay	Methanol extract	The extracts showed anticancer property	Stikić et al. (2020)
Antiinflammatory, anticancer	In vitro	MCF-7 cells (human breast cancer cell)	Bound and free phenolics extracts of white, red, and black quinoa seeds	Black and red quinoa seed extracts inhibited the proliferation of MCF-7 cells, while downregulating the NO production. Their extracts were more efficient than that of white seed one	Liu et al. (2020)
Antimicrobial	In vitro	Three Gram positive, three Gram negative bacteria and six fungi were tested	80% ethanol extract	The quinoa extract showed a range of antibacterial and antifungal activities. The extract had higher antifungal activity against *P. ochrochloron* as compared to ketoconazole (a fungicide)	Sampaio et al. (2020)

Antioxidant	In vivo	Rats with oxidative stress induced by CCl_4	70% ethanol extracts of germinated seeds	Germination increased the phenolic content and in vitro antioxidant activity. Feeding the rats with the sprouts reduced fasting blood glucose, aspartate aminotransferase, alanine aminotransferase, and total bilirubin, while improving the conditions of liver inflammation. It decreased the level of low-density lipoproteins and very low-density lipoproteins, while increasing that of high-density lipoproteins. The feeding of the spouts decreased the level of malonaldehyde (MDA) and increased the activity of superoxide dismutase and glutathione level	Al-Qabba et al. (2020)
Liver protection	In vivo	Mice with liver injury and fibrosis induced by CCl_4	*Chenopodium formosanum*: 50% ethanol extracts or water extracts of bran and whole seeds, seed powder	Compared with rutin, the seed powder extract better protected the mice from CCl_4-induced fibrosis, proinflammatory factor expression, and oxidative stress. The ethanol extract of the bran had the highest antifibrosis and liver protection effect. Mechanistically, the tumor necrosis factor α (TNF-)/ interleukin 6 and transforming growth factor β1 pathways were blocked	Lin, Han, et al. (2019), Lin, Ke, et al. (2019)
Controlling postprandial hyperglycemia and antidiabetic	In vivo	ICR mice	Extracts of free and bound phenolics of red quinoa seeds	Extracts of bound phenolics were more efficient in suppressing postprandial glucose increases in the mice than those of free phenolics. The controlling effect of the extracts of bound phenolics at 50 mg/kg were similar to that acarbose (an antidiabetic drug) at 20 mg/kg	Zhang, Bai, et al. (2022)

(*Continued*)

Table 8.2 (Continued)

Biological property	Mode	Method	Sample form	Major findings	References
Saponins related					
Antibacterial	In vitro	Halitosis-related bacteria *Porphyromonas gingivalis, Fusobacterium nucleatum, Clostridium perfringens*	Alkaline treated saponins were fractioned using AB-2 resin	Specific saponin fraction showed strong inhibitory activity against *F. nucleatum* by disrupting the bacterial membrane	Sun et al. (2019)
Antibacterial	In vitro	*Listeria ivanovii, Staphylococcus aureus, S. epidermidis, Bacillus cereus, Pseudomonas aeruginosa, Salmonella enteritidis*	Six types of saponins (Fig. 8.4) isolated from quinoa seed husk were used	All the saponins showed antibacterial activities in a dose-dependent manner. The minimum inhibitory and minimum bactericidal concentrations were 0.0625 and 0.125 mg/mL, respectively. For saponins with specific structure, the addition increased the release of proteins and nucleic acids from *B. cereus*, while disrupting the membrane structure	Dong et al. (2020)
Antiobesity	In vitro	3T3-L1 preadipocytes were used as model cells	Quinoa flour was extracted using MeOH before partition with EtOAc and n-BuOH. The resulting filtrates were passed through resin column to obtain saponin fraction	Quinoa saponin fraction decreased 3T3-L1 adipocyte differentiation and viability. At the molecular level, the saponin fraction downregulated a range of adipogenesis-related transcriptional factors	Yao et al. (2015)
Immunoadjuvant activity	In vivo	Mice immunized by ovalbumin were used as model system	Saponin fractions	Saponin fractions significantly increased the humoral and cellular immune responses to ovalbumin in the mice	Verza et al. (2012)
Gut microbiota regulation	In vitro	Fecal slurry for fermentation	Methanol (10%) and 1-butanol extracts rich in saponins	The major sapogenins after the fermentation were hederagenin, phytolaccagenic acid, oleanolic acid, and serjanic acid. Fermentation of quinoa extracts led to the formation of more sapogenins than that of fenugreek and lentil. Quinoa extracts significantly inhibited the *Lactobacillus* spp. and lactic acid bacteria of human gut microbiota	del Hierro, Cueva et al. (2020)

		gastrointestinal digestion and fermentation using human feces		the in vitro antioxidant activity of the extract but had no hydrolytic effect of releasing sapogenins. Gut microbiota fermentation released hederagenin and phytolaccagenin from the extracts. Saponin-rich extract reduced the growth of *Shigella* spp., while enhancing that of *Bifidobacterium* spp. and *Lactobacillus* spp. Extracts of different quinoa varieties showed different fermentation properties	
Anticancer	In vitro	Colorectal cancer cells were used	Saponin-rich extracts	Quinoa seed extracts showed no anticancer property, whereas the extracts from fenugreek had strong anticancer capacity	de Cedrón et al. (2020)
Pancreatic lipase inhibition, hypocholesterolemic	In vitro	Pancreatic lipase; in vitro gastrointestinal digestion of cholesterol	Saponin-rich extracts	Acid hydrolysis increased the lipase inhibition activity of quinoa extract, while significantly reducing the bioaccessible cholesterol	Del Hierro et al. (2021)
Cellular toxicity	In vitro	Human gastric mucosal cells	Saponin-rich extracts	Quinoa saponin extracts showed toxicity towards human gastric mucosal cells. Soaking reduced the toxicity by degrading the saponins	Xue et al. (2020)
Toxicity	In vitro and in vivo	Rats	Saponin-rich extracts	The LD_{50} (the amount of a material, given all at once, leads to 50% death of tested animals) was over 10 g/kg body weight. Food intake of the rats decreased with some developing diarrhea. Overall, quinoa saponins had no mutagenicity. Quinoa saponins were excreted in the form of aglycones	Lin et al. (2021)

(*Continued*)

Table 8.2 (Continued)

Biological property	Mode	Method	Sample form	Major findings	References
Toxicity	In vivo	Healthy adult Sprague-Dawley rats (males and females)	Saponin-rich extracts	Metabonomics analysis showed that the number of metabolites in male rats (23) was lower than that in female rats (32). High-dose saponin treatment increased the diversity of gut microbiota, while inducing slight necrosis intestinal mucosa in female rats. This effect was sex-dependent. The level of saponins intake to cause no adverse effect was <50 mg/kg body/day for females and males	Zhang, Zhai et al. (2022)
Skin penetration enhancer	In vitro	Intercellular lipid mixture mimicking the outermost layer of skin (stratum corneum)	Saponin-rich extracts	Saponins got incorporated into lipid monolayer and could be used as enhancers for active components to penetrate into skin	Jurek et al. (2019)
Membrane permeability	In vitro	Parallel artificial membrane was used for the permeability tests. In vitro gastrointestinal digestion	Saponins and sapogenins in quinoa extracts	Saponins and sapogenins had low to moderate permeability, and gastrointestinal digestion increased it. The presence of lipids contributed to the permeability of the saponins and sapogenins	del Hierro, Piazzini et al. (2020)

obtained (Escribano et al., 2017). The extracts from red, violet, and yellow quinoa grains tended to have higher antioxidant activities than those of black and white grains. Red quinoa samples had higher antioxidant activities than black and white ones (Liu et al., 2020). The extracts of bound phenolics (mainly containing ferulic acid and its derivatives) had higher antioxidant activities than those of free phenolics (mainly containing hydroxybenzoic acid and its derivatives) (Zhang, Bai, et al., 2022). The antioxidant activity of quinoa extract was less than that of buckwheat and more than that of amaranth grains (Paśko et al., 2019). The antioxidant activities of quinoa extracts were related to the contents of phenolics. For example, antioxidant activity was positively correlated with the content of quercetin derivatives (Antognoni et al., 2021). Extracts analyzed in different studies were mostly mixtures of phenolics and other components such as betalains and saponins. Indeed, together with free and bound phenolics, betalains in quinoa contributed to the antioxidant activity (Escribano et al., 2017). The extracts of different parts of quinoa plant including stem, leaves, seeds, roots, and sprouts were studied for in vitro antioxidant activities (Lim et al., 2019). Quinoa root had the highest antioxidant capacity, indicating its potential for functional food applications.

A range of different in vitro chemical model systems (QUENCHER-CUPRAC, oxygen radical absorbance capacity (ORAC), ferric reducing antioxidant power assay (FRAP), 2,2-diphenyl-1-picrylhydrazyl (DPPH), 2,2′-azino-bis(3-ethylbenzothiazoline-6-sulfonic acid) (ABTS), and ferric reducing power) were used to characterize the antioxidant and radical scavenging properties of quinoa extracts (Table 8.2). There has been increasing pressure to develop other types of methods for studying antioxidant activity. This is because these in vitro chemical assays may have little relevance towards to the applications of their results in human nutrition. Unfortunately, the research community continues to use these "easy" methods out of a common trap. Other methods such as cell-based approaches may be developed and used to define the antioxidant activity (Martinelli et al., 2021). For example, Caco-2 cell line could be a suitable model for cellular antioxidant activity and bioavailability measurements (Martinelli et al., 2021).

Antioxidant activity of germinated quinoa seeds in the form of sprouts was studied using in vivo rat model (Al-Qabba et al., 2020). The oxidative stress in rats was induced using CCl_4. Feeding the rats with the sprouts reduced fasting blood glucose level, levels of aspartate aminotransferase, alanine aminotransferase, and total bilirubin, while improving the conditions of liver inflammation. It decreased the levels of low-density lipoproteins and very low-density lipoproteins, while increasing that of high-density lipoproteins. The feeding of the spouts decreased the level of malonaldehyde (MDA) and increased the activity of superoxide dismutase and glutathione level (Al-Qabba et al., 2020). Such results should be translated into human applications.

8.3.1.2 Anticancer

In vitro anticancer properties of quinoa extracts were studied (Liu et al., 2020; Paśko et al., 2019; Stikić et al., 2020). Paśko et al. (2019) used human cervical carcinoma cell lines (C4-I, HTB-35 and, HTB-34) and normal human fibroblasts to study the anticancer effect of quinoa methanol extract. The quinoa extract had moderate inhibitory activity against the cancer cells, while having little effect on normal human fibroblasts. The anticancer effect of the quinoa extract was less than that of buckwheat extract and more than that of amaranth extract (Paśko et al., 2019). Stikić et al. (2020) showed that methanol extract of quinoa had anticancer activity against human colorectal cancer cell line HCT-116. Liu et al. (2020) indicated that black and red quinoa seed extracts inhibited the proliferation of MCF-7 cells (human breast cancer cell line), while downregulating the NO production. Their extracts were more efficient than that of white seed one (Liu et al., 2020). In these studies, no commercial anticancer drugs were employed for comparison. It is not clear if these quinoa extracts have any practical anticancer potential due to the lack of any meaningful comparison.

8.3.1.3 Antiobesity and antihypoglycemia

Antiobesity and antihypoglycemia capacities could be related to the inhibition of starch degradation enzymes such as α-glucosidase and α-amylase (Herrera et al., 2019; Tang et al., 2016; Zhang, Bai, et al., 2022). Tang et al. (2016) showed that bound phenolics from black, red, and white quinoa seeds had a range of α-glucosidase inhibition activities. Han, Chi, Zhang, Zhang, Fan, Dong, et al. (2019) and Han, Chi, Zhang, Zhang, Fan, Huang, et al. (2019) revealed that the extracts of free phenolics had higher inhibition on α-glucosidase than those of bound phenolics of quinoa seeds. In contrast, Zhang, Bai, et al. (2022) showed that extracts of bound phenolics in red quinoa seeds had higher inhibitory activity against α-glucosidase activities than those of free phenolics. The differences between the two studies may be due to the different quinoa varieties used in different studies. In a comparative study of Herrera et al. (2019), quinoa extract had higher α-amylase inhibition activity than fenugreek extract. In vitro intestinal digestion had little effect on the α-amylase inhibition of the extracts.

Antiobesity could be partially related to the inhibition of lipase (Herrera et al., 2019). Tang et al. (2016) showed that bound phenolics from black, red, and white quinoa seeds had a range of lipase inhibition activities. In a comparative study of Herrera et al. (2019), quinoa extract had lower lipase inhibition activity than fenugreek extract. In vitro intestinal digestion decreased the lipase inhibition activity of the extracts (Herrera et al., 2019).

Extracts of free and bound phenolics of red quinoa seeds were studied for controlling postprandial hyperglycemia and antidiabetic properties using in vivo ICR (Institute of Cancer Research) mice model (Zhang, Bai, et al., 2022). Extracts of bound phenolics were more efficient in suppressing the increases of postprandial glucose level in the mice than those of free phenolics. The controlling effects of the extracts of bound phenolics at 50 mg/kg were similar to that of acarbose (an antidiabetic drug) at 20 mg/kg (Zhang, Bai, et al., 2022).

8.3.1.4 Liver protection

The liver protective effect of the extracts of *Chenopodium formosanum* was studied using mouse model in vivo (Lin, Han, et al., 2019; Lin, Ke, et al., 2019). Mice were treated by CCl_4 to induce liver injury and fibrosis. Compared with rutin, the seed powder better protected the mice from CCl_4-induced fibrosis, proinflammatory factor expression, and oxidative stress. The ethanol extract of the bran had the highest antifibrosis and liver protection effect. Mechanistically, the tumor necrosis factor α (TNF-)/interleukin 6 and transforming growth factor β1 pathways were blocked (Lin, Han, et al., 2019; Lin, Ke, et al., 2019).

8.3.2 Saponins-related biological activities

Biological functions of quinoa grains could be significantly related to the presence of saponins apart from phenolics (Table 8.2). Saponins are not only found in quinoa, but also in many species of the Amaranthaceae family (Mroczek, 2015). Biological activities of these saponins were identified. Cytotoxicity, antidiabetes, immunomodulation, hypolipidemia, antiosteoporosis, hepatoprotection, anthelmintic, antifungal, and antiviral activities were the claimed biological functions of these saponins (Melini & Melini, 2021c; Mroczek, 2015).

8.3.2.1 Antibacterial activity

Sun et al. (2019) tested the in vitro antibacterial activity of fractions of alkaline treated saponins against halitosis-related bacteria (*Porphyromonas gingivalis*, *Fusobacterium nucleatum*, and *Clostridium perfringens*). Specific saponin fraction with less polarity showed stronger inhibitory activity against *F. nucleatum* by disrupting the bacterial membrane (Sun et al., 2019). Dong et al. (2020) analyzed six different types of saponins as antibacterial agents against *Listeria ivanovii*, *Staphylococcus aureus*, *S. epidermidis*, *Bacillus cereus*, *Pseudomonas aeruginosa*, and *Salmonella enteritidis*. All the saponins showed antibacterial activities in a dose-dependent manner. The minimum inhibitory and minimum bactericidal concentrations were 0.0625 and 0.125 mg/mL, respectively. For saponins with specific structure, its addition increased the release of proteins and nucleic acids from *B. cereus*, while disrupting the membrane structure (Fig. 8.6). Yao et al. (2015) analyzed the in vitro antiobesity capacity of quinoa saponin-rich fraction using 3T3-L1 preadipocytes cells. Quinoa saponin fraction decreased 3T3-L1 adipocyte differentiation and viability. At the molecular level, the saponin fraction downregulated a range of adipogenesis-related transcriptional factors (Yao et al., 2015). However, systematic analysis of literature showed that the efficiency of saponins for antiobesity application was inconclusive (Marrelli et al., 2016).

8.3.2.2 Gut microbiota regulation

Gut microbiota is related to a range of health effects of humans (Fan & Pedersen, 2021). Human fecal slurry was used for fermentation studies on the effect of saponin-rich extracts from quinoa (del Hierro, Cueva, et al., 2020). After the fermentation, the major sapogenins formed were hederagenin, phytolaccagenic acid, oleanolic acid, and serjanic acid. Fermentation of quinoa extracts led to the

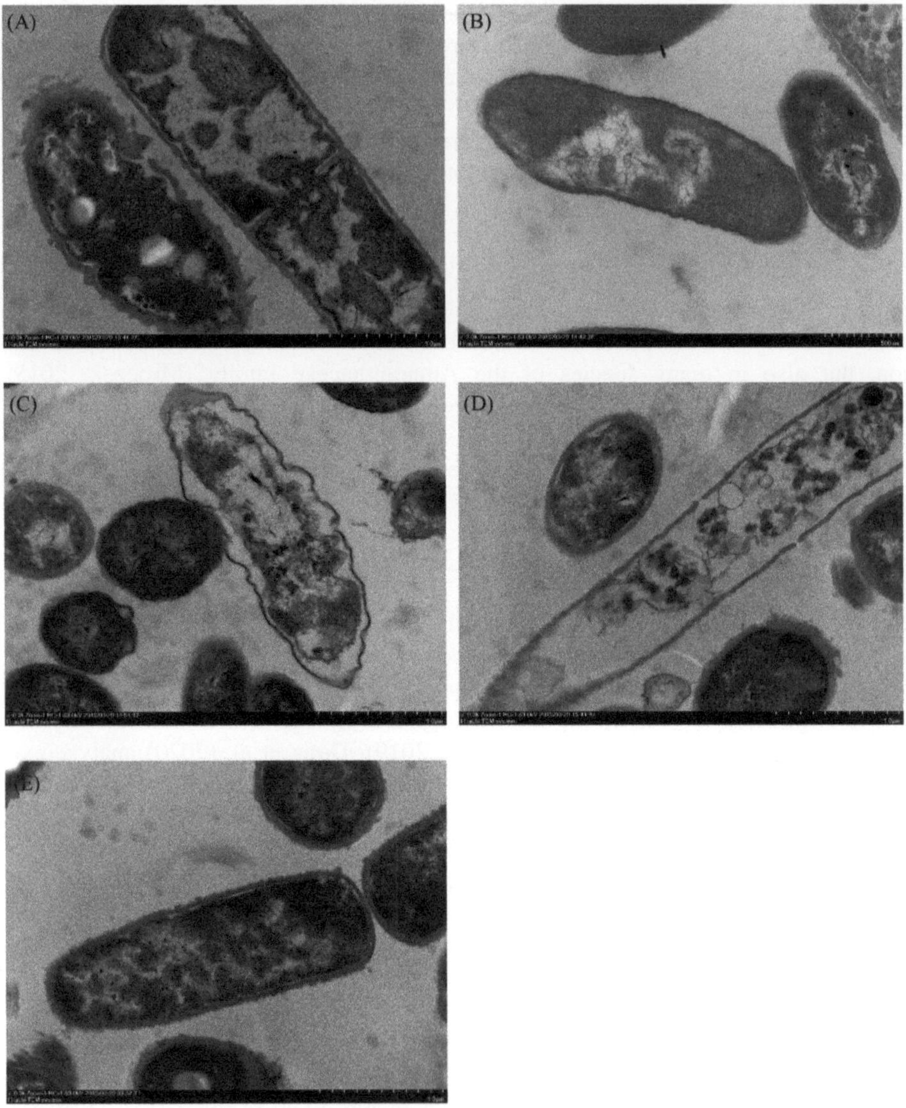

Figure 8.6 *Bacillus cereus* under transmission electron microscopy. (A) *Bacillus cereus* treated with 1/2 minimum inhibitory concentration (MIC) of quinoa saponins. (B) *Bacillus cereus* treated with MIC. (C) *Bacillus cereus* treated with minimum bactericidal concentration (MBC). (D) Positive control (treated with cefixime, an antibiotic medication). (E) Negative control (untreated *Bacillus cereus*) (Dong et al., 2020).
Source: Reprinted with permission from Elsevier.

formation of more sapogenins than that of fenugreek and lentil extracts. Quinoa extracts significantly inhibited the *Lactobacillus* spp. and lactic acid bacteria (del Hierro, Cueva, et al., 2020). Chen et al. (2022) found that human feces-based fermentation of saponins-rich extracts of quinoa brans produced sapogenins including phytolaccagenin and hederagenin. The growth of *Shigella* spp. (harmful bacteria) was suppressed, whereas that of beneficial *Bifidobacterium* spp. and *Lactobacillus* spp. were enhanced by the extract of black quinoa (Chen et al., 2022). Such results were in contradiction with those of del Hierro, Cueva, et al. (2020). Overall, it appeared that the effect of saponins in quinoa on gut microbiota remains unclear and inconclusive.

8.3.2.3 Anticancer

Saponin-rich extracts from quinoa seeds were tested for in vitro anticancer property against colorectal cancer cells (de Cedrón et al., 2020). The quinoa seed extracts showed no anticancer property, whereas the extracts from fenugreek had strong anticancer capacity.

8.3.2.4 Pancreatic lipase inhibition and hypocholesterolemia

Saponin-rich extracts were tested for in vitro inhibitory activity on pancreatic lipase and hypocholesterolemia capacity (Del Hierro et al., 2021). Only after acid hydrolysis, the extracts had increased lipase inhibition activity of quinoa extract, and significantly reduced the bioaccessible cholesterol. The hydrolyzed extracts had similar hypocholesterolemia effects compared to β-sitosterol (Del Hierro et al., 2021).

8.3.2.5 Immunoadjuvant activity

Quinoa saponin fractions showed in vivo immunoadjuvant activity (Verza et al., 2012). Mice immunized by ovalbumin were used as model system. Saponin fractions significantly increased the humoral and cellular immune responses to ovalbumin in the mice (Verza et al., 2012).

8.3.2.6 Membrane and skin permeability

The capacity of quinoa saponins as skin penetration enhancer was tested in vitro (Jurek et al., 2019). An intercellular lipid mixture mimicking the outermost layer of skin (stratum corneum) was created and tested against saponin-rich extracts from quinoa. Saponins got incorporated into the lipid monolayer and could be used as enhancers for active components to penetrate into skin (Jurek et al., 2019). Parallel artificial membrane was used for the permeability tests to analyze the effect of in vitro gastrointestinal digestion of saponins and sapogenins in quinoa extracts (del Hierro, Piazzini, et al., 2020). Saponins and sapogenins had low-to-moderate permeability, and gastrointestinal digestion increased it. The presence of lipids contributed to the permeability of the saponins and sapogenins (del Hierro, Piazzini, et al., 2020).

8.3.2.7 Toxicity

A major concern of quinoa has been the safety profile due to the presence of saponins (Lin et al., 2021; Xue et al., 2020; Zhang, Zhai, et al., 2022). Cellular toxicity of saponin-rich extracts from quinoa seeds was tested in vitro using human gastric mucosal cells (Xue et al., 2020). Quinoa saponin extracts showed toxicity towards human gastric mucosal cells. Soaking reduced the toxicity by degrading and removing the saponins (Xue et al., 2020). Studies on *Quillaja saponaria* showed that saponin parent compounds were 3 – 7 times more toxic towards aquatic zebrafish and crustacean than their degraded products (Jiang et al., 2018). In vivo rat-based tests showed that quinoa saponins had rather limited acute toxicity effects (Lin et al., 2021). The LD_{50} (median lethal dose) was over 10 g/kg body weight. Food intake of the rats decreased with some of them developing diarrhea. Quinoa saponins were excreted in the form of aglycone (Lin et al., 2021). Zhang, Zhai, et al. (2022) analyzed subchronic toxicity of saponins from quinoa husk using healthy adult Sprague-Dawley rats (males and females). Metabolomics analysis showed that the number of metabolites in male rats (23) was lower than that in female rats (32). High-dose saponin treatment increased the diversity of gut microbiota, while inducing slight necrosis intestinal mucosa in female rats. This effect was sex-dependent. The level of saponins intake to cause no adverse effect was <50 mg/kg body/day for females and males (Zhang, Zhai, et al., 2022). Overall, quinoa saponins had no mutagenicity. There has been no data on the toxicity of saponins in humans.

8.3.3 Biological activities related to other bioactive compounds (phytoecdysteroids, betalains, and alkylresorcinols)

Apart from phenolics and saponins, other minor compounds in quinoa including phytoecdysteroids and betalains showed biological activities. Previous studies on phytoecdysteroids from other sources showed that they had antiinflammatory, anabolic, metabolism-modulatory, adaptogenic, antidiabetic, antioxidant, hypolipidemic, tissue differentiation, anticancer, and hepatoprotective effects (Das et al., 2021). Similarly, previous studies on betalains from other sources showed that they had antimicrobial, antioxidant, anticancer, antilipidemic, hepatoprotective, neuroprotective, antiinflammatory, and hypoglycemic activities (Sadowska-Bartosz & Bartosz, 2021). The molecular mechanisms of action of these components have been analyzed for the biological functions (Das et al., 2021; Sadowska-Bartosz & Bartosz, 2021). Even though these biological activities of phytoecdysteroids and betalains have not been studied on quinoa, it may be expected that these two types of components also contribute to the biofunctions of quinoa. For example, Foucault et al. (2014) evaluated the effect of a quinoa extract rich in 20-hydroxyecdysone (the most abundant phytoecdysteroid in quinoa) on the development of obesity in mice. The intake of 20-hydroxyecdysone decreased the lipid absorption, while increasing the energy expenditure in the mice. Overall, 20-hydroxyecdysone in quinoa demonstrated antiobesity effect (Foucault et al., 2014). Graf et al. (2014) showed that quinoa seed leachate containing a significant amount of 20-hydroxyecdysone largely decreased the level of fasting blood glucose in hyperglycemic and obese mice. The leachate showed antidiabetic potential (Graf et al., 2014). On the other hand, the levels of phytoecdysteroids and betalains can be very low in whole quinoa seeds, depending on the genotypes and processing conditions. Therefore the biological significance of these components in quinoa remains to be studied. Novel quinoa varieties with increased levels of these minor bioactive components may be developed through breeding program and genetic means.

Alkylresorcinols were found in quinoa seeds (Ross et al., 2017). A total of 40 different types of alkylresorcinols in quinoa included even- and odd- alkyl chain alkylresorcinols, methylalkylresorcinols and branched-chain alkylresorcinols. The composition of the alkylresorcinols was related to the geographic origins of the quinoa samples. Alkylresorcinols were found in the plasma of humans after quinoa consumption. These alkylresorcinols have the potential as biomarkers to indicate quinoa intake in humans (Ross et al., 2017). Previous studies also showed that dietary alkylresorcinols have biological functions such as reducing the risks of cancer development in humans (Kruk et al., 2017).

8.3.4 General discussion on biological activities of quinoa minor components

The above-mentioned biological activities of quinoa seed extracts could be attributed to the components present in these extracts. A major type of the components is polyphenols (Abderrahim et al., 2015; Antognoni et al., 2021; Liu et al., 2020; Paśko et al., 2019; Stikić et al., 2020; Tang et al., 2015). Quinoa seeds contain a range of different polyphenols and their conjugates as described in the Section 8.2 above. Many studies have shown that polyphenols and their conjugates from different types of plant sources have different biological functions and health effects (Boccellino & D'Angelo, 2020; Cirillo et al., 2016; Koch, 2019; Wan et al., 2021). Polyphenols may have potential in the prevention of different chronic diseases such as cardio-vascular diseases, cancers, obesity, diabetes, and neurodegenerative diseases via different mechanisms such as being antiinflammatory and modulation of gut microbiota (Koch, 2019; Wan et al., 2021). Phenolics may positively influence the gut microbiome of humans through interactions with other food components especially dietary fibers through covalent and noncovalent bonds (Rocchetti et al., 2022). Thus it is naturally expected that quinoa extracts rich in polyphenols tend to have a range of biological functions as described in this section. Similar observations could be obtained on other minor components of quinoa including saponins, phytoecdysteroids, betalains, and alkylresorcinols. Many studies have shown that these components have a range of health benefits, though they have been understudied in the case of quinoa. Lack of research in these biological aspects of quinoa suggests future research opportunities. A major criticism of the reports published on quinoa minor components and biological functions is related to dose effect. The results of the reports may have no meaning in relation to human intake of minor components. The amount of in-taking these minor components when eating quinoa products may be too low to have a health effect. Unfortunately, our research community has continued to ignore such an important and difficult question.

8.4 Effect of processing on chemical composition and biological activities of polyphenols and saponins in quinoa

A range of different types of processing methods including both thermal and non-thermal treatments on the chemical composition and biological functions of quinoa products containing phenolics and saponins have been studied (Table 8.3).

8.4.1 Drying

Quinoa sprouts were subjected to drying at different temperatures of up to 60°C and simulated gastrointestinal digestion (Zhang et al., 2020; Złotek et al., 2019). The effect of drying on phenolics of quinoa sprouts depended on the quinoa variety and the drying conditions (Złotek et al., 2019). The sprout of red quinoa seeds had more phenolics than that of white quinoa seeds. The phenolics in the spout of red quinoa seeds were more stable under the drying conditions than those of white quinoa seeds. The phenolic contents of the sprouts and leaves decreased at 60°C. Drying at 30°C had little effects on the phenolics of both quinoa sprouts and leaves (Złotek et al., 2019). Germination and drying at 30°C improved the in vitro antioxidant activity of quinoa seeds (Złotek et al., 2019). Therefore to maximize the nutritional capacity and phenolic content of sprouts, suitable drying temperatures and quinoa varieties should be chosen.

8.4.2 Simulated gastrointestinal digestion

In vitro simulated gastrointestinal digestion (oral, gastric, and intestinal phases) of quinoa sprout, seeds, and flakes increased the percentages of bioavailable phenolics for adsorption (Balakrishnan & Schneider, 2020; Pellegrini et al., 2017; Zhang et al., 2020). The recovery of both flavonoids and phenolics was increased after oral and gastric phases, but was decreased after the intestinal phase. Vanillic acid, syringic acid, and caffeic acid were the main phenolics found during the digestion (Zhang et al., 2020). Seven of the 11 identified flavonoids were not structurally affected by the digestion. Flavonoids (mostly kaempferol and quercetin glycosides) were released from the matrix after the digestion (Balakrishnan & Schneider, 2020). The digestion at different phases increased in vitro antioxidant activity and bioaccessibility of the flavonoids of the samples as more phenolics were released from the process (Balakrishnan & Schneider, 2020; Zhang et al., 2020). After the digestion, the percentages of the bioavailable flavonoids and phenolics were 13% and 73%, respectively (Pellegrini et al., 2017). Such data should be related to their significance on gut microbiota, which remains to be studied.

8.4.3 Germination and fermentation

Germination of quinoa seeds for 72 h increased the contents of tocopherols and ascorbic acid, whereas the fermentation process with two types of *Saccharomyces*

Table 8.3 Effect of processing on phenolics and saponins of quinoa grains

Sample types	Processing types	Processing conditions	Major findings			References
			Content and composition	Biological activity	Other properties	
Phenolics related						
Seeds	Boiling, autoclaving, roasting, microwave heating	Boiling for up to 20 min: seed to water ratio at 1:3; roasting for up to 6.5 min at 180°C; microwave heating up to 5 min (900 w, 2450 MHz); autoclaving at 121°C for 15 min	Roasting, microwave heating, and autoclaving increased the total contents of phenolics (by 30%–105%) and flavonoids (by 5%–26%), whereas boiling decreased that of phenolics by 11%	Roasting, microwave heating and autoclaving increased the in vitro antioxidant activities with the microwave heating being the most efficient, whereas boiling decreased that	The thermal treatments increased the water absorption capacity by up to 60%. Roasting and microwave heating increased the swelling power by up to 22%, whereas boiling reduced that by 3%. Flour dispersibility and emulsification capacities were decreased by the thermal treatments	Sharma et al. (2022)
Sprout and leaves	Drying	Drying temperatures of 30°C, 45°C, and 60°C	The phenolics in the spouts of red quinoa seeds were more stable under the drying conditions than those of white quinoa seeds. The phenolic content of the sprouts and leaves decreased at 60°C. Drying at 30°C had little effects on the phenolics of both quinoa sprouts and leaves	Germination and drying at 30°C improved the in vitro antioxidant activity of quinoa seeds		Złotek et al. (2019)

(*Continued*)

Table 8.3 (Continued)

Sample types	Processing types	Processing conditions	Major findings			References
			Content and composition	Biological activity	Other properties	
Sprout	In vitro simulated gastrointestinal digestion	Oral, gastric, and intestinal phases	The digestion at different phases increased the percentages of bioavailable phenolics for adsorption. The recovery of both flavonoids and phenolics were increased after oral and gastric phases, but was decreased after the intestinal phase. Vanillic acid, syringic acid, and caffeic acid were the main phenolics found during the digestion	The digestion at different phases increased in vitro antioxidant activity of the samples		Zhang et al. (2020)
Seeds, flakes, sprouts	In vitro gastrointestinal digestion	Oral, gastric, and intestinal phases	Seven of the eleven identified flavonoids were not structurally affected by the digestion. Flavonoids (mostly kaempferol and quercetin glycosides) were released from the matrix after the digestion	The digestion increased the in vitro antioxidant activity and bioaccessibility of the flavonoids in the samples		Balakrishnan and Schneider (2020)
Seed	In vitro simulated gastrointestinal digestion	Oral, gastric, and intestinal phases	The recovery indexes of the contents of total phenolics and flavonoids increased after the digestion of the three phases. After the digestion, the percentages of the bioavailable flavonoids and phenolics were 13% and 73%, respectively	The in vitro antioxidant activity increased during the digestion process due to the release of the phenolics		Pellegrini et al. (2017)

Seed	Fermentation	Different fermentation conditions in different studies	Fermentation in general can increase phenolic content. Fermentation conditions and quinoa varieties should be suitable	Fermentation in general can increase the antioxidant activity. Fermentation conditions and quinoa varieties should be suitable		Melini and Melini (2021a)
Seed	Germination, fermentation	72 h of germination; fermentation with two types of *Saccharomyces cerevisiae*	Germination of 72 h increased the contents of tocopherols and ascorbic acid, whereas the fermentation process reduced them. Both of the germination and fermentation increased the in vitro antioxidant activity and phenolic contents of quinoa seeds. The germination was more efficient for the increases			Carciochi et al. (2016)
Seed	Fermentation, roasting and boiling	Lactic acid bacteria (*Pediococcus pentosaceus* GS · B and *Lactobacillus paracasei* A1 2.6) were used for the fermentation (30°C, 48 h)	The fermentation increased the phenolic content of quinoa seeds. The fermentation, roasting, and boiling released tyrosols and phenolic acids. Overall, flavonoids were more affected by the processing operations than the other phenolics	The fermentation as well as roasting and boiling increased the in vitro antioxidant activity of quinoa seeds		Rocchetti et al. (2019)

(*Continued*)

Table 8.3 (Continued)

Sample types	Processing types	Processing conditions	Major findings			References
			Content and composition	Biological activity	Other properties	
Seed	Soaking, germination	Soaking for 24 h, germination at 20°C for 72 h	Soaking and germination increased the phenolic content of the seeds	Soaking and germination increased the in vitro antioxidant activity of the seeds		Kaur et al. (2016)
Seed	Malting	100% of relative humidity, 48 h, 25°C, 45%–55% moisture content of the grains	Malting increased the contents of total phenolics and flavonoids. The extents of the increase were dependent on the quinoa variety	Malting increased the in vitro antioxidant activity. The extents of the increase were dependent on the quinoa variety		Aguilar et al. (2019)
Seed	Milling	Bran and hull fractions were obtained	Milling of whole grains reduced the phenolic content by 30%. In different fractions, the major phenolic acids were ferulic and vanillic acids, whereas quercetin and rutin were the main flavonoids	The fractions (extract of 80% methanol) showed a range of in vitro antioxidant activities (DPPH, H_2O_2 scavenging, FRAP, ferrous ion chelating). The metal chelating activity of dehulled and milled fraction was higher than that of bran and hull fractions. The extracts of hull and bran fractions had stronger inhibitory activities against α-amylase and α-glucosidase		Hemalatha et al. (2016)
Seed	Ultrasound	Flour was treated with ultrasound (20 kHz, 250 W, up to 19 h)	Ultrasound treatment decreased the total phenolics content of the flour in a treatment length dependent manner	Ultrasound treatment decreased the in vitro antioxidant activity of the flour in a treatment length-dependent manner		Zhu and Li (2019a)

Seed	High hydrostatic pressure	Flour was treated with hydrostatic pressure of up to 600 MPa	The pressure treatment had little effect on the total phenolic content of the three types of flours			Zhu and Li (2019b)
Seed	Electron beam irradiation	Irradiation at up to 8 kGy	The irradiation at 8 kGy increased the content of free phenolics while decreasing that of bound phenolics			Luo et al. (2021)
Seed	Washing, washing and hydration, toasting, cooking with or without pressure	Toasting at 180°C for 15 min, pressure cooker with a water to sample ratio of 3:1, washing and hydration with a water to sample ratio of 3:1 at 60°C for 30 min	Cooking under pressure retained the highest amount of phenolics, whereas toasting led to the highest amount of loss of the phenolics. The content of saponins was decreased by washing and was not affected by toasting and cooking	Cooking under pressure retained the highest in vitro antioxidant activity, whereas toasting led to the highest loss of the in vitro antioxidant activity		Nickel et al. (2016)
Seed	Germination, normal steaming, high-pressure steaming, baking	Germination at 25°C for 1 day; normal steaming: soaked seeds cooked at 100°C for 20 min; pressure steaming, steamed at 110°C and for 10 min under 50 kPa; baking at 145°C for 30 min	Germination increased the phenolic content. The normal and pressure steaming decreased the phenolic content by disrupting the grain structure of quinoa (Fig. 8.8). Baking had little effect on the phenolics and tocopherols	Germination increased the in vitro antioxidant activity. The normal and pressure steaming decreased the antioxidant activity. Under in vitro simulated digestion, both the steaming types increased the bioaccessibility of tocopherols, whereas germination increased that of ferulic acid		Gu et al. (2021)
Seed	Fermentation with *Monascus anka*	Quinoa seeds were treated at 121°C (20 min) before fermentation (culture at 5%, v/w)	Ferulic acid, *p*-hydroxybenzoic acid and protocatechuic acid were the major phenolics in the fermented samples.	The fermentation increased the absorption of free phenolics in the intestine, in vitro antioxidant activity (especially that of		Yang et al. (2022)

(*Continued*)

Table 8.3 (Continued)

Sample types	Processing types	Processing conditions	Major findings			References
			Content and composition	Biological activity	Other properties	
		at 80% (relative humidity) and 32°C for 10 days	*M. anka*-based fermentation for 6 days gave the seeds the highest total phenolic content.	the free phenolics), and inhibition activities against α-glucosidase and α-amylase. The free phenolics were largely responsible for the observed activities		
Saponins related						
Seed	Ultrasound-assisted extraction	Sonication amplitude of 60%, water, ethanol, or their mixtures (1:1)	The extraction process using ethanol gave the highest saponin yield. Quinoa had less yield of saponins than red lentil and fenugreek and had higher yield than soybean and lupin	Saponins in the extract obtained using water and ethanol mixture had a bioaccessibility of ~100%. The presence of phenolics and lipids decreased and increased the bioaccessibility of saponins, respectively		del Hierro et al. (2018)
Seed	Milling	Methanol for saponin extraction. 80% acetone for extraction of free phenolics and NaOH solution for bound phenolics	Increasing degree of milling from 0% to 27% decreased the total phenolic and flavonoid contents by 32% and 41%, respectively. Increasing degree of milling from 0% to 27% decreased the total saponin content by 42%	Increasing degree of milling decreased the in vitro antioxidant activity		Han et al. (2019)
Seed	Fermentation	35°C, 4 days, *Lactobacillus plantarum* 1	Fermentation reduced the saponin content of quinoa seeds		The fermentation process affected the pasting properties of the flour	Bolívar-Monsalve et al. (2018)

Seed	Sprouting	At 22°C up to 72 h before drying at 55°C for 6 h	Sprouting decreased the amount of saponins. The changes occurred mostly within 48 h and the saponin content reached 0.086 mg/100 g		The astringency and bitterness of quinoa decreased by the sprouting largely due to the decreased content of saponins. The changes of phenolics had little effect on the taste of quinoa during sprouting	Suárez-Estrella et al. (2021)
Seed	Electron beam irradiation	Irradiation up to 8 kGy			The irradiation had no effect on saponin content	Luo et al. (2021)
Seed	Boiling, autoclaving, roasting, microwave heating	Boiling for up to 20 min: seed to water ratio at 1:3; roasting for up to 6.5 min at 180°C; microwave heating up to 5 min (900 w, 2450 MHz); autoclaving at 121°C for 15 min	Autoclaving and boiling reduced the content of saponins by up to 64%, whereas microwave heating and roasting enhanced that by up to 25%			Sharma et al. (2022)
Quinoa husk	Microwave-assisted hydrolysis of saponins	Hydrolysis was under alkaline conditions, 150°C, 5 min for microwave	Monodesmosidic saponins were produced from bidesmosidic saponins under microwave-assisted hydrolysis	The membranolytic activity of saponins increased drastically after the hydrolysis. The membranolytic activity related to their actions against numerous organisms such as fungi, bacteria, and cancer cells		Colson et al. (2020)

cerevisiae reduced them (Carciochi et al., 2016). Both of the germination and fermentation increased the in vitro antioxidant activity and phenolic contents of quinoa seeds. The germination was more efficient to obtain the increases than the fermentation (Carciochi et al., 2016). Similarly, malting (100% of relative humidity, 48 h, 25°C, 45%–55% moisture content of the grains) increased the contents of total phenolics and flavonoids. The extents of the increase were dependent on the quinoa variety (Aguilar et al., 2019). Soaking (24 h) and germination (20°C for 72 h) increased the phenolic content of quinoa seeds (Kaur et al., 2016). Soaking and germination increased the in vitro antioxidant activity of the seeds (Kaur et al., 2016). In the study of Rocchetti et al. (2019), lactic acid bacteria (*Pediococcus pentosaceus* GS · B and *Lactobacillus paracasei* A1 2.6) were used for the fermentation (30°C, 48 h) of quinoa seeds. The fermentation increased the phenolic content of quinoa seeds and released tyrosols and phenolic acids. Overall, flavonoids were more affected by the processing operations than the other phenolics. The fermentation increased the in vitro antioxidant activity of quinoa seeds (Rocchetti et al., 2019). In the study of Gu et al. (2021), germination (25°C for 1 day) increased the in vitro antioxidant activity. Alvarez-Jubete et al. (2010) found that the polyphenol content of quinoa seeds (especially kaempferol and quercetin glycosides) was significantly increased during the sprouting. Compared with amaranth and wheat sprouts, and the sprouts of quinoa and buckwheat seeds contained more polyphenols (Alvarez-Jubete et al., 2010). In the study of Yang et al. (2022), quinoa seeds were treated at 121°C (20 min) before fermentation (culture at 5%, v/w) at 80% (relative humidity) and 32°C for 10 days using *Monascus anka*. Ferulic acid, *p*-hydroxybenzoic acid, and protocatechuic acid were the major phenolics in the fermented samples. *M. anka*-based fermentation for 6 days gave the seeds the highest total phenolic content. The fermentation increased the absorption of free phenolics in the intestine, in vitro antioxidant activity (especially that of the free phenolics), and inhibition activities against α-glucosidase and α-amylase. The free phenolics were largely responsible for the observed activities (Yang et al., 2022). Different types of microorganisms can transform phenolics into a range of products through sulfate conjugation, methylation, hydroxylation, carboxylation, and oxidation (Gulsunoglu-Konuskan & Kilic-Akyilmaz, 2022) (Fig. 8.7). The bound phenolics may also be released by microbial enzymes through fermentation. The transformation of quinoa phenolics through different types of fermentation and microorganisms remains to be studied for value added processing.

Fermentation (35°C, 4 days) of quinoa seeds with *Lactobacillus plantarum* 1 reduced the saponin content of quinoa seeds (Bolívar-Monsalve et al., 2018). Biotransformation of saponins by hydrolysis was obtained (del Hierro, Cueva, et al., 2020). The fermentation process also affected the pasting properties of the flour probably due to the effect on starch (Bolívar-Monsalve et al., 2018). Sprouting of quinoa seeds (at 22°C up to 72 h before drying at 55°C) for 6 h decreased the amount of saponins (Suárez-Estrella et al., 2021). The changes occurred mostly within 48 h and the saponin content decreased to 0.086 mg/100 g. The astringency and bitterness of quinoa were decreased by the sprouting, largely due to the decreased content of saponins. The changes of phenolics had little effect on the taste of quinoa during sprouting (Suárez-Estrella et al., 2021).

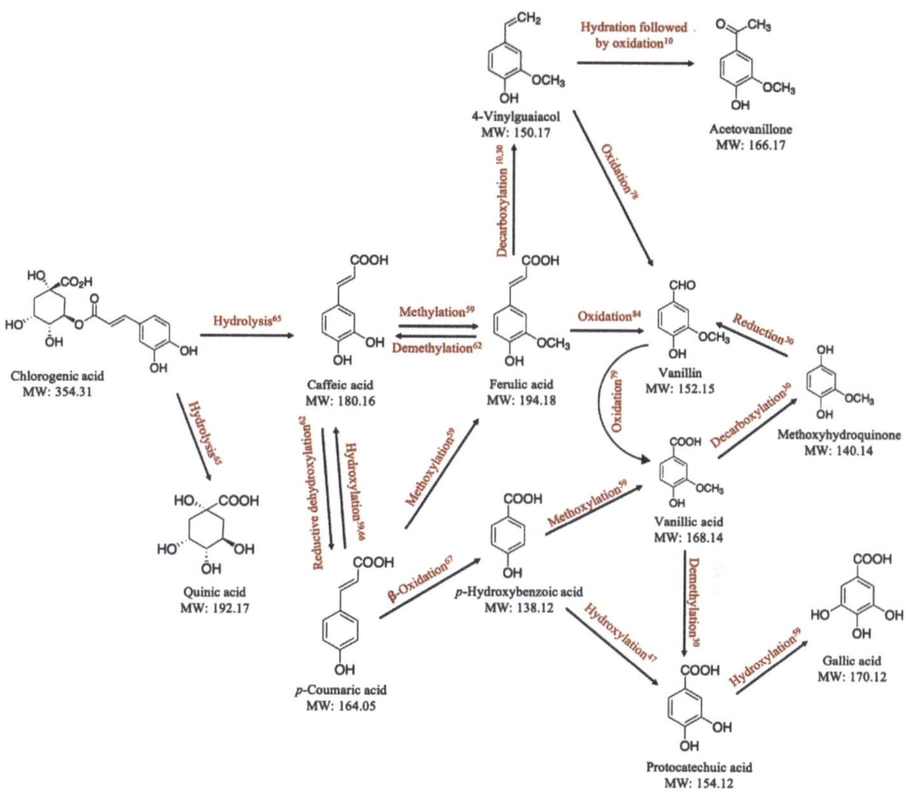

Figure 8.7 Schematic representation of bioconversions of phenolic acids (Gulsunoglu-Konuskan & Kilic-Akyilmaz, 2022).
Source: Reprinted with permission from American Chemical Society.

8.4.4 Home cooking methods

A range of home cooking methods including washing, hydration, microwave heating, toasting, baking, and cooking with or without pressure were used to process quinoa (Gu et al., 2021; Nickel et al., 2016). Among different processing methods, cooking with pressure affected the phenolic and saponin composition and the antioxidant activities to greater extents (Nickel et al., 2016). Cooking under pressure retained the highest amount of phenolics, whereas toasting led to the highest loss of the phenolics. Cooking under pressure retained the highest in vitro antioxidant activity, whereas toasting led to the highest loss of the in vitro antioxidant activity (Nickel et al., 2016). Baking of quinoa seeds had little effect on the phenolics and tocopherols (Gu et al., 2021). The normal and pressure steaming of the seeds decreased the antioxidant activity. Under in vitro simulated digestion, both the steaming types increased the bioaccessibility of tocopherols (Gu et al., 2021). The normal and pressure steaming decreased the phenolic content by disrupting the grain structure of quinoa (Fig. 8.8). Sharma et al. (2022) compared the effects of boiling, microwave heating, autoclaving, and roasting on the

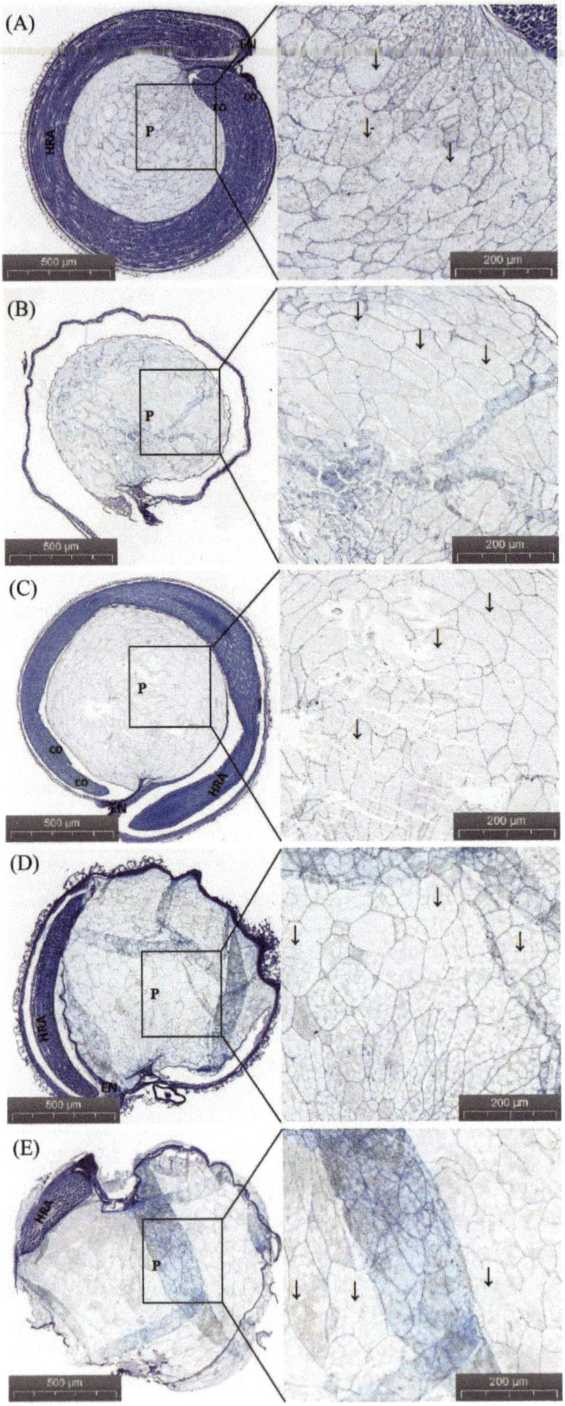

Figure 8.8 Light microscopy of a median longitudinal section of a seed of raw quinoa (A), germinating quinoa (B), quinoa after baking (C), quinoa after normal pressure steaming (NPS) (D), and quinoa after high-pressure steaming (HPS) (E); germination (Ger), perisperm (P), endosperm (EN), hypocotyl-radicle axis (HRA), and cotyledon (co) (Gu et al., 2021). *Source*: Reprinted with permission from Elsevier.

contents of phenolics and saponins of quinoa seeds. Roasting, microwave heating, and autoclaving increased the total contents of phenolics (by 30%−105%) and flavonoids (by 5%−26%), whereas boiling decreased that of phenolics by 11%. Autoclaving and boiling reduced the content of saponins by up to 64%, whereas microwave heating and roasting enhanced that by up to 25% (Sharma et al., 2022). The increased contents of the phenolics were attributed to the release of bound phenolics from the treatments. Another study showed that the content of saponins in quinoa seeds was decreased by washing and was not affected by toasting and cooking (Nickel et al., 2016). The differences may be due to the different experimental conditions used in different studies. Roasting, microwave heating, and autoclaving increased the in vitro antioxidant activities with the microwave heating being the most efficient, whereas boiling decreased that. Apart from the release of bound phenolics, the formation of Maillard reaction products also contributed to the increased antioxidant activities (Sharma et al., 2022). The thermal treatments increased the water absorption capacity by up to 60%. Roasting and microwave heating increased the swelling power by up to 22%, whereas boiling reduced that by 3%. Flour dispersibility and emulsification capacities were decreased by the thermal treatments due to the denaturation and gelatinization of components (Sharma et al., 2022).

8.4.5 Milling

The effect of milling on properties of quinoa seed fractions is described in Chapter 3. The milling had an impact on composition and in vitro antioxidant activities of quinoa flour fractions including bran and hull fractions (Hemalatha et al., 2016). Milling of whole grain reduced the phenolic content by 30%. In different fractions, the major phenolic acids were ferulic and vanillic acids, whereas quercetin and rutin were the main flavonoids. The fractions (in the form of extracts of 80% methanol) showed a range of in vitro antioxidant activities (DPPH, H_2O_2 scavenging, FRAP, ferrous ion chelating). The metal chelating activity of dehulled and milled fraction was higher than that of bran and hull fractions. The extracts of hull and bran fractions had stronger inhibitory activities against α-amylase and α-glucosidase (Hemalatha et al., 2016). Han, Chi, Zhang, Zhang, Fan, Dong, et al. (2019) and Han, Chi, Zhang, Zhang, Fan, Huang, et al. (2019) showed that increasing degree of milling from 0% to 27% decreased the total phenolic and flavonoid contents by 32% and 41%, respectively. Increasing degree of milling from 0% to 27% decreased the total saponin content by 42%. Increasing degree of milling decreased the in vitro antioxidant activity due to the decreased contents of the bioactive compounds (Han, Chi, Zhang, Zhang, Fan, Dong, et al., 2019; Han, Chi, Zhang, Zhang, Fan, Huang, et al., 2019). Milling could be used to obtain fractions with desired nutritional properties.

8.4.6 Ultrasound, high hydrostatic pressure, and electron beam irradiation treatments

Novel nonthermal processing methods including ultrasound, high hydrostatic pressure, and electron beam irradiation were used to treat quinoa flour to obtain a range of functional properties (Luo et al., 2021; Zhu & Li, 2019a,b). Three types of quinoa flours were treated with ultrasound processing (20 kHz, 250 W, up to 19 h). The ultrasound treatment decreased the total phenolic contents of the flours in a treatment length-dependent manner. The treatment decreased the in vitro antioxidant activity of the flours in a treatment length-dependent manner. The extents of decreases in the phenolic contents and antioxidant activities depended also on the type of quinoa flours (Zhu & Li, 2019a). In another study, quinoa flour was treated with high hydrostatic pressure of up to 600 MPa. The pressure treatment had little effect on the total phenolic contents of the three types of quinoa flours (Zhu & Li, 2019b). Luo et al. (2021) studied the effects of electron beam irradiation at up to 8 kGy on the quinoa phenolics. The irradiation at 8 kGy increased the content of free phenolics, while decreasing that of bound phenolics (Luo et al., 2021). It would be interesting to study the effect of irradiation (e.g., γ-ray) at higher doses (e.g., 100 kGy) on the phenolic composition of quinoa flours. Electron beam irradiation up to 8 kGy had no effect on saponin content in the quinoa samples (Luo et al., 2021).

8.4.7 Ultrasound-assisted extraction and microwave-assisted hydrolysis

Ultrasound was used to assist the extraction of saponins from quinoa seeds with different solvents including water, ethanol, or their mixtures (1:1) (del Hierro et al., 2018). The extraction process using ethanol gave the highest saponin yield. Quinoa had lower yield of saponins than red lentil and fenugreek and had higher yield than soybean and lupin. Saponins in the extracts obtained using ethanol mixture had an in vitro bioaccessibility of ~100%. The presence of phenolics and lipids decreased and increased the bioaccessibility of saponins, respectively (del Hierro et al., 2018). Phenolics may form less soluble complexes with saponins, reducing the bioavailability. Lipids may contribute to the formation of micelles, which may improve the adsorption and bioavailability of saponins. These hypotheses remain to be studied.

Quinoa husk was subjected to microwave (150°C, 5 min)-assisted hydrolysis of saponins under alkaline conditions (Colson et al., 2020). Monodesmosidic saponins were produced from bidesmosidic saponins. Membranolytic activity relates to their actions against numerous organisms such as fungi, bacteria, and cancer cells. The membranolytic activity of saponins increased drastically after the hydrolysis due to the structural changes, which favored the bacteria-saponin binding (Colson et al., 2020).

8.5 Conclusions

Different parts of quinoa plant including seeds, leaves, husk, and roots contain a range of different structural types of polyphenols, saponins, phytoecdysteroids, betalains, and alkylresorcinols. Most of the studies have been focused on the composition of these bioactive components in quinoa seeds. The major free and bound polyphenols of quinoa seeds included phenolic acids such as ferulic acid and flavonoids such as kaempferol and quercetin and their glycosides. Quinoa contains different types of saponins with their sapogenins being hederagenin, phytolaccagenic acid, and oleanolic acid. Quinoa roots contain a significant amount of saponins. Phytoecdysteroids (e.g., 20-hydroxyecdysone) and betalains were also found in quinoa. These components contributed to a range of biological functions of quinoa. These functions included antioxidant, antiobesity, antihypoglyemia, anticancer, antiinflammatory, antimicrobial, immunoadjuvant activity, gut microbiota regulation, and hypocholesterolemia properties. A range of different types of processing methods changed the composition of these bioactive components and their biological functions to various degrees, depending on the processing conditions and quinoa genotypes. Quinoa genotypes of increased amounts of specific bioactive compounds for targeted nutritional applications should be developed. Fractionation and extraction processes to isolate these bioactive compounds should contribute to the value addition of quinoa. There has been a lack of confirmative studies of the published reports using clinical trials. The data published on the composition and biological functions of quinoa bioactive components remain to be made meaningful to help the rationalization of diet for nutritional applications and disease prevention.

References

Abderrahim, F., Huanatico, E., Segura, R., Arribas, S., Gonzalez, M. C., & Condezo-Hoyos, L. (2015). Physical features, phenolic compounds, betalains and total antioxidant capacity of coloured quinoa seeds (*Chenopodium quinoa* Willd.) from Peruvian Altiplano. *Food Chemistry*, *183*, 83–90.

Aguilar, J., Miano, A. C., Obregón, J., Soriano-Colchado, J., & Barraza-Jáuregui, G. (2019). Malting process as an alternative to obtain high nutritional quality quinoa flour. *Journal of Cereal Science*, *90*, 102858.

Ahmed, I. A. M., Al Juhaimi, F., & Özcan, M. M. (2021). Insights into the nutritional value and bioactive properties of quinoa (*Chenopodium quinoa*): Past, present and future prospective. *International Journal of Food Science & Technology*, *56*, 3726–3741.

Al-Qabba, M. M., El-Mowafy, M. A., Althwab, S. A., Alfheeaid, H. A., Aljutaily, T., & Barakat, H. (2020). Phenolic profile, antioxidant activity, and ameliorating efficacy of *Chenopodium quinoa* sprouts against CCl_4-induced oxidative stress in rats. *Nutrients*, *12*, 2904.

Alvarez-Jubete, L., Wijngaard, H., Arendt, E. K., & Gallagher, E. (2010). Polyphenol composition and in vitro antioxidant activity of amaranth, quinoa buckwheat and wheat as affected by sprouting and baking. *Food Chemistry*, *119*, 770–778.

Antognoni, F., Potente, G., Biondi, S., Mandrioli, R., Marincich, L., & Ruiz, K. B. (2021). Free and conjugated phenolic profiles and antioxidant activity in quinoa seeds and their relationship with genotype and environment. *Plants*, *10*, 1046.

Balakrishnan, G., & Schneider, R. G. (2020). Quinoa flavonoids and their bioaccessibility during in vitro gastrointestinal digestion. *Journal of Cereal Science*, *95*, 103070.

Boccellino, M., & D'Angelo, S. (2020). Anti-obesity effects of polyphenol intake: Current status and future possibilities. *International Journal of Molecular Sciences*, *21*, 5642.

Bolívar-Monsalve, J., Ceballos-González, C., Ramírez-Toro, C., & Bolívar, G. A. (2018). Reduction in saponin content and production of gluten-free cream soup base using quinoa fermented with *Lactobacillus plantarum*. *Journal of Food Processing and Preservation*, *42*, e13495.

Buitrago, D., Buitrago-Villanueva, I., Barbosa-Cornelio, R., & Coy-Barrera, E. (2019). Comparative examination of antioxidant capacity and fingerprinting of unfractionated extracts from different plant parts of quinoa (*Chenopodium quinoa*) grown under greenhouse conditions. *Antioxidants*, *8*, 238.

Cabanillas, B., Espichán, F., Estrada, R., Neyra, E., & Rojas, R. (2021). Metabolomic profile and discrimination of white quinoa seeds from Peru based on UHPLC-HRMS and multivariate analysis. *Journal of Cereal Science*, *101*, 103307.

Carciochi, R. A., Galván-D'Alessandro, L., Vandendriessche, P., & Chollet, S. (2016). Effect of germination and fermentation process on the antioxidant compounds of quinoa seeds. *Plant Foods for Human Nutrition*, *71*, 361–367.

Chen, J., Zhang, Y., Guan, X., Cao, H., Li, L., Yu, J., & Liu, H. (2022). Characterization of saponins from differently colored quinoa cultivars and their in vitro gastrointestinal digestion and fermentation properties. *Journal of Agricultural and Food Chemistry*, *70*, 1810–1818.

Cirillo, G., Curcio, M., Vittorio, O., Iemma, F., Restuccia, D., Spizzirri, U. G., et al. (2016). Polyphenol conjugates and human health: A perspective review. *Critical Reviews in Food Science and Nutrition*, *56*, 326–337.

Colson, E., Savarino, P., Claereboudt, E. J. S., Cabrera-Barjas, G., Deleu, M., Lins, L., et al. (2020). Enhancing the membranolytic activity of *Chenopodium quinoa* saponins by fast microwave hydrolysis. *Molecules*, *25*, 1731.

Das, N., Mishra, S. K., Bishayee, A., Ali, E. S., & Bishayee, A. (2021). The phytochemical, biological, and medicinal attributes of phytoecdysteroids: An updated review. *Acta Pharmaceutica Sinica B*, *11*, 1740–1766.

de Cedrón, M. G., del Hierro, J. N., Reguero, M., Wagner, S., Bouzas, A., Quijada-Freire, A., Reglero, G., Martín, D., & de Molina, A. R. (2020). Saponin-rich extracts and their acid hydrolysates differentially target colorectal cancer metabolism in the frame of precision nutrition. *Cancers*, *12*, 3399.

Del Hierro, J. N., Casado-Hidalgo, G., Reglero, G., & Martin, D. (2021). The hydrolysis of saponin-rich extracts from fenugreek and quinoa improves their pancreatic lipase inhibitory activity and hypocholesterolemic effect. *Food Chemistry*, *338*, 128113.

del Hierro, J. N., Cueva, C., Tamargo, A., Núñez-Gómez, E., Moreno-Arribas, M. V., Reglero, G., & Martin, D. (2020). *In vitro* colonic fermentation of saponin-rich extracts from quinoa, lentil, and fenugreek. Effect on sapogenins yield and human gut microbiota. *Journal of Agricultural and Food Chemistry*, *68*, 106–116.

del Hierro, J. N., Piazzini, V., Reglero, G., Martin, D., & Bergonzi, M. C. (2020). *In vitro* permeability of saponins and sapogenins from seed extracts by the parallel artificial membrane permeability assay: Effect of in vitro gastrointestinal digestion. *Journal of Agricultural and Food Chemistry*, *68*, 1297–1305.

del Hierro, J. N., Herrera, T., García-Risco, M. R., Fornari, T., Reglero, G., & Martin, D. (2018). Ultrasound-assisted extraction and bioaccessibility of saponins from edible seeds: Quinoa, lentil, fenugreek, soybean and lupin. *Food Research International*, *109*, 440–447.

Di Lorenzo, C., Colombo, F., Biella, S., Stockley, C., & Restani, P. (2021). Polyphenols and human health: The role of bioavailability. *Nutrients*, *13*, 273.

Dong, S., Yang, X., Zhao, L., Zhang, F., Hou, Z., & Xue, P. (2020). Antibacterial activity and mechanism of action saponins from *Chenopodium quinoa* Willd. husks against food-borne pathogenic bacteria. *Industrial Crops and Products, 149*, 112350.

El Hazzam, K., Hafsa, J., Sobeh, M., Mhada, M., Taourirte, M., EL Kacimi, K., & Yasri, A. (2020). An insight into saponins from quinoa (*Chenopodium quinoa* Willd): A review. *Molecules, 25*, 1059.

Escribano, J., Cabanes, J., Jiménez-Atiénzar, M., Ibañez-Tremolada, M., Gómez-Pando, L. R., et al. (2017). Characterization of betalains, saponins and antioxidant power in differently colored quinoa (*Chenopodium quinoa*) varieties. *Food Chemistry, 234*, 285–294.

Fan, Y., & Pedersen, O. (2021). Gut microbiota in human metabolic health and disease. *Nature Reviews Microbiology, 19*, 55–71.

Fiallos-Jurado, J., Pollier, J., Moses, T., Arendt, P., Barriga-Medina, N., et al. (2016). Saponin determination, expression analysis and functional characterization of saponin biosynthetic genes in *Chenopodium quinoa* leaves. *Plant Science, 250*, 188–197.

Foucault, A. S., Even, P., Lafont, R., Dioh, W., Veillet, S., Tomé, D., Huneau, J. F., Hermier, D., & Quignard-Boulangé, A. (2014). Quinoa extract enriched in 20-hydroxyecdysone affects energy homeostasis and intestinal fat absorption in mice fed a high-fat diet. *Physiology & Behavior, 128*, 226–231.

Gianna, V., Montes, J. M., Calandri, E. L., & Guzmán, C. (2012). Impact of several variables on the microwave extraction of *Chenopodium quinoa* willd saponins. *International Journal of Food Science & Technology, 47*, 1593–1597.

Gil-Ramirez, A., Salas-Veizaga, D. M., Grey, C., Karlsson, E. N., Rodriguez-Meizoso, I., & Linares-Pastén, J. A. (2018). Integrated process for sequential extraction of saponins, xylan and cellulose from quinoa stalks (*Chenopodium quinoa* Willd.). *Industrial Crops & Products, 121*, 54–65.

Góral, I., Stochmal, A., & Wojciechowski, K. (2021). Surface activity of the oat, horse chestnut, cowherb, soybean, quinoa and soapwort extracts – Is it only due to saponins? *Colloid and Interface Science Communications, 42*, 100400.

Graf, B. L., Poulev, A., Kuhn, P., Grace, M. H., Lila, M. A., & Raskin, I. (2014). Quinoa seeds leach phytoecdysteroids and other compounds with anti-diabetic properties. *Food Chemistry, 163*, 178–185.

Graf, B. L., Rojo, L. E., Delatorre-Herrera, J., Poulev, A., Calfio, C., & Raskin, I. (2016). Phytoecdysteroids and flavonoid glycosides among Chilean and commercial sources of *Chenopodium quinoa*: Variation and correlation to physico-chemical characteristics. *Journal of the Science of Food and Agriculture, 96*, 633–643.

Gu, R., Chang, X., Bai, G., Li, X., Di, Y., Liu, X., Sun, L., & Wang, Y. (2021). Effects of household cooking methods on changes of tissue structure, phenolic antioxidant capacity and active component bioaccessibility of quinoa. *Food Chemistry, 350*, 129138.

Gulsunoglu-Konuskan, Z., & Kilic-Akyilmaz, M. (2022). Microbial bioconversion of phenolic compounds in agro-industrial wastes: A review of mechanisms and effective factors. *Journal of Agricultural and Food Chemistry, 70*, 6901–6910.

Han, Y., Chi, J., Zhang, M., Zhang, R., Fan, S., Dong, L., Huang, F., & Liu, L. (2019). Changes in saponins, phenolics and antioxidant activity of quinoa (*Chenopodium quinoa* willd) during milling process. *LWT-Food Science and Technology, 114*, 108381.

Han, Y., Chi, J., Zhang, M., Zhang, R., Fan, S., Huang, F., Xue, K., & Liu, L. (2019). Characterization of saponins and phenolic compounds: Antioxidant activity and inhibitory effects on α-glucosidase in different varieties of colored quinoa (*Chenopodium quinoa* Willd). *Bioscience, Biotechnology, and Biochemistry, 83*, 2128–2139.

Hemalatha, P., Bomzan, D. P., Rao, B. V. S., & Sreerama, Y. N. (2016). Distribution of phenolic antioxidants in whole and milled fractions of quinoa and their inhibitory effects on α-amylase and α-glucosidase activities. *Food Chemistry*, *199*, 330–338.

Herrera, T., del Hierro, J. N., Fornari, T., Reglero, G., & Martin, D. (2019). Inhibitory effect of quinoa and fenugreek extracts on pancreatic lipase and α-amylase under in vitro traditional conditions or intestinal simulated conditions. *Food Chemistry*, *270*, 509–517.

Jiang, X., Cao, Y., Jørgensen, L. V. G., Strobel, B. W., Hansen, H. C., & Cedergreen, N. (2018). Where does the toxicity come from in saponin extract? *Chemosphere*, *204*, 243–250.

Jurek, I., Góral, I., Gęsiński, K., & Wojciechowski, K. (2019). Effect of saponins from quinoa on a skin-mimetic lipid monolayer containing cholesterol. *Steroids*, *147*, 52–57.

Kaur, I., Tanwar, B., Reddy, M., & Chauhan, A. (2016). Vitamin C, total polyphenols and antioxidant activity in raw, domestically processed and industrially processed Indian *Chenopodium quinoa* seeds. *Journal of Applied Pharmaceutical Science*, *6*, 139–145.

Koch, W. (2019). Dietary polyphenols—Important non-nutrients in the prevention of chronic noncommunicable diseases. A systematic review. *Nutrients*, *11*, 1039.

Kruk, J., Aboul-Enein, B., Bernstein, J., & Marchlewicz, M. (2017). Dietary alkylresorcinols and cancer prevention: A systematic review. *European Food Research and Technology*, *243*, 1693–1710.

Laqui-Vilca, C., Aguilar-Tuesta, S., Mamani-Navarro, W., Montaño-Bustamante, J., & Condezo-Hoyos, L. (2018). Ultrasound-assisted optimal extraction and thermal stability of betalains from colored quinoa (*Chenopodium quinoa* Willd) hulls. *Industrial Crops & Products*, *111*, 606–614.

Lazo-Vélez, M. A., Guajardo-Flores, D., Mata-Ramírez, D., Gutiérrez-Uribe, J. A., & Serna-Saldivar, S. O. (2016). Characterization and quantitation of triterpenoid saponins in raw and sprouted *Chenopodium berlandieri* spp. (huauzontle) grains subjected to germination with or without selenium stress conditions. *Journal of Food Science*, *81*, C19–C26.

Lee, M. J., & Sim, K. H. (2018). Nutritional value and the kaempferol and quercetin contents of quinoa (*Chenopodium quinoa* Willd.) from different regions. *Korean Journal of Food Science and Technology*, *50*, 680–687.

Letelier, M. E., Rodríguez-Rojas, C., Sánchez-Jofré, S., & Aracena-Parks, P. (2011). Surfactant and antioxidant properties of an extract from *Chenopodium quinoa* Willd seed coats. *Journal of Cereal Science*, *53*, 239–243.

Lim, J., Park, H.-M., & Yoon, K. S. (2019). Analysis of saponin composition and comparison of the antioxidant activity of various parts of the quinoa plant (*Chenopodium quinoa* Willd.). *Food Science & Nutrition*, *8*, 694–702.

Lin, B., Qi, X., Fang, L., Zhao, L., Zhang, R., Jing, J., Zhang, S., et al. (2021). In vivo acute toxicity and mutagenic analysis of crude saponins from *Chenopodium quinoa* Willd husks. *RSC Advances*, *11*, 4829.

Lin, M., Han, P., Li, Y., Wang, W., Lai, D., & Zhou, L. (2019). Quinoa secondary metabolites and their biological activities or functions. *Molecules*, *24*, 2512.

Lin, T. A., Ke, B. J., Cheng, C. S., Wang, J. J., Wei, B. L., & Lee, C. L. (2019). Red quinoa bran extracts protects against carbon tetrachloride-induced liver injury and fibrosis in mice via activation of antioxidative enzyme systems and blocking TGF-1 pathway. *Nutrients*, *11*, 395.

Liu, M., Zhu, K., Yao, Y., Chen, Y., Guo, H., Ren, G., Yang, X., & Li, J. (2020). Antioxidant, anti-inflammatory, and antitumor activities of phenolic compounds from white, red, and black *Chenopodium quinoa* seed. *Cereal Chemistry*, *97*, 703–713.

Liu, Y., Liu, J., Kong, Z., Huan, X., Li, L., Zhang, P., & Wang, Q. (2022). Transcriptomics and metabolomics analyses of the mechanism of flavonoid synthesis in seeds of differently colored quinoa strains. *Genomics*, *114*, 138–148.

Luo, X., Du, Z., Yang, K., Wang, J., Zhou, J., Liu, J., & Chen, Z. (2021). Effect of electron beam irradiation on phytochemical composition, lipase activity and fatty acid of quinoa. *Journal of Cereal Science*, *98*, 103161.

Marrelli, M., Conforti, F., Araniti, F., & Statti, G. A. (2016). Effects of saponins on lipid metabolism: A review of potential health benefits in the treatment of obesity. *Molecules*, *21*, 1404.

Martinelli, E., Granato, D., Azevedo, L., Gonçalves, J. E., Lorenzoe, J. M., Munekata, P. E. S., et al. (2021). Current perspectives in cell-based approaches towards the definition of the antioxidant activity in food. *Trends in Food Science & Technology*, *116*, 232–243.

Medina-Meza, I. G., Aluwi, N. A., Saunders, S. R., & Ganjyal, G. M. (2016). GC – MS profiling of triterpenoid saponins from 28 quinoa varieties (*Chenopodium quinoa* Willd.) grown in Washington State. *Journal of Agricultural and Food Chemistry*, *64*, 8583–8591.

Melini, F., & Melini, V. (2021a). Impact of fermentation on phenolic compounds and antioxidant capacity of quinoa. *Fermentation*, *7*, 20.

Melini, V., & Melini, F. (2021b). Modelling and optimization of ultrasound-assisted extraction of phenolic compounds from black quinoa by response surface methodology. *Molecules*, *26*, 3616.

Melini, V., & Melini, F. (2021c). Functional components and anti-nutritional factors in gluten-free grains: A focus on quinoa seeds. *Foods*, *10*, 351.

Mokrani, A., & Madani, K. (2016). Effect of solvent, time and temperature on the extraction of phenolic compounds and antioxidant capacity of peach (*Prunus persica* L.) fruit. *Separation and Purification Technology*, *162*, 68–76.

Mroczek, A. (2015). Phytochemistry and bioactivity of triterpene saponins from Amaranthaceae family. *Phytochemistry Reviews*, *14*, 577–605.

Mufari, J. R., Rodríguez-Ruiz, A. C., Bergesse, A. E., Miranda-Villa, P. P., Nepote, V., & Velez, A. R. (2021). Bioactive compounds extraction from malted quinoa using water-ethanol mixtures under subcritical conditions. *LWT-Food Science and Technology*, *138*, 110574.

Nickel, J., Spanier, L. P., Botelho, F. T., Gularte, M. A., & Helbig, E. (2016). Effect of different types of processing on the total phenolic compound content, antioxidant capacity, and saponin content of *Chenopodium quinoa* Willd grains. *Food Chemistry*, *209*, 139–143.

Nowak, V., Du, J., & Charrondière, U. R. (2016). Assessment of the nutritional composition of quinoa (*Chenopodium quinoa* Willd.). *Food Chemistry*, *193*, 47–54.

Palombini, S., Claus, T., Maruyama, S., Gohara, A., Souza, A., Souza, N., Visentainer, J., Gomes, S., & Matsushita, M. (2013). Evaluation of nutritional compounds in new amaranth and quinoa cultivars. *Food Science and Technology*, *33*, 339–344.

Paśko, P., Tyszka-Czochara, M., Namieśnik, J., Jastrzębskic, Z., Leontowicz, H., et al. (2019). Cytotoxic, antioxidant and binding properties of polyphenols from the selected gluten-free pseudocereals and their by-products: In vitro model. *Journal of Cereal Science*, *87*, 325–333.

Pellegrini, M., Lucas-Gonzalez, R., Fernández-López, J., Ricci, A., Pérez-Álvarez, J. A., Sterzo, C. L., & Viuda-Martos, M. (2017). Bioaccessibility of polyphenolic compounds of six quinoa seeds during in vitro gastrointestinal digestion. *Journal of Functional Foods*, *38*, 77–88.

Präger, A., Munz, S., Nkebiwe, P. M., Mast, B., & Graeff-Hönninger, S. (2018). Yield and quality characteristics of different quinoa (*Chenopodium quinoa* Willd.) cultivars grown under field conditions in southwestern Germany. *Agronomy*, *8*, 197.

Rocchetti, G., Gregorio, R. P., Lorenzo, J. M., Barba, F. J., Oliveira, P. G., Prieto, M. A., et al. (2022). Functional implications of bound phenolic compounds and phenolics–food interaction: A review. *Comprehensive Reviews in Food Science and Food Safety*, *21*, 811–842.

Rocchetti, G., Miragoli, F., Zacconi, C., Lucini, L., & Rebecchi, A. (2019). Impact of cooking and fermentation by lactic acid bacteria on phenolic profile of quinoa and buckwheat seeds. *Food Research International*, *119*, 886–894.

Ross, A. B., Svelander, C., Karlsson, G., & Savolainen, O. I. (2017). Identification and quantification of even and odd chained 5-n alkylresorcinols, branched chain-alkylresorcinols and methylalkylresorcinols in quinoa (*Chenopodium quinoa*). *Food Chemistry*, *220*, 344–351.

Ruiza, K. B., Khakimov, B., Engelsen, S. B., Bak, S., Biondi, S., & Jacobsen, S. E. (2017). Quinoa seed coats as an expanding and sustainable source of bioactive compounds: An investigation of genotypic diversity in saponin profiles. *Industrial Crops & Products*, *104*, 156–163.

Sadowska-Bartosz, I., & Bartosz, G. (2021). Biological properties and applications of betalains. *Molecules*, *26*, 2520.

Sampaio, S. L., Fernandes, Â., Pereira, C., Calhelha, R. C., Sokovic, M., Santos-Buelga, C., Barros, L., & Ferreira, I. C. F. R. (2020). Nutritional value, physicochemical characterization and bioactive properties of the Brazilian quinoa BRS Piabiru. *Food & Function*, *11*, 2969.

Segura, R., Vásquez, G., Colson, E., Gerbaux, P., Frischmon, C., Nesic, A., García, D. E., & Cabrera-Barjas, G. (2020). Phytostimulant properties of highly stable silver nanoparticles obtained with saponin extract from *Chenopodium quinoa*. *Journal of the Science of Food and Agriculture*, *100*, 4987–4994.

Sharma, S., Kataria, A., & Singh, B. (2022). Effect of thermal processing on the bioactive compounds, antioxidative, antinutritional and functional characteristics of quinoa (*Chenopodium quinoa*). *LWT-Food Science and Technology*, *160*, 113256.

Stanschewski, C. S., Rey, E., Fiene, G., Craine, E. B., Wellman, G., Melino, V. J., Patiranage, D. S. R., et al.On behalf of the quinoa phenotyping consortium. (2021). Quinoa phenotyping methodologies: An international consensus. *Plants*, *10*, 1759.

Stikić, R. I., Milinčić, D. D., Kostić, A. Ž., Jovanović, Z. B., Gašić, U. M., et al. (2020). Polyphenolic profiles, antioxidant, and in vitro anticancer activities of the seeds of Puno and Titicaca quinoa cultivars. *Cereal Chemistry*, *97*, 626–633.

Suárez-Estrella, D., Borgonovo, G., Buratti, S., Ferranti, P., Accardo, F., Pagani, M. A., & Marti, A. (2021). Sprouting of quinoa (*Chenopodium quinoa* Willd.): Effect on saponin content and relation to the taste and astringency assessed by electronic tongue. *LWT-Food Science and Technology*, *144*, 111234.

Suárez-Estrella, D., Torri, L., Pagani, M. A., & Marti, A. (2018). Quinoa bitterness: Causes and solutions for improving product acceptability. *Journal of the Science of Food and Agriculture*, *98*, 4033–4041.

Sun, X., Yang, X., Xue, P., Zhang, Z., & Ren, G. (2019). Improved antibacterial effects of alkali-transformed saponin from quinoa husks against halitosis-related bacteria. *BMC Complementary and Alternative Medicine*, *19*, 46.

Tang, Y., & Tsao, R. (2017). Phytochemicals in quinoa and amaranth grains and their antioxidant, anti-inflammatory, and potential health beneficial effects: A review. *Molecular Nutrition & Food Research, 61*, 1600767.

Tang, Y., Li, X., Zhang, B., Chen, P. X., Liu, R., & Tsao, R. (2015). Characterisation of phenolics, betanins and antioxidant activities in seeds of three *Chenopodium quinoa* Willd. genotypes. *Food Chemistry, 166*, 380–388.

Tang, Y., Zhang, B., Li, X., Chen, P. X., Zhang, H., Liu, R., & Tsao, R. (2016). Bound phenolics of quinoa seeds released by acid, alkaline, and enzymatic treatments and their antioxidant and α-glucosidase and pancreatic lipase inhibitory effects. *Journal of Agricultural and Food Chemistry, 64*, 1712–1719.

Verza, S. G., Silveira, F., Cibulski, S., Kaiser, S., Ferreira, F., Gosmann, G., Roehe, P. M., & Ortega, G. G. (2012). Immunoadjuvant activity, toxicity assays, and determination by UPLC/Q-TOF-MS of triterpenic saponins from *Chenopodium quinoa* seeds. *Journal of Agricultural and Food Chemistry, 60*, 3113–3118.

Villacrés, E., Quelal, M., Galarza, S., Iza, D., & Silva, E. (2022). Nutritional value and bioactive compounds of leaves and grains from quinoa (*Chenopodium quinoa* Willd.). *Plants, 11*, 213.

Wan, M. L. Y., Co, V. A., & El-Nezami, H. (2021). Dietary polyphenol impact on gut health and microbiota. *Critical Reviews in Food Science and Nutrition, 61*, 690–711.

Wang, D., Cao, D., Yao, Y., Wang, J., Li, Z., & Liu, B. (2020). Understanding the chemical foundation and genetic mechanism of the black grain trait in quinoa by integrating metabolome and transcriptome analyses. *Biotechnology & Biotechnological Equipment, 34*, 1095–1103.

Woldemichael, G., & Wink, M. (2001). Identification and biological activities of triterpenoid saponins from *Chenopodium quinoa*. *Journal of Agricultural and Food Chemistry, 49*, 2327–2332.

Xue, P., Zhao, L., Wang, Y., Hou, Z., Zhang, F., & Yang, X. (2020). Reducing the damage of quinoa saponins on human gastric mucosal cells by a heating process. *Food Science & Nutrition, 8*, 500–510.

Yang, A., Akhtar, S. S., Iqbal, S., Qi, Z., Alandia, G., Saddiq, M. S., & Jacobsen, S. E. (2018). Saponin seed priming improves salt tolerance in quinoa. *Journal of Agronomy and Crop Science, 204*, 31–39.

Yang, P., Huang, K., Zhang, Y., Li, S., Cao, H., Song, H., Zhang, Y., & Guan, X. (2022). Biotransformation of quinoa phenolic compounds with *Monascus anka* to enhance the antioxidant capacity and digestive enzyme inhibitory activity. *Food Bioscience, 46*, 101568.

Yao, Y., Zhu, Y., Gao, Y., Shi, Z., Hu, Y., & Ren, G. (2015). Suppressive effects of saponin-enriched extracts from quinoa on 3T3-L1 adipocyte differentiation. *Food & Function, 6*, 3282.

Yousif, A. M., Snowball, R., D'Antuono, M. F., Dhammu, H. S., & Sharma, D. L. (2021). Water droplet surface tension method – An innovation in quantifying saponin content in quinoa seed. *Food Chemistry, 343*, 128483.

Zhang, Q., Xing, B., Sun, M., Zhou, B., Ren, G., & Qin, P. (2020). Changes in bio-accessibility, polyphenol profile and antioxidants of quinoa and djulis sprouts during in vitro simulated gastrointestinal digestion. *Food Science and Nutrition, 8*, 4232–4241.

Zhang, Y., Bai, B., Yan, Y., Liang, J., & Guan, X. (2022). Bound polyphenols from red quinoa prevailed over free polyphenols in reducing postprandial blood glucose rises by inhibiting α-glucosidase activity and starch. *Nutrients, 14*, 728.

Zhang, R., Zhai, Q., Yu, Y., Li, X., Zhang, F., Hou, Z., Gao, Y., Feng, J., & Xue, P. (2022). Safety assessment of crude saponins from *Chenopodium quinoa* willd. husks: 90-day oral toxicity and gut microbiota & metabolomics study in rats. *Food Chemistry*, *375*, 131655.

Zhu, F. (2018). Interactions between cell wall polysaccharides and polyphenols. *Critical Reviews in Food Science and Nutrition*, *58*, 1808–1831.

Zhu, F., & Li, H. (2019a). Modification of quinoa flour functionality using ultrasound. *Ultrasonics-Sonochemistry*, *52*, 305–310.

Zhu, F., & Li, H. (2019b). Effect of high hydrostatic pressure on physicochemical properties of quinoa flour. *LWT- Food Science and Technology*, *114*, 108367.

Złotek, U., Gawlik-Dziki, U., Dziki, D., Świeca, M., Nowak, R., & Martinez, E. (2019). Influence of drying temperature on phenolic acids composition and antioxidant activity of sprouts and leaves of white and red quinoa. *Journal of Chemistry*, *2019*. Article ID 7125169.

Health and nutritional effects of quinoa

9

9.1 Introduction

The major reason for the rise of quinoa as a "superfood" is its attractive nutritional qualities such as being gluten free. The gluten-free nature of quinoa was described in Chapter 6. Quinoa consumption has a positive influence on metabolic syndrome, cardiovascular diseases, and general health and wellbeing (Caeiro et al., 2022; Graf et al., 2015a,b; Karimian et al., 2021; Lin et al., 2019; Navruz-Varli & Sanlier, 2016; Simnadis et al., 2015). The chemistry and composition of quinoa seeds were described in Chapters 2, 4–8. The components in quinoa such as dietary fiber polysaccharides, polyphenols, and vitamins are well known to have a range of health benefits. Thus it is natural for researchers to look into the health and nutritional effects of quinoa in relation to the chemistry of its components. Significant progress in understanding the health effects of quinoa in humans has been made during the last decade. For example, Simnadis et al. (2015) reviewed the literature (published up to the year 2015) on the health effects associated with quinoa consumption. Weight control, enhanced response towards oxidative stress, and improved lipid profile were the reported health benefits of quinoa consumption. However, they concluded that the quality of the published research was not of the highest standard and that clinical trials are needed for confirmation (Simnadis et al., 2015). Lin et al. (2019) reviewed the biological functions of diverse secondary metabolites (phenolic compounds, nitrogen-containing compounds steroids, and terpenoids) of quinoa from the published literature up to 2019. These metabolites showed different functions including molluscicidal, insecticidal, cytotoxic, antimicrobial, antidiabetic, antioxidant, and antiinflammatory activities (Lin et al., 2019). However, quinoa has also been found to have negative health effects such as allergy.

This chapter summarizes the in vitro, in vivo, and clinical studies on the diverse health effects of quinoa and related products such as protein isolate, polysaccharide fractions, and extracts. The relationships between chemical components and biological functions of quinoa seeds are discussed. The methodology used to study these effects is also analyzed.

9.2 In vitro studies on health and nutritional effects of quinoa

In vitro studies have mostly focused on the positive health effects of quinoa in different forms including whole seeds, flours, extracts, and isolated components

Quinoa. DOI: https://doi.org/10.1016/B978-0-323-99909-0.00004-0

(Table 9.1). Some of the health benefits of quinoa and related products included its prebiotic, antiinflammatory, antioxidant, cyto-protective, immunonutritional, anticancer, antiproliferative, antiviral, antihypertensive, antidiabetic activity, as well as gut microbiota modulation and prevention of skin aging (Alarcón et al., 2020; Ballegaard et al., 2021; Capraro et al., 2020; Graf et al., 2015a,b; Gu et al., 2021; Gullón et al., 2016; Hur et al., 2018; Mudgil et al., 2020; Obaroakpo, Liu, et al., 2020; Ong et al., 2020; Rangaswamy et al., 2021; Rotondo et al., 2021; Srdić et al., 2020; Tan et al., 2020; Teng et al., 2021; Ujiroghene et al., 2019a,b; Usman et al., 2022; Vilcacundo et al., 2018; Zeyneb et al., 2021).

9.2.1 Whole quinoa seeds

9.2.1.1 Fermentation property

Gullón et al. (2016) analyzed the prebiotic property of whole quinoa seeds. The seeds were cooked and digested in vitro. The digesta were used as culture for fecal human inocula. Incubation for 2 days increased the concentration of short-chain fatty acids (propionate, acetate, and butyrate being dominant) of the system, while decreasing the pH. The effect of quinoa seeds on the growth of *Lactobacillus–Enterococcus*, *Bifidobacterium* spp., *Atopobium*, *Clostridium coccoides*, *Eubacterium rectale*, *Bacteroides–Prevotella*, *Roseburia intestinalis*, and *Faecalibacterium prausnitzii* was different from that of amaranth seeds (Gullón et al., 2016). The prebiotic effect of quinoa could be attributed to its components such as dietary fiber (Xu et al., 2022). Zeyneb et al. (2021) used human fecal microbiota to ferment raw and cooked quinoa and quinoa polysaccharide extract after in vitro human digestion. The anaerobic fermentation of the samples for 24 h increased the contents of short-chain fatty acids and decreased the pH, while increasing the amounts of *Bacteroides* and *Prevotella* species. Cooked quinoa seeds increased the level of short-chain fatty acids to a larger extent than raw seeds, probably due to the easier access of the nutrients such as dietary fiber polysaccharide in the cooked samples.

9.2.2 Quinoa extracts

9.2.2.1 Skin care function

Graf, Cheng, et al. (2015) analyzed the potential application of quinoa ethanol extract for skin care and prevention of skin aging. Solid phase extraction column was used to fractionate the extract into three fractions (carbohydrate-rich (fraction 1), oil-rich (fraction 2), and polyphenol-, phytoecdysone-, and protein-rich (fraction 3) fractions). Quinoa extract had no effect on cell viability of four mammalian cell lines. The extract, oil-rich fraction, and polyphenol-, phytoecdysone-, and protein-rich fraction (5 μg/mL) inhibited the mRNA expression in human dermal fibroblasts. The extract and fractions containing polyphenols and phytoecdysone inhibited the activities of MMP-9 and tyrosinase. The reactive oxygen species were inhibited by the extract,

Table 9.1 Health effects of quinoa seeds and related fractionated products: In vitro studies.

Quinoa form	Targeted health and nutritional effects	Experimental	Major findings	References
Whole food				
Seeds	Prebiotic	Seeds were cooked and digested in vitro. The digesta were used as culture for fecal human inocula	Incubation for 2 days increased the concentrations of short-chain fatty acids (propionate, acetate and butyrate being dominant), while decreasing the pH. The effect of quinoa seeds on the growth of *Lactobacillus–Enterococcus*, *Bifidobacterium* spp., *Atopobium*, *Clostridium coccoides–Eubacterium rectale*, *Bacteroides–Prevotella*, *Roseburia intestinalis* and *Faecalibacterium prausnitzii* was different from that of amaranth	Gullón et al. (2016)
Raw and cooked seeds	Gut microbiota fermentation	Human fecal microbiota was used to ferment raw, cooked quinoa or quinoa polysaccharide extract after in vitro human digestion	Anaerobic fermentation of the samples for 24 h increased the contents of short-chain fatty acids and decreased the pH, while increasing the amounts of *Bacteroides* and *Prevotella* species. Cooked quinoa seeds increased the level of short-chain fatty acids to a larger extent that raw seeds	Zeyneb et al. (2021)
Extracts				
Seed extracts (70% ethanol)	Prevention of skin aging	Solid phase extraction column was used to fractionate the extract into three fractions (carbohydrate-rich, oil-rich and polyphenol, phytoecdysones, and protein-rich fractions). Their effects on the viability of four mammalian cell lines, intracellular reactive oxygen species, metalloproteinase and tyrosinase activity, and matrix metalloproteinase mRNA expression were studied	Quinoa extract had no effect on cell viability. The extract, oil-rich fraction and polyphenol, phytoecdysone and protein-rich fraction (5 μg/mL) inhibited the mRNA expression in human dermal fibroblasts. The extract and fractions containing polyphenols and phytoecdysones inhibited activities of MMP-9 and tyrosinase. The reactive oxygen species were inhibited by the extracts, 20-hydroxyecdysone and fractions-rich in polyphenols and phytoecdysones and oil	Graf, Cheng, et al. (2015), Graf, Rojas-Silva, et al. (2015)
Ethanol extracts of *Rhizopus oligosporus* fermented seeds	Antioxidant	DPPH (2,2-diphenyl-1-picrylhydrazyl) and ABTS (2,2′-azino-bis(3-ethylbenzothiazoline-6-sulfonic acid)) based chemical assays	Fermentation increased the contents of GABA, L-carnitine, vanillic acid, gallic acid total phenolics, and in vitro antioxidant activity	Hur et al. (2018)

(*Continued*)

Table 9.1 (Continued)

Quinoa form	Targeted health and nutritional effects	Experimental	Major findings	References
Extracts of quinoa seeds with three different colors (pressurized hot water extraction)	Antioxidant	ABTS (2,2′-azino-bis(3-ethylbenzothiazoline-6-sulfonic acid)), DPPH based chemical assays	The extracts of black and red quinoa seeds showed higher in vitro antioxidant activities than that of white seeds	Ong et al. (2020)
Extracts	Antioxidant	DPPH, ABTS based chemical assays	Different household processing methods affected the antioxidant activity of phenolic extracts to different extents. Germination increased the antioxidant activity, baking had small effect, whereas steaming decreased that. Steaming and germination increased the in vitro bioaccessibility of tocopherol and ferulic acid, respectively	Gu et al. (2021)
Ethanol extracts of *Rhizopus oligosporus* fermented seeds	Antiinflammatory	RAW264.7 macrophages	Fermentation increased the contents of GABA (gamma-aminobutyric acid), L-carnitine, vanillic acid, gallic acid, total phenolics, and inhibition of nitric oxide production in RAW264.7 macrophages	Hur et al. (2018)
Ethanol extracts	Antiplatelet effect	Human platelet aggregation test using flow cytometry in ADP-stimulated human platelets	Extracts of different lupin varieties were significantly antiaggregatory, whereas quinoa extracts had no effect	Alarcón et al. (2020)
Seed extracts containing aglycones of saponins	Intestinal barrier function	Caco-2/TC7 cells were used as model of epithelial monolayer to study the effect of quinoa extract (in comparison with capsaicin and cholera toxin) on the transport through intestinal barrier	Quinoa extract, capsaicin, or cholera toxin had no effect on intestinal permeability for protein intake	Ballegaard et al. (2021)
Extracts of quinoa seeds with three different colors (pressurized hot water extraction)	Cyto-protective	Dermal microvascular endothelium (HMEC-1) cells was treated with H_2O_2 to induce damage, while being treated with quinoa extract (0.001 g/mL)	The extracts of black and red quinoa seeds showed higher in vitro cytoprotective activities than that of white seeds	Ong et al. (2020)

Protein				
Protein fraction (salt soluble fraction of <30 kDa)	Immuno-nutritional property	Human-like macrophage cells (HB-8902) were used as model system	The protein fraction had potential to enhance the CD68/CD206 ratios as TLR4 agonists, and increased nonmitochondrial respiration and oxygen consumption rates. The protein fraction had potential to intervene immune-metabolic-based diseases	Srdić et al. (2020)
Chenopodin isoforms	Antiinflammatory	Caco-2 cells were used as model system	Both of low charge and high charge chenopodin Isoforms showed antiinflammatory capacity with different mechanisms of action (Fig. 9.1)	Capraro et al. (2020)
Protein isolate affected by Maillard reactions	Antiinflammatory	Mouse RAW264.7 macrophage cells were treated with protein isolate or its Maillard reaction products (up to 4 mg/mL) in the presence of lipopolysaccharide for 24 h	Maillard reactions of quinoa protein isolates with mannose and xylose increased antiinflammatory activity of the protein. The ratio of protein-to-sugar at 2 to 1 gave the highest activity	Teng et al. (2021)
Peptides	Anticancer	Peptides released from in vitro gastrointestinal digestive process were tested for inhibition on colon cancer cells	Digesta from the intestinal phase showed the highest anticancer effects, and the fraction of the peptides of >5 kDa had the most effective anticancer effects	Vilcacundo et al. (2018)
Quinoin	Anticancer	Human continuous glioblastoma cells (U87Mg), two primary cells (ZAR and NULU) established from gliomas of patients	Quinoin (5 and 2.5 nM) significantly inhibited the growth of glioblastoma cells. It can be incorporated into glioblastoma therapy combined with temozolomide (a chemotherapeutic agent)	Rotondo et al. (2021)
Peptides in the hydrolysates from simulated gastrointestinal digestion using trypsin and pancreatin	Anticancer	Anticancer properties of the peptides against Caco-2 cells of colon cancer	Three peptides of NWFPLPR, FHPFPR, and HYNPYFPG showed significant anticancer property. Molecular simulation showed that the peptides were docked into the active "pocket" of histone deacetylase 1. The peptides decreased the expressions of histone deacetylase 1, Bcl-2, NFκB, IL-8, and IL-6, while increasing that of caspase3	Fan et al. (2022)
Protein isolate affected by Maillard reactions	Antiproliferative	MDA-MB-231 cells (human breast cancer) were treated with protein isolate or its Maillard reaction products (up to 4 mg/mL) for 48 h	Maillard reactions of quinoa protein isolates with mannose and xylose increased the antiproliferative activity of the protein. The ratio of protein to sugar at 2 to 1 gave the highest activity	Teng et al. (2021)
Peptides	Antiviral	Peptides from protein hydrolysates were studied for inhibitory activity against SARS-CoV-2	Quinoa peptide NWRTVKYG showed effective inhibitory activity against SARS-CoV-2 through molecular simulation	Rangaswamy et al. (2021)

(*Continued*)

Table 9.1 (Continued)

Quinoa form	Targeted health and nutritional effects	Experimental	Major findings	References
Peptides	Antiviral	Peptides screened from protein hydrolysates were studied for inhibitory activity against SARS-CoV-2 using a computational approach	Proteins of five quinoa varieties were hydrolyzed using subtilisin and papain to generate a total of 1465 peptides in silico. Seven of these peptides (molecular mass from 547 to 3975 g/mol) interacted with catalytic and binding residues of the virus and they were nonallergenic and nontoxic. VEDKGMMH QQRMMEK AMNIPRM CGTMQRKCRMS bound the virus with the highest numbers of binding sites	Wong et al. (2021)
Peptides	Antihypertensive, antidiabetic	Peptides were produced using chymotrypsin, bromelain, and pronase E. The inhibitory activities against angiotensin converting enzyme, α-glucosidase, dipeptidyl peptidase-IV were tested	The peptides obtained from the hydrolysis using different enzymes showed a range of activities against angiotensin converting enzyme, α-glucosidase, and dipeptidyl peptidase-IV. A total of 35 peptides in the hydrolysates from the peptides obtained using chymotrypsin for the hydrolysis for 6 h were found to be bioactive. QHPHGL GALCAAPPST had the highest inhibitory activities against the targeted enzymes	Mudgil et al. (2020)
Protein hydrolysates	Antihypertensive, antidiabetic	Sprouted quinoa yogurt was fermented using *Lactobacillus casei*. The protein and peptides were isolated from the yogurt α-Glucosidase and angiotensin converting enzyme were tested enzymes for inhibition by the protein	QLCZ and QLCSY13 had the highest inhibitory activities against angiotensin converting enzyme and α-glucosidase, respectively. QLCSY13 had higher hydrophobicity than the others. Other peptides having inhibitory activities included VAHPVF and LAHMIVAGA	Ujiroghene et al. (2019a)

Polysaccharide				
Polysaccharide extracts	Antidiabetic and antioxidant	Five fractions of polysaccharide extract were obtained using gradient ethanol precipitation and tested for the inhibitory activities against DPPH and ABTS radicals and α-amylase and α-glucosidase	The five fractions of polysaccharide extract showed antioxidant and antidiabetic activities to different extents	Tan et al. (2020)
Polysaccharide extracts	Gut microbiota fermentation	Human fecal microbiota was used to ferment raw, cooked quinoa, or quinoa polysaccharide extract after in vitro human digestion	Anaerobic fermentation of the samples for 24 h increased contents of short-chain fatty acids and decreased the pH, while increasing the amounts of *Bacteroides* and *Prevotella* species. In particular, quinoa polysaccharide fraction increased the amounts of *Collinsella* and *Bifidobacterium* species with prebiotic function	Zeyneb et al. (2021)
Polyphenol-polysaccharide conjugates	Antidiabetic property and glucagon-like peptide-1 release	Human endothelial NCI-H716 cells were treated by the polyphenol-polysaccharide conjugates, polyphenol, or polysaccharide extracts obtained from sprouted quinoa-based yogurt	Glucagon-like peptide-1 release from the NCI-H716 cells was dependent on the germination conditions of quinoa and the dose of the polyphenol-polysaccharide conjugates. The treatment of NCI-H716 cells with the conjugates increased the expression of proglucagon mRNA. Sprouted quinoa-based yogurt can regulate energy and glucose homeostasis	Obaroakpo, Liu, et al. (2020)

20-hydroxyecdysone, and fractions rich in polyphenols and phytoecdysones and oil (Graf, Cheng, et al., 2015; Graf, Rojas-Silva, et al., 2015).

9.2.2.2 Antioxidant activity

In vitro antioxidant activities of different quinoa extracts were studied using chemical assays including DPPH (2,2-diphenyl-1-picrylhydrazyl) and ABTS (2,2′-azino-bis(3-ethylbenzothiazoline-6-sulfonic acid)) free radical scavenging capacity (Gu et al., 2021; Hur et al., 2018; Ong et al., 2020). All the extracts of quinoa showed antioxidant capacities to different extents, depending on the processing methods and quinoa genotypes. For example, Hur et al. (2018) obtained the ethanol extracts of *Rhizopus oligosporus* fermented quinoa seeds. The fermentation (30°C, up to 5 days) increased the contents of GABA (γ-aminobutyric acid), L-carnitine, vanillic acid, gallic acid, total phenolics, and in vitro antioxidant activity (Hur et al., 2018). Different household processing methods affected the antioxidant activity of phenolic extracts to different extents (Gu et al., 2021). Germination increased the antioxidant activity, baking had a small effect, and steaming decreased it. Steaming and germination increased the in vitro bioaccessibility of tocopherol and ferulic acid, respectively (Gu et al., 2021). Ong et al. (2020) showed that the extracts of black and red quinoa seeds had higher in vitro antioxidant activities than that of white seeds. Overall, processing conditions and quinoa varieties should be optimized to maximize the antioxidant activity.

9.2.2.3 Antiinflammatory property

Ethanol extracts of *Rhizopus oligosporus* fermented seeds were studied for antiinflammatory activity (Hur et al., 2018). The fermentation (30°C, up to 5 days) increased the contents of GABA, L-carnitine, vanillic acid, gallic acid and total phenolics, and inhibition capacity of nitric oxide production in RAW264.7 macrophages (Hur et al., 2018). Indeed, previous studies showed that different food components such as phenolics were antiinflammatory (Li et al., 2021).

9.2.2.4 Antiplatelet effect

Human platelet aggregation tests were done on ethanol extracts of both quinoa seeds and lupin varieties using flow cytometry in ADP-stimulated human platelets (Alarcón et al., 2020). Antiplatelet effect has been associated with reduced risk of cardiovascular diseases. Extracts of different lupin varieties were significantly anti-aggregatory, whereas quinoa extracts had no effect. The difference may be due to the different phenolic composition of the two extracts. Indeed, Hrubša et al. (2022) showed that only phenolic derivatives with specific structures can be antiplatelet in human blood. The results of Alarcón et al. (2020) did not support the claim that quinoa is a “superfood”.

9.2.2.5 Intestinal barrier function

Ballegaard et al. (2021) studied the intestinal barrier function affected by quinoa seed extract containing aglycones of saponins. Caco-2/TC7 cells were used as model of epithelial monolayer to study the effect of quinoa extract (in comparison with capsaicin and cholera toxin) on the transport through intestinal barrier. Quinoa extract, capsaicin, or cholera toxin had no effect on intestinal permeability for protein intake (Ballegaard et al., 2021). Compromised intestinal permeability may allow pathogens and toxins to pass through into circulation increasing the risk of developing diseases (Snelson et al., 2021). Intestinal barrier function protects the entrance of these harmful substances, while selectively taking in useful nutrients from foods. The results of Ballegaard et al. (2021) suggested that quinoa seed extract had no effect on increasing the risk of health issues.

9.2.2.6 Cyto-protection

Extracts of three quinoa varieties with different colors were obtained using pressurized hot water extraction (Ong et al., 2020). Dermal microvascular endothelium (HMEC-1) cells were treated with H_2O_2 to induce damage. Cyto-protective function of the quinoa extracts on these cells was studied by treating them with quinoa extract (0.001 g/mL). The extracts of black and red quinoa seeds showed higher in vitro cyto-protective activities than that of white seeds (Ong et al., 2020). The damage to the cells was largely due to the oxidative stress created by the H_2O_2. Quinoa extracts contained a range of antioxidant compounds including polyphenols, which alleviated the oxidative stress on the cells and protected them from damaging. In Section 9.2.2.2, the antioxidant activities of quinoa were described.

9.2.3 Quinoa proteins

The gluten-free nature of quinoa protein can aid people with celiac diseases (Caeiro et al., 2022). Proteins and peptides were isolated from quinoa seeds and showed different biological functions (Capraro et al., 2020; Mudgil et al., 2020; Rangaswamy et al., 2021; Rotondo et al., 2021; Srdić et al., 2020; Teng et al., 2021; Ujiroghene et al., 2019a; Vilcacundo et al., 2018; Wong et al., 2021; Usman et al., 2022).

9.2.3.1 Immuno-nutritional property

A protein fraction (fraction of salt-soluble fraction of <30 kDa) was obtained from quinoa seeds (Srdić et al., 2020). The immuno-nutritional property of the protein fraction was studied using human-like macrophage cells (HB-8902) as the model system. The protein fraction had the potential function as TLR4 agonists to enhance the CD68/CD206 ratios, increasing nonmitochondrial respiration and oxygen consumption rates. The protein fraction also had the potential to intervene immune-metabolic-based diseases (Srdić et al., 2020).

9.2.3.2 Antiinflammatory property

Chenopodin isoforms and quinoa protein isolate affected by Maillard reactions were tested for antiinflammatory activity (Capraro et al., 2020; Teng et al., 2021). Caco-2 cells were used as the model system for testing chenopodin isoforms. Both low-charged and high-charged chenopodin isoforms showed antiinflammatory capacity with different mechanisms of action (Fig. 9.1) (Capraro et al., 2020). In the study of Teng et al. (2021), mouse RAW264.7 macrophage cells were treated with protein isolate or its Maillard reaction products (up to 4 mg/mL) in the presence of lipopolysaccharide for 24 h. Maillard reactions of quinoa protein isolates with mannose and xylose increased antiinflammatory activity of the protein. The ratio of protein to sugar at 2 to 1 gave the highest antiinflammatory activity (Teng et al., 2021). The Maillard reaction changed the hydrophilicity of proteins, which may change the interactions of the products with the RAW264.7 macrophage cells and improve their antiinflammatory capacity. Previous studies showed that other protein sources can also be antiinflammatory (Concha et al., 2022). It remains to test if quinoa protein-based compounds have higher antiinflammatory activity than the other proteins.

9.2.3.3 Anticancer property

Quinoa peptides, quinoin, and quinoa protein isolate affected by Maillard reactions showed anticancer and antiproliferative capacity against different types of cancer cells (Fan et al., 2022; Rotondo et al., 2021; Teng et al., 2021; Vilcacundo et al., 2018). Peptides released from in vitro gastrointestinal digestive process were tested for inhibition of colon cancer cells (Vilcacundo et al., 2018). The digesta from the intestinal phase showed the highest anticancer effects, and the fraction of the peptides of >5 kDa had the most effective anticancer effect (Vilcacundo et al., 2018).

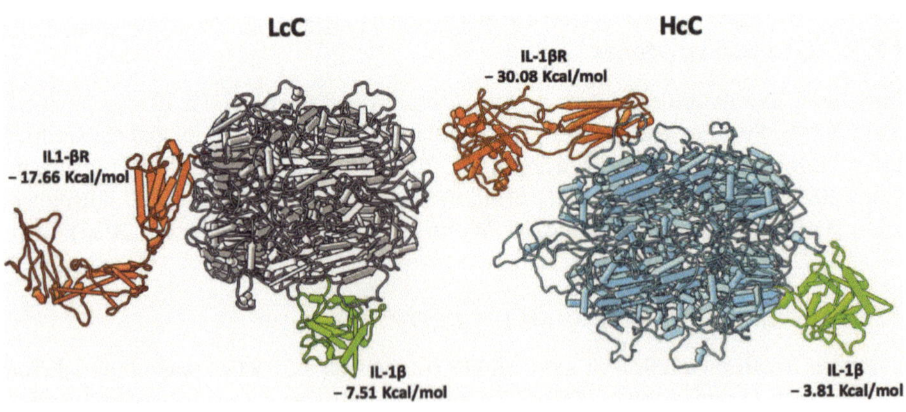

Figure 9.1 Molecular docking models. Pipe and plank representations of IL-1β (green) and the soluble interleukin 1β (IL-1β) receptor (orange) binding to low charge chenopodin (LcC) (gray) and high charge chenopodin (HcC) (light blue). The binding energy for each complex is reported (Capraro et al., 2020).
Source: Reprinted with permission under Creative Commons Attribution License.

Quinoin, a type 1 ribosome-inactivating protein from quinoa seeds, showed anticancer capacity against human continuous glioblastoma cells (U87Mg) and two primary cells (ZAR and NULU) established from gliomas of patients. Quinoin (5 and 2.5 nM) significantly inhibited the growth of glioblastoma cells. It can be incorporated into glioblastoma therapy combined with temozolomide (a chemotherapeutic agent) (Rotondo et al., 2021). The anticancer properties of quinoa peptides against Caco-2 cells of colon cancer were studied (Fan et al., 2022). Three peptides of NWFPLPR, FHPFPR, and HYNPYFPG showed significant anticancer properties. Molecular simulation showed that these peptides were docked into the active "pocket" of histone deacetylase 1. On the molecular level, the peptides decreased the expressions of histone deacetylase 1, Bcl-2, NFκB, IL-8, and IL-6, while increasing that of caspase3 (Fan et al., 2022). Teng et al. (2021) treated MDA-MB-231 cells (human breast cancer) with protein isolate or its Maillard reaction products (up to 4 mg/mL) for 48 h. The Maillard reactions of quinoa protein isolates with mannose and xylose increased the antiproliferative activity of the protein. The ratio of protein to sugar at 2 to 1 gave the highest antiproliferative activity (Teng et al., 2021).

Different mechanisms of action have been associated with anticancer property of proteins and peptides (Chalamaiah et al., 2018) (Fig. 9.2). Proteins and peptides may arrest cell cycle progression and stimulate apoptosis, inhibit cell adhesion,

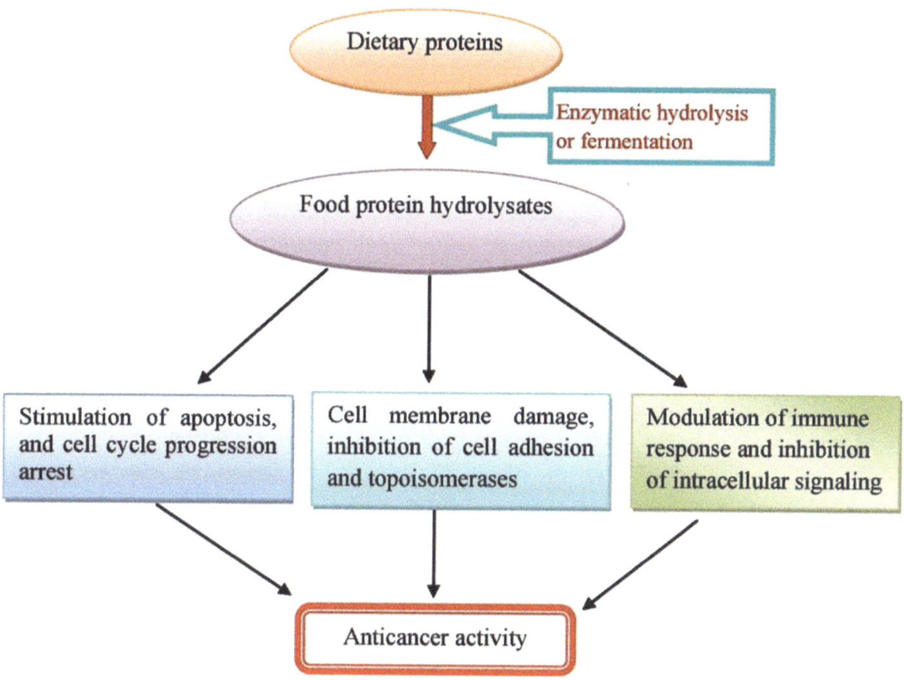

Figure 9.2 Mechanisms of action of food-derived protein hydrolysates and peptides with anticancer property (Chalamaiah et al., 2018).
Source: Reprinted with permission from Elsevier.

topoisomerases and intracellular signaling, damage cell membrane, and modulate immune response (Chalamaiah et al., 2018). The anticancer mechanisms of action of quinoa proteins and peptides remain to be better studied and compared with that of commonly used proteins and peptides.

9.2.3.4 Antiviral property

Peptides from quinoa protein hydrolysates were studied for inhibitory activity against SARS-CoV-2 (Rangaswamy et al., 2021; Wong et al., 2021). Quinoa peptide NWRTVKYG showed effective inhibitory activity against SARS-CoV-2 through molecular simulation (Rangaswamy et al., 2021). Similarly, peptides screened from quinoa protein hydrolysates were studied for inhibitory activity against SARS-CoV-2 using a computational approach (Wong et al., 2021). Proteins of five quinoa varieties were hydrolyzed using subtilisin and papain to generate a total of 1465 peptides in silico. Seven of these peptides (molecular mass from 547 to 3975 g/mol) interacted with catalytic and binding residues of the virus and they were nonallergenic and nontoxic. VEDKGMMHQQRMMEKAMNIPRMCGTMQRKCRMS bound the virus with the highest numbers of binding sites (Wong et al., 2021). However, this kind of research may have little practical significance considering its in vitro and in silico nature.

9.2.3.5 Antihypertensive property

Quinoa peptides were produced using chymotrypsin, bromelain, and pronase E (Mudgil et al., 2020). The inhibitory activities of the peptides against angiotensin converting enzyme varied depending on the type of enzymes. A total of 35 peptides in the hydrolysates from the peptides obtained using chymotrypsin for the hydrolysis for 6 h were found to be bioactive. QHPHGLGALCAAPPST had the highest inhibitory activities against angiotensin converting enzyme (Mudgil et al., 2020). Sprouted quinoa yogurt was fermented using *Lactobacillus casei*. The protein and peptides were isolated from the yogurt for testing their antihypertensive property (Ujiroghene et al., 2019a). Among the different peptides identified, QLCZ had the highest inhibitory activity against angiotensin converting enzyme. Other peptides with inhibitory activity included VAHPVF and LAHMIVAGA (Ujiroghene et al., 2019a). Peptides from other sources such as soybean can be antihypertensive (Li et al., 2022). It is not clear if quinoa peptides are more effective compared with the peptides of other sources.

9.2.3.6 Antidiabetic property

Quinoa peptides were produced using chymotrypsin, bromelain, and pronase E (Mudgil et al., 2020). The inhibitory activities of the peptides against α-glucosidase and dipeptidyl peptidase-IV varied depending on the type of enzymes. A total of 35 peptides in the hydrolysates from the peptides obtained from the hydrolysis using chymotrypsin for 6 h were found to be bioactive. QHPHGLGALCAAPPST had the highest inhibitory activities against the targeted enzymes (Mudgil et al., 2020). Sprouted quinoa yogurt was fermented with *Lactobacillus casei*. The protein and peptides were isolated from the yogurt for testing their antidiabetic property (Ujiroghene et al., 2019a,b). Among the different peptides identified, QLCSY13 with the highest hydrophobicity had the highest inhibitory activities against α-glucosidase. Other peptides having the inhibitory activity included VAHPVF and LAHMIVAGA (Ujiroghene et al., 2019a). Overall, peptides derived from quinoa protein had the capacity to control type 2 diabetes and hypertension via inhibition of α-glucosidase and dipeptidyl peptidase-IV as well as angiotensin-converting enzyme (Mudgil et al., 2020). There are many different types of peptides from other sources such as soybean showing promising antidiabetic capacity (Casey et al., 2021; Das et al., 2022). It is not clear if quinoa-based peptides have any technical advantages over those based on other sources.

9.2.4 Quinoa polysaccharide

Nonstarch polysaccharides of quinoa seeds in different forms showed antioxidant, antidiabetic, and gut microbiota modulating effects (Obaroakpo, Liu, et al., 2020; Tan et al., 2020; Zeyneb et al., 2021). Tan et al. (2020) obtained five different fractions of polysaccharide extract of quinoa seeds using gradient ethanol precipitation. The five fractions of the polysaccharide extract showed antioxidant activities of different degrees (measured using chemical assays including DPPH and ABTS-based methods). These fractions also showed inhibitory activities against α-amylase and α-glucosidase to different degrees, indicating their antidiabetic property (Tan et al., 2020). Zeyneb et al. (2021) analyzed the fermentative property of in vitro digestive products of quinoa polysaccharide extract and raw and cooked quinoa seeds using human fecal microbiota. Anaerobic fermentation of the samples for 24 h increased contents of short-chain fatty acids and decreased the pH, while increasing the amounts of *Bacteroides* and *Prevotella* species. In particular, quinoa polysaccharide fraction increased the amounts of *Collinsella* and *Bifidobacterium* species with prebiotic function (Zeyneb et al., 2021). This indicated the modulating capacity of the polysaccharide extract on human gut microbiota. Obaroakpo, Liu, et al. (2020) analyzed the antidiabetic property of polyphenol-polysaccharide conjugates obtained from sprouted quinoa-based yogurt. The conjugates stimulated the release glucagon-like peptide-1 from human endothelial NCI-H716 cells. The glucagon-like peptide-1 release from the NCI-H716 cells was dependent on the germination conditions of quinoa and the dose of the polyphenol-polysaccharide conjugates. The treatment of NCI-H716 cells with the conjugates increased the expression of proglucagon mRNA. Sprouted quinoa-based yogurt can regulate energy and glucose homeostasis (Obaroakpo, Liu, et al., 2020). Many other studies showed that polysaccharides from other sources such as fruit seeds and cereals had various biological functions similar to those of quinoa polysaccharides (Sivieri et al., 2022; Wei et al., 2022). Comparison between quinoa polysaccharides and those of other sources remains to be done to reveal if the former is unique in terms of biological functions and efficiency.

9.3 In vivo studies on health and nutritional effects of quinoa

Different in vivo studies on quinoa health effects showed that intake of quinoa in different forms had different health effects (Table 9.2). These health effects included hypercholesterolemia, hepato-protection, antiobesity, antidiabetes, antihypertension, neuroprotection, mitigating chemical toxicity, modulating gut microbiota, immunonutritional property, and lifespan improvement (An et al., 2021; Bahnasy & Ragheb, 2020; Cisneros-Yupanqui et al., 2020; Fotschki et al., 2020; Halaby et al., 2017; Hamed, 2021; Mohamed et al., 2019; Noratto et al., 2019; Obaroakpo et al., 2020; Abdel-Wahhab et al., 2021; Cao et al., 2020; Fan et al., 2019; Graf et al., 2017; Guo et al., 2021; Hashem et al., 2021; Llopis et al., 2020; Selma-Gracia et al., 2020; Selma-Gracia et al., 2021; Souza et al., 2020).

9.3.1 Quinoa seeds and flours

9.3.1.1 Hypercholesterolemia

Halaby et al. (2017) fed male albino rats with a diet containing cholesterol (2%) and quinoa flour (up to 40%) for 60 days. Feeding rats with diet containing cholesterol (2%) increased risk of hypercholesterolemia related diseases, whereas feeding quinoa flour to these rats significantly reduced the adverse effects of hypercholesterolemia by improving feed consumption, feed efficiency ratio, body weight gain, and kidney and liver functions and reducing blood cholesterol level (Halaby et al., 2017).

9.3.1.2 Hepatoprotection

Male rats were fed on high fructose diet containing 20% of quinoa flour for 4 weeks (Mohamed et al., 2019). The high fructose diet created dyslipidemia and increased lipid peroxidation and levels of liver triglycerides and total cholesterol. Incorporation of quinoa flour into the diet improved these physiological conditions in the rats against nonalcoholic fatty liver disease (Mohamed et al., 2019). Mustafa et al. (2019) also reported that feeding chicken with quinoa improved liver and kidney functions.

9.3.1.3 Antiobesity and antidiabetes

Obese-diabetic (db/db) mice were fed with a diet containing quinoa (165 g/kg) in comparison with AIN-93G diet of similar calorie density for 8 weeks (Noratto et al., 2019). These mice were compared to lean mice in their physiological parameters. The quinoa-based diet largely decreased the levels of plasma total cholesterol, oxidized-low-density lipoprotein, and low-density lipoprotein cholesterol, hepatic steatosis, and total-cholesterol accumulation to be similar to those of lean

Table 9.2 Health effects of quinoa seeds and related fractionated products: In vivo and clinical trials.

Quinoa form	Targeted health and nutritional effects	Experimental model	Experimental	Major findings	References
Quinoa flakes	Metabolic improvement	Postmenopausal women	Postmenopausal overweight women (n = 35) had corn or quinoa flakes (25 g per day) for 4 weeks	Eating quinoa or corn flakes decreased the levels of serum triglycerides, TBARS (thiobarbituric acid reactive substances), vitamin E, while increasing the urinary excretion of enterolignans. Eating quinoa but not corn flakes decreased levels of LDL (low-density lipoprotein)-cholesterol and total cholesterol, while increasing that of reduced glutathione	De Carvalho et al. (2014)
Seeds	Glycemic control	Human	Patients with type 2 diabetes and healthy persons were fed with buckwheat and quinoa for breakfast. The glycemic properties of the subjects after breakfast and lunch were studied	Healthy subjects with buckwheat as breakfast had lower incremental area under the curve of glucose after breakfast than those with quinoa or white breads. Diabetic subjects with buckwheat or quinoa as breakfast had lower incremental area under the curve of glucose after breakfast than those that had white breads. After lunch, the subjects with buckwheat and quinoa as breakfast showed flatter glycemic response than those that had white bread. Quinoa gave a much higher glycemic index (89) than buckwheat (27). Overall, quinoa only had moderate potential to modulate glycemic response in humans	Gabrial et al. (2016)

(*Continued*)

Table 9.2 (Continued)

Quinoa form	Targeted health and nutritional effects	Experimental model	Experimental	Major findings	References
Seeds	Glycemic properties	Human	Humans (n = 10) were fed with cooked quinoa. The blood glucose levels of the subjects were measured up to 2 h after the quinoa meals	The ranges of glycemic indexes and glycemic loads were 51 − 78 and 26 − 39, respectively. The averages of glycemic indexes and glycemic loads were 63 and 32, respectively. Quinoa was considered to have medium glycemic index and high glycemic load	Priyanka et al. (2018)
Seeds	Plasma lipid profile improvement	Human	In a parallel-designed, randomized, controlled, and single-blind trial, obese and overweight participants (n = 50) were fed with 25 and 50 g quinoa per day for 12 weeks	Quinoa intake (50 g per day) after 12 weeks decreased the occurrence of metabolic syndrome (by 70%) and the concentration of serum triglycerides (from 1.1 to 0.72 mmol/L). Quinoa intake (25 g per day) after 12 weeks decreased the occurrence of metabolic syndrome (by 40%) and had no effect on the concentration of serum triglycerides	Navarro-Perez et al. (2017)
Seeds	Cardiovascular disease	Human	Humans were fed with quinoa containing (20 g flour) breads for 4 weeks in a crossover intervention	During intervention, the area under the curve of blood glucose level was lower in the groups eating quinoa-containing breads than that of the groups eating white breads. After intervention, the levels of low-density lipoprotein (LDL) cholesterol, blood glucose and other blood metabolites as well as anthropometric data were similar in humans having breads containing quinoa or not containing quinoa	Li et al. (2018)

Seeds	Cardiovascular disease	Human (n = 40)	Healthy adults daily consumed quinoa-containing biscuits (15 g, 60% quinoa flour) in comparison with wheat flour-containing biscuits for 4 weeks in a randomized-controlled, double-blind crossover trial	Eating quinoa biscuits decreased the weight and body mass index, the levels of total cholesterol and low-density lipoprotein cholesterol, and the ratio of total cholesterol to high-density lipoprotein cholesterol, while having little effect on the levels of triglycerides, high-density lipoprotein cholesterol, polyunsaturated fatty acid, or C-reactive protein	Pourshahidi et al. (2020)
Seeds and different forms	Cardiovascular disease	Human	Meta-analysis of controlled clinical tests on the effect of quinoa addition on the development and risk of cardiovascular diseases	The review paper analyzed the literature published up to February of 2020 and selected five randomized controlled trials involving 206 tested subjects. Adding quinoa to the diet decreased body weight, fat mass, waist circumference, levels of insulin, low-density lipoprotein, triglycerides, and total cholesterol in humans. The results suggested that quinoa had potential to reduce the risk of cardiovascular diseases in humans	Karimian et al. (2021)
Flour	Hypercholesterolemia	Male albino rats	Rats were fed with a diet containing cholesterol (2%) and quinoa flour (up to 40%) for 60 days	Feeding rats with diet containing cholesterol (2%) increased risk of hypercholesterolemia-related disease, whereas feeding quinoa flour to these rats significantly reduced the adverse effects of hypercholesterolemia by improving feed consumption, feed efficiency ratio, body weight gain, and kidney and liver functions, and reducing blood cholesterol level	Halaby et al. (2017)

(*Continued*)

Table 9.2 (Continued)

Quinoa form	Targeted health and nutritional effects	Experimental model	Experimental	Major findings	References
Flour	Hepato-protection	Male rats	Rats were fed on high fructose diet containing 20% of quinoa flour for 4 weeks	High fructose diet created dyslipidemia and increased lipid peroxidation and levels of liver triglycerides and total cholesterol. Incorporation of quinoa flour into the diet improved these physiological conditions in the rats against nonalcoholic fatty liver disease	Mohamed et al. (2019)
Flour	Antiobesity and antidiabetic	Obese-diabetic (db/db) mice	Mice were fed with diet containing quinoa (165 g/kg) in comparison with AIN-93G diet of similar calorie density for 8 weeks. These mice were compared to lean mice	Quinoa-based diet largely decreased the levels of plasma total-cholesterol, oxidized-low-density lipoprotein and low-density lipoprotein cholesterol, hepatic steatosis, total-cholesterol accumulation to be similar to those of lean mice. Quinoa-based diet increased plasma insulin in the obese-diabetic mice and had no protection against pathophysiological conditions	Noratto et al. (2019)
Sprouted quinoa yogurt	Antidiabetes	Mice (C57BL/6 N) with type 2 diabetes induced by streptozotocin (50 mg/kg body weight) and high fat diet	Diabetic mice were orally fed with diet containing sprouted quinoa yogurt for 6 weeks	Diet containing quinoa yogurt decreased the lipid peroxidation, levels of fasting blood glucose, total cholesterol, low-density lipoprotein-cholesterol and triglycerides, and those of proinflammatory cytokines (IL-6,	Obaroakpo, Nan, et al. (2020)

				TNF-α and IL-1β) in the diabetic mice, while increasing their body weights, activities of catalase, superoxide dismutase, and glutathione peroxidase and the levels of hepatic glycogen and IL-10 (antiinflammatory cytokine), and protecting tissue structure from damage. AKT/AMPK/PI3K signaling pathways were involved in the hypoglycemic and antihyperlipidemic effects in the diabetic mice obtained for diet containing sprouted quinoa yogurt	
Flour	Antiobesity	C57BL/6 J female mice	Mice were fed with high- fat diet to induce obesity before feeding with a diet containing quinoa (2 g quinoa/day) for 12 weeks	Incorporation of quinoa into the diet of obese mice gave them less triglycerides, cholesterol, low-density lipoprotein, and blood glucose, and oxidative stress, while increasing glucose tolerance and histological data of liver tissues. On the molecular level, in the liver, the mRNA expression of endoplasmic reticulum stress markers (including GRP78, eIF2α, and CHOP) was reduced by adding quinoa into the diet	An et al. (2021)
Flour	Antihypertension	Rats	Rats were treated by N (ω)-nitro-L-arginine methyl ester to induce hypertension. The rats were also fed with a diet containing cooked quinoa flour (30%) or captopril (a commercial antihypertensive drug)	Adding quinoa flour into the diet enhanced the level of high-density lipoprotein, blood glucose, and increased activities of catalase and superoxide dismutase. Overall, adding quinoa flour into the diet improved the medical conditions of hypertensive rats	Cisneros-Yupanqui et al. (2020)

(*Continued*)

Table 9.2 (Continued)

Quinoa form	Targeted health and nutritional effects	Experimental model	Experimental	Major findings	References
Flour-rich in protein	Intestinal microbiota	Wistar rats	Protein-rich fraction of a protein content of 36% was obtained by using suitable wheel speed of impact mill. The fraction was incorporated in a diet which was fed to rats for 2 weeks in comparison to a protein-rich fraction from buckwheat and a soy protein isolate	Diet containing quinoa or buckwheat protein-rich fraction more increased the body weight gain and biological value of protein in the rats compared to that containing soy protein isolate. In the cecum of the rats, the quinoa or buckwheat flour increased the concentrations of short-chain fatty acids and the activities of α-glucosidase, α-galactosidase and α-glucosidase. The flour-containing diet decreased the levels of plasma low density lipoprotein (LDL) cholesterol and total cholesterol. Adding quinoa fraction into the diet increased the levels of alanine transaminase and plasma interleukin 6 and inflammatory activity	Fotschki et al. (2020)
Flour	Lipid profile protection against synthetic colorant induced changes	Sprague Dawley male albino rats	Rats were fed with sunset yellow (2.5 mg/kg body weight) or chocolate brown (0.3 mg/kg body weight). They were also fed with diet containing quinoa flour of up to 10% for 8 weeks	Quinoa intake by the rats decreased the side effects induced by intake of the synthetic colorants, while regulating the lipid profile	Bahnasy and Ragheb (2020)

Seeds	Mitigating chemical toxicity	Rats	Rats were fed with chlorfenapyr (180 mg/kg) to induce testicular toxicity, quinoa seeds (30% of the diet) were also fed to the rats for 4 weeks	Chlorfenapyr treatment led to sagging of testicles, increased the testis weights, decreased sex hormone level, sperm count, viability and motility, and increased sperm abnormality. Feeding quinoa seeds mitigated these symptoms in the rats induced by chlorfenapyr	Hamed (2021)
Flour (red and white varieties)	Immunonutritional property	Mice (C57BL/6) with hyperglycemia induced by streptozotocin	Mice were fed with a high fat diet for 3 weeks. Breads containing 25% quinoa flour and 20% chia flour were also fed to mice (150 g/day/70 kg body weight) 3 times per week (150 g/day/70 kg body weight)	Feeding mice with breads containing quinoa and chia largely decreased the value of homeostatic model assessment of insulin resistance in hyperglycaemic mice compared to that with white breads. This decreased value had little relation to the reduced starch content in the formulated breads. Feeding mice with breads containing red quinoa and chia increased the levels of triglycerides in the mouse plasma and reduced hepatic lipid peroxidation induced by high fat diet. Feeding mice with breads containing white quinoa and chia increased the levels of hepatic IL-6 and TNF-α in the mice. Mice fed with breads containing red quinoa and chia had much less IL-17 than those fed with samples containing white quinoa and chia. The levels of hepatic granulocyte–monocyte colony stimulation factor in the mice were increased by the breads containing quinoa and chia	Selma-Gracia et al. (2020)

(*Continued*)

Table 9.2 (Continued)

Quinoa form	Targeted health and nutritional effects	Experimental model	Experimental	Major findings	References
Flour	Immunonutritional property	C57BL/6 Mice	Breads containing 25% quinoa flour and 25% chia flour were fed to mice (150 g/day/70 kg body weight) for up to 3 weeks in different models	Compared to the mice fed with white wheat breads, those with breads containing quinoa and chia had normalized/regulated insulin resistance glycemia, and lipid and glucose metabolism, glucose homeostasis, hepatic lipid accumulation expression of coronary risk and lipogenic markers via controlling the hepatic myeloid cells	Selma-Gracia et al. (2021)
Protein fraction (fraction of salt soluble fraction of <30 kDa)	Immunonutritional property	Mice	Mice were fed with high fat diet for 1 week before receiving the protein fraction solution and diethylnitrosoamine injection	The protein fraction increased the resistance of the mice to hepato-carcinogenesis. The proportions of F4/80^{+} cells and the level of immune mediators (CD68/CD206 ratio) in intestinal tissue were increased. The analysis of fecal microbiota showed that the fraction intake led to the pattern of microbial composition to be similar to that with less potential of developing obesity	Llopis et al. (2020)
Protein isolate	Antihypertensive property and gut microbiota	Spontaneously hypertensive rats	Quinoa protein isolate was fed to rats (up to 400 mg quinoa protein/kg body weight) up to 5 weeks	The protein feeding significantly decreased the blood pressure level of the hypertensive rats, enhanced alpha diversity, and shifted the microbial composition to that of normal rats. The increased abundance of genera including	Guo et al. (2021)

				Allobaculum and *Turicibacter* in the fecal microbiota was related to decreased blood pressure level. The composition of fecal microbiota in the hypertensive rats fed with quinoa protein was more similar to that in normal rats than that in captopril treated rats. Captopril is an antihypertensive drug	
Ethanol extract	Hypolipidemic	Rats	Rats were treated with Triton X-100 (intraperitoneal injection) to induce hyperlipidemia. The rats also received quinoa extracts for 40 days	The feeding of extract to the hyperlipidemic rats had hepato-renal protective effect, while decreasing lipid peroxidation, enhancing growth performance and antioxidant activity, and regulating the levels of leptin and adiponectin hormone	Hashem et al. (2021)
Ethanol extract	Hepatoprotective	Rats	Rats were treated with cyclophosphamide and/or the ethanol extract	The extract reduced the cyclophosphamide-induced hepatotoxicity in the rats. The levels/activities of serum alanine aminotransferase, aspartate aminotransferase, gamma-glutamyl transferase, alkaline phosphatase, low density lipoprotein (LDL)-cholesterol, total cholesterol, triglycerides, and TNF-α IL 1β and IL6, hepatic malondialdehyde, nitric oxide levels, and DNA fragmentation in the rats were decreased by the extract feeding	Abdel-Wahhab et al. (2021)

(*Continued*)

Table 9.2 (Continued)

Quinoa form	Targeted health and nutritional effects	Experimental model	Experimental	Major findings	References
Ethanol extract of seeds	Neuroprotective	Mice	Declarative memory deficits in mice were induced by scopolamine. The mice were fed with the extracts (up to 300 mg/day/kg) up to 9 days	The memory deficit in the mice was significantly reduced by the extracts. The scopolamine-induced changes in the activities of acetylcholinesterase and superoxide dismutase, and the levels of reactive oxygen species and thiobarbituric acid reactive substance in the cerebral cortex and hippocampus of the mice were reduced by the extracts	Souza et al. (2020)
Polysaccharide	Immunity-improving	ICR mice	A polysaccharide fraction was purified from quinoa crude polysaccharide extract using chromatography. Mice were treated with cyclophosphamide for immunosuppression. The mice were fed with the polysaccharide fraction (100 mg/kg/day) for 3 weeks	The polysaccharide fraction increased the phagocytic function of mononuclear macrophages, the concentrations of lysozyme, IFN-γ, IFN-α, IL-6, and IgM (immunological factors) in serum of the mice, while decreasing the delayed allergy	Fan et al. (2019)
Soluble polysaccharide	Gut microbiota regulation	Rats	Rats were fed with high-fat diet and also the polysaccharide (up to 600 mg/kg/day) for 8 weeks	The levels/activities of serum low-density lipoprotein cholesterol, total triglyceride, malondialdehyde, hepatic lipid accumulation, total glutamic pyruvic transaminase, and glutamic oxaloacetic transaminase in the rats fed with high fat diet were decreased by the polysaccharide	Cao et al. (2020)

Soluble polysaccharide	Hyperlipidemia regulation	Rats	Rats were fed with high-fat diet and also the polysaccharide (up to 600 mg/kg/day) for 8 weeks	High-fat diet disrupted the gut microbiota composition. The quinoa polysaccharide intake increased the microbial diversity, decreased the percentages of *Proteobacteria*, *Desulfovibrio*, and *Allobaculum* and the ratio of *Firmicutes* to *Bacteroides*	Cao et al. (2020)
Phytoecdysteroid extract	Healthspan improvement	*Caenorhabditis elegans*	Red quinoa seed leachate was obtained using 70% ethanol for extraction (4 h, 80°C). *Caenorhabditis elegans* (wild type N2) was treated with the extract (1.0 mg/mL) or metformin (6.5 mg/mL, as control)	The quinoa extract treatment enhanced the basal respiration rate (by 37%), locomotory performance (from 104 to 115 head thrashes/min), and lifespan by 2 days (from 9 to 11 days), while reducing the formation of advanced glycation end-product pigments, body fat and reactive oxygen species by 24, 14, and 20%, respectively. Treating the worms with 20-hydroxyecdysone at equivalent dosage gave similar biological effects	Graf et al. (2017)

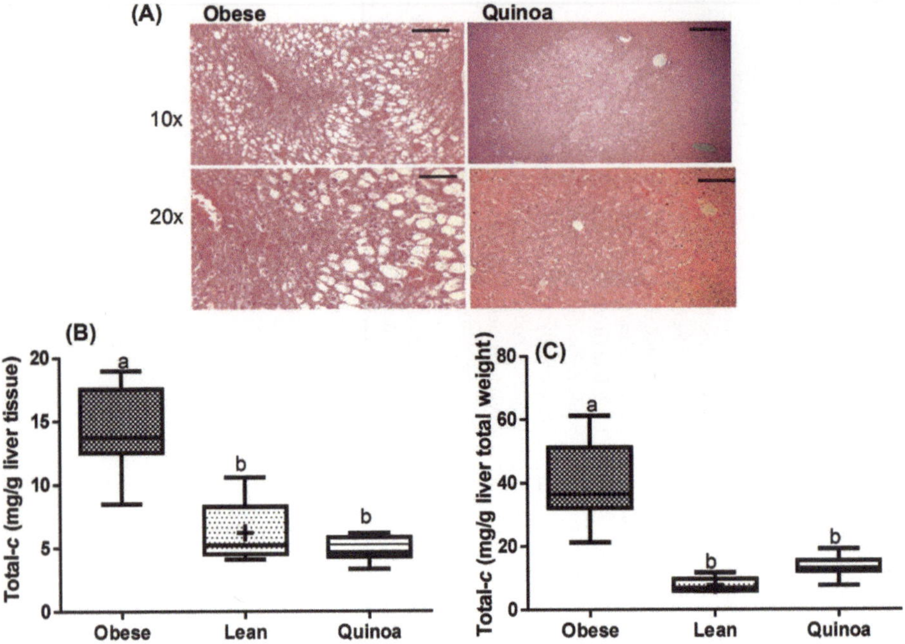

Figure 9.3 Liver fat accumulation and plasma total-cholesterol level (total-c) in mice were reduced by quinoa intake. (A) Fat accumulation in liver. Representative photomicrographs of liver tissues stained with H & E at magnification of 10 × and 20 × [bar = 200 μm (10 ×) or 100 μm (20 ×)]. (B) Total-c (mg/g liver tissue). (C) Total-c (mg/liver total weight). Boxplot and whiskers represent min to max, lines inside box represent median. Data collected from n = 11 (lean group), n = 10 (obese group), n = 9 (quinoa fed group of obese mice) (Noratto et al., 2019).
Source: Reprinted with permission from Elsevier.

mice (Fig. 9.3). Quinoa-based diet increased plasma insulin in the obese-diabetic mice and had no protection against pathophysiological conditions (Noratto et al., 2019).

An et al. (2021) fed C57BL/6 J female mice with a high-fat diet to induce obesity before feeding them with a diet containing quinoa (2 g quinoa/day) for 12 weeks. The incorporation of quinoa into the diet of the obese mice resulted in lower triglycerides, cholesterol, low-density lipoprotein and blood glucose, and oxidative stress, while increasing glucose tolerance and histological data of liver tissues. On the molecular level, in the liver, the mRNA expression of endoplasmic reticulum stress markers (including GRP78, eIF2α, and CHOP) was reduced by adding quinoa into the diet (An et al., 2021). Quinoa fortification decreased obesity-related conditions in mice with obesity induced by high-fat diet via regulating the stress in endoplasmic reticulum and improving the metabolism of glycolipids (An et al., 2021).

Obaroakpo, Nan, et al. (2020) tested the antidiabetic property of sprouted quinoa yogurt. Mice (C57BL/6 N) were treated with streptozotocin (50 mg/kg body weight) and a high-fat diet to induce type 2 diabetes. The diabetic mice were orally fed with a diet containing sprouted quinoa yogurt for 6 weeks. The diet containing quinoa yogurt decreased the lipid peroxidation, levels of fasting blood glucose, total cholesterol, low-density lipoprotein-cholesterol and triglycerides, and those of proinflammatory cytokines (IL-6, TNF-α and IL-1β) in the diabetic mice, while increasing their body weights, activities of catalase, superoxide dismutase, and glutathione peroxidase and the levels of hepatic glycogen and IL-10 (antiinflammatory cytokine), and protecting tissue structure from damage. AKT/AMPK/PI3K signaling pathways were involved in the hypoglycemic and antihyperlipidemic effects in the diabetic mice obtained with a diet containing sprouted quinoa yogurt (Obaroakpo, Nan, et al., 2020).

9.3.1.4 Antihypertension

Rats were treated by N (ω)-nitro-L-arginine methyl ester to induce hypertension (Cisneros-Yupanqui et al., 2020). The rats were also fed with a diet containing cooked quinoa flour (30%) or captopril (a commercial antihypertensive drug). Adding quinoa flour into the diet enhanced the level of high-density lipoprotein, blood glucose, and increased activities of catalase and superoxide dismutase. Overall, adding quinoa flour into the diet improved the medical conditions of the hypertensive rats (Cisneros-Yupanqui et al., 2020).

9.3.1.5 Lipid profile and immune regulating

A protein-rich fraction with a protein content of 36% was obtained using a suitable wheel speed of impact mill (Fotschki et al., 2020). The fraction was incorporated in a diet that was fed to Wistar rats for 2 weeks in comparison to a protein-rich fraction from buckwheat and a soy protein isolate. The diet containing quinoa or buckwheat protein-rich fraction increased the body weight gain and biological value of protein in the rats to larger extents compared to that containing soy protein isolate. In the cecum of the rats, the addition of quinoa or buckwheat flour increased the concentration of short-chain fatty acids and the activities of α-glucosidase, α-galactosidase, and α-glucosidase. The flour-containing diet decreased the levels of plasma low density lipoprotein (LDL) cholesterol and total cholesterol. Adding quinoa fraction into the diet increased the levels of alanine transaminase and plasma interleukin 6 and inflammatory activity (Fotschki et al., 2020).

9.3.1.6 Immunonutritional property

The immunonutritional property of quinoa flour was studied using C57BL/6 mice (Selma-Gracia et al., 2020, 2021). The mice were treated with streptozotocin to induce hyperglycemia. The mice were fed with a high fat diet for 3 weeks. Breads containing 25% of quinoa flour and 20% of chia flour were also fed to mice (150 g/day/70 kg body weight) at a frequency of three times per week (150 g/day/70 kg body weight). Feeding mice with breads containing quinoa and chia largely decreased the value of homeostatic model assessment of insulin resistance in hyperglycaemic mice compared to that with white breads. This decreased value had little relation to the reduced starch content in the formulated breads. Feeding mice with breads containing red quinoa and chia increased the levels of triglycerides in the mouse plasma and reduced hepatic lipid peroxidation induced by the high-fat diet. Feeding mice with breads containing white quinoa and chia increased the levels of hepatic IL-6 and TNF-α in the mice. Mice fed with breads containing red quinoa and chia had much less IL-17 than those fed with samples containing white quinoa and chia. The levels of hepatic granulocyte–monocyte colony stimulation factor in the mice were increased by the breads containing quinoa and chia (Selma-Gracia et al., 2020). Feeding mice with breads containing quinoa and chia enhanced the metabolic imbalances in mice with chemically induced hyperglycemia (Selma-Gracia et al., 2020). In another study of Selma-Gracia et al. (2021), compared to the mice fed with white wheat breads, those fed with breads containing quinoa and chia had normalized/regulated insulin resistance, lipid and glucose metabolism, glucose homeostasis, hepatic lipid accumulation expression of coronary risk, and lipogenic markers via controlling the hepatic myeloid cells (Selma-Gracia et al., 2021).

9.3.1.7 Mitigating chemical toxicity

Sprague-Dawley male albino rats were fed with sunset yellow (an orange azo dye) (2.5 mg/kg body weight) or chocolate brown (0.3 mg/kg body weight). They were also fed with a diet containing quinoa flour of up to 10% for 8 weeks. The quinoa intake by the rats decreased the side effects induced by the intake of the synthetic colorants, while regulating the lipid profile (Bahnasy & Ragheb, 2020). In the study of Hamed (2021), rats were fed with chlorfenapyr (180 mg/kg) to induce testicular toxicity. Quinoa seeds (30% of the diet) were also fed to the rats for 4 weeks. Chlorfenapyr treatment led to sagging of testicles, increased testis weights, decreased sex hormone level, sperm count, viability and motility, and increased sperm abnormality. Feeding quinoa seeds mitigated these symptoms in the rats induced by chlorfenapyr (Hamed, 2021). The effects of mitigating chemical toxicity may be related to the bioactive components of quinoa such as polyphenol antioxidants.

9.3.2 Quinoa protein

9.3.2.1 Immunonutritional, antihypertensive, and gut microbiota modulating properties

Protein fractions and isolate from quinoa seeds were immunonutritional, antihypertensive, antiviral, and gut microbiota modulating as studied in vivo (Guo et al., 2021; Llopis et al., 2020). A protein fraction of salt-soluble fraction of <30 kDa was isolated (Llopis et al., 2020). Mice were fed with a high-fat diet for 1 week before receiving the protein fraction solution and diethylnitrosoamine injection. The protein fraction increased the resistance of the mice to hepato-carcinogenesis. The proportions of F4/80$^+$ cells and the level of immune mediators (CD68/CD206 ratio) in the intestinal tissue were increased. The analysis of fecal microbiota showed that the fraction intake led to the pattern of microbial composition to be similar to that with less potential of developing obesity (Llopis et al., 2020). In the study of Guo et al. (2021), spontaneously hypertensive rats were fed with quinoa protein isolate (up to 400 mg quinoa protein/kg body weight) for up to 5 weeks. The protein feeding significantly decreased the blood pressure level of the hypertensive rats, enhanced alpha diversity, and shifted the microbial composition to that of normal rats. The increased abundance of genera including *Allobaculum* and *Turicibacter* in the fecal microbiota was related to decreased blood pressure level. The composition of fecal microbiota in the hypertensive rats fed with quinoa protein was more similar to that in normal rats than that in captopril (an antihypertensive drug) treated rats. Overall, quinoa protein isolate can lower blood pressure and regulate gut dysbiosis due to hypertension (Guo et al., 2021). Similarly, quinoa protein hydrolysates showed different biological functions such as inhibition of angiotensin I converting enzyme and didpetidyl peptidase-IV (López et al., 2019).

9.3.3 Quinoa extracts

Ethanol extracts of quinoa seeds showed hypolipidemic, hepatoprotective, and neuroprotective properties in vivo (Abdel-Wahhab et al., 2021; Hashem et al., 2021; Souza et al., 2020).

9.3.3.1 Hypolipidemia

Rats were treated with Triton X-100 (intraperitoneal injection) to induce hyperlipidemia. The rats also received quinoa extracts for 40 days. The feeding of the extract to the hyperlipidemic rats had hepato-renal protective effect, while decreasing lipid peroxidation, enhancing growth performance and antioxidant activity, and regulating the levels of leptin and adiponectin hormone (Hashem et al., 2021).

9.3.3.2 Hepatoprotection

Rats were treated with cyclophosphamide and/or the ethanol extract (Abdel-Wahhab et al., 2021). The extract reduced the cyclophosphamide-induced hepatotoxicity in the rats. The levels/activities of serum alanine aminotransferase, aspartate aminotransferase, gamma-glutamyl transferase, alkaline phosphatase, low density lipoprotein (LDL)-cholesterol, total cholesterol, triglycerides, and TNF-α IL 1β and IL6, hepatic malondialdehyde, nitric oxide levels, and DNA fragmentation in the rats were decreased by the extract feeding (Abdel-Wahhab et al., 2021).

9.3.3.3 Neuroprotection

Declarative memory deficits in mice were induced by scopolamine (Souza et al., 2020). The mice were fed with the extracts (up to 300 mg/day/kg) for up to 9 days. The memory deficit in the mice was significantly prevented by the extracts. The scopolamine-induced changes in the activities of acetylcholinesterase and superoxide dismutase, and the levels of reactive oxygen species and thiobarbituric acid reactive substance in the cerebral cortex and hippocampus of the mice were reduced by the extracts (Souza et al., 2020). The extracts of quinoa seeds are rich in phenolic antioxidants, which contributed to their neuroprotective effect.

9.3.4 Quinoa polysaccharide

Nonstarch polysaccharide fractions from quinoa seeds showed immunity-improving, gut microbiota regulation, and hyperlipidemia regulation properties in vivo (Cao et al., 2020; Fan et al., 2019).

9.3.4.1 Immunity-improving

A polysaccharide fraction was purified from quinoa crude polysaccharide extract using chromatography (Fan et al., 2019). Mice were treated with cyclophosphamide to induce immunosuppression. The mice were fed with the polysaccharide fraction (100 mg/kg/day) for 3 weeks. The polysaccharide fraction increased the phagocytic function of mononuclear macrophages and the concentrations of lysozyme, IFN-γ, IFN-α, IL-6, and IgM (immunological factors) in the serum of the mice, while decreasing the delayed allergy (Fan et al., 2019).

9.3.4.2 Gut microbiota and hyperlipidemia regulation

Soluble polysaccharide fraction of quinoa seeds (up to 600 mg/kg/day) was fed to rats that also received a high-fat diet for 8 weeks (Cao et al., 2020). The levels/activities of serum low-density lipoprotein cholesterol, total triglyceride, malondialdehyde, hepatic lipid accumulation, total glutamic pyruvic transaminase, and glutamic oxaloacetic transaminase in the rats fed with a high-fat diet were decreased by the polysaccharide. The high-fat diet disrupted the gut microbiota composition. The quinoa polysaccharide intake increased the microbial diversity and decreased the percentages of *Proteobacteria*, *Desulfovibrio*, and *Allobaculum* and the ratio of *Firmicutes* to *Bacteroides* (Cao et al., 2020). Quinoa fraction of soluble polysaccharide positively regulated the serum lipid profile in rats fed with the high-fat diet via regulating gut microbiota (Cao et al., 2020). Studies have shown that gut microbiota composition can be related to dysfunctional immune responses and disease severity in COVID-19 patients (Yeoh et al., 2021). Since quinoa fiber fractions can positively regulate the gut microbiome and reduce the inflammation, intake of quinoa seeds may help alleviate COVID-19-related symptoms in patients.

9.3.5 Quinoa phytoecdysteroid extract

9.3.5.1 Healthspan improvement

Red quinoa seed leachate was obtained using 70% ethanol for extraction (4 h, 80°C) (Graf et al., 2017). *Caenorhabditis elegans* (wild type N2) was treated with the extract (1.0 mg/mL) or metformin (6.5 mg/mL) as control. The quinoa extract treatment enhanced the basal respiration rate (by 37%), locomotory performance (from 104 to 115 head thrashes/min), and lifespan by 2 days (from 9 to 11 days), while reducing the formation of advanced glycation end-product pigments, body fat, and reactive oxygen species by 24%, 14%, and 20%, respectively. Treating the worms with 20-hydroxyecdysone at a equivalent dosage gave similar biological effects (Graf et al., 2017). Ecdysteroids have been proven to have a range of biological functions including antiinflammatory and antioxidant activities and regulation of lipid and carbohydrate metabolism (Savchenko et al., 2022). These properties could be readily linked to the healthspan improvement capacity of the quinoa phytoecdysteroid extract reported by Graf et al. (2017).

9.4 Human trials on health and nutritional effects of quinoa

A few clinical trials on the health effects of quinoa seeds were conducted (Table 9.2). The health effects tested included improving metabolic and serum lipid profiles, glycemic control, and decreasing the risk of cardiovascular diseases (De Carvalho et al., 2014; Gabrial et al., 2016; Graf, Cheng, et al., 2015; Graf, Rojas-Silva, et al., 2015; Karimian et al., 2021; Li et al., 2018; Navarro-Perez et al., 2017; Priyanka et al., 2018).

9.4.1 Metabolic improvement

Postmenopausal overweight women (n = 35) had corn or quinoa flakes (25 g per day) for 4 weeks (De Carvalho et al., 2014). Eating quinoa or corn flakes decreased the levels of serum triglycerides, TBARS (thiobarbituric acid reactive substances), and vitamin E, while increasing the urinary excretion of enterolignans. Eating quinoa but not corn flakes decreased the levels of low density lipoprotein (LDL) cholesterol and total cholesterol, while increasing that of reduced glutathione (De Carvalho et al., 2014). Overall, the intake of quinoa flakes on a daily basis resulted in metabolic improvement in postmenopausal women.

9.4.2 Glycemic control

Patients with type 2 diabetes and healthy persons were fed with buckwheat and quinoa for breakfast (Gabrial et al., 2016). The glycemic properties of the subjects after breakfast and lunch were studied. Healthy subjects who had buckwheat for breakfast had lower incremental area under the curve of glucose after breakfast than those with quinoa or white breads. Diabetic subjects with buckwheat or quinoa as breakfast had lower incremental area under the curve of glucose after breakfast than those with white breads. After lunch, the subjects with buckwheat and quinoa as breakfast showed flatter glycemic response than those with white bread. Quinoa gave a much higher glycemic index (89) than buckwheat (27). Overall, quinoa had the potential to modulate glycemic response in humans (Gabrial et al., 2016). In another study, humans (n = 10) were fed with cooked quinoa (Priyanka et al., 2018). The blood glucose levels of the subjects were measured up to 2 h after the quinoa meals. The ranges of glycemic indexes and glycemic loads were 51 − 78 and 26 − 39, respectively. The averages of glycemic indexes and glycemic loads were 63 and 32, respectively. Quinoa is considered to have medium glycemic index and high glycemic load (Priyanka et al., 2018).

9.4.3 Plasma lipid profile improvement

In a parallel-designed, randomized, controlled, and single-blind trial, obese and overweight participants (n = 50) were fed with 25 and 50 g quinoa per day for 12 weeks (Navarro-Perez et al., 2017). Quinoa intake (50 g per day) after 12 weeks decreased the occurrence of metabolic syndrome (by 70%) and the concentration of serum triglycerides (from 1.1 to 0.72 mmol/L). Quinoa intake (25 g per day) after 12 weeks decreased the occurrence of metabolic syndrome (by 40%) and had no effect on the concentration of serum triglycerides (Navarro-Perez et al., 2017).

9.4.4 Reduction of cardiovascular disease risk

Humans were fed with quinoa-containing (20 g flour) breads for 4 weeks in a cross-over intervention (Li et al., 2018). During intervention, the area under the curve of blood glucose level was lower in the groups eating quinoa-containing breads than that of the groups eating white breads. After intervention, the levels of low density lipoprotein (LDL) cholesterol, blood glucose, and other blood metabolites as well as anthropometric data were similar in humans having breads containing quinoa or not containing quinoa (Li et al., 2018). Overall, the clinical studies showed that daily intake of quinoa (20 g) for 4 weeks decreased the glycemic response but had no effect on other cardiovascular disease-related parameters (Li et al., 2018). In the study of Pourshahidi et al. (2020), healthy adults daily consumed quinoa-containing biscuits (15 g, 60% quinoa flour) in comparison with wheat flour-containing biscuits for 4 weeks in a randomized-controlled, double-blind crossover trial. Eating quinoa biscuits decreased weight and body mass index, the levels of total cholesterol and low-density lipoprotein cholesterol, and the ratio of total cholesterol to high-density lipoprotein cholesterol, while having little effect on the levels of triglycerides, high-density lipoprotein cholesterol, or polyunsaturated fatty acids, or C-reactive protein. Overall, quinoa biscuit consumption by older adults (aged 50−75) lowered their cardiovascular disease risk (Pourshahidi et al., 2020).

Karimian et al. (2021) did a meta-analysis of controlled clinical tests on the effect of quinoa addition on the development and risk of cardiovascular disease in humans. The review paper analyzed the literature published up to February of 2020 and selected five randomized controlled trials involving 206 tested subjects. Adding quinoa to the diet decreased body weight, fat mass, waist circumference, levels of insulin, low-density lipoprotein, triglyceride, and total cholesterol in humans. The results suggested that quinoa had the potential to reduce the risk of cardiovascular diseases in humans (Karimian et al., 2021). Different types of cereals and pseudocereals such as whole grain wheat and buckwheat also showed different health benefits such as reduction of cardiovascular disease risk and glycemic control (Seal et al., 2021; Zhu, 2016). Comparative clinical studies between quinoa and other food grains remain to be done to reveal if there are any nutritional advantages of the former.

9.5 Component-biofunction relationships of quinoa

The various biological functions of quinoa seeds obtained using in vitro, in vivo, and clinical trials (Tables 9.1 and 9.2) could be attributed to the different functional components. For example, the health effects of quinoa proteins, dietary fiber polysaccharides, polyphenols, vitamins, and phytoecdysteroids have been well documented using in vitro and in vivo methods (Tables 9.1 and 9.2). Each type of chemical components in quinoa tends to be multifunctional. For example, the ethanol extracts of quinoa seeds rich in polyphenols and phytoecdysteroids were hypolipidemic, hepatoprotective, and neuroprotective as revealed using mouse and rat models (Abdel-Wahhab et al., 2021; Hashem et al., 2021; Savchenko et al., 2022; Souza et al., 2020). Dietary fiber polysaccharides of quinoa seeds showed immunity improvement and gut microbiota and hyperlipidemia regulation capacities based on mouse and rat models (Cao et al., 2020; Fan et al., 2019). Quinoa protein isolate had antihypertensive and gut microbiota regulating properties (Guo et al., 2021; Llopis et al., 2020). Quinoa contains a significant amount of phytoestrogens, which are effective in preventing some chronic diseases and gynecological diseases (De Carvalho et al., 2014). A comparative study of eight different types of whole grains including buckwheat, proso millet, foxtail millet, barley, coix, oat, sorghum, and quinoa suggested that different grains were nutritionally suitable for different populations (Zhang et al., 2022). Based on in vitro digestibility and nutritional composition of the eight porridges, quinoa porridge along with those of buckwheat and oats had higher dietary fiber content and protein quality, and lower glycemic loads and glycemic indexes compared to those of other grains. Quinoa porridge was recommended for the elderly, obese, athletic, and constipated subjects (Zhang et al., 2022) (Fig. 9.4).

Chapters 5–8 of this book summarized the diverse biological functions of the individual components. Therefore it is known that quinoa seeds or flours in whole form have a range of biological properties. Indeed, in vitro, in vivo, and clinical trials showed that quinoa seeds or flours had a range of different biological properties (Tables 9.1 and 9.2). It remains to be revealed if the claimed health effects of individual quinoa components have any relevance to those of whole quinoa seeds or flours. The interactions between the chemical components of quinoa may affect the biological functions. Many previous studies showed that phytochemical interactions can affect biological activities and bioavailability (Phan et al., 2018;

Figure 9.4 Dietary recommendations of different grains for different populations. F1, lower glycemic load and glycemic index; F2, bioaccessibility of essential amino acid, F3, digestible protein content, F4, dietary fiber content (Zhang et al., 2022).
Source: Reprinted with permission from Elsevier.

Rocchetti et al., 2022). However, the effect of the interactions of quinoa components on these biological activities of whole seeds and flour remains to be studied. The most common way of eating quinoa is whole form instead of its fractionated form such as protein isolate or dietary fiber fraction. In the studies on the isolated components or fractions, the doses used in the experiments usually did not reflect the amount used in quinoa-based foods consumed on a daily basis. Overall, it is very challenging to establish mathematical precision to describe the relationships between the biological activities of individual components based on the studies of isolated compounds and those of whole seeds and flours. On the other hand, the results on the individual components may be useful in value-added processing of quinoa and new product development (Chapter 13).

9.6 Negative health effects of quinoa

9.6.1 Quinoa allergy revealed by clinical reports

A major health concern associated with quinoa consumption has been allergy (Table 9.3). Clinical reports during the last decade or so have shown that some people are allergic to quinoa (Astier et al., 2009; Cuervo-Pardo et al., 2015; Guarnieri et al., 2019; Hong et al., 2013; Kesh et al., 2018; López et al., 2014). For example, a 52-year-old man had quinoa with bread and fish. The man developed symptoms of dysphonia, dysphagia, angio-edema, and urticaria after eating quinoa containing

Table 9.3 Negative health effects of quinoa.

Quinoa form	Sample status	Targeted health effects	Type of experiment	Experimental	Major findings	References
Seed	Grains including quinoa collected in South America and Northern Europe	Mycotoxins	Analytical	Multi-analyte liquid chromatography–tandem mass spectrometry (LC–MS/MS)	Mycotoxin and fungal metabolites (a total of 101) of *Penicillium* spp., *Fusarium* spp., and *Aspergillus* spp. were detected. Their composition varied among different grains, depending on the grain type, agricultural practices, and geographical locations. Cleaning significantly decreased the concentrations of the mycotoxins. The profiles of mycotoxins of quinoa were different from those of wheat, barley, and oats collected from the same region. The concentrations of mycotoxins in quinoa collected from South America were lower than those of quinoa samples from Northern Europe	Ramos-Diaz et al. (2021)
Seed	Andean grains (quinoa grains were not specified) from the market in Peru	Mycotoxins	Analytical	Multi-analyte liquid chromatography–tandem mass spectrometry (LC–MS/MS)	Mycotoxins of *Fusarium* spp. were detected at moderate levels in quinoa samples	Ducos et al. (2021)

(*Continued*)

Table 9.3 (Continued)

Quinoa form	Sample status	Targeted health effects	Type of experiment	Experimental	Major findings	References
Seed	Quinoa cultivated in arsenic-contaminated soils with the presence of biochar from cotton shell. The soil was saline or nonsaline	Cancer and noncancer risk	In vitro	Human health risk was calculated using math equations	The addition of biochar in the soils improved the plant growth and reduced the Na and As uptake. Both noncancer and cancer risk related to As in the quinoa seeds were significantly reduced by the presence of the biochar	Shabbir et al. (2021)
Seed	Flour	Intestinal barrier function	In vivo	Quinoa flour (300 mg) in the presence/absence of capsaicin (12 mg) or cholera toxin (10 μg) was fed to Brown Norway rats for 1 week	Quinoa flour in the diet of rats enhanced intestinal permeability and uptake of compartment-specific proteins. The mechanism of quinoa flour in enhancing the intestinal permeability in rats was different from that of capsaicin or cholera toxin	Ballegaard et al. (2021)
Seed	Cooked quinoa	Allergy (anaphylaxis)	Clinical	A 52-year-old man had quinoa with bread and fish. Immunoblotting test was done	The man developed dysphonia, dysphagia, angio-edema and urticaria after eating a quinoa containing dish. IgE reacted with protein band of 35 kDa in both heated (95°C) and crude quinoa extract	Astier et al. (2009)
Seed	Cooked quinoa in a salad	Allergy (anaphylaxis)	Clinical	A 29-year-old woman was subjected to skin prick testing. Quinoa allergens were removed through extraction using Coca's solution	The woman developed urticaria on arms and chest, "choking" sensation, lip angioedema, and pruritis of soles and palms within a few minutes after eating a salad containing quinoa. Skin prick testing was positive towards quinoa. Quinoa extracts with the allergens removed became negative towards skin prick testing	Hong et al. (2013)

Seed	Cooked quinoa	Allergy	Clinical	A 38-year-old woman with eosinophilic esophagitis was subjected to skin prick tests and serum specific IgE measurement. Quinoa extracts were subjected to SDS-PAGE immunoblotting test	Cross-reactivity between quinoa, wheat, and buckwheat flours was found in the patient. The patient followed flour free diet and eosinophilic inflammation was gone afterward	El-Qutob López et al. (2014)
Seed	Dinner containing cooked quinoa	Allergy	Clinical	A 29-year-old man had a dish containing a variety of different ingredients including quinoa and mustard seeds	The man developed lower extremity urticaria, facial flushing and angio-edema, chest tightness and vomiting in half hour of eating the quinoa dish. Medication with diphenhydramine and loratadine solved the anaphylaxis, which was attributed to mustard	Cuervo-Pardo et al. (2015)
Flour	Pasta containing quinoa flour	Occupational asthma	Clinical	39-year-old male had pasta containing quinoa flour	The man had wheezing of mild-moderate degree when he worked at a plant of quinoa flour processing for 2 years. After eating the pasta, the patient developed urticaria, dyspnea, and angioedema. Prick test using quinoa seed extract showed positive result. By avoiding the contact with quinoa for 1 year, the man's asthma condition improved	Guarnieri et al. (2019)
Seed	Unspecified	Quinoa sensitization	Clinical	A total of 212 tested subjects for IgE of soy, wheat, apple, and quinoa were subjected to cluster analysis	Cluster analysis showed that sensitization to quinoa and that to apple are statistically associated	Kesh et al. (2018)

dish. Immunoblotting test was done to analyze the allergy. IgE reacted with protein band of 35 kDa in both heated (95°C) and crude quinoa extract (Astier et al., 2009). A 29-year-old woman developed urticaria on arms and chest, a "choking" sensation, lip angioedema, and pruritis of soles and palms within a few minutes after eating a salad containing quinoa (Hong et al., 2013). The woman was subjected to skin prick testing for confirmation. The results of the skin prick testing were positive for quinoa. Quinoa allergens were removed through extraction using Coca's solution. The quinoa extracts with the allergens being removed became negative in skin prick testing (Hong et al., 2013). Occupational asthma related to quinoa was reported for a 39-year-old male who had pasta containing quinoa flour (Guarnieri et al., 2019). The man had wheezing of mild-moderate degree when he worked at the operation plant of quinoa flour processing for 2 years. After eating the pasta, the patient developed symptoms of urticaria, dyspnea, and angioedema. Prick test using quinoa seed extract showed a positive result. After avoiding contact with quinoa for 1 year, the man's asthma condition improved (Guarnieri et al., 2019). A total of 212 tested subjects for IgE of soy, wheat, apple, and quinoa were subjected to cluster analysis (Kesh et al., 2018). The cluster analysis showed that sensitization to quinoa and that to apple are statistically associated (Kesh et al., 2018). The exact nature of allergens in quinoa and apple should be studied as the incidents of food anaphylaxis continue to increase. Cross-reacting allergens with quinoa should be studied (Skypala, 2019). The occurrence and epidemiology of quinoa allergy have not been studied well. Breeding programs aimed at developing quinoa varieties with less allergens should be created.

9.6.2 Contaminated quinoa

Quinoa was cultivated in arsenic-contaminated soils (saline or nonsaline) with the presence of biochar from cotton shells (Shabbir et al., 2021). The cancer and noncancer risk of the harvested quinoa seeds for humans were calculated using math equations. The addition of biochar in the soils improved the plant growth and reduced the Na and As uptake. Both noncancer and cancer risk related to As in the quinoa seeds was significantly reduced by the presence of the biochar due to absorption (Shabbir et al., 2021). A range of fungal metabolites and mycotoxins of *Penicillium* spp., *Aspergillus* spp., and *Fusarium* spp. were detected in quinoa grains (Ducos et al., 2021; Ramos-Diaz et al., 2021). The concentrations of these compounds were dependent on climatic and geographic conditions, crop physiology, and cleaning methods. Cleaning can significantly decreased the concentrations of mycotoxins (by ~50%), though it did not eliminate them (Ramos-Diaz et al., 2021). The presence of similar mycotoxins was also found in other food grains such as wheat and kãniwa (*Chenopodium pallidicaule*) (Ramos-Diaz et al., 2021). The presence of toxins in quinoa grains available in the international market and their influencing factors remain to be better studied for effective prevention methods to be developed.

9.6.3 *Intestinal permeability*

Intestinal barrier function of rats affected by quinoa intake was studied in vivo (Ballegaard et al., 2021). Quinoa flour (300 mg) in the presence/absence of capsaicin (12 mg) or cholera toxin (10 μg) was fed to Brown Norway rats for 1 week. Quinoa flour in the diet of the rats enhanced intestinal permeability and uptake of compartment-specific proteins. The mechanism of quinoa flour in enhancing the intestinal permeability in rats was different from that of capsaicin or cholera toxin (Ballegaard et al., 2021). Increased intestinal permeability of cells in humans is associated with several chronic diseases (Odenwald & Turner, 2013). The quinoa flour increased the intestinal permeability and protein uptake in rats, indicating increased risk of developing chronic diseases.

9.7 Research gaps on health effects of quinoa

The usefulness of in vitro studies for practical applications in humans has been under question. For example, the in vitro chemical assays for measuring the antioxidant activities of plant materials have been under question for their relevance for human nutrition (Granato et al., 2018). The results of the in vitro antioxidant activities of quinoa-related products were mostly based on chemical assays such as ABTS and DPPH radical scavenging capacities (Gu et al., 2021; Hur et al., 2018; Ong et al., 2020). It is known that any type of food materials or components can show some degree of antioxidant activity using these chemical assays. Interpretation of these in vitro results such as anticancer property can be misleading in some studies on biological functions of food materials (Wang & Zhu, 2017). In some cases, the results of in vitro tests contradicted with those of in vivo animal-based tests (Ballegaard et al., 2021). The fundamental issue with these kinds of studies is that the relevance of these in vitro models to the practical applications of the results in humans has not been substantiated in the first place. Researchers tend to simply follow the initially proposed model without being able to tell if these in vitro models were truly useful and relevant in the first place when they began their studies. After all, these chemical assays are easy and cheap to perform.

Animal-based studies aim to extrapolate their data to include humans. However, the animal models used to study specific biological functions such as anticancer property may have pitfalls (Murayama & Gotoh, 2019). In the studies on the biological properties of quinoa, different animal models were used. Once again, the fundamental issue with these kinds of studies is that the relevance of these in vivo models to the practical applications of the results in humans has not been substantiated in the first place. Researchers tend to simply follow the initially proposed animal model without being able to tell if these in vivo models were truly useful and relevant in the first place when they begin with their studies. The research community at large needs to go back to the origin of the problem to meticulously identify the suitability of the models for humans. This will be revolutionary in our research community.

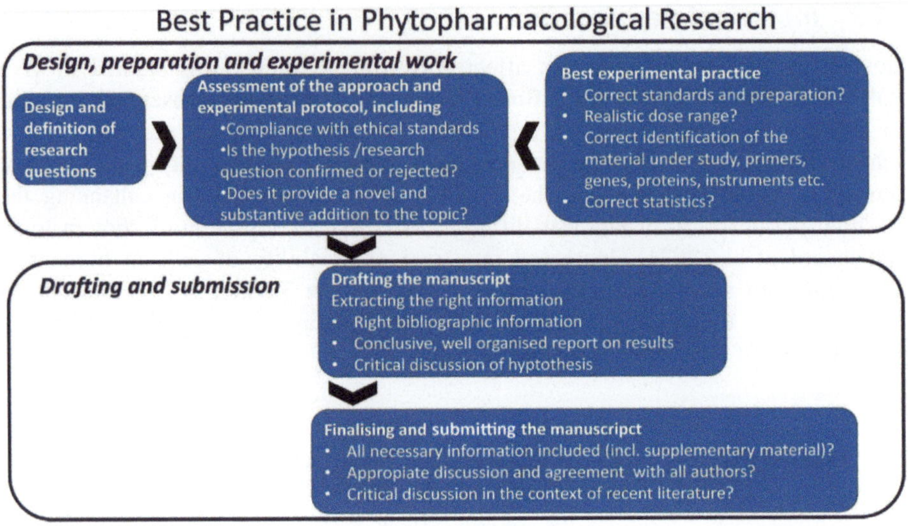

Figure 9.5 Scientific research and publishing as an iterative process of project development, data generation and assessment, evaluation, and manuscript development (Heinrich et al., 2020).
Source: Reprinted with permission from Elsevier.

There has been great inconsistency and bad practice in experimental design in the research of food science and related areas such as phyto-pharmacological research (Heinrich et al., 2020; Karimian et al., 2021). Such a lack of carefulness in experimental design, meticulous execution, and detailed reports of the studies limits the usefulness and validity of these reports on the health effects of quinoa seeds and related products (Fig. 9.5). In the different studies summarized in this chapter, the intervention duration of in vivo and clinical studies varied widely from 9 days to 2 months (Table 9.2). Long-term health effects of quinoa consumption are not clear as the studies were short in duration. Previous studies showed that participants involved in health interventions often change habits for healthier lifestyles. This is a limitation as the influence of these changes may affect our understanding of the health benefits of quinoa (Van den Driessche et al., 2018). Furthermore, many of the studies included a small sample size in both animal and human (e.g., $n = 10$) studies. Larger sample sizes are needed to make the studies more meaningful and results useful. There are many studies showing that different cereals and pseudocereals in whole grain form have a range of health functions (Seal et al., 2021; Ugural & Akyol, 2022). These health functions are similar to those reported for quinoa. Comparison of quinoa with other food grains should be done to reveal the "uniqueness" of quinoa among food grains. Epidemiological studies of quinoa consumption in relation to human health are needed.

9.8 Conclusions

In vitro, in vivo, and clinical studies have shown a range of health effects of quinoa and related products in a consistent manner. Quinoa seeds or whole flours had prebiotic and gut microbiota modulating, glycemic-controlling, and lipid profile regulating, immunonutritional control, hepato-protection, antiobesity, and antidiabetic properties, and had the potential to mitigate chemical toxicity. Isolated quinoa components or fractions also showed similar biological functions, supporting the health claims of whole quinoa seeds and flours. These biological properties were dependent on the quinoa variety and experimental conditions. Negative health effects of eating quinoa included allergy and anaphylaxis, the presence of toxins, and intestinal barrier disruption. However, this area remains to be better studied to have a balanced view on the nutritional quality of quinoa. The validity of the in vitro and in vivo methods used to obtain the claimed health effects of quinoa is being questioned for their practical clinical applications. This chapter clearly showed that quinoa can be applied in food product development for enhanced nutritional quality. The processing and food product development of quinoa were described in Chapters 10–12 of this book.

References

Abdel-Wahhab, K. G., Mannaa, F. A., Ashry, M., Khaled, D. M., Hassan, L. K., & Gomaa, H. F. (2021). *Chenopodium quinoa* ethanolic extract ameliorates cyclophosphamide®-induced hepatotoxicity in male rats. *Comparative Clinical Pathology, 30*, 267–276.

Alarcón, M., Bustos, M., Mendez, D., Fuentes, E., Palomo, I., & Lutz, M. (2020). In vitro assay of quinoa (*Chenopodium quinoa* Willd.) and lupin (*Lupinus* spp.) extracts on human platelet aggregation. *Plant Foods for Human Nutrition, 75*, 215–222.

An, T., Liu, J. X., Yang, X. Y., Lv, B. H., Wu, Y. X., & Jiang, G. J. (2021). Supplementation of quinoa regulates glycolipid metabolism and endoplasmic reticulum stress in the high-fat diet-induced female obese mice. *Nutrition & Metabolism, 18*, 95.

Astier, C., Moneret-Vautrin, D. A., Puillandre, E., & Bihain, B. E. (2009). First case report of anaphylaxis to quinoa, a novel food in France. *Allergy, 64*, 819–820.

Bahnasy, R. M., & Ragheb, E. M. (2020). Effect of quinoa (*Chenopodium quinoa*) on lipid profile in rats exposed to two synthetic food colors. *Egyptian Journal of Nutrition and Health, 15*, 53–70.

Ballegaard, A. S. R., Larsen, J. M., Rasmussen, P. H., Untersmayr, E., Pilegaard, K., & Bøgh, K. L. (2021). Quinoa (*Chenopodium quinoa* Willd.) seeds increase intestinal protein uptake. *Molecular Nutrition & Food Research, 65*, 2100102.

Caeiro, C., Pragosa, C., Cruz, M. C., Pereira, C. D., & Pereira, S. G. (2022). The role of pseudocereals in celiac disease: Reducing nutritional deficiencies to improve well-being and health. *Journal of Nutrition and Metabolism, 2022*, Article ID 8502169.

Cao, Y., Zou, L., Li, W., Song, Y., Zhao, G., & Hu, Y. (2020). Dietary quinoa (*Chenopodium quinoa* Willd.) polysaccharides ameliorate high-fat diet-induced hyperlipidemia and modulate gut microbiota. *International Journal of Biological Macromolecules, 163*, 55–65.

Capraro, J., De Benedetti, S., Di Dio, M., Bona, E., Abate, A., Corsetto, P. A., & Scarafoni, A. (2020). Characterization of chenopodin isoforms from quinoa seeds and assessment of their potential anti-inflammatory activity in Caco-2 cells. *Biomolecules*, *10*, 795.

Casey, R., Adelfio, A., Connolly, M., Wall, A., Holyer, I., & Khaldi, N. (2021). Discovery through machine learning and preclinical validation of novel anti-diabetic peptides. *Biomedicines*, *9*, 276.

Chalamaiah, M., Yu, W., & Wu, J. (2018). Immunomodulatory and anticancer protein hydrolysates (peptides) from food proteins: A review. *Food Chemistry*, *245*, 205–222.

Cisneros-Yupanqui, M., Lante, A., Mihaylova, D., Krastanov, A. I., & Vílchez-Perales, C. (2020). Impact of consumption of cooked red and black *Chenopodium quinoa* Willd. over blood lipids, oxidative stress, and blood glucose levels in hypertension-induced rats. *Cereal Chemistry*, *97*, 1254–1262.

Concha, D. R. M., Martínez, J. E. B., Velázquez, T. G. G., Martínez, C. J., & Ruiz, J. C. R. (2022). Impact of germination time on protein solubility and anti-inflammatory properties of *Pisum sativum* L grains. *Food Chemistry: X*, *13*, 100219.

Cuervo-Pardo, L., Gonzalez-Estrada, A., & Fernandez, J. (2015). Anaphylaxis after homemade quinoa dinner: Hold the mustard. *BMJ Case Reports*, *2015*, bcr2014208980.

Das, D., Kabir, M. E., Sarkar, S., Wann, S. B., Kalita, J., & Manna, P. (2022). Antidiabetic potential of soy protein/peptide: A therapeutic insight. *International Journal of Biological Macromolecules*, *194*, 276–288.

De Carvalho, F. G., Ovídio, P. P., Padovan, G. J., Jordão, J. A. A., Marchini, J. S., & Navarro, A. M. (2014). Metabolic parameters of postmenopausal women after quinoa or corn flakes intake-a prospective and double-blind study. *International Journal of Food Sciences and Nutrition*, *65*, 380–385.

Ducos, C., Pinson-Gadais, L., Chereau, S., Richard-Forget, F., Vásquez-Ocmín, P., Cerapio, J. P., Casavilca-Zambrano, S., et al. (2021). Natural occurrence of mycotoxin-producing *Fusaria* in market-bought Peruvian cereals: A food safety threat for Andean populations. *Toxins*, *13*, 172.

Fan, S., Li, J., & Bai, B. (2019). Purification, structural elucidation and in vivo immunity-enhancing activity of polysaccharides from quinoa (*Chenopodium quinoa* Willd.) seeds. *Bioscience, Biotechnology, and Biochemistry*, *83*, 2334–2344.

Fan, X., Guo, H., Teng, C., Zhang, B., Blecker, C., & Ren, G. (2022). Anti-colon cancer activity of novel peptides isolated from in vitro digestion of quinoa protein in Caco-2 cells. *Foods*, *11*, 194.

Fotschki, B., Juśkiewicz, J., Jurgoński, A., Amarowicz, R., Opyd, P., Bez, J., Muranyi, I., Petersen, I. L., & Llopis, M. L. (2020). Protein-rich flours from quinoa and buckwheat favourably affect the growth parameters, intestinal microbial activity and plasma lipid profile of rats. *Nutrients*, *12*, 2781.

Gabrial, S. G. N., Shakib, M. C. R., & Gabrial, G. N. (2016). Effect of pseudocereal-based breakfast meals on the first and second meal glucose tolerance in healthy and diabetic subjects. *Open Access Macedonian Journal of Medical Sciences*, *4*, 565–573.

Graf, B. L., Cheng, D. M., Esposito, D., Shertel, T., Poulev, A., Plundrich, N., Itenberg, D., Dayan, N., Lila, M. A., & Raskin, I. (2015). Compounds leached from quinoa seeds inhibit matrix metalloproteinase activity and intracellular reactive oxygen species. *International Journal of Cosmetic Science*, *37*, 212–221.

Graf, B. L., Kamat, S., Cheong, K. Y., Komarnytsky, S., Driscoll, M., & Di, R. (2017). Phytoecdysteroid-enriched quinoa seed leachate enhances healthspan and mitochondrial metabolism in *Caenorhabditis elegans*. *Journal of Functional Foods*, *37*, 1–7.

Graf, B. L., Rojas-Silva, P., Rojo, L. E., Delatorre-Herrera, J., Baldeón, M. E., & Raskin, I. (2015). Innovations in health value and functional food development of quinoa (*Chenopodium quinoa* Willd.). *Comprehensive Reviews in Food Science and Food Safety*, *14*, 431–445.

Granato, D., Shahidi, F., Wrolstad, R., Kilmartin, P., Melton, L. D., Hidalgo, F. J., Miyashita, K., Camp, J. V., Alasalvar, C., Ismail, A. B., Elmore, S., Birch, G. G., Charalampopoulos, D., Astley, S. B., Pegg, R., Zhou, P., & Finglas, P. (2018). Antioxidant activity, total phenolics and flavonoids contents: Should we ban in vitro screening methods? *Food Chemistry*, *264*, 471–475.

Gu, R., Chang, X., Bai, G., Li, X., Di, Y., Liu, X., Sun, L., & Wang, Y. (2021). Effects of household cooking methods on changes of tissue structure, phenolic antioxidant capacity and active component bioaccessibility of quinoa. *Food Chemistry*, *350*, 129138.

Guarnieri, G., Bonadonna, P., Olivieri, E., & Schiappoli, M. (2019). Occupational asthma induced by exposure to quinoa. *Journal of Investigational Allergology and Clinical Immunology*, *29*, 451–452.

Gullón, B., Gullón, P., Tavaria, F. K., & Yáñez, R. (2016). Assessment of the prebiotic effect of quinoa and amaranth in the human intestinal ecosystem. *Food & Function*, *7*, 3782.

Guo, H., Hao, Y., Fan, X., Richel, A., Everaert, N., Yang, X., & Ren, G. (2021). Administration with quinoa protein reduces the blood pressure in spontaneously hypertensive rats and modifies the fecal microbiota. *Nutrients*, *13*, 2446.

Halaby, M. S., Abdel-Rahman, M. K., & Hassan, R. A. (2017). Protective influence of quinoa on hypercholesterolemia in male rats. *Current Science International*, *6*, 259–270.

Hamed, N. A. (2021). Quinoa mitigates chlorfenapyr-induced testicular toxicity in rats. *Alexandria Science Exchange Journal*, *42*, 475–486.

Hashem, M. A., Mahmoud, E. A., & Abd-Allah, N. A. (2021). Hypolipidemic activity of an ethanolic extract of quinoa seeds in Triton X-100-induced hyperlipidemic rats. *Comparative Clinical Pathology*, *30*, 473–482.

Heinrich, M., Appendino, G., Efferth, T., Fürst, R., Izzo, A. A., Kayser, O., et al. (2020). Best practice in research—overcoming common challenges in phytopharmacological research. *Journal of Ethnopharmacology*, *246*, 112230.

Hong, J., Convers, K., Reeves, N., & Temprano, J. (2013). Anaphylaxis to quinoa. *Annals of Allergy, Asthma & Immunology*, *110*, 60–61.

Hrubša, M., Alva, R., Parvin, M. S., Macáková, K., Karlíčková, J., Fadraersada, J., Konečný, L., et al. (2022). Comparison of antiplatelet effects of phenol derivatives in humans. *Biomolecules*, *12*, 117.

Hur, J., Nguyen, T. T. H., Park, N., Kim, J., & Kim, D. (2018). Characterization of quinoa (*Chenopodium quinoa*) fermented by *Rhizopus oligosporus* and its bioactive properties. *AMB Express*, *8*, 143.

Karimian, J., Abedi, S., Shirinbakhshmasoleh, M., Moodi, F., Moodi, V., & Ghavami, A. (2021). The effects of quinoa seed supplementation on cardiovascular risk factors: A systematic review and meta-analysis of controlled clinical trials. *Phytotherapy Research*, *35*, 1688–1696.

Kesh, S., Barnes, C. S., Noel-Macdonnell, J., & Patel, N. N. (2018). Association of quinoa and apple sensitization. *The Journal of Allergy and Clinical Immunology*, *141*, AB243.

Li, L., Lietz, G., Bal, W., Watson, A., Morfey, B., & Seal, C. (2018). Effects of quinoa (*Chenopodium quinoa* Willd.) consumption on markers of CVD risk. *Nutrients*, *10*, 777.

Li, M., Xu, T., Zheng, W., Gao, B., Zhu, H., Xu, R., Deng, H., Wang, B., et al. (2021). Triacylglycerols compositions, soluble and bound phenolics of red sorghums, and their radical scavenging and anti-inflammatory activities. *Food Chemistry*, *340*, 128123.

Li, T., Zhang, X., Ren, Y., Zeng, Y., Huang, Q., & Wang, C. (2022). Antihypertensive effect of soybean bioactive peptides: A review. *Current Opinion in Pharmacology*, *62*, 74–81.

Lin, M., Han, P., Li, Y., Wang, W., Lai, D., & Zhou, L. (2019). Quinoa secondary metabolites and their biological activities or functions. *Molecules*, *24*, 2512.

Llopis, J. M. L., Brown, D., & Saiz, B. (2020). *Chenopodium quinoa* and *Salvia hispanica* provide immunonutritional agonists to ameliorate hepatocarcinoma severity under a high-fat diet. *Nutrients*, *12*, 1946.

López, D. E. Q., Bartolomé, Z. B., & Ortiz, I. (2014). Cross-reactivity between buckwheat and quinoa in a patient with eosinophilic esophagitis caused by wheat. *Journal of Investigational Allergology and Clinical Immunology*, *24*, 56–57.

López, D. N., Galante, M., Raimundo, G., Spelzini, D., & Boeris, V. (2019). Functional properties of amaranth, quinoa and chia proteins and the biological activities of their hydrolyzates. *Food Research International*, *116*, 419–429.

Mohamed, D. A., Fouda, K. A., & Mohamed, R. S. (2019). In vitro anticancer activity of quinoa and safflower seeds and their preventive effects on non-alcoholic fatty liver. *Pakistan Journal of Biological Sciences*, *22*, 383–392.

Mudgil, P., Kilari, B. P., Kamal, H., Olalere, O. A., FitzGerald, R. J., Gan, C. Y., & Maqsood, S. (2020). Multifunctional bioactive peptides derived from quinoa protein hydrolysates: Inhibition of α-glucosidase, dipeptidyl peptidase-IV and angiotensin I converting enzymes. *Journal of Cereal Science*, *96*, 103130.

Murayama, T., & Gotoh, N. (2019). Patient-derived xenograft models of breast cancer and their application. *Cells*, *8*, 621.

Mustafa, S., Tariq, Z., Riaz, A., Masoud, M. S., Riaz, M. A., Ilyas, M., Nazar, S., Haq, A. U., & Sabir, Z. (2019). Laboratory scale study on the effect of feeding quinoa (*Chenopodium quinoa*) as meal on serum biochemistry of broiler. *Pure and Applied Biology*, *8*, 2326–2332.

Navarro-Perez, D., Radcliffe, J., Tierney, A., & Jois, M. (2017). Quinoa seed lowers serum triglycerides in overweight and obese subjects: A dose-response randomized controlled clinical trial. *Current Developments in Nutrition*, *1*, e001321.

Navruz-Varli, S., & Sanlier, N. (2016). Nutritional and health benefits of quinoa (*Chenopodium quinoa* Willd.). *Journal of Cereal Science*, *69*, 371–376.

Noratto, G. D., Murphy, K., & Chew, B. P. (2019). Quinoa intake reduces plasma and liver cholesterol, lessens obesity associated inflammation, and helps to prevent hepatic steatosis in obese db/db mouse. *Food Chemistry*, *287*, 107–114.

Obaroakpo, J. U., Liu, L., Zhang, S., Lu, J., Liu, L., Pang, X., & Lv, J. (2020). In vitro modulation of glucagon-like peptide release by DPP-IV inhibitory polyphenol-polysaccharide conjugates of sprouted quinoa yoghurt. *Food Chemistry*, *324*, 126857.

Obaroakpo, J. U., Nan, W., Hao, L., Liu, L., Zhang, S., Lu, J., Pang, X., & Lv, J. (2020). The hyperglycemic regulatory effect of sprouted quinoa yoghurt in high-fat-diet and streptozotocin-induced type 2 diabetic mice via glucose and lipid homeostasis. *Food & Function*, *11*, 8354.

Odenwald, M. A., & Turner, J. R. (2013). Intestinal permeability defects: Is it time to treat? *Clinical Gastroenterology and Hepatology*, *11*, 1075–1083.

Ong, E. S., Pek, C. J. N., Tan, J. C. W., & Leo, C. H. (2020). Antioxidant and cytoprotective effect of quinoa (*Chenopodium quinoa* Willd.) with pressurized hot water extraction (PHWE). *Antioxidants*, *9*, 1110.

Phan, M. A. T., Paterson, J., Bucknall, M., & Arcot, J. (2018). Interactions between phytochemicals from fruits and vegetables: Effects on bioactivities and bioavailability. *Critical Reviews in Food Science and Nutrition*, *58*, 1310–1329.

Pourshahidi, L. K., Caballero, E., Osses, A., Hyland, B. W., Ternan, N. G., & Gill, C. I. (2020). Modest improvement in CVD risk markers in older adults following quinoa (*Chenopodium quinoa* Willd.) consumption: A randomised-controlled crossover study with a novel food product. *European Journal of Nutrition*, *59*, 3313–3323.

Priyanka, M., Suneetha, W. J., Maheswari, K. U., Suneetha, K. B., & Kumari, B. A. (2018). Glycemic index profiling of quinoa (*Chenopodium quinoa* Willd) variety. *Journal of Pharmaceutical Research International*, *21*, JPRI37915.

Ramos-Diaz, J. M., Sulyok, M., Jacobsen, S. E., Jouppila, K., & Nathanail, A. V. (2021). Comparative study of mycotoxin occurrence in Andean and cereal grains cultivated in South America and North Europe. *Food Control*, *130*, 108260.

Rangaswamy, A. N., Ashok, A., Hanumanthappa, P., Chandrashekaramurthy, A. S., Kumbaiah, M., Hiregouda, P., Sharma, V., & Sosalegowda, A. H. (2021). Identification of potential peptide inhibitors of ACE-2 target of SARS-CoV-2 from buckwheat & quinoa. *International Journal of Peptide Research and Therapeutics*, *27*, 1799–1813.

Rocchetti, G., Gregorio, R. P., Lorenzo, J. M., Barba, F. J., Oliveira, P. G., Prieto, M. A., et al. (2022). Functional implications of bound phenolic compounds and phenolics–food interaction: A review. *Comprehensive Reviews in Food Science and Food Safety*, *21*, 811–842.

Rotondo, R., Ragucci, S., Castaldo, S., Oliva, M. A., Landi, N., Pedone, P. V., Arcella, A., & Maro, A. D. (2021). Cytotoxicity effect of quinoin, type 1 ribosome-inactivating protein from quinoa seeds, on glioblastoma cells. *Toxins*, *13*, 684.

Savchenko, R. G., Veskina, N. A., Odinokov, V. N., Benkovskaya, G. V., & Parfenova, L. V. (2022). Ecdysteroids: Isolation, chemical transformations, and biological activity. *Phytochemistry Reviews*. Available from https://doi.org/10.1007/s11101-021-09792-y.

Seal, C. J., Courtin, C. M., Venema, K., & de Vries, J. (2021). Health benefits of whole grain: Effects on dietary carbohydrate quality, the gut microbiome, and consequences of processing. *Comprehensive Reviews in Food Science and Food Safety*, *20*, 2742–2768.

Selma-Gracia, R., Haros, C. M., & Llopis, J. M. L. (2020). Inclusion of *Salvia hispanica* L. and *Chenopodium quinoa* into bread formulations improves *meta*bolic imbalances derived from a high-fat intake in hyperglycaemic mice. *Food & Function*, *11*, 7994.

Selma-Gracia, R., Megušar, P., Haros, C. M., & Llopis, J. M. L. (2021). Immunonutritional bioactives from *Chenopodium quinoa* and *Salvia hispanica* L. flour positively modulate insulin resistance and preserve alterations in peripheral myeloid population. *Nutrients*, *13*, 1537.

Shabbir, A., Saqib, M., Murtaza, G., Abbas, G., Imran, M., Rizwan, M., Naeem, M. A., Ali, S., & Javeed, H. M. R. (2021). Biochar mitigates arsenic-induced human health risks and phytotoxicity in quinoa under saline conditions by modulating ionic and oxidative stress responses. *Environmental Pollution*, *287*, 117348.

Simnadis, T. G., Tapsell, L. C., & Beck, E. J. (2015). Physiological effects associated with quinoa consumption and implications for research involving humans: A review. *Plant Foods for Human Nutrition*, *70*, 238–249.

Sivieri, K., de Oliveira, S. M., de Souza Marquez, A., Pérez-Jiménez, J., & Diniz, S. N. (2022). Insights on β-glucan as a prebiotic coadjuvant in the treatment of diabetes mellitus: A review. *Food Hydrocolloids for Health*, *2*, 100056.

Skypala, I. J. (2019). Food-induced anaphylaxis: Role of hidden allergens and cofactors. *Frontiers in Immunology*, *10*, 673.

Snelson, M., de Pasquale, C., Ekinci, E. I., & Coughlan, M. T. (2021). Gut microbiome, prebiotics, intestinal permeability and diabetes complications. *Best Practice & Research: Clinical Endocrinology & Metabolism*, *35*, 101507.

Souza, S. P., Roos, A. A., Gindri, A. L., Domingues, V. O., Ascari, J., & Guerra, G. P. (2020). Neuroprotective effect of red quinoa seeds extract on scopolamine-induced declarative memory deficits in mice: The role of acetylcholinesterase and oxidative stress. *Journal of Functional Foods*, *69*, 103958.

Srdić, M., Ovčina, I., Fotschki, B., Haros, C. M., & Llopis, J. M. L. (2020). *C. quinoa* and *S. hispanica* L. seeds provide immunonutritional agonists to selectively polarize macrophages. *Cells*, *9*, 593.

Tan, M., Chang, S., Liu, J., Li, H., Xu, P., Wang, P., Wang, X., Zhao, M., Zhao, B., Wang, L., & Zhao, Q. (2020). Physicochemical properties, antioxidant and antidiabetic activities of polysaccharides from quinoa (*Chenopodium quinoa* Willd.) seeds. *Molecules*, *25*, 3840.

Teng, C., Xing, B., Fan, X., Zhang, B., Li, Y., & Ren, G. (2021). Effects of Maillard reaction on the properties and anti-inflammatory, anti-proliferative activity in vitro of quinoa protein isolates. *Industrial Crops & Products*, *174*, 114165.

Ugural, A., & Akyol, A. (2022). Can pseudocereals modulate microbiota by functioning as probiotics or prebiotics? *Critical Reviews in Food Science and Nutrition*, *62*, 1725−1739.

Ujiroghene, O. J., Liu, L., Zhang, S., Lu, J., Zhang, C., Lv, J., Pang, X., & Zhang, M. (2019a). Antioxidant capacity of germinated quinoa-based yoghurt and concomitant effect of sprouting on its functional properties. *LWT-Food Science and Technology*, *116*, 108592.

Ujiroghene, O. J., Liu, L., Zhang, S., Lu, J., Pang, X., & Lv, J. (2019b). α-Glucosidase and ACE dual inhibitory protein hydrolysates and peptide fractions of sprouted quinoa yoghurt beverages inoculated with *Lactobacillus casei*. *Food Chemistry*, *299*, 124985.

Usman, M., Patil, P. J., Mehmood, A., Rehman, A., Shah, H., Haider, J., Xu, K., Zhang, C., & Li, X. (2022). Comparative evaluation of pseudocereal peptides: A review of their nutritional contribution. *Trends in Food Science & Technology*, *122*, 287−313.

Van den Driessche, J. J., Plat, J., & Mensink, R. P. (2018). Effects of superfoods on risk factors of metabolic syndrome: A systematic review of human intervention trials. *Food & Function*, *9*, 1944−1966.

Vilcacundo, R., Miralles, B., Carrillo, W., & Hernández-Ledesma, B. (2018). In vitro chemopreventive properties of peptides released from quinoa (*Chenopodium quinoa* Willd.) protein under simulated gastrointestinal digestion. *Food Research International*, *105*, 403−411.

Wang, S., & Zhu, F. (2017). Dietary antioxidant synergy in chemical and biological systems. *Critical Reviews in Food Science and Nutrition*, *57*, 2343−2357.

Wei, M., Hu, Y., Zou, W., Li, Y., Cao, Y., Li, S., et al. (2022). Physicochemical property and antioxidant activity of polysaccharide from the seed cakes of *Camellia oleifera* Abel. *Food Science & Nutrition*, *10*, 1667−1682.

Wong, F. C., Ong, J. H., Kumar, D. T., & Chai, T. T. (2021). *In silico* identification of multi-target anti-SARS-CoV-2 peptides from quinoa seed proteins. *International Journal of Peptide Research and Therapeutics*, *27*, 1837−1847.

Xu, T., Wu, X., Liu, J., Sun, J., Wang, X., Fan, G., Meng, X., Zhang, J., & Zhang, Y. (2022). The regulatory roles of dietary fibers on host health via gut microbiota-derived short chain fatty acids. *Current Opinion in Pharmacology*, *62*, 36−42.

Yeoh, Y. K., Zuo, T., Lui, G. C., Zhang, F., Liu, Q., Li, A. Y. L., et al. (2021). Gut microbiota composition reflects disease severity and dysfunctional immune responses in patients with COVID-19. *Gut*, *70*, 698−706.

Zeyneb, H., Pei, H., Cao, X., Wang, Y., Win, Y., & Gong, L. (2021). In vitro study of the effect of quinoa and quinoa polysaccharides on human gut microbiota. *Food Science & Nutrition*, *9*, 5735–5745.

Zhang, S., Hu, J., Sun, Y., Ji, H., Liu, F., Peng, X., Zhong, Y., Geng, F., & Nie, S. (2022). In vitro digestion of eight types of wholegrains and their dietary recommendations for different populations. *Food Chemistry*, *370*, 131069.

Zhu, F. (2016). Chemical composition and health effects of Tartary buckwheat. *Food Chemistry*, *203*, 231–245.

Domestic cooking and sensory properties of quinoa

10

10.1 Introduction

Understanding consumers' attitude towards quinoa is essential to its sustainable consumption and production. Consumer preferences are not only determined by quinoa's sensory quality but also by a range of social, political, and cultural factors (Medina et al., 2020; Nosi et al., 2020). Purchasing behaviors and attitudes of consumers towards quinoa and its organics influence its preferences (Gosine & McSweeney, 2019; Nosi et al., 2020; Higuchi & Maehara, 2021; Muziri et al., 2021). However, this area is much understudied.

Quinoa seeds have been formulated into a range of cuisines and commercial food products such as salads, gluten free cakes, cookies, and vegetable meatballs (Sezgina & Sanlier, 2019). Raw quinoa seeds are subjected to domestic cooking. There are different types of domestic cooking (Goh & Lee, 2017; Jeong & Sim, 2020; Chen et al., 2022; Gu et al., 2021; Repo-Carrasco-Valencia et al., 2010; Carciochi et al., 2016; Castro-Alba et al., 2019; Mota et al., 2016). Diverse domestic cooking can significantly affect the chemical composition and properties of diverse phytochemicals in plant foods (Zhao et al., 2019). It is essential to understand the effects of these cooking methods on quality attributes of quinoa. Quinoa seeds may undergo biochemical and chemical changes during storage (Kibar et al., 2021). Thermal processing can lead to partial degradation and loss of different bioactive components to various extents, depending on the type of processing and its conditions (Mhada et al., 2020). Suitable processing conditions should be employed to minimize the loss of nutrients. On the other hand, the sensory quality of quinoa is essential for its consumption and commercial success. The taste, texture, and aroma of cooked quinoa seeds have been studied (Wu, 2016). However, the sensory quality is affected by the interplay among these parameters, which is still to be better studied. To be successful in commercialization of quinoa, its nutritional, chemical, and sensory attributes should be balanced to meet consumers' requirements. The interplay between sensory and nutritional properties of cooked quinoa affected by diverse cooking methods has not been well understood. This hinders the commercial development of quinoa on the world stage.

The processing of quinoa by a range of methods including extrusion, fermentation, and germination are described in Chapter 11. The formulations and product development of quinoa have been summarized in Chapter 12. This chapter summarizes the factors affecting consumer preferences, effects of different domestic and household cooking methods on physicochemical and nutritional properties, and sensory properties of quinoa seeds. This content of this chapter can be useful to culinary practice and research.

Quinoa. DOI: https://doi.org/10.1016/B978-0-323-99909-0.00009-X

10.2 Consumer preference for quinoa

Consumer preference to food products including quinoa can be influenced by a range of factors such as sensory acceptance, political values, cultural associations, ecological welfare, nutritional claims, and perceived corporate social responsibility (Gosine & McSweeney, 2019; Nosi et al., 2020; Higuchi & Maehara, 2021; Medina et al., 2020; Muziri et al., 2021). A few studies reported on these factors. For example, among products formulated with millet, sorghum, quinoa, amaranth, buckwheat, and spelt, those with quinoa had the highest consumer acceptance (Gosine & McSweeney, 2019). The products that were nutty and sweet were preferred. Health claims including reduction in the risk of type 2 diabetes and/or cardiovascular diseases and as a source of dietary fiber were associated with higher consumer acceptance (Gosine & McSweeney, 2019). Medina et al. (2020) surveyed the consumer behaviors of 365 households towards quinoa at Latacunga in Cotopaxi Province, Ecuador. Over half of the households consumed quinoa flour or grains. Increased awareness of the nutritional benefits of quinoa and a tradition associated with quinoa were correlated with increased quinoa consumption (Medina et al., 2020). Higuchi and Maehara (2021) used factor-cluster analysis of the quinoa consumption behaviors of 381 consumers from Lima, Peru. The consumer preference towards quinoa was dominated by two patterns including the ethnical and attitudinal identity as well as the past experience and subjective norm identity (Higuchi & Maehara, 2021). Nosi et al. (2020) studied consumers' tendency to buy organic quinoa. Overall, consumers' purchase of organic quinoa was positively driven by the ecological welfare and corporate social responsibility of retailers, and was negatively driven by political values. The political values denote that the product comes from a country the consumer approves of politically, from a country in which human rights are not violated, has the country of origin clearly marked, and has been prepared in a way that does not conflict with the consumers' political values (Lindeman & Väänänen, 2000). Muziri et al. (2021) used principal component analysis to study consumers' perception of health claims of quinoa in Zimbabwe. Age, income, and education level were found to be the factors affecting consumers' perceptions (Muziri et al., 2021). Overall, consumers' attitude towards quinoa is affected by a range of different factors that not only included the sensory acceptance but also social, political, and cultural aspects. In this chapter, the sensory properties of cooked quinoa seeds are described in detail.

10.3 Domestic cooking of quinoa

Quinoa seeds were cooked using different household cooking methods including frying, steaming, atmospheric pressure cooking, roasting, baking, and boiling (Goh & Lee, 2017; Jeong & Sim, 2020; Chen et al., 2022; Gu et al., 2021; Repo-Carrasco-Valencia et al., 2010; Carciochi et al., 2016; Castro-Alba et al., 2019; Mota et al., 2016). Thermal processing can alter the structure of quinoa seeds

(Gu et al., 2021). Different chemical and physicochemical parameters of quinoa seeds affected by different cooking methods at different conditions have been studied. Quinoa has been used in various different recipes for culinary uses (Sonoma Press, 2015). Understanding how domestic cooking methods can affect the nutritional and sensory properties of quinoa is essential to understand the nutritional quality of quinoa-based dishes and to expand its applications and consumption.

10.3.1 Boiling

Mota et al. (2016) found that cooking types had an effect on the mineral composition of different pseudocereals including quinoa. True retention rates of minerals ranged from 83% for copper in boiled quinoa seeds to 100% for manganese, iron, and phosphorus in steamed seeds. Therefore the data on the mineral composition of quinoa described in Chapter 2 of this book should be treated with caution for nutritional applications. The mineral retention in quinoa seeds depended on the cooking as well as the mineral type (Mota et al., 2016). Boiled quinoa seeds had more free amino acids than roasted and steamed samples (Jeong & Sim, 2020). Repo-Carrasco-Valencia et al. (2010) found that boiling increased in vitro zinc dialyzability of quinoa. Goh and Lee (2017) showed that boiling had little effect on the contents of proteins, lipids, ash, and starch. Boiling reduced the contents of flavonoids and phenolics to a larger extent than steaming or roasting. Heating methods including boiling, roasting, and steaming increased water absorption index, yellowness, starch digestibility, and reduced the lightness, water solubility index, and in vitro antioxidant activity. Boiling reduced the in vitro antioxidant activity to a larger extent than steaming and roasting (Goh & Lee, 2017). Boiling increased the protein and starch hydrolysis of quinoa as revealed by in vitro oral and gastric digestion model (Tagle-Freire et al., 2022). Cooking with heat such as boiling, steaming, and roasting in general decreased the contents of unsaturated fatty acids and vitamins and increased those of saturated fatty acids and amino acids (Jeong & Sim, 2020).

10.3.2 Steaming

Jeong and Sim (2020) showed that steamed quinoa seeds had higher carbohydrate and dry matter contents, more iron, magnesium and zinc, and manganese, and less amino acids, water-soluble vitamins, free sugars, and fatty acids than boiled and roasted samples. Goh and Lee (2017) showed that roasting and steaming had little effect on the contents of proteins, lipids, ash, and starch, while reducing those of phenolics and flavonoids. Gu et al. (2021) showed that steaming reduced the contents of flavonoids and phenolics and the in vitro antioxidant activity of quinoa seeds. The steaming process significantly disrupted the tissue structure of quinoa seeds (Fig. 8.8). The steaming increased the bioaccessibility of tocopherol (Gu et al., 2021). Mhada et al. (2020) compared the effect of boiling (sample-to-water ratio at 1–3, boiled for 15 min), steaming (steamed for 45 min), and grain polishing on some nutritional aspects of quinoa samples. Cooking (steaming and boiling) and polishing decreased the contents of proteins, phenolics, and minerals. The effect of polishing was more significant than that of cooking. Steaming better retained polyphenols and minerals than boiling. Boiling led up to 88% of the loss of phenolics and 40% of the loss of some minerals due to the leaching effects (Mhada et al., 2020).

10.3.3 Baking and roasting

Gu et al. (2021) showed that baking had little effect on the contents of flavonoids and phenolics and the in vitro antioxidant activity of quinoa seeds. The hypocotyl-radicle axis of the seeds protected the phenolics from degradation (Fig. 8.8). Repo-Carrasco-Valencia et al. (2010) found that roasting had little effect on in vitro mineral dialyzability of quinoa. Maillard reaction products were produced from the roasting process on quinoa seeds (Carciochi et al., 2016). Increasing the heating temperature increased the amount of Maillard reaction products and in vitro antioxidant activity. Roasting time had less impact on the products and the antioxidant function. A maximum of the in vitro antioxidant activity was obtained at a roasting condition of 190°C for 30 min. The antioxidant activity was mostly contributed by the Maillard reaction products rather than by the phenolics (Carciochi et al., 2016). A comparative study showed that roasted quinoa seeds had more dietary fibers and glucose than boiled and steamed samples (Jeong & Sim, 2020). Castro-Alba et al. (2019) found that combinations of fermentation and dry roasting at different conditions significantly reduced the phytate content of quinoa by 30%−73%. Fermentation followed by the roasting of the raw quinoa flour most effectively increased the estimated bioavailability of iron and zinc. Dry roasting increased the sensory attributes of the fermented quinoa flour. The treatment conditions were important in affecting the outcomes of the processing. Fermentation of quinoa flour for 4 h followed by dry roasting gave higher sensory acceptance of the resulting porridge than that for 10 h (Castro-Alba et al., 2019). Overall, baking and roasting conditions are important to determine the sensory and nutritional quality of quinoa products.

10.3.4 High-pressure assisted cooking

Quinoa seeds were subjected to heating (3−5 min) at 0.6 MPa (Wu et al., 2021). The maximal amounts of flavonoid and saponin contents, and the maximal in vitro antioxidant activities were obtained at the treatment conditions. The saponin and flavonoid contents were better retained in pressure-heated samples than in boiled, extruded, and baked samples (Wu et al., 2021).

10.3.5 Combined cooking methods

Pressure cooking (117°C, 80 kPa, 10 min), microwave cooking (15 min at 490 W), or atmospheric pressure cooking (in a stainless steel pot, 35 min) in combination with freeze drying assisted by infrared were used to make instant quinoa (Chen et al., 2022). The quinoa seed-to-water ratio was set at 1−1.5. The cooking gelatinized starch in quinoa, while the infrared assisted drying kept the gelatinized state of quinoa. The highest rehydration capacity and porosity were observed in pressure-cooked samples. The rehydrated products of microwaved and pressure-cooked samples were thicker and softer than that of atmospheric pressure-cooked samples. The freeze drying process with the assistance of infrared retained the flavor of the samples (Chen et al., 2022). On another note, the freeze drying equipment coupled with infrared is not readily available on the commercial market. Thus this cooking method may not become readily domestic.

It is apparent that the physicochemical and nutritional properties of cooked quinoa seeds are largely dependent on the cooking methods and conditions. The ultimate choice of domestic cooking methods and conditions is not only dependent on the nutritional composition, but also largely on the taste and sensory preference of the consumers. In most of the studies summarized in this section, the sensory properties of the cooked quinoa seeds were not studied. Future studies should be comprehensive in considering both the nutritional and sensory outcomes of the cooking methods.

10.4 Sensory properties of cooked quinoa

A few studies have explored the taste, color, aroma, flavor, and texture of cooked quinoa seeds (Bonfiglio et al., 2020; Gamboa et al., 2018; Suárez-Estrella et al., 2018, 2021, Wu et al., 2014, 2017b, 2017a; Zhang et al., 2019; Wu, 2016) (Table 10.1).

10.4.1 Taste

The flavor or taste of cooked quinoa seeds was described as toasty, grain-like, nutty, bitter, earthy, and sweet (Wu et al., 2017a). A major issue with the taste of cooked quinoa has been the bitterness (Koziol, 1991; Suárez-Estrella et al., 2018; Gamboa et al., 2018; Bonfiglio et al., 2020; Suárez-Estrella et al., 2021; León-Roque et al., 2019; Kavali et al., 2019). The bitter taste of quinoa could be largely attributed to the presence of saponins (León-Roque et al., 2019). A literature survey by Suárez-Estrella et al. (2018) showed that saponins are bitter in taste and negatively affect the taste of the quinoa products. A survey of smallholders of farming from Peruvian Andes showed that smallholders preferred to cultivate quinoa varieties with mid-to-low saponin contents (Gamboa et al., 2018). The threshold and dynamics of bitter taste of saponins in quinoa were studied using sensory and chemical approaches (Bonfiglio et al., 2020). Different degrees of sensitivity and perception dynamics towards bitterness perception were obtained among the panelists. The stimulus concentrations were positively correlated with three types of saponins identified. The area under the curve, maximum intensity, rising slope, and total duration were strongly affected by the three types of saponins. The dynamics of bitterness perception were more related to the three types of saponin fractions than to the total saponin content (Bonfiglio et al., 2020).

Various methods were used to decrease the bitterness and the level of saponins in quinoa. Washing, mechanical pearling, and/or breeding sweet varieties are the strategies

Table 10.1 Sensory and textural properties of quinoa seeds.

Parameters	Number of varieties	Experimental	Major findings	References
Texture	13	Texture profile analysis of cooked quinoa seeds of 13 different varieties	Great diversity in textural properties of these samples was obtained. Depending on the specific varieties, the cooked seeds can be harder, softer, more adhesive, or stickier in texture. These textural parameters were positively correlated with the protein content in the seeds. Hardness, chewiness, and gumminess of the cooked seeds were positively correlated with seed hardness. Hardness, gumminess, cohesiveness, and chewiness were positively correlated with seed coat proportion and were negatively with seed density. Increasing cooking length enhanced the values of these four textural parameters. Increasing water intake decreased the hardness, chewiness, and gumminess. Pasting and thermal parameters of quinoa flour were correlated with different textural parameters	Wu et al. (2014)
Texture	13	Texture profile analysis of cooked quinoa seeds in relation to starch properties	Significant variations in the properties of the starches and textural parameters of texture profile analysis of the cooked seeds of 13 quinoa varieties were obtained. More cohesive and more adhesive, chewier, and firmer textures of cooked seeds were correlated with higher enthalpy change (ΔH) of their starches. Higher amylose leaching and more amylose–lipid complexes of the starch components gave the cooked seeds less chewy and softer texture. Higher amylose contents gave the cooked seeds stickier, harder, more gummy, more chewy, and more cohesive texture	Wu et al. (2017b)
Aroma	30	Volatile compounds of quinoa porridge	A total of 53 volatile compounds were detected. They included 14 alkanals, 10 alkanes, 10 acids and esters, 8 heterocycles, 7 ketones, and 4 alcohols. Acids and esters (44.3%) and alkanes (23.0%) dominated the weight of the volatiles. The porridges of the 30 quinoa the varieties had similar numbers and types of the volatiles, but they varied in the concentrations. Hexanal, 1-octen-3-ol, 2-pentylfuran, nonanal, (E,E)-2,4-decadienal and 6,10-dimethyl-5,9-undecadien-2-one, heptanal, benzeneacetaldehyde and decanal were the major aroma contributors. All the porridges had a relatively high content of 6,10,14-trimethyl-2-pentadecanone	Zhang et al. (2019)
Aroma	11	Raw and cooked seeds	Yellowish, white and black quinoa seeds had caramel, grassy, and fruity notes of aroma, respectively. A total of 61 compounds were identified in the 11 quinoa varieties. Raw white quinoa seeds had a high amount of "grassy" volatiles (e.g., n-hexanol). The quinoa seeds (2 g) were mixed with water (2.6 mL) and were heated at 100°C for 20 min to stimulate cooking process. "Green" aromatic compounds such as (E)-2-heptenal and (E)-2-octenal were developed by the cooking. Cooking led to the formation of caramel and sweet aromatic notes with volatiles such as decanal, 5-methyl-furfural, and 2-furfural. Cooking also led to the formation of fruity aromatic notes with compounds such as phenylacetaldehyde and methyl hexanoate	Yang et al. (2021)

Aroma, taste, flavor, texture, color	21	Cooked seeds	Lexicon to describe aroma, texture, taste, flavor, and color of cooked quinoa seeds was developed. The lexicon with sensory descriptors effectively grouped the cooked seeds of 21 quinoa varieties based on the sensory categories. An adhesive texture was observed for three cooked seeds. Consumer evaluation (100 participants) identified the most acceptable and least acceptable varieties. Crunchy and firm texture and grassy aroma contributed to high consumer acceptance, though specific groups of consumers had different extents of acceptance in different sensory attributes	Wu et al. (2017a), Wu (2016)
Taste	Not applicable	Literature survey	Saponins are bitter in taste and negatively affect the taste of the quinoa products. Washing, mechanical pearling, and/or breeding sweet varieties are the strategies to reduce the saponin content and the bitter taste	Suárez-Estrella et al. (2018)
Taste	Not applicable	Survey of smallholders of farming from Peruvian Andes	Smallholders preferred to cultivate quinoa varieties with mid-to-low saponins contents	Gamboa et al. (2018)
Taste	1	Sensory and chemical analysis	The threshold and dynamics of bitterness taste of saponins in quinoa were studied. Different degrees of sensitivity and dynamics towards bitterness perception were obtained among the panelists. The stimulus concentrations were positively correlated with three types of saponins identified. The area under the curve, maximum intensity, rising slope, and total duration were strongly affected by the three types of saponins. The dynamics of bitterness perception were more related to the three types of saponins fractions than the total saponin content	Bonfiglio et al. (2020)
Taste	1	Seed sprouting for up to 72 h at 22°C	Astringency and bitterness of the quinoa seeds were decreased with increasing sprouting time by decreasing the amount of saponins. The polyphenols in quinoa played little role in affecting the taste of quinoa	Suárez-Estrella et al. (2021)

used to reduce the saponin content and the bitter taste (Suárez-Estrella et al., 2018). Koziol (1991) showed that quinoa varieties containing less than 0.11% of saponins were considered "sweet" in taste. Thus a breeding program could be developed to make new "sweet" varieties. Kavali et al. (2019) found that soaking quinoa seeds in sodium bicarbonate (1%) was more efficient at removing saponins and the bitter taste than citric acid soaking and cold/hot water washing. Suárez-Estrella et al. (2021) studied seed sprouting for up to 72 h at 22°C in relation to saponin content and quinoa taste. Astringency and bitterness of the quinoa seeds were decreased with increasing sprouting time by decreasing the amount of saponins. The polyphenols in quinoa had little role in affecting the taste of quinoa (Suárez-Estrella et al., 2021). On the other hand, saponins from quinoa showed positive health effects such as regulating human gut microbiota (Chen et al., 2022). Therefore a balanced view of the role of saponins in quinoa quality is needed to advise consumers on suitable portion of serving.

10.4.2 Aroma

The aroma profiles of cooked quinoa seeds were described as grassy, nutty, caramel, woody, earthy, and buttery (Wu et al., 2017a, 2017b). Yellowish, white, and black quinoa seeds had caramel, grassy, and fruity notes of aroma, respectively (Yang et al., 2021). A total of 61 compounds were identified in the 11 quinoa varieties. Raw white quinoa seeds had a high amount of "grassy" volatiles (e.g., *n*-hexanol). The quinoa seeds (2 g) were mixed with water (2.6 mL) and were heated at 100°C for 20 min to stimulate cooking process (Yang et al., 2021). "Green" aromatic compounds such as (E)-2-heptenal and (E)-2-octenal were developed by the cooking. Cooking led to the formation of caramel and sweet aromatic notes with volatiles such as decanal, 5-methyl-furfural, and 2-furfural. Cooking also led to the formation of fruity aromatic notes with compounds such as phenylacetaldehyde and methyl hexanoate (Yang et al., 2021). Zhang et al. (2019) studied the volatile compounds of porridge samples of 30 quinoa varieties. A total of 53 volatile compounds were detected. They included 14 alkanals, 10 alkanes, 10 acids and esters, 8 heterocycles, 7 ketones, and 4 alcohols. Acids and esters (44.3%) and alkanes (23.0%) dominated weight of the volatiles. The porridges of the 30 quinoa varieties had similar numbers and types of the volatiles, but they varied in the concentrations. Hexanal, 1-octen-3-ol, 2-pentylfuran, nonanal, (E,E)-2,4-decadienal and 6,10-dimethyl-5,9-undecadien-2-one, heptanal, benzeneacetaldehyde, and decanal were the major aroma contributors. All the porridges had a relatively high content of 6,10,14-trimethyl-2-pentadecanone (Zhang et al., 2019). The molecular basis of the aromatic profiles of quinoa remains to be better studied.

10.4.3 Texture

The descriptors for texture of cooked quinoa seeds included "astringent", "firm", "cohesive", "crunchy", "pasty", "adhesive", "moist", and "chewy" (Wu et al., 2017a, 2017b). Wu et al. (2014) analyzed the textural attributes of cooked quinoa seeds of 13 different varieties using texture profile analysis (TPA). Great diversity in the textural properties of these samples was obtained. Depending on the specific varieties, the cooked seeds can be harder, softer, more adhesive, or stickier in texture. These textural parameters were positively correlated with the protein content in the seeds. Hardness, chewiness, and gumminess of the cooked seeds were positively correlated with seed hardness. Hardness, gumminess, cohesiveness, and chewiness were positively correlated with seed coat proportion and were negatively correlated with seed density. Increasing cooking length enhanced the values of these four textural parameters. Increasing water intake decreased the hardness, chewiness, and gumminess. Pasting and thermal parameters of quinoa flour were correlated with different textural parameters of cooked quinoa seeds (Wu et al., 2014). For example, adhesiveness was negatively correlated with setback viscosities of the flour. Setback and final viscosities were negatively correlated with the four textural parameters. The breakdown values and peak viscosities of flour pasting were inversely correlated with hardness, chewiness, and gumminess of the cooked quinoa seeds. The four textural parameters were positively correlated with onset gelatinization temperature (T_o), but they had no correlations with enthalpy change (ΔH) nor conclusion temperature (T_c) of starch gelatinization (Wu et al., 2014). Starch is the major component of quinoa seeds. TPA of cooked quinoa seeds in relation to starch properties was studied among 13 quinoa varieties (Wu et al., 2017b). Significant variations in the properties of the starches and textural parameters of TPA of cooked seeds of the 13 quinoa varieties were recorded. More cohesive and more adhesive, chewier, and firmer textures of cooked seeds were correlated with higher enthalpy change (ΔH) of their starches. Higher amylose leaching and more amylose–lipid inclusion complexes of the starch components gave the cooked seeds a less chewy and softer texture. Higher amylose contents gave the cooked seeds stickier, harder, gummier, chewier, and more cohesive texture (Wu et al., 2017b). There has been increasing concerns about using the TPA to analyze food texture due to its inability to accurately reflect food textural system (Peleg, 2019). Such a concern remains to be applied in the case of quinoa texture. Unfortunately, most of the researchers continue to use TPA for food texture analysis without asking if this method is appropriate for their subject of research.

10.4.4 Interplay among different sensory attributes of quinoa

A lexicon to describe aroma, texture, taste, flavor, and color of cooked quinoa seeds was developed (Wu et al., 2017a; Wu, 2016). The lexicon with sensory descriptors effectively grouped the cooked seeds of 21 quinoa varieties based on the sensory categories. An adhesive texture was observed for three cooked seeds. Consumer evaluation (100 participants) identified the most acceptable and the least acceptable varieties. Crunchy and firm texture together with grassy aroma contributed to high consumer acceptance, though specific groups of consumers had different extents of acceptance in different sensory attributes (Wu et al., 2017a; Wu, 2016). Overall, the sensory acceptance of cooked quinoa is determined by multiple factors including texture and aroma. However, the interplay between the texture and aroma profiles affecting the sensory properties of quinoa remains to be studied. Such interplay in relation to culinary innovation should be considered.

10.5 Conclusions

The consumer choice in purchasing quinoa is affected by a range of factors including sensory acceptance, nutritional claims, social, political, and cultural influences. Quinoa commercial development should take the social and political factors into consideration to make it successful on a world stage especially in Western countries. Quinoa seeds were subjected to different cooking methods including frying, steaming, atmospheric pressure cooking, roasting, baking, and boiling. These cooking methods changed the physicochemical and nutritional properties of quinoa to different extents. The extents of changes depended on the type of cooking and cooking conditions such as heating length and temperatures. Thermal processing such as boiling tended to degrade certain nutrients such as polyphenols in quinoa. Suitable thermal processing conditions should be used to minimize the loss of nutrients. Sensory descriptors of cooked quinoa seeds were developed. The bitterness of cooked quinoa seeds may be reduced using processing methods such as washing and mechanical pearling. The sensory acceptance of cooked quinoa is determined by the interplay between taste, texture, and aroma. To maximize the eating quality and commercial viability of quinoa, sensory, nutritional, and social aspects should be the points of considerations.

References

Bonfiglio, G. V., Wierna, R. V., Bonini, N. A., Armada, M., & Goldner, M. C. (2020). Study of bitterness perception of quinoa (*Chenopodium quinoa* Willd.) saponin extracts. *Journal of Cereal Science*, *95*, 103032.

Carciochi, R. A., D'Alessandro, L. G., & Manrique, G. D. (2016). Effect of roasting conditions on the antioxidant compounds of quinoa seeds. *International Journal of Food Science and Technology*, *51*, 1018–1025.

Castro-Alba, V., Lazarte, C. E., Perez-Rea, D., Sandberg, A. S., Carlsson, N. G., Almgren, A., Bergenståhl, B., & Granfeldt, Y. (2019). Effect of fermentation and dry roasting on the nutritional quality and sensory attributes of quinoa. *Food Science & Nutrition*, *7*, 3902–3911.

Chen, J., Zhang, Y., Guan, X., Cao, H., Li, L., Yu, J., & Liu, H. (2022). Characterization of saponins from differently colored quinoa cultivars and their in vitro gastrointestinal digestion and fermentation properties. *Journal of Agricultural and Food Chemistry*, *70*, 1810–1818.

Chen, K., Zhang, M., Bhandari, B., & Chen, J. (2022). Instant quinoa prepared by different cooking methods and infrared-assisted freeze drying: Effects of variables on the physicochemical properties. *Food Chemistry*, *370*, 131091.

Gamboa, C., Van den Broeck, G., & Maertens, M. (2018). Smallholders' preferences for improved quinoa varieties in the Peruvian Andes. *Sustainability*, *10*, 3735.

Goh, H. K., & Lee, Y. T. (2017). Effects of heat treatments on physicochemical properties and in vitro biological activities of quinoa (*Chenopodium quinoa* Willd.). *Journal of the Korean Society of Food Science and Nutrition*, *46*, 688–694.

Gosine, L., & McSweeney, M. B. (2019). Consumers' attitudes towards alternative grains: A conjoint analysis study. *International Journal of Food Science and Technology*, *54*, 1588–1596.

Gu, R., Chang, X., Bai, G., Li, X., Di, Y., Liu, X., Sun, L., & Wang, Y. (2021). Effects of household cooking methods on changes of tissue structure, phenolic antioxidant capacity and active component bioaccessibility of quinoa. *Food Chemistry*, *350*, 129138.

Higuchi, A., & Maehara, R. (2021). A factor-cluster analysis profile of consumers. *Journal of Business Research*, *123*, 70–78.

Jeong, K. Y., & Sim, K. H. (2020). Comparison of the nutritional composition of quinoa seeds cultivated in Korea depending on different cooking nethods. *Korean Journal of Food and Nutrition*, *33*, 117–130.

Kavali, S., Shobha, D., Shekhara, N. R., & Brundha, A. R. (2019). Development of value added products from quinoa using different cooking methods. *The Pharma Innovation Journal*, *8*, 548–554.

Kibar, H., Sönmez, F., & Temel, S. (2021). Effect of storage conditions on nutritional quality and color characteristics of quinoa varieties. *Journal of Stored Products Research*, *91*, 101761.

Koziol, M. J. (1991). Afrosimetric estimation of threshold saponin concentration for bitterness in quinoa (*Chenopodium quinoa* Willd). *Journal of the Science of Food and Agriculture*, *54*, 211–219.

León-Roque, N., Aguilar-Tuesta, S., Quispe-Neyra, J., Mamani-Navarro, W., Alfaro-Cruz, S., & Condezo-Hoyos, L. (2019). A green analytical assay for the quantitation of the total saponins in quinoa (*Chenopodium quinoa* Willd.) based on macro lens-coupled smartphone. *Talanta*, *204*, 576–585.

Lindeman, M., & Väänänen, M. (2000). Measurement of ethical food choice motives. *Appetite*, *34*, 55–59.

Medina, P. H., Maqueda, R. H., Gavilanes, P., & Quevedo, K. P. M. (2020). Culture and healthy lifestyle: Factors influencing the decision to buy quinoa in the city of Latacunga in Cotopaxi province, Ecuador. *Journal of Food Products Marketing*, *26*, 440–455.

Mhada, M., Metougui, M. L., El Hazzam, K., El Kacimi, K., & Yasri, A. (2020). Variations of saponins, minerals and total phenolic compounds due to processing and cooking of quinoa (*Chenopodium quinoa* Willd.) seeds. *Foods*, *9*, 660.

Mota, C., Nascimento, A. C., Santos, M., Delgadoa, I., Coelho, I., Rego, A., Matos, A. S., Torres, D., & Castanheira, I. (2016). The effect of cooking methods on the mineral content of quinoa (*Chenopodium quinoa*), amaranth (*Amaranthus* sp.) and buckwheat (*Fagopyrum esculentum*). *Journal of Food Composition and Analysis*, *49*, 57–64.

Muziri, T., Chaibva, P., Chofamba, A., Madanzi, T., Mangeru, P., Mudada, N., et al. (2021). Using principal component analysis to explore consumers' perception toward quinoa health and nutritional claims in Gweru, Zimbabwe. *Food Science & Nutrition*, *9*, 1025–1033.

Nosi, C., Zollo, L., Rialti, R., & Ciappei, C. (2020). Sustainable consumption in organic food buying behavior: The case of quinoa. *British Food Journal*, *122*, 976–994.

Peleg, M. (2019). The instrumental texture profile analysis revisited. *Journal of Texture Studies*, *50*, 362–368.

Repo-Carrasco-Valencia, R. A., Encina, C. R., Binaghi, M. J., Greco, C. B., & Ronayne de Ferrer, P. A. (2010). Effects of roasting and boiling of quinoa, kiwicha and kañiwa on composition and availability of minerals *in vitro*. *Journal of the Science of Food and Agriculture*, *90*, 2068–2073.

Sezgina, A. C., & Sanlier, N. (2019). A new generation plant for the conventional cuisine: Quinoa (*Chenopodium quinoa* Willd.). *Trends in Food Science & Technology*, *86*, 51–58.

Sonoma Press. (2015). *Quinoa: The everyday superfood: 150 Gluten-free recipes to delight every kind of eater*. Sonoma Press.

Suárez-Estrella, D., Borgonovo, G., Buratti, S., Ferranti, P., Accardo, F., Pagani, M. A., & Marti, A. (2021). Sprouting of quinoa (*Chenopodium quinoa* Willd.): Effect on saponin content and relation to the taste and astringency assessed by electronic tongue. *LWT-Food Science and Technology*, *144*, 111234.

Suárez-Estrella, D., Torri, L., Pagani, M. A., & Marti, A. (2018). Quinoa bitterness: Causes and solutions for improving product acceptability. *Journal of the Science of Food and Agriculture*, *98*, 4033–4041.

Tagle-Freire, D., Mennah-Govela, Y., & Bornhorst, G. M. (2022). Starch and protein hydrolysis in cooked quinoa (*Chenopodium quinoa* Willd.) during static and dynamic *in vitro* oral and gastric digestion. *Food & Function*, *13*, 920–932.

Wu, G. (2016). *Quinoa seed quality and sensory evaluation* (Ph.D. thesis). Washington State University.

Wu, G., Morris, C. F., & Murphy, K. M. (2014). Evaluation of texture differences among varieties of cooked quinoa. *Journal of Food Science*, *79*, S2337.

Wu, G., Morris, C. F., & Murphy, K. M. (2017b). Quinoa starch characteristics and their correlations with the texture profile analysis (TPA) of cooked quinoa. *Journal of Food Science*, *82*, 2387–2395.

Wu, G., Ross, C. F., Morris, C. F., & Murphy, K. M. (2017a). Lexicon development, consumer acceptance, and drivers of liking of quinoa varieties. *Journal of Food Science*, *82*, 993–1005.

Wu, L. G., Wang, A., Shen, R., & Qu, L. (2021). Effect of heating under pressure treatment on the antioxidant of quinoa. *International Journal of Food Engineering*, *17*, 795–804.

Yang, X., Zhu, K., Guo, H., Geng, Y., Lv, W., Wang, S., Guo, Y., et al. (2021). Characterization of volatile compounds in differently coloured *Chenopodium quinoa* seeds before and after cooking by headspace-gas chromatography-ion mobility spectrometry. *Food Chemistry*, *348*, 129086.

Zhang, Y., Zhang, S., Fan, W., Duan, M., Han, Y., & Li, H. (2019). Identification of volatile compounds and odour activity values in quinoa porridge by gas chromatography–mass spectrometry. *Journal of the Science of Food and Agriculture*, *99*, 3957–3966.

Zhao, C., Liu, Y., Lai, S., Cao, H., Guan, Y., Cheang, W. S., et al. (2019). Effects of domestic cooking process on the chemical and biological properties of dietary phytochemicals. *Trends in Food Science & Technology*, *85*, 55–66.

Thermal, nonthermal, and biotechnological processing of quinoa

11

11.1 Introduction

It is essential to cook and process quinoa for food and nutritional applications. Quinoa seeds and flour can be processed using a range of different operations to create different functional and physicochemical properties and new food products (Ahmed & Thomas, 2020; Zhu & Li, 2019a; Song et al., 2022; Mandhare et al., 2020; Zapana et al., 2020; Darwish et al., 2021). Different types of thermal processes used for quinoa processing included extrusion, drying, heating under pressure, microwave treatment, popping, boiling, steaming, and roasting (Wu et al., 2021; Motta et al., 2017; Duliński et al., 2017; Multari et al., 2018; Wu et al., 2021; Zapana et al., 2020; Maqueda et al., 2018). Some of the thermal processes have been among the most commonly used as they are home-based cooking methods. These thermal processes transform quinoa into edible products, while affecting the nutritional quality. Thermal processing may lead to the degradation of certain heat-sensitive nutrients and loss of nutritional quality. Nonthermal processing has gained research interest in recent years due to its capacity to create novel food properties and to retain a significant proportion of nutrients (Jadhav et al., 2021; Li et al., 2021). Different methods of nonthermal processing of quinoa include high hydrostatic pressure, washing, plasma activation, and ultrasound (Zhu & Li, 2019b; Ahmed & Thomas, 2020; Lalaleo et al., 2020; Gómez-Ramírez et al., 2017; Zhu & Li, 2019a; Zare et al., 2022). Even though some of the nonthermal processing methods are still under investigation at the laboratory level and are far from being practical for any commercial applications, they have been tested in quinoa-based food systems and showed some promising results. Bioprocesses of quinoa processing included fermentation, germination, and malting (Zhang et al., 2021; Aguilar et al., 2019; Ma et al., 2021; Cizeikiene et al., 2021; Motta et al., 2019). These bioprocesses have been used in many traditional cooking methods (e.g., sourdough breadmaking) to improve the nutritional and sensory quality of different food products. Their applications have been extended to include quinoa-based products.

Different processing methods lead to different nutritional composition and physicochemical properties of quinoa seeds (Mezzatesta et al., 2020; Zhu & Li, 2019b). This chapter is focused on the effects of nonthermal (Table 11.1), thermal (Table 11.2), and biotechnological (Table 11.3) processing of quinoa seeds and flours on the physicochemical and nutritional properties of related products (mostly intermediate products). The effects of different processing methods on the

Quinoa. DOI: https://doi.org/10.1016/B978-0-323-99909-0.00005-2

Table 11.1 Effect of nonthermal processing on properties of quinoa seeds and flours.

Treatment type	Quinoa form	Treatment conditions	Targeted properties and functions	Major findings	References
High pressure	Flour	High hydrostatic pressure of up to 600 MPa	Physicochemical properties	Pressure treatments at 400 and 200 MPa had little effects on the physicochemical properties. The treatments at 600 MPa reduced the in vitro starch digestibility, gel hardness, gelatinization enthalpy change, and pasting viscosity, while increasing the solubility. The extents of the changes were dependent on the type of quinoa flour used and the pressure level	Zhu and Li (2019b)
High pressure	Flour	Water to flour ratio of 2:1, pressure of up to 600 MPa	Physicochemical and in vitro antioxidant properties	Amylopectin crystallites of starch granules were completely gelatinized at 600 MPa. Increasing pressure level increased the mechanical rigidity of the resulting gels from pressure treatment. Heating of the pressure-treated gels increased the rigidity. Pressure treatments had no effect on protein denaturation property. Under 450 MPa, the antioxidant activity of the flour was increased	Ahmed and Thomas (2020)

Washing	Seed	The grains of three varieties were washed to remove the foam	Metabolomics profile (nontargeted)	Washing decreased the saponin levels of quinoa seeds and affected the concentrations of most of the metabolites including soluble proteins. The metabolomics profile depended on the quinoa genotype	Lalaleo et al. (2020)
Plasma activation	Seed	Dielectric barrier discharge and radio frequency plasma treatment at 500 and 0.1 mbar	Germination capacity	The plasma treatments increased the germination capacity. The treatments depleted the nitrate species and potassium ions of the seeds when exposing to water, contributing to the increased germination	Gómez-Ramírez et al. (2017)
Atmospheric pressure cold plasma	Flour	Dielectric barrier discharge plasma was applied on quinoa flour for 5 or 10 min at 50 or 60 kV	Physicochemical and structural properties	The plasma treatments induced cross-linking in the proteins, which was increased with increasing treatment time. The treatment increased the gelatinization temperatures of the flours and decreased the storage modulus (G') of flour gels during frequency sweep. The water-holding capacity and swelling power of the flours were either increased or decreased, depending on the treatment length and voltage	Zare et al. (2022)
Ultrasound	Quinoa flour	Ultrasound treatments at 20 kHz and 250 W up to 19 h	Physicochemical properties	Ultrasound treatment for at least 5 h decreased the pasting viscosity, gelation, in vitro antioxidant activity, gelatinization enthalpy change and temperatures, while increasing starch solubility and digestibility in a treatment time dependent manner	Zhu and Li (2019a)

(*Continued*)

Table 11.1 (Continued)

Treatment type	**Quinoa form**	**Treatment conditions**	**Targeted properties and functions**	**Major findings**	**References**
Electron beam irradiation	Quinoa seed	Up to 8 kGy	Phytochemical and fatty acid composition, lipase activity, bacterial count	The irradiation inhibited the increase of the free fatty acid content and the lipase activity, while inhibiting the increase of the bacterial count. The fatty acid composition was not significantly affected by the irradiation of up to 8 kGy. The irradiation decreased the phytic acid content (19% at 8 kGy) without affecting the saponin content. The content of free phenolics was increased and that of bound phenolics was decreased	Luo et al. (2021)

physicochemical and structural properties of isolated components of quinoa seeds such as starch and protein have been summarized in Chapter 13 of this book.

11.2 Nonthermal processing and modifications of quinoa seeds and flours

11.2.1 Washing

Washing is a common processing method and is critical for many types of home cooking (Table 11.1). Quinoa seeds of three different varieties were washed to remove the foam. The metabolomics profile was analyzed using a nontargeted approach (Lalaleo et al., 2020). The washing decreased the saponin levels of quinoa seeds and affected the concentrations of most of the metabolites including soluble proteins. The metabolomic profiles depended on the quinoa genotype (Lalaleo et al., 2020). The metabolomics profiles, however, should be related to any chemical and nutritional properties of quinoa seeds and related food products.

11.2.2 High hydrostatic pressure treatment

Quinoa flour suspensions were treated using high hydrostatic pressure of up to 600 MPa to modify the physicochemical and nutritional properties (Zhu & Li, 2019b; Ahmed & Thomas, 2020). The water content of the suspensions was high enough to induce full starch gelatinization. Zhu and Li (2019b) showed that pressure treatments at 400 and 200 MPa had little effect on the physicochemical properties of three different types of quinoa flour. The treatments at 600 MPa reduced the in vitro starch digestibility, gel hardness, gelatinization enthalpy change, and pasting viscosity of all the quinoa flours, while increasing their water solubility. The extents of the changes in these physicochemical and functional properties were seen to be dependent on the type of quinoa flour used and the pressure level (Zhu & Li, 2019b). The changes were largely attributed to the pressure-induced effects on the quinoa components including starch and protein (Li & Zhu, 2018; Akharume et al., 2021). There have been many reports on the effects of high pressure on food macromolecules including starch and proteins (Castro et al., 2020; Hall & Moraru, 2022). The knowledge accumulated could be feasibly used to explain the changes in the properties of quinoa flour affected by high-pressure processing. Amylopectin crystallites of quinoa starch granules were completely gelatinized at 600 MPa (Ahmed & Thomas, 2020). Increasing pressure level increased the mechanical rigidity of the resulting gels from the pressure treatment. Heating of the pressure-treated gels increased the rigidity. Pressure treatments had no effect on protein denaturation property. Under the pressure of 450 MPa, the in vitro antioxidant activity of the flour was increased, probably due to the release of the phenolics from the disrupted cell structure (Ahmed & Thomas, 2020). There are other types of pressure-based processing methods such as dynamic pressure treatments (e.g., high-pressure microfluidization and high-pressure homogenization) (Zhu, 2021). These dynamic pressure processing methods remain to be used to include quinoa flour systems to create "novel" functionalities.

11.2.3 Cold plasma modification

Cold plasma technology has different applications in seeds to improve their agricultural performances and in food products to improve quality (Gómez-Ramírez et al., 2017; Starič et al., 2020). The reactive species (free radicals and unionized molecules) of the cold plasma interact with the molecules of the products, changing the properties. This technology could be used to modify the genetics, physiology, and morphology of the seeds (Starič et al., 2020). Quinoa seeds were modified using dielectric barrier discharge (DBD) and radio frequency plasma-based treatments at 500 and 0.1 mbar (Gómez-Ramírez et al., 2017). The germination capacity of the plasma-treated quinoa seeds was studied. The plasma treatments increased the germination capacity. The treatments depleted the nitrate species and potassium ions of the seeds when exposed to water, contributing to the increased germination (Gómez-Ramírez et al., 2017). Nonthermal plasma has been used to modify food properties and increase food quality and safety. The plasma technology is considered environmentally friendly (Pankaj et al., 2018). For example, DBD plasma significantly modified the physicochemical properties of quinoa starch (Gao et al., 2019). Such an approach was used to include quinoa-based products such as quinoa flour (Zare et al., 2022). Atmospheric pressure cold plasma in the form of DBD plasma was applied on quinoa flour for 5 or 10 min at 50 or 60 kV (Zare et al., 2022). The plasma treatments induced crosslinking in the proteins, which was increased with increasing treatment time. The treatment increased the gelatinization temperatures of the flours, and decreased the storage modulus (G') of flour gels during frequency sweep. The water-holding capacity and swelling power of the flours were either increased or decreased, depending on the treatment length and voltage (Zare et al., 2022).

11.2.4 Ultrasound modification

Low-intensity ultrasound could be used as a nondestructive analytical tool to evaluate food composition and structure, whereas high-intensity ultrasound is applied in food structure modifications and to promote extraction and reactions (Ortega-Rivas, 2012). Quinoa flour was subjected to ultrasound processing at 20 kHz and 250 W up to 19 h (Zhu & Li, 2019a). Physicochemical properties of the quinoa flour affected by the ultrasound treatments were studied. Ultrasound treatment for at least 5 h decreased the pasting viscosity, gelation, in vitro antioxidant activity, and gelatinization enthalpy change (ΔH) and temperatures of the quinoa flour, while increasing starch solubility and digestibility in a treatment time-dependent manner (Zhu & Li, 2019a). Ultrasound in food systems can induce cavitation, which creates high temperature and shear force (Cui & Zhu, 2021). The cavitation led to physical and molecular degradation of food molecules such as starch in the system and changed the physicochemical properties of the quinoa flour (Zhu & Li, 2019a). Indeed, many previous studies have shown the degradation effect of ultrasound on starch and protein at the supramolecular and molecular level (Zhu, 2015; Mir et al., 2019). Overall, these results obtained are largely experimental on a laboratory scale. it is very questionable if ultrasound processing of quinoa samples for functionality modifications can be meaningful at all for industrial uses.

11.2.5 Electron beam irradiation

Quinoa seeds were subjected to electron beam irradiation of up to 8 kGy before storage (Luo et al., 2021). The irradiation inhibited the increase of the free fatty acid content and the lipase activity of the quinoa seeds, while retarding the increase of the bacterial count. The fatty acid composition was not significantly affected by the irradiation of up to 8 kGy. The irradiation decreased the phytic acid content (19% at 8 kGy) without affecting the saponin content. The content of free phenolics was increased and that of bound phenolics was decreased (Luo et al., 2021).

11.3 Thermal processing of quinoa seeds and flours

Different types of heat-based processing methods have been used to treat quinoa in the forms of seeds and flours (Table 11.2). Extrusion, drying, heating under pressure, microwave treatment, popping, boiling, steaming, and roasting were used to treat quinoa samples under different conditions.

11.3.1 Extrusion

Extrusion is a combination of mixing, heating, shearing, compressing, and forcing a melted sample through a narrow die (Leonard et al., 2020). Kowalski et al. (2016) used corotating twin screw extruder to process quinoa flour under a range of different extrusion parameters (feed moisture of 15%–25%, screw speed of 300–500 rpm, and feed zone set (22°C), following zones at 45°C–100°C and last zones at 120°C–160°C). The resulting extrudates had a range of physicochemical parameters including water absorption index (2.3 – 3.1 g/g), expansion ratio (1.2 – 1.6 g/cm^3), water solubility index (14.5% – 15.9%), and unit density (0.45 – 1.0 g/cm^3). The expansion capacity of the quinoa extrudates was lower than that of major cereals such as maize and wheat. Overall, the quinoa variety was not suitable for making extruded products (Kowalski et al., 2016). Therefore quinoa was modified in different forms for extrusion processing. Aluwi et al. (2016) compared the applicability of unscarified, scarified, and degermed quinoa for extrusion processing. The scarified variety with more proteins gave a lower expansion rate than the other varieties. The degermed quinoa seeds with less fiber, lipids, and proteins gave higher expansion than the other varieties (Aluwi et al., 2016). Gearhart and Rosentrater (2017) employed quinoa seeds and milled seeds of different sizes

Table 11.2 Thermal processing of quinoa seeds and flours.

Process	Quinoa form	Extrusion conditions	Targeted properties	Major findings	References
Extrusion	Flour	Corotating twin screw extruder, feed moisture of 15%–25%, screw speed of 300–500 rpm, feed zone set temperature (22°C), following zones at 45°C–100°C and last zones at 120°C–160°C	Response surface method was used to analyze the effect of changing process parameters on the physicochemical properties of the extrudates	A range of physicochemical parameters including water absorption index (2.3 – 3.1 g/g), expansion ratio (1.2 – 1.6 g/cm^3), water solubility index (14.5% – 15.9%), and unit density (0.45 – 1.0 g/cm^3) were recorded. The expansion capacity of the quinoa extrudates was lower than that of major cereals such as maize and wheat. Overall, the quinoa variety was not suitable for making extruded products	Kowalski et al. (2016)
Extrusion	Unscarified, scarified, and degermed varieties	Corotating twin-screw extruder, barrel temperature at 140°C, feed moisture of 15%–25%, screw speed of 100–200 rpm	Physicochemical properties of extruded products and process response parameters	The scarified variety with more proteins gave lower expansion than the other varieties. The degermed quinoa seeds with less fiber, lipids, and proteins gave higher expansion than the other varieties	Aluwi et al. (2016)
Extrusion	White quinoa seeds or seeds ground to have sizes of 1 and 2 mm	Moisture content of samples (20% and 40%), screw speeds (50 and 100 rpm), temperature was not controlled	Color and physicochemical properties of extrudates	Increasing moisture content decreased the expansion ratio of the extrudates, while increasing the a^* value. Decreasing particle size of the raw materials decreased the a^* and b^* values. Increasing screw speed increased the b^* value, while decreasing the L value	Gearhart and Rosentrater (2017)

Extrusion	Germinated flour of *Chenopodium album*	Barrel temperature (115°C–135°C), feed moisture (18%–20%), screw speed (225–275 rpm)	Nutritional, antioxidant, physicochemical and sensory properties of extrudates	Antioxidant activity (DPPH radical scavenging: 11.6%–15.9%), total phenolic (56.8–81.3 mg/100 g) and fiber (18.7%–22.1%) contents, lightness (64.0–68.4), and hardness (12.3–19.5 N) of the extrudates. The barrel temperature was the most significant factor affecting the properties (acceptability, antioxidant activity, hardness, total phenolics, and expansion ratio) of the extrudates	Jan et al. (2017)
Extrusion	Composite flour of quinoa:corn (oats):rice at a ratio of 60:25:15	Corotating twin-screw extruder with feed rate (7 kg/h), screw speed (300 rpm) and barrel temperature (100 to 170°C)	Physicochemical and nutritional properties of extrudates	Changing barrel temperature from 100°C to 170°C led to the formation of products with a range of parameters including unit density (0.12–0.20 g/cm^3), expansion ratios (3.3–3.5), water absorption index (5.77–6.40 g/g), and water solubility index (12.59%–20.93%). The optimized nutritional composition for products with the highest sensory acceptance included carbohydrates (78%), protein (16%), lipids (4.2%), and ash (1.9%)	Alajil et al. (2020)

(*Continued*)

Table 11.2 (Continued)

Process	Quinoa form	Extrusion conditions	Targeted properties	Major findings	References
Extrusion	Seeds	Corotating twin screw extruder; screw speed (300–500 rpm), feed zone set at 22°C, following zones at 45°C, 70°C, 100°C and last zone ranged from 120°C to 160°C, feed rate at 3.0 kg/h	In vitro antioxidant activity and phenolic, flavonoid and saponins contents	Extrusion cooking decreased the antioxidant activities, and the contents of phenolics and saponins	Wu et al. (2021)
Extrusion	Flour	Self-wiping, corotating and twin-screw extruder; the temperatures of the second zone were set at 120°C, 140°C, 160°C, and 180°C, whereas that of first and three zones was at 80°C. The speeds of cutting and feeding were 30 and 8 Hz, respectively	Bound and free phenolic compound composition	The extrusion process significantly decreased the amount of bound phenolics from 155 down to 77 mg/kg, while increasing that of free phenolics from 22 up to 84 mg/kg (at 160°C)	Song et al. (2022)
Extrusion	Flour	Extruding temperature (69°C–95°C), screw velocity 149 rpm, feeding speed 50 rpm. The extruded samples were dried at 130°C for 10 min	Composition, structural, and physicochemical properties	Extrusion processing increased the level of soluble fiber and protein aggregation and cross-link formation. The amino acid composition was little affected. The protein cross-linking was promoted by the final drying step. The extrusion increased the level of V-type structure of the starch, while decreasing the crystallinity	Kuktaite et al. (2022)

Boiling and steaming	Raw seed	Boiling (15 min, 100°C) and steaming (30 min)	Folate profile	Raw quinoa seeds had more total folate content (310 μg/100 g dw) than buckwheat (150 μg/100 g dw) and amaranth (230 μg/100 g dw). 5-Methyltetrahydrofolate (260 μg/100 g dw) was dominant followed by 10-formyltetrahydrofolate (53 μg/100 g dw) and folic acid (9 μg/100 g dw). Boiling and steaming had little effect on the total folate content	Motta et al. (2017)
Boiling and steaming	Raw seed	Steaming (120°C, 25 min) and boiling (15 min, 100°C)	Amino acid profile	Steaming and boiling had little effect on the amino acid (except for sulfur-containing ones) composition	Motta et al. (2019)
Boiling	Raw seed of three varieties	15 min of boiling	Metabolite profile	Boiling affected the concentrations of most of the metabolites	Lalaleo et al. (2020)
Boiling and roasting	Raw seed	15 min of boiling, roasting conditions were not given in the study	Composition of myo-inositol phosphates	Cooking reduced the phytate content (e.g., from 14 to 9 μmol/g dry matter in red quinoa seeds) and the phytate bioavailability (e.g., from 43 to 26 in white quinoa seeds). The extents of reduction were dependent on the quinoa genotype	Duliński et al. (2017)
Roasting	Seed	Seeds were roasted at different temperatures (100°C–190°C) and for different time length (15–120 min)	Maillard reaction products, phenolic content, and in vitro antioxidant activity	The highest antioxidant activities were obtained when the seeds were roasted for 0.5 h at 190°C. Maillard reaction products had a higher contribution to the antioxidant activity than the phenolic content. Overall, roasting temperature had a higher influence on the outcomes than roasting time length	Carciochi et al. (2016b)

(*Continued*)

Table 11.2 (Continued)

Process	Quinoa form	Extrusion conditions	Targeted properties	Major findings	References
Roasting	Seed	Seeds were parched on a frying pan with hot sand at 160°C–190°C for up to 30 s	Physicochemical and nutritional properties of quinoa seeds	The roasting increased the bulk density, expansion volume and water absorption, in vitro starch, and protein digestibility. Roasting at 160°C for 30 s was recommended for the maximal effects	Kumar et al (2022)
Drying	Seed	Soaked and washed seeds were dried with an air velocity of 0.8 m/s at 80°C, 100°C, 120°C, and 140°C	A coupled energy and mass model was developed to describe the drying curves	The relationships between drying time, temperature, and seed moisture content were described using a coupled energy and mass math model	Irigoyen and Giner (2017)
Drying	Seed	Drying temperatures at 40°C, 50°C, 60°C, or 70°C. The seeds were dried to have a moisture content of 94%–95%	Recovery and composition of phenolics and carotenoids	Drying at 60°C–70°C increased the extractability of phenolics (maximum extractability at 70°C) and carotenoids (maximum extractability at 60°C) from the seeds. Heating may loosen the tissue and inactivate phytochemical-degrading enzymes	Multari et al (2018)
Heating under pressure	Seed	Seeds were heated under pressure of up to 0.6 MPa and for 3–5 min	In vitro antioxidant activity and phenolic, flavonoid, and saponin contents	The process parameters were optimized using response surface methodology to retain the antioxidant activity and the bioactive components including phenolics, flavonoids and saponins. The heating under pressure method better retained these parameters than other heat-based processing methods including extrusion, boiling, and baking	Wu et al. (2021)

Microwave	Raw seed	Power density of 1.75–7 W/g. The seeds were then dried for seedling and germination tests	Seedling and germination capacities of the seeds	Microwave treatment decreased the germination capacity by over 75%. To increase the germination rates of quinoa seeds, lower power density (<1.75 W/g), and temperature should be tested	Maqueda et al. (2018)
Popping	Raw seed coated with starch-chitosan	Popping was done using extrusion, gun and microwave-based techniques with conditions varying in pressure, energy consumption, power, and moisture content	Physicochemical and nutritional properties of the popped quinoa	Gun and extrusion puffing gave better nutritional quality of the puff with less loss in the nutrients and bioavailability of proteins and organic matter compared with microwave-based puffing. The optimized conditions for gun puffing were at 1.31 MPa with a time length of 780 s, and for extrusion puffing at 180 s and 500 rpm	Zapana et al. (2020)
Popping	Raw seed	For traditional popping, the temperatures were tested for a range of 200°C–280°C and for a length of 13–90 s. For microwave popping, the treatment time of 135–167 s and power level of 60%–100% were tested	Volume and expansion ratio of the puffed grain as well as the percentage of puffed grain (number of grains puffed out of 100 grains after popping)	A range of volumes (7.9 – 9.1 mL) and expansion ratios (2.0 – 4.4) of puffed grains and the percentages of puffed grain (59%–87%) were obtained by the traditional popping method. A range of volumes (6.9 – 8 mL) and expansion ratios (1.9 – 3.1) of puffed grains and the percentages of puffed grain (49%–65%) were obtained by microwave-based popping	Mandhare et al. (2020)

(average of 1 or 2 mm) under different extrusion parameters. Increasing moisture content (20% and 40%) decreased the expansion ratio of the extrudates, while increasing the a^* value. Decreasing particle size of the raw materials decreased the a^* and b^* values. Increasing screw speed increased the b^* value, while decreasing the L value (Gearhart & Rosentrater, 2017). To make quinoa more suitable for making extruded products, composite flour of quinoa:corn (oats):rice at a ratio of 60:25:15 was used for extrusion (Alajil et al., 2020). Changing barrel temperature from 100°C to 170°C led to the formation of products with a range of parameters including unit density (0.12–0.20 g/cm^3), expansion ratios (3.3–3.5), water absorption index (5.77–6.40 g/g), and water solubility index (12.59%–20.93%). The optimized nutritional composition for the products with the highest sensory acceptance included carbohydrates (78%), protein (16%), lipids (4.2%), and ash (1.9%) (Alajil et al., 2020). The effect of extrusion on in vitro antioxidant activity and composition of bioactive components including phenolics and saponins was studied (Wu et al., 2021; Song et al., 2022). Extrusion cooking decreased the in vitro antioxidant activities and the contents of phenolics and saponins (Wu et al., 2021). Song et al. (2022) found that the extrusion process significantly decreased the amount of bound phenolics from 155 down to 77 mg/kg, while increasing that of free phenolics from 22 up to 84 mg/kg (at 160°C) (Song et al., 2022). The increased amount of free phenolics may indicate increased bioavailability of phenolics in the extruded quinoa samples. Kuktaite et al. (2022) used extrusion to produce different types of breakfast cereals of quinoa flour (Fig. 11.1). Extrusion processing increased the level of soluble fiber and protein aggregation and crosslink formation. The amino acid composition was little affected. The protein crosslinking was promoted by the final drying step. The extrusion increased the level of V-type structure of the starch due to inclusion complex formation with lipids, while decreasing the crystallinity (Kuktaite et al., 2022). Many previous studies have shown that extrusion process can affect the structure and properties of food components including starch and protein (Ye et al., 2018; Mosibo et al., 2022). The mechanism of action has been reviewed in detail. The effect of extrusion cooking on the chemical and physicochemical properties of quinoa flour and seeds largely could be related to the effects on the individual components.

11.3.2 Boiling, steaming, and roasting

Different household cooking methods including boiling, steaming, and roasting were used to treat quinoa and the changes in nutritional properties were obtained (Motta et al., 2017; Motta et al., 2019; Lalaleo et al., 2020; Duliński et al., 2017; Carciochi et al., 2016b; Kumar et al., 2022). Motta et al. (2017) showed that raw quinoa seeds had more folates (310 μg/100 g dw (dry weight)) than buckwheat (150 μg/100 g dw) and amaranth (230 μg/100 g dw). 5-Methyltetrahydrofolate (260 μg/100 g dw) was dominant followed by 10-formyltetrahydrofolate (53 μg/100 g dw) and folic acid (9 μg/100 g dw). Boiling and steaming had little effect on the total folate content. Steaming and boiling had little effect on the amino acid (except for sulfur-containing ones) composition (Motta et al., 2019). Lalaleo et al. (2020) analyzed the

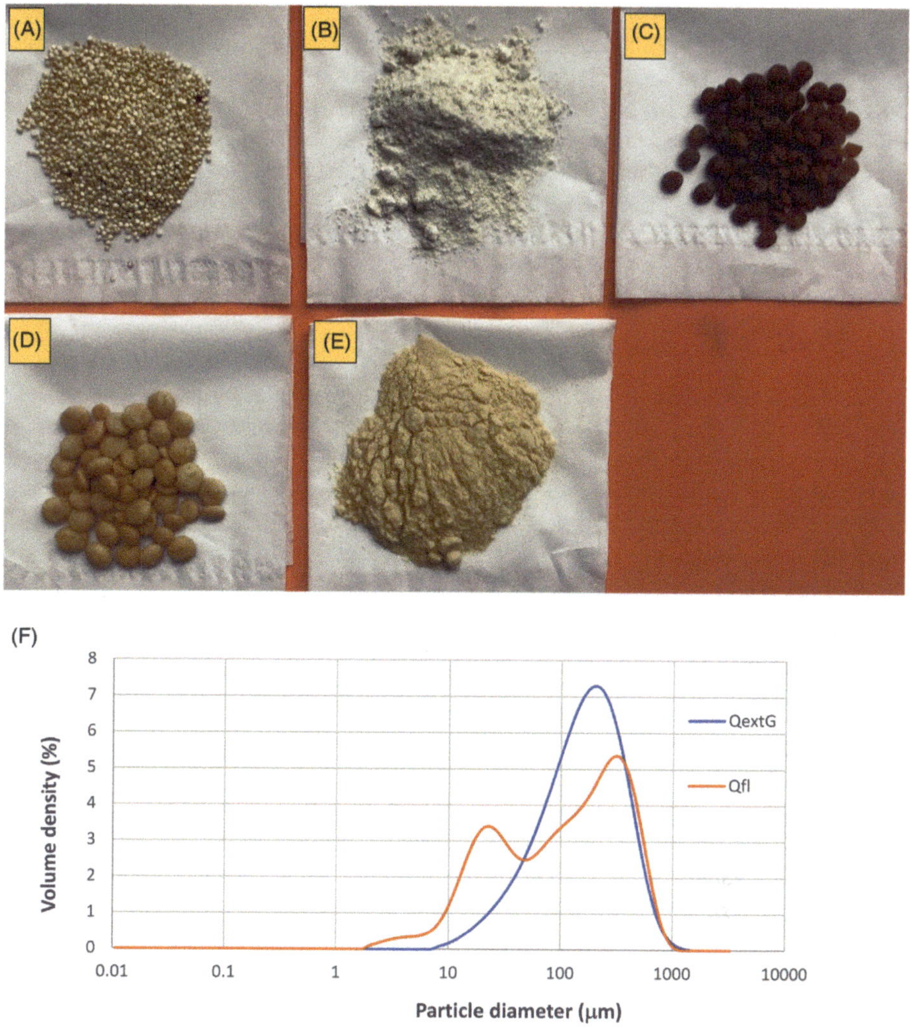

Figure 11.1 Processed quinoa samples (A) pearl quinoa seeds; (B) milled quinoa flour (Qfl); (C) extruded quinoa breakfast cereals (Qext); (D) extruded quinoa breakfast cereals post-dried at high temperature (QextD); (E) QextD sample ground into flour (QextG); (F) volume density (%) and particle diameter (μm) of Qfl and QextG samples (Kuktaite et al., 2022). *Source*: Reprinted with permission under Creative Commons CC BY License.

metabolite profile of quinoa seeds and found that boiling (15 min) affected the concentrations of most of the metabolites. Duliński et al. (2017) found that boiling and roasting decreased the phytate content in quinoa seeds. Carciochi et al. (2016b) studied the effect of roasting on the in vitro antioxidant properties of quinoa seeds. The highest antioxidant activities were obtained when the seeds were roasted for 0.5 h at 190°C. Maillard reaction products had a higher contribution to the antioxidant activity than the phenolic content. Overall, roasting temperature had a higher influence on the outcomes than roasting time length (Carciochi et al., 2016b). Quinoa seeds were parched on a frying pan with hot sand at 160°C–190°C for up to 30 s (Kumar et al., 2022). The roasting increased the bulk density, expansion volume, and water absorption, in vitro starch and protein digestibility. Roasting at 160°C for 30 s was recommended for the maximal effects (Kumar et al., 2022). Overall, the effect of different home cooking methods on the physicochemical and nutritional properties not only depended on the processing conditions but also on the quinoa variety.

11.3.3 Heating under pressure

Quinoa seeds were heated under pressure of up to 0.6 MPa and for 3–5 min and the effect of the heating on in vitro antioxidant activity and phenolic, flavonoid, and saponin contents was studied (Wu et al., 2021). The process parameters were optimized using response surface methodology to retain the antioxidant activity and the bioactive components. The heating under pressure method better retained these parameters than other heat-based processing methods including extrusion, boiling, and baking (Wu et al., 2021).

11.3.4 Drying

Soaked and washed seeds were dried with an air velocity of 0.8 m/s at 80°C, 100°C, 120°C, and 140°C for preservation purpose (Irigoyen & Giner, 2017). A coupled energy and mass model was developed to describe the drying curves. The relationships between drying time, temperature, and seed moisture content were described using a coupled energy and mass math model (Irigoyen & Giner, 2017). In another study, Multari et al. (2018) dried quinoa seeds at different temperatures of 40°C, 50°C, 60°C, or 70°C. The seeds were dried to have a moisture content of 94%–95%. Recovery and composition of phenolics and carotenoids of the seeds affected by the drying were studied. Drying at 60°C–70°C increased the extractability of phenolics (maximum extractability obtained at 70°C) and carotenoids (maximum extractability obtained at 60°C) from the seeds. Heating may loosen the tissue and inactivate phytochemical-degrading enzymes (Multari et al., 2018).

11.3.5 Microwave heating

Raw quinoa seeds were treated using microwave at a range of power density of 1.75–7 W/g (Maqueda et al., 2018). The seeds were then dried for seedling and germination tests. The microwave treatment decreased the germination capacity of the quinoa seeds by over 75%. To increase the germination rates of quinoa seeds, lower power density ($<$1.75 W/g) and temperature should be tested. Microwave could also be used to modify the physicochemical properties of starch rich systems (Tao et al., 2020). For example, microwave at a level of heating energy high enough can gelatinize starch in the presence of enough amount of water. This kind of application of microwave treatment on quinoa remains to be studied. Microwave-based popping of raw quinoa seeds is described in the section below.

11.3.6 Popping

The popping property of quinoa seeds was studied (Zapana et al., 2020; Mandhare et al., 2020). Raw quinoa seeds were popped using extrusion, gun, and microwave-based techniques with conditions varying in pressure, energy consumption, power, and moisture (Zapana et al., 2020). Popping process gelatinized starch. Gun and extrusion puffing gave better nutritional quality of the puff with less loss in the nutrients and bioavailability of proteins and organic matter compared with microwave-based puffing. The optimized conditions for gun puffing were at 1.31 MPa with a time length of 780 s, and for extrusion puffing at 180 s and 500 rpm (Zapana et al., 2020). The physicochemical and expansion properties of puffed quinoa made by traditional and microwave puffing were studied (Mandhare et al., 2020). Varying heating temperatures resulted in the formation of popped products with a range of volumes (7.9 − 9.1 mL) and expansion ratios (2.0 − 4.4) of puffed grains and the percentages of puffed grain (59%−87%). Changing the microwave power levels and temperatures produced popped quinoa seeds with a range of volumes (6.9 − 8 mL), expansion ratios (1.9 − 3.1), and the percentages of puffed grain (49%−65%) (Mandhare et al., 2020). The sensory and textural properties of the popped products should be further studied for commercial application purposes.

11.4 Bioprocessing of quinoa seeds and flours

Quinoa seeds and flours were subjected to biotechnological processing operations including fermentation, malting, and germination (Table 11.3).

11.4.1 Fermentation

Different fermentation conditions with different bacterial species were used for quinoa seed and flour treatments (Starzyńska-Janiszewska et al., 2016; Lopes et al., 2019; Ruiz Rodríguez et al., 2016; Rizzello et al., 2017; Ayyash et al., 2018; Ayyasha et al., 2019; Paz et al., 2020; Cizeikiene et al., 2021; Yang et al., 2022; Kewuyemi et al., 2022). Different aspects of the fermentation processing on various properties of quinoa were studied. For example, Ruiz Rodríguez et al. (2016) analyzed the biodiversity of the lactic acid bacteria in quinoa flour-based sourdough made from daily backslopping (30°C, 10 days). *Lactobacillus brevis* and *L. plantarum* were dominant in the lactic acid bacteria of the sourdough. *Leuconostoc mesenteroides* CRL1907 and *L. plantarum* CRL1905 showed potential in terms of antibiotic susceptibility, acidification, ropy phenotype, and antimicrobial capacity to be starters to make quinoa sourdough (Ruiz Rodríguez et al., 2016). Paz et al. (2020)

Table 11.3 Bioprocessing of quinoa seeds and flours.

Processing methods	Sample form	Experimental/ processing methods	Targeted properties/ functions	Major findings	References
Fermentation	Flour	Fermentation was done by daily backslopping (30°C, 10 days)	Biodiversity of the lactic acid bacteria, selection of suitable strains for making quinoa sourdough	*Lactobacillus brevis* and *L. plantarum* were dominant in the lactic acid bacteria of the sourdough. *Leuconostoc mesenteroides* CRL1907 and *L. plantarum* CRL1905 showed potential in terms of antibiotic susceptibility, acidification, ropy phenotype, and antimicrobial capacity to be starters to make quinoa sourdough	Ruiz Rodríguez et al. (2016)
Fermentation	Flour	Fermentation of quinoa flour was done (30°C, 16 h) using *Lactobacillus plantarum* T6B10 and *L. rossiae* T0A16	Nutritional properties of the sourdough, physicochemical and sensory properties of breads fortified with the quinoa sourdough (20%)	The fermentation process increased the levels of free amino acids, organic acids, total phenolics, soluble fibers, phytase activity, and in vitro protein digestibility and antioxidant activities of quinoa flour. The use of the quinoa sourdough improved the nutritional (e.g., increased protein digestibility and decreased starch digestibility) and sensory quality of the resulting breads	Rizzello et al. (2016)
Fermentation	Seed	*Rhizopus oligosporus* was used for solid state fermentation of quinoa seeds at 31°C for up to 40 h	In vitro antioxidant activities and phenolic composition	Fermentation significantly increased in vitro antioxidant activity of quinoa. The fermentation of 40 h increased the levels of free rutin, vanillic acid and protocatechuic acid, total protein, free amino acids, riboflavin and thiamine in quinoa	Starzyńska-Janiszewska et al. (2016)

Sprouting and fermentation	Flour	Sprouting and natural fermentation were done at 25°C and 26°C, respectively	Glycemic index and physiological conditions in Wistar rats fed with high carbohydrate diet	Incorporation of sprouted and fermented quinoa flour into the diet reduced the glycemic index and total cholesterol and triacylglycerol level in the rats	Lopes et al. (2019)
Fermentation	Flour	Quinoa flour was fermented (37°C, 24 h) using autochthonous lactic acid bacteria	In vitro antioxidant activities (both chemical and cellular assays)	Fermentation increased the antioxidant activities of both salt-soluble and water extracts. The extract obtained from the dough fermented using *Lactobacillus plantarum* T0A10 had the highest prevention against linoleic acid autoxidation. A total of 5 peptides (5 − 9 amino acid residues) were obtained from the protein of the fermented dough and showed strong antioxidant activity. The fractions showed cellular antioxidant capacity on human keratinocytes NCTC 2544 model	Rizzello et al. (2017)
Fermentation	Raw seed	Solid state fermentation using *Rhizopus oligosporus* (2 h at 35°C and then 38 h at 31°C)	Composition of myo-inositol phosphates	Fermentation decreased the percentages of phytate, while increasing that of inositol triphosphates due to the presence of microbial 3-phytase. The total phytate content of quinoa seeds was reduced (e.g., from 14 to 5 μmol/g dry matter in red quinoa seeds). The reduction was dependent on the quinoa genotypes	Duliński et al. (2017)

(*Continued*)

Table 11.3 (Continued)

Processing methods	Sample form	Experimental/ processing methods	Targeted properties/ functions	Major findings	References
Fermentation	Seed	Cooked quinoa was subjected to solid-state fermentation with *Neurospora intermedia* (oncom-type treatments)	Nutritional and bioactive components	The fermentation increased the contents of protein, fiber, carotenoids (six times), riboflavin (five times), in vitro antioxidant activities, while decreasing the level of inositol phosphates. Fungal glucosamine was found in the final product of fermented quinoa seeds	Starzyńska-Janiszewska et al. (2017)
Fermentation	Seed	Quinoa seeds were subjected to solid state fermentation (37°C, 72 h) with *Bifidobacterium animalis*, *B. longum*, and *B. breve*	Phenolic content, in vitro anticancer (breast cancer cell (MCF-7) and colon cancer cell (Caco-2)), and antioxidant properties, α-glucosidase, α-amylase, angiotensin-converting-enzyme (ACE) inhibition	Fermentation significantly increased, to different extents, the total phenolic content and in vitro antioxidant activity, α-glucosidase inhibition, ACE-inhibition and anticancer activities of quinoa seeds, depending on the type of bacterial strain. The fermentation induced increases in the anticancer functions were seen to be greater in lupin than in quinoa	Ayyash et al. (2018)
Fermentation	Flour	Quinoa flour was fermented using *Weissella confusa* DSM 20194 (exopolysaccharide-producing), *Lactobacillus rhamnosus* SP1 or *Lactobacillus plantarum* T6B10 (20 h, 30°C)	Physicochemical and sensory properties of the resulting yogurt-like drinks	The fermentation process with any of the strains increased protein digestibility, the levels of free amino acids and γ-aminobutyric acid, polyphenol bioavailability, and in vitro antioxidant activity. The water-holding capacity and viscosity of the resulting products fermented with *Weissella confusa* were not affected by storage (4°C). The sensory acceptance of the products was good	Lorusso et al. (2018)

Fermentation	Seed	Quinoa seeds were fermented (30°C, up to 5 days)	Composition of GABA, L-carnitine, and phenolics, in vitro antioxidant activity	Fermentation increased the contents of GABA, L-carnitine, vanillic acid, gallic acid, total phenolics, in vitro antioxidant activity and inhibition of nitric oxide production in RAW264.7 macrophages	Hur et al. (2018)
Fermentation	Seed	Quinoa seeds were subjected to solid state fermentation (37°C, 72 h) with *Lactobacillus reuteri* K777 and *L. plantarum* K779	In vitro anticancer (breast cancer cell (MCF-7) and colon cancer cell (Caco-2)), and antioxidant properties, α-glucosidase, α-amylase, angiotensin-converting-enzyme (ACE) inhibition	Fermentation significantly increased, to different extents, the total phenolic content and in vitro antioxidant activity, α-glucosidase inhibition, ACE-inhibition and anticancer activities of quinoa seeds, depending on the bacterial strain. The fermentation induced increases in the anticancer functions were seen to be greater in lupin than in quinoa	Ayyasha et al. (2019)
Fermentation	Seed	Quinoa seeds were subjected to solid state fermentation using *Rhizopus oligosporus* (31°C, up to 6 days), *Aspergillus oryzae* (25°C, up to 6 days) or *Neurospora intermedia* (30°C, 72 h)	In vitro antioxidant activity and phenolic composition	Fermentation with *Rhizopus oligosporus* more drastically increased the antioxidant activity and the content of free vanillin compared to that with the other two strains. All the fermentation increased the contents of free amino acids and peptides	Starzyńska-Janiszewska et al. (2019)

(*Continued*)

Table 11.3 (Continued)

Processing methods	Sample form	Experimental/ processing methods	Targeted properties/ functions	Major findings	References
Fermentation	Seed	Quinoa seeds were treated at 121°C (20 min) before fermentation (culture at 5%, v/w) at 80% (relative humidity) and 32°C for 10 days using *Monascus anka* for solid state fermentation	Composition of phenolic compounds and in vitro biological activities	Ferulic acid, *p*-hydroxybenzoic acid and protocatechuic acid were the major phenolics in the fermented samples. *M. anka*-based fermentation for 6 days gave the seeds the highest total phenolic content. The fermentation increased the absorption of free phenolics in the intestine, in vitro antioxidant activity (especially that of the free phenolics), and inhibition activities against α-glucosidase and α-amylase. The free phenolics were largely responsible for the observed activities	Yang et al. (2022)
Fermentation	Flour	Quinoa flour with starch and xanthan was fermented at 30°C for up to 6 h to obtain a pH of <4 using *Lactobacillus plantarum* Q823. The fermented sample was cooled at 6°C for storage of 28 days. A range of different ingredients including date, banana, and bilberry were added to make spoonable products	Lactic acid bacteria count, sensory property, nutritional composition	The fermented quinoa flour-based products after 28 days of storage had lactic acid bacteria counts comparable to commercial products. The nutritional quality of the quinoa-based products was superior than commercial products to meet recommended human nutritional requirements. The drawback of the quinoa-based products was related to unpleasant aftertaste	Väkeväinen et al. (2020)

Fermentation	Flour	Red quinoa flour was subjected to steaming, α-amylolysis, complex enzymatic hydrolysis, and/or lactic acid bacteria fermentation	Phenolic composition and in vitro antioxidant activity	Combination of complex enzymatic hydrolysis and lactic acid bacteria fermentation led to much higher increases in the total phenolic (both free and bound forms) content and antioxidant activities. The combined treatment increased the contents of catechin, protocatechuic acid, quercetin, and procyanidin B2 by 102%, 126%, 296%, and 524%, respectively	Zhang et al. (2021)
Fermentation	Flour	Quinoa flour was fermented (37°C, 20 h) using *Lactobacillus plantarum* ATCC 8014 or spontaneously	The properties of gluten-free muffins with the addition of quinoa sourdough	The use of *Lactobacillus plantarum* ATCC 8014 fermented sourdough increased the level of folic acid and organic acids, flavonoids and total phenolics of the formulated muffins, while increasing the in vitro antioxidant activity (DPPH assay)	Chiş et al. (2020)
Fermentation	Flour	Quinoa flour was fermented (35°C, up to days) using *Lactobacillus plantarum* 1 and *L. plantarum* JCM1149	The fermented dough was made into gluten free bread. The properties of the bread were studied	The fermentation increased soluble protein level. The fermentation of >2 days gave the breads homogeneous porous microstructure and higher sensory quality	Ceballos-González et al. (2018)

(*Continued*)

Table 11.3 (Continued)

Processing methods	Sample form	Experimental/ processing methods	Targeted properties/ functions	Major findings	References
Fermentation	Quinoa juice suspension	The suspension was fermented (30°C, 2 days) using *Lactobacillus plantarum* DSM 9843. The resulting product was stored at 4°C for 28 days	Microbial composition	Without fermentation, a range of different microbial species were found in the drinks including *Escherichia*, *Enterobacter*, *Leclercia*, *Pantoea*, *Klebsiella*, *Enterococcus*, and *Salmonella*. With *Lactobacillus plantarum*-based fermentation, the counts of *Enterobacteriaceae* became negligible, whereas the *Lactobacillus* count became prominent (11 log CFU/mL). After the storage, *Pediococcus pentosaceous* (6%) and *Enterococcus mudtii* (19%) were found in the drinks. Increasing the inoculum load or using a different *Lactobacillus* strain may decrease the growth of *E. mudtii*	Paz et al. (2020)
Fermentation	Flour	Spontaneous fermentation of quinoa flour at 30°C to make type 1 sourdough back-slopped up to 8 days	Identification and extraction of exopolysaccharide-making yeast and lactic acid bacteria	The microbes were consisted of *Lactobacillus*, *Lactococcus*, *Enterococcus*, *Pediococcus*, *Leuconostoc*, and *Weissella*. In a starch-based medium, *W. cibaria*, *P. pentosaceous*, and *L. citreum* produced exopolysaccharide. *P. pentosaceous* showed the highest efficiency in producing exopolysaccharide and decreasing the pH of the dough after 24 h of fermentation	Franco et al. (2020)

Fermentation	Flour	Quinoa flour was fermented using *L. acidophilus* DSM 20079 (37°C, 72 h), *L. brevis* R26 (37°C, 72 h), or *L. plantarum* MR24 (30 °C, 72 h)	Nutritional and physicochemical properties of fermented flour and breads incorporated with the fermented flour	Fermentation decreased the pH of the quinoa flour dough, while increasing the contents of lactic acid and phenolics, and the activities of amylase, cellulase, protease, and phytase in the dough. The extents of these changes were dependent on the specific strain. The addition of the fermented flour with *L. plantarum* MR24 at up to 10% gave desirable sensory quality of the resulting wheat-based breads	Cizeikiene et al. (2021)
Fermentation and malting	Whole grain flour	Spontaneous fermentation was done at 28°C for 48 h to obtain sourdough. Malting was done on soaked grains at 28°C for 48 h. The malted grains were dried at 35°C for 24 h and milled into flours. Fermented and malted quinoa flours were compared with those of cowpea flours	Physicochemical and chemical properties	Fermented quinoa flour sourdough had lower contents of flavonoids, phenolics, dietary fibers, and proteins, in vitro antioxidant activities, and lower oil and water adsorption and higher acidity than fermented cowpea flour sourdough. Malted quinoa flour had higher dispersibility, pasting viscosities, and swelling power than malted cowpea flour	Kewuyemi et al. (2022)

(*Continued*)

Table 11.3 (Continued)

Processing methods	Sample form	Experimental/ processing methods	Targeted properties/ functions	Major findings	References
Malting and roasting	Seed	Quinoa seeds were germinated at 23°C for 3 days before kilning (50°C, 24 h). The resulting malt was roasted at different temperatures (100°C, 145°C, 190°C) for 30 min	Phenolics, in vitro antioxidant activity, Maillard reaction products	The highest amounts of Maillard reaction products and phenolics in malted and roasted quinoa seeds were obtained at a roasting temperature of 145°C. At a higher roasting temperature of 190°C, degradation of phenolics and Maillard reaction products and reduction in the antioxidant activity	Carciochi et al. (2016a)
Malting	Seed	Seeds were steeped at 30°C for 48 h before germination (23°C, 48 h)	Composition of folate vitamers	The malting had little effect on the folate composition, while increasing the total folate content of buckwheat and amaranth	Motta et al. (2017)
Malting	Flour	Seeds were steeped at 20°C for 2 h before germination (25°C, up to 72 h). The sprouted seeds were ground	The malted flour (30%) was incorporated into rice flour to make gluten free muffins. Physicochemical and sensory properties of the muffins were studied	Adding quinoa flour increased the levels of protein, minerals and amino acids of the muffins. Adding quinoa flour malted for 24 h gave the best scores of smell and taste	Miranda-Villa et al. (2019)
Malting	Seed	Seeds were steeped at 30°C for 48 h before germination (23°C, 48 h)	Amino acid composition and protein quality	Malting increased the protein content as well as the ratio of essential amino acids to nonessential amino acids	Motta et al. (2019)

Malting	Seed	Seeds of three quinoa varieties were steeped in water for 4 h at 25°C. The hydrated seeds were germinated at 25°C for 48 h before drying at 55°C for 24 h. The radicle of the resulted malted seeds was removed	Nutritional and chemical composition	Malting process increased the levels of flavonoids, phenolics, reducing sugars and ascorbic acid as well as in vitro antioxidant activities, while reducing the lipid and ash contents. The protein content of quinoa seeds was either increased or decreased, depending on the specific variety	Aguilar et al. (2019)
Germination	Seed	Germination conditions varied at 12°C–28°C and 12–72 h	Production of bioactive compounds and in vitro antioxidant activity using response surface optimization of the germination conditions	The maximum contents of bioactive compounds (increased by 80%) and antioxidant activity (increased by 30%) of the seeds were obtained at 20°C for 42 h. The germination significantly increased the contents of GABA, quercetin-*O*-glucuronide and kaempferol-*O*-dirhamnosyl-galactopyranose	Paucar-Menacho et al. (2018)
Germinated and fermented	Flour	Quinoa flour was fermented (30°C, 16 h) using *Pediococcus pentosaceus* QB17. Quinoa seeds were germinated at 25°C for 24 h before drying and milling	The germinated and fermented flours (20%) were incorporated into rice flour to make gluten free breads. Physicochemical and sensory properties of the breads were studied	The addition of germinated or fermented (sourdough) quinoa flours increased the levels of proteins and amino acids. The addition of quinoa sourdough increased the shelf life (to 6 days) and improved the texture of the resulting breads, whereas the addition of germinated quinoa flour gave the breads a decreased shelf life (4 days)	Franco et al. (2021)

(*Continued*)

Table 11.3 (Continued)

Processing methods	Sample form	Experimental/ processing methods	Targeted properties/ functions	Major findings	References
Sprouting and storage	Seed	Quinoa seeds were germinated for 4 days (20°C, 70% humidity) before storage at 5°C for 7 days	In vitro antioxidant activity, vitamin C and total phenolic contents	Germination of 4 days increased the antioxidant activity and total phenolic content by over two times. Storage increased the total phenolic and vitamin C contents	Laus et al. (2017)
Germination	Seed	Quinoa seeds were germinated at 10°C for 24 h	Metabolomics profile	Germination increased the concentrations of most of the metabolites including soluble proteins. The metabolomics profile depended on the quinoa genotype. The germination increased, had no effect or decreased the saponin levels of quinoa seeds, depending on the quinoa genotypes	Lalaleo et al. (2020)
Germination and kilning	Seed	Germination: 22°C, 72 h, 95% of relative humidity; kilning of the above germinated seeds: 90°C, 5 min	Phenolic profile and chemical composition	Both kilning and germination increased the protein and total phenolic levels in the seeds. Germination increased significantly the levels of a kaempferol tri-glycoside and coumaric acid	Pilco-Quesada et al. (2020)
Germination	Seed	Quinoa seeds were germinated at 20°C for 4 days	Chemical composition, starch structure, and properties	Chemical composition, starch structure, and properties were affected to different extents by the germination, depending on the time. However, no clear pattern of changes was obtained	Ma et al. (2021)

Germination	Seed	Quinoa seeds were soaked at 25°C for 12 h before germination for 72 h. The germinated seeds were dried at 50°C for 24 h	Nutritional composition and in vivo biological functions (albino rat model)	Germination increased the concentrations of vitamin C, Fe, Ca, Zn, and carotenoids, while reducing those of saponins and phytic acid. Incorporation of 10% of quinoa sprout in the diet of the rats improved the blood cell counts, serum profile and body weight	Darwish et al. (2021)

analyzed the microbial composition of quinoa juice suspension fermented (30°C, 2 days) using *L. plantarum* DSM 9843. The resulting product was stored at 4°C for 28 days. Without the fermentation, a range of different microbial species were found in the drinks including *Escherichia*, *Enterobacter*, *Leclercia*, *Pantoea*, *Klebsiella*, *Enterococcus*, and *Salmonella*. With *L. plantarum*-based fermentation, the counts of *Enterobacteriaceae* became negligible, whereas the *Lactobacillus* count became prominent (11 log CFU/mL). After the storage, *Pediococcus pentosaceous* (6%) and *Enterococcus mudtii* (19%) were found in the drinks. Increasing the inoculum load or using a different *Lactobacillus* strain may decrease the growth of *E. mudtii* (Paz et al., 2020). Franco et al. (2020) analyzed exopolysaccharide-making yeast and lactic acid bacteria in spontaneous fermented (30°C, back-slopped up to 8 days) quinoa flour. The microbes consisted of *Lactobacillus*, *Lactococcus*, *Enterococcus*, *Pediococcus*, *Leuconostoc*, and *Weissella*. In a starch-based medium, *W. cibaria*, *P. pentosaceous*, and *L. citreum* produced exopolysaccharide. *P. pentosaceous* showed the highest efficiency in producing exopolysaccharide and decreasing the pH of the dough after 24 h of fermentation (Franco et al., 2020). Quinoa seeds were treated at 121°C (20 min) before fermentation (culture at 5%, v/w) at 80% (relative humidity) and 32°C for 10 days using *Monascus anka* (Yang et al., 2022). Ferulic acid, *p*-hydroxybenzoic acid, and protocatechuic acid were the major phenolics in the fermented samples. *M. anka*-based fermentation for 6 days gave the seeds the highest total phenolic content. The fermentation increased the absorption of free phenolics in the intestine, in vitro antioxidant activity (especially that of the free phenolics), and inhibition activities against α-glucosidase and α-amylase. The free phenolics were largely responsible for the observed activities (Yang et al., 2022). Overall, selection of suitable strains for making quinoa sourdough may be done for improved physicochemical profile.

The effect of fermentation on nutritional and chemical properties of quinoa has been studied (Table 11.3). For example, fermentation of food grains using lactic acid bacteria can be used to improve the nutritional values (Rollán et al., 2019). The fermentation can decrease the levels of antinutritional nutrients (e.g., phytic acid), while producing certain nutrients (e.g., some B vitamins) and increasing those of certain phytochemicals (e.g., phenolic compounds) (Rollán et al., 2019). Duliński et al. (2017) studied the composition of myo-inositol phosphates in quinoa seeds affected by solid state fermentation using *Rhizopus oligosporus*. Rizzello et al. (2017) analyzed the in vitro antioxidant activities of quinoa flour fermented (37°C, 24 h) using autochthonous lactic acid bacteria. Fermentation increased the antioxidant activities of both salt-soluble and water extracts. The extract obtained from the dough fermented using *L. plantarum* T0A10 had the highest prevention capacity against linoleic acid autoxidation. A total of 5 peptides (5 − 9 amino acid residues) were obtained from the protein of the fermented dough and showed strong in vitro antioxidant activity. The fractions showed cellular antioxidant capacity on human keratinocytes NCTC 2544. The fermentation decreased the percentage of phytate, while increasing that of inositol triphosphate due to the presence of microbial 3-phytase. The total phytate content of quinoa seeds was reduced (e.g., from 14 to 5 μmol/g dry matter in red quinoa seeds). The reduction was dependent on the quinoa genotypes (Rizzello et al., 2017). Ayyasha et al. (2019) analyzed the biological properties of quinoa seeds subjected to solid-state fermentation (37°C, 72 h) with *Lactobacillus reuteri* K777 and *L. plantarum* K779. The fermentation significantly increased, to different extents, the total phenolic content and in vitro antioxidant activity, α-glucosidase inhibition, ACE-inhibition, and anticancer activities of quinoa seeds, depending on the type of bacterial strain. The fermentation induced increases in the anticancer functions were seen to be greater in lupin than in quinoa (Ayyasha et al., 2019). Cizeikiene et al. (2021) analyzed nutritional and physicochemical properties of quinoa flour fermented with *L. acidophilus* DSM 20079 (37°C, 72 h), *L. brevis* R26 (37°C, 72 h), or *L. plantarum* MR24 (30°C, 72 h) and breads incorporated with the fermented flour. The fermentation decreased the pH of the quinoa flour dough, while increasing the contents of lactic acid and phenolics, and the activities of amylase, cellulase, protease, and phytase in the dough. The extents of these changes were dependent on the specific strain. The addition of the fermented flour with *L. plantarum* MR24 at up to 10% gave desirable sensory quality of the resulting wheat-based breads (Cizeikiene et al., 2021). Kewuyemi et al. (2022) compared the chemical and physicochemical properties of quinoa flour sourdough with those of cowpea sourdough. The quinoa sourdough had lower contents of flavonoids, phenolics, dietary fibers, and proteins, in vitro antioxidant activities, and lower oil and water adsorption and higher acidity than the fermented cowpea flour sourdough (Kewuyemi et al., 2022). Overall, the effects of fermentation on the chemical and nutritional properties of quinoa products depend on a range of factors including the fermentation conditions and microbial strains used and the genotypes of quinoa and quinoa forms (flour vs seed). Fermentation can be used efficiently to enhance the nutritional quality of quinoa-based products.

11.4.2 Malting

Quinoa seeds were malted, and the changes in the nutritional properties and food applications of malted quinoa flour were studied (Carciochi et al., 2016a; Motta et al., 2017; Miranda-Villa et al., 2019; Motta et al., 2019; Aguilar et al., 2019; Kewuyemi et al., 2022). For example, Motta et al. (2017) found that malting process had little effect on the folate vitamer composition, while increasing the total folate content of buckwheat and amaranth. The malting increased the protein content as well as the ratio of essential amino acids to nonessential amino acids (Motta et al., 2019). Another study showed that malting process increased the levels of flavonoids, phenolics, reducing sugars, and ascorbic acid as well as in vitro antioxidant activities, while reducing the lipid and ash contents. The protein content of quinoa seeds was either increased or decreased, depending on the specific variety (Aguilar et al., 2019). Overall, the effect of malting on the chemical and nutritional properties depends on the malting conditions and quinoa genotypes and forms. Kewuyemi et al. (2022) compared the chemical and physicochemical properties of malted quinoa and cowpea flours. Malted quinoa flour had higher dispersibility, pasting viscosities, and swelling power than malted cowpea flour. Malted quinoa seeds were mixed with rice flour to make gluten-free muffins (Miranda-Villa et al., 2019). Adding the quinoa flour increased the levels of proteins, minerals, and amino acids of the muffins. Adding the quinoa flour malted for 24 h gave the best scores of smell and taste (Miranda-Villa et al., 2019). It may be expected in the future that malted quinoa flour can be used in the formulations of other food products.

11.4.3 Germination

Quinoa seeds were germinated at different conditions. The effect of germination on chemical and nutritional properties of quinoa sprouts was studied (Paucar-Menacho et al., 2018; Franco et al., 2021; Lalaleo et al., 2020; Laus et al., 2017; Pilco-Quesada et al., 2020; Ma et al., 2021; Darwish et al., 2021). For example, Paucar-Menacho et al. (2018) showed that the maximum contents of bioactive compounds (increased by 80%) and antioxidant activity (increased by 30%) of the seeds were obtained at 20°C for 42 h. The germination significantly increased the contents of γ-aminobutyric acid (GABA), quercetin-*O*-glucuronide, and kaempferol-*O*-dirhamnosyl-galactopyranose (Paucar-Menacho et al., 2018). Lalaleo et al. (2020) showed that germination increased the concentrations of most of the metabolites including soluble proteins. The metabolomics profile depended on the quinoa genotype. The germination increased, had no effect, or decreased the saponin levels of quinoa seeds, depending on the quinoa genotypes (Lalaleo et al., 2020). Darwish et al. (2021) showed that germination increased the concentrations of vitamin C, Fe, Ca, Zn, and carotenoids, while reducing those of saponins and phytic acid. Incorporation of 10% of the quinoa sprout in the diet of the rats improved the blood cell counts, serum profile, and body weight (Darwish et al., 2021). Therefore germination at suitable conditions can be used to improve the nutritional quality of quinoa sprouts. Germinated quinoa flour was incorporated into rice flour to make gluten-free breads (Franco et al., 2021). The addition of the germinated quinoa flours increased the levels of proteins and amino acids. However, the addition gave the breads a decreased shelf life (Franco et al., 2021). Therefore the process of using germinated quinoa flour for product development remains to be optimized. The suitability of germinated quinoa flour for specific food formulations remains to be tested.

11.5 Conclusions

Diverse thermal, nonthermal, and biotechnological methods have been used to process quinoa flour and seeds. These treatments created a range of different functional, physicochemical, sensory and nutritional properties of quinoa-based products. Depending on the application purpose and targeted functionalities, the types of processing methods and process parameters as well as the quinoa varieties should be carefully selected. Nonthermal processes and bioprocessing methods are promising to better retain or improve the nutritional quality of the quinoa products. The treated intermediate products including flour and seeds should be further tested to formulate food products.

Quinoa-based food product development is described in Chapter 12 of this book. Sensory properties of the resulting products remain to be studied. Irradiation with gamma rays, ultraviolet rays, and electron beam may be used to process quinoa to create different functional properties, while increasing the microbiological shelf life of quinoa products. Combination of different types of process operations remains to be used to treat quinoa. There has been a lack of studies on the interplay between processing conditions and quinoa variety. Some of the processing methods such as ultrasound and cold plasma treatments are too far away from being practical for industrial uses. These represent research opportunities for quinoa processing researchers.

References

Aguilar, J., Miano, A. C., Obregón, J., Soriano-Colchado, J., & Barraza-Jáuregui, G. (2019). Malting process as an alternative to obtain high nutritional quality quinoa flour. *Journal of Cereal Science*, *90*, 102858.

Ahmed, J., & Thomas, L. (2020). Changes in structural, functional and antioxidant properties induced by high pressure on quinoa flour. *Journal of Food Measurement and Characterization*, *14*, 401−410.

Akharume, F. U., Aluko, R. E., & Adedeji, A. A. (2021). Modification of plant proteins for improved functionality: A review. *Comprehensive Reviews in Food Science and Food Safety*, *20*, 198−224.

Alajil, O., Hymavathi, T. V., Maheswari, K. U., & Rudra, S. G. (2020). Nutritional, physicochemical and sensory attributes of quinoa based extrudates. *Vegetos*, *33*, 390−400.

Aluwi, N. A., Gu, B. J., Dhumal, G. S., Medina-Meza, I. G., Murphy, K. M., & Ganjyal, G. M. (2016). Impacts of scarification and degermination on the expansion characteristics of select quinoa varieties during extrusion processing. *Journal of Food Science*, *81*, E2939−E2949.

Ayyash, M., Johnson, S. K., Liu, S. Q., Al-Mheiri, A., & Abushelaibi, A. (2018). Cytotoxicity, antihypertensive, antidiabetic and antioxidant activities of solid-state fermented lupin, quinoa and wheat by *Bifidobacterium* species: In vitro investigations. *LWT-Food Science and Technology*, *95*, 295−302.

Ayyasha, M., Johnson, S. K., Liu, S. Q., Mesmari, N., Dahmani, S., Dhaheri, A. S. A., & Kizhakkayil, J. (2019). In vitro investigation of bioactivities of solid-state fermented lupin, quinoa and wheat using *Lactobacillus* spp. *Food Chemistry*, *275*, 50−58.

Carciochi, R. A., Dimitrov, K., & D Alessan, L. G. (2016a). Effect of malting conditions on phenolic content, Maillard reaction products formation, and antioxidant activity of quinoa seeds. *Journal of Food Science and Technology*, *53*, 3978−3985.

Carciochi, R. A. D., Alessandro, L. G., & Manrique, G. D. (2016b). Effect of roasting conditions on the antioxidant compounds of quinoa seeds. *International Journal of Food Science and Technology*, *51*, 1018−1025.

Castro, L. M. G., Alexandre, E. M. C., Saraiva, J. A., & Pintado, M. (2020). Impact of high pressure on starch properties: A review. *Food Hydrocolloids*, *106*, 105877.

Ceballos-González, C., Bolívar-Monsalve, J., Ramírez-Toro, C., & Bolívar, G. A. (2018). Effect of lactic acid fermentation on quinoa dough to prepare gluten-free breads with high nutritional and sensory quality. *Journal of Food Processing and Preservation*, *42*, e13551.

Chiş, M. S., Păucean, A., Man, S. M., Vodnar, D. C., Teleky, B. E., Pop, C. R., Stan, L., Borsai, O., et al. (2020). Quinoa sourdough fermented with *Lactobacillus plantarum* ATCC 8014 designed for gluten-free muffins – A powerful tool to enhance bioactive compounds. *Journal of Food Processing and Preservation*, *10*, 7140.

Cizeikiene, D., Gaide, I., & Basinskiene, L. (2021). Effect of lactic acid fermentation on quinoa characteristics and quality of quinoa-wheat composite bread. *Foods*, *10*, 171.

Cui, R., & Zhu, F. (2021). Ultrasound modified polysaccharides: A review of structure, physicochemical properties, biological activities and food applications. *Trends in Food Science & Technology*, *107*, 491–508.

Darwish, A. M. G., Al-Jumayi, H. A. O., & Elhendy, H. A. (2021). Effect of germination on the nutritional profile of quinoa (*Cheopodium quinoa* Willd.) seeds and its anti-anemic potential in Sprague–Dawley male albino rats. *Cereal Chemistry*, *98*, 315–327.

Duliński, R., Starzyńska-Janiszewska, A., Byczyński, Ł., & Błaszczyk, U. (2017). Myo-inositol phosphates profile of buckwheat and quinoa seeds: Effects of hydrothermal processing and solid-state fermentation with *Rhizopus oligosporus*. *International Journal of Food Properties*, *20*, 2088–2095.

Franco, W., Evert, K., & Van Nieuwenhove, C. (2021). Quinoa flour, the germinated grain flour, and sourdough as alternative sources for gluten-free bread formulation: Impact on chemical, textural and sensorial characteristics. *Fermentation*, *7*, 115.

Franco, W., Pérez-Díaz, I. M., Connelly, L., & Diaz, J. T. (2020). Isolation of exopolysaccharide-producing yeast and lactic acid bacteria from quinoa (*Chenopodium quinoa*) sourdough fermentation. *Foods*, *9*, 337.

Gao, S., Liu, H., Sun, L., Liu, N., Wang, J., Huang, Y., Wang, F., Cao, J., Fan, R., Zhang, X., & Wang, M. (2019). The effects of dielectric barrier discharge plasma on physicochemical and digestion properties of starch. *International Journal of Biological Macromolecules*, *138*, 819–830.

Gearhart, C., & Rosentrater, K. A. (2017). Extrusion processing of amaranth and quinoa into gluten free snack foods for celiac and gluten free diets. *Journal of Food Research*, *6*, 107–116.

Gómez-Ramírez, A., López-Santos, C., Cantos, M., García, J. L., Molina, R., Cotrino, J., et al. (2017). Surface chemistry and germination improvement of quinoa seeds subjected to plasma activation. *Scientific Reports*, *7*, 5924.

Hall, A. E., & Moraru, C. I. (2022). Comparative effects of high pressure processing and heat treatment on in vitro digestibility of pea protein and starch. *NPJ Science of Food*, *6*, Article number: 2 (2022).

Hur, J., Nguyen, T. T. H., Park, N., Kim, J., & Kim, D. (2018). Characterization of quinoa (*Chenopodium quinoa*) fermented by *Rhizopus oligosporus* and its bioactive properties. *AMB Express*, *8*, 143.

Irigoyen, R. M. T., & Giner, S. A. (2017). Modeling thin layer drying-roasting kinetics of soaked quinoa. Coupled mass and energy transfer. *Biosystems Engineering*, *157*, 99–108.

Jadhav, H. B., Annapure, U. S., & Deshmukh, R. R. (2021). Non-thermal technologies for food processing. *Frontiers in Nutrition*, *8*, 657090.

Jan, R., Saxena, D. C., & Singh, S. (2017). Effect of extrusion variables on antioxidant activity, total phenolic content and dietary fibre content of gluten-free extrudate from germinated Chenopodium (*Chenopodium album*) flour. *International Journal of Food Science and Technology*, *52*, 2623–2630.

Kewuyemi, Y. O., Kesa, H., & Adebo, O. A. (2022). Biochemical properties, nutritional quality, colour profile and techno-functional properties of whole grain sourdough and malted cowpea and quinoa flours. *International Journal of Food Science and Technology*, *57*, 1527–1543.

Kowalski, R. J., Medina-Meza, I. G., Thapa, B. B., Murphy, K. M., & Ganjyal, G. M. (2016). Extrusion processing characteristics of quinoa (*Chenopodium quinoa* Willd.) var. Cherry Vanilla. *Journal of Cereal Science*, *70*, 91–98.

Kuktaite, R., Repo-Carrasco-Valencia, R., Mendoza, C. C., Plivelic, T. S., Hall, S., & Johansson, E. (2022). Innovatively processed quinoa (*Chenopodium quinoa* Willd.) food: Chemistry, structure and end-use characteristics. *Journal of the Science of Food and Agriculture*, *102*, 5065–5076.

Kumar, M., Singh, A., Bharti, A., & Upadhyay, A. (2022). Effect of parching process on nutritional and physical characteristics of quinoa. *The Pharma Innovation Journal*, *11*, 806–811.

Lalaleo, L., Hidalgo, D., Valle, M., Calero-Cáceres, W., Lamuela-Raventós, R. M., & Becerra-Martínez, E. (2020). Differentiating, evaluating, and classifying three quinoa ecotypes by washing, cooking and germination treatments, using 1H NMR-based metabolomic approach. *Food Chemistry*, *331*, 127351.

Laus, M. N., Cataldi, M. P., Robbe, C., D'Ambrosio, T., & Amodio, M. L. (2017). Antioxidant capacity, phenolic and vitamin C contents of quinoa (*Chenopodium quinoa* Willd.) as affected by sprouting and storage conditions. *Italian Journal of Agronomy*, *12*, 816.

Leonard, W., Zhang, P., Ying, D., & Fan, Z. (2020). Application of extrusion technology in plant food processing byproducts: An overview. *Comprehensive Reviews in Food Science and Food Safety*, *19*, 218–246.

Li, G., & Zhu, F. (2018). Effect of high pressure on rheological and thermal properties of quinoa and maize starches. *Food Chemistry*, *241*, 380–386.

Li, S., Zhang, R., Lei, D., Huang, Y., Cheng, S., Zhu, Z., Wu, Z., & Cravotto, G. (2021). Impact of ultrasound, microwaves and high-pressure processing on food components and their interactions. *Trends in Food Science & Technology*, *109*, 1–15.

Lopes, C. D. O., Barcelos, M. D. F. P., Vieira, C. N. D. G., Abreu, W. C. D., Ferreira, E. B., Pereira, R. C., & Angelis-Pereira, M. C. D. (2019). Effects of sprouted and fermented quinoa (*Chenopodium quinoa*) on glycemic index of diet and biochemical parameters of blood of Wistar rats fed high carbohydrate diet. *Journal of Food Science and Technology*, *56*, 40–48.

Lorusso, A., Coda, R., Montemurro, M., & Rizzello, C. G. (2018). Use of selected lactic acid bacteria and quinoa flour for manufacturing novel yogurt-like beverages. *Foods*, *7*, 51.

Luo, X., Du, Z., Yang, K., Wang, J., Zhou, J., Liu, J., & Chen, Z. (2021). Effect of electron beam irradiation on phytochemical composition, lipase activity and fatty acid of quinoa. *Journal of Cereal Science*, *98*, 103161.

Ma, Z., Guan, X., Gong, B., & Li, C. (2021). Chemical components and chain-length distributions affecting quinoa starch digestibility and gel viscoelasticity after germination treatment. *Food & Function*, *12*, 4060.

Mandhare, L. L., More, D. R., & Nagulwar, M. M. (2020). Effect of popping methods on popping characteristics of quinoa seed. *Journal of Pharmacognosy and Phytochemistry*, *9*, 1943–1945.

Maqueda, R. H., Redondo, I. B., Jácome, S. J., & Moreno, Á. H. (2018). Microwave drying of amaranth and quinoa seeds: Effects of the power density on the drying time, germination rate and seedling vigour. *Journal of Microwave Power and Electromagnetic Energy*, *52*, 299–311.

Mezzatesta, P., Farah, S., Di Fabio, A., & Emilia, R. (2020). Variation of the nutritional composition of quinoa according to the processing used. *Proceedings*, *53*, 4.

Mir, N. A., Riar, C. S., & Singh, S. (2019). Physicochemical, molecular and thermal properties of high-intensity ultrasound (HIUS) treated protein isolates from album (*Chenopodium album*) seed. *Food Hydrocolloids*, *96*, 433–441.

Miranda-Villa, P. P., Mufari, J. R., Bergesse, A. E., & Calandri, E. L. (2019). Effects of whole and malted quinoa flour addition on gluten-free muffins quality. *Journal of Food Science*, *84*, 147–153.

Mosibo, O. K., Ferrentino, G., Alam, M. R., Morozova, K., & Scampicchio, M. (2022). Extrusion cooking of protein-based products: Potentials and challenges. *Critical Reviews in Food Science and Nutrition*, *62*, 2526–2547.

Motta, C., Castanheira, I., Gonzales, G. B., Delgado, I., Torres, D., Santos, M., & Matosf, A. S. (2019). Impact of cooking methods and malting on amino acids content in amaranth, buckwheat and quinoa. *Journal of Food Composition and Analysis*, *76*, 58–65.

Motta, C., Delgado, I., Matos, A. S., Gonzales, G. B., Torres, D., Santos, M., Chandra-Hioef, M. V., et al. (2017). Folates in quinoa (*Chenopodium quinoa*), amaranth (*Amaranthus* sp.) and buckwheat (*Fagopyrum esculentum*): Influence of cooking and malting. *Journal of Food Composition and Analysis*, *64*, 181–187.

Multari, S., Marsol-Vall, A., Keskitalo, M., Yang, B., & Suomela, J. P. (2018). Effects of different drying temperatures on the content of phenolic compounds and carotenoids in quinoa seeds (*Chenopodium quinoa*) from Finland. *Journal of Food Composition and Analysis*, *72*, 75–82.

Ortega-Rivas, E. (2012). *Non-thermal food engineering operations*. Springer Science & Business Media.

Pankaj, S. K., Wan, Z., & Keener, K. M. (2018). Effects of cold plasma on food quality: A review. *Foods*, *7*, 4.

Paucar-Menacho, L. M., Martínez-Villaluenga, C., Dueñas, M., Frias, J., & Peñas, E. (2018). Response surface optimisation of germination conditions to improve the accumulation of bioactive compounds and the antioxidant activity in quinoa. *International Journal of Food Science and Technology*, *53*, 516–524.

Paz, P. C., Janny, R. J., & Håkansson, Å. (2020). Safeguarding of quinoa beverage production by fermentation with *Lactobacillus plantarum* DSM 9843. *International Journal of Food Microbiology*, *324*, 108630.

Pilco-Quesada, S., Tian, Y., Yang, B., Repo-Carrasco-Valencia, R., & Suomela, J. P. (2020). Effects of germination and kilning on the phenolic compounds and nutritional properties of quinoa (*Chenopodium quinoa*) and kiwicha (*Amaranthus caudatus*). *Journal of Cereal Science*, *94*, 102996.

Rizzello, C. G., Lorusso, A., Montemurro, M., & Gobbetti, M. (2016). Use of sourdough made with quinoa (*Chenopodium quinoa*) flour and autochthonous selected lactic acid bacteria for enhancing the nutritional, textural and sensory features of white bread. *Food Microbiology*, *56*, 1–13.

Rizzello, C. G., Lorusso, A., Russo, V., Pinto, D., Marzani, B., & Gobbetti, M. (2017). Improving the antioxidant properties of quinoa flour through fermentation with selected autochthonous lactic acid bacteria. *International Journal of Food Microbiology*, *241*, 252–261.

Rollán, G. C., Gerez, C. L., & LeBlanc, J. G. (2019). Lactic fermentation as a strategy to improve the nutritional and functional values of pseudocereals. *Frontiers in Nutrition*, *6*, 98.

Ruiz Rodríguez, L., Pingitore, E. V., Rollan, G., Cocconcelli, P. S., Fontana, C., Saavedra, L., Vignolo, G., & Hebert, E. M. (2016). Biodiversity and technological-functional potential of lactic acid bacteria isolated from spontaneously fermented quinoa sourdoughs. *Journal of Applied Microbiology*, *120*, 1289–1301.

Song, J., Shao, Y., Chen, X., & Li, X. (2022). Release of characteristic phenolics of quinoa based on extrusion technique. *Food Chemistry*, *374*, 128780.

Starič, P., Vogel-Mikuš, K., Mozetič, M., & Junkar, I. (2020). Effects of nonthermal plasma on morphology, genetics and physiology of seeds: A review. *Plants*, *9*, 1736.

Starzyńska-Janiszewska, A., Duliński, R., Stodolak, B., Mickowska, B., & Wikiera, A. (2016). Prolonged tempe-type fermentation in order to improve bioactive potential and nutritional parameters of quinoa seeds. *Journal of Cereal Science*, *71*, 116–121.

Starzyńska-Janiszewska, A., Stodolak, B., Duliński, R., Mickowska, B., & Sabat, R. (2017). Fermentation of colored quinoa seeds with *Neurospora intermedia* to obtain oncom-type products of favorable nutritional and bioactive characteristics. *Cereal Chemistry*, *94*, 619–624.

Starzyńska-Janiszewska, A., Stodolaka, B., Gómez-Caravaca, A. M., Mickowska, B., Martin-Garcia, B., & Byczyński, Ł. (2019). Mould starter selection for extended solid-state fermentation of quinoa. *LWT-Food Science and Technology*, *99*, 231–237.

Tao, Y., Yan, B., Fan, D., Zhang, N., Ma, S., & Wang, L. (2020). Structural changes of starch subjected to microwave heating: A review from the perspective of dielectric properties. *Trends in Food Science & Technology*, *99*, 593–607.

Väkeväinen, K., Ludena-Urquizo, F., Korkala, E., Lapveteläinen, A., Peräniemi, S., von Wright, A., & Plumed-Ferrera, C. (2020). Potential of quinoa in the development of fermented spoonable vegan products. *LWT-Food Science and Technology*, *120*, 108912.

Wu, L. G., Wang, A., Shen, R., & Qu, L. (2021). Effect of heating under pressure treatment on the antioxidant of quinoa. *International Journal of Food Engineering*, *17*, 795–804.

Yang, P., Huang, K., Zhang, Y., Li, S., Cao, H., & Song, H. (2022). Biotransformation of quinoa phenolic compounds with *Monascus anka* to enhance the antioxidant capacity and digestive enzyme inhibitory activity. *Food Bioscience*, *46*, 101568.

Ye, J., Hu, X., Luo, S., Liu, W., Chen, J., Zeng, Z., & Liu, C. (2018). Properties of starch after extrusion: A review. *Starch/Stärke*, *70*, 1700110.

Zapana, F., de Bruijn, J., Vidal, L., Melín, P., González, M. E., Cabrera, G., et al. (2020). Physical, chemical and nutritional characteristics of puffed quinoa. *International Journal of Food Science and Technology*, *55*, 313–322.

Zare, L., Mollakhalili-Meybodi, N., Fallahzadeh, H., & Arab, M. (2022). Effect of atmospheric pressure cold plasma (ACP) treatment on the technological characteristics of quinoa flour. *LWT-Food Science and Technology*, *155*, 112898.

Zhang, J., Li, M., Cheng, J., Zhang, X., Li, K., Li, B., Wang, C., & Liu, X. (2021). Viscozyme L hydrolysis and *Lactobacillus* fermentation increase the phenolic compound content and antioxidant properties of aqueous solutions of quinoa pretreated by steaming with α-amylase. *Journal of Food Science*, *86*, 1726–1736.

Zhu, F. (2015). Impact of ultrasound on structure, physicochemical properties, modifications, and applications of starch. *Trends in Food Science and Technology*, *43*, 1–17.

Zhu, F. (2021). Structure and physicochemical properties of starch affected by dynamic pressure treatments: A review. *Trends in Food Science & Technology*, *116*, 639–654.

Zhu, F., & Li, H. (2019a). Modification of quinoa flour functionality using ultrasound. *Ultrasonics-Sonochemistry*, *52*, 305–310.

Zhu, F., & Li, H. (2019b). Effect of high hydrostatic pressure on physicochemical properties of quinoa flour. *LWT-Food Science and Technology*, *114*, 108367.

Quinoa-based food product development

12

12.1 Introduction

Quinoa popularity has increased significantly due to the nutritional benefits related to the consumption of the seeds as whole grain used in different food products (Ismail & Rayan, 2022; Zoumpopoulou & Tsakalidou, 2019; Franco et al., 2021). Quinoa has special nutritional composition and health potential as described in Chapters 2 and 9 of this book. Starch is the major component of quinoa grain (Li & Zhu, 2018). Thus quinoa has similar processing properties to major grains such as wheat and rice. There have been increasing number and varieties of quinoa-based food products that have become available in the market in recent years. This trend is largely driven by consumer preferences. Quinoa seeds can be made as porridge or baked products such as biscuits, noodles, pasta, breads, tortillas, and crisps. The seeds are fermented to make beer or traditional alcoholic beverages (e.g., chicha) (Navruz-Varli & Sanlier, 2016).

A major attractive quality attribute of quinoa is being gluten-free (Guo et al., 2022). There has been increasing diagnosis of digestive disorders in humans due to gluten intake. This is caused by the gluten-forming storage protein prolamin in gluten-containing staple grains such as wheat, barley, and rye (Azizi et al., 2020). These digestive disorders include conditions such as gluten sensitivity, allergies to wheat, and celiac disease. Epidemiological studies have shown that approximately 1% of the world's population is suffering from these disorders (Zoumpopoulou & Tsakalidou, 2019). Awareness of these gluten-related medical conditions has caused growing demand from consumers for more gluten-free food and beverage products, while providing high nutritional value and taste and texture standards that gluten-containing foods tend to lack. There is growing interest in quinoa application to commercial gluten-free food product development to meet this consumer demand (Zoumpopoulou & Tsakalidou, 2019). Some food products of high interest in the food industry include gluten-free pastas, breads and other baked goods, snacks, and beverages such as beer (Navruz-Varli & Sanlier, 2016; Linares-García et al., 2019; Cerdá-Bernad et al., 2022; Aguiar et al., 2022). Dairy-free and gluten-free products such as milk substitute and cheese analogs have been developed (Huang et al., 2021; Chen et al., 2020). There has been keen interest in using quinoa-related products to improve the nutritional, sensory, and shelf-life quality of animal-based products such as dairy-, fish-, and meat-based foods (Fernández-López et al., 2021; Romero et al., 2018). This chapter aims to outline how quinoa grain and related products such as flour and extract have been applied to formulate food and beverage products including both gluten-containing and gluten-free products. Physicochemical,

Quinoa. DOI: https://doi.org/10.1016/B978-0-323-99909-0.00010-6

chemical, nutritional, and sensory properties of quinoa-based products are described in relation to quinoa addition and fortification. Regulatory and consumer-related issues on quinoa-based gluten-free food products are discussed, and research opportunities in using quinoa for food product development are suggested.

12.2 Gluten-containing products with quinoa

Most of the gluten-containing products formulated with quinoa addition have been bakery products including cupcakes, tarhana, biscuit and cookies, breads, steamed bread, instant noodles, kishk, and beef burgers (Table 12.1). In the formulations of these products, wheat flour was partially replaced with quinoa. Different properties including physicochemical, chemical, nutritional, textural, and sensory aspects of these gluten-containing products as affected by quinoa addition were studied.

12.2.1 Cupcake

Quinoa protein isolate was obtained and modified by freeze-drying or heating (López-Alarcón et al., 2019). The quinoa protein isolate of up to 30% replaced wheat flour to make cupcakes containing egg, sugar, vanilla extract, cinnamon, yeast, and oil as the ingredients. Sensory, texture, physicochemical, and microbiological properties of the cupcakes as affected by the quinoa protein isolate addition were studied. The addition of the quinoa protein isolate in different forms increased the hardness and chewiness of the cupcakes, while increasing the water activity. The cupcakes containing raw, heat-treated, or freeze-dried quinoa protein isolate samples had higher acceptance of sensory quality and longer microbiological shelf life than the control samples. Mechanistically, the freeze-drying or heating-modified proteins had more nano-pores and higher water retention capacity, decreasing the water availability to the fungal growth and increasing the shelf life (López-Alarcón et al., 2019).

12.2.2 Tarhana and kishk

Quinoa products have been formulated into traditional food products including tarhana and kishk (Ismail & Rayan, 2022; Çevik & Ertaş, 2019).

Tarhana is a traditional fermented food product. Yogurt, wheat flour, yeast, spices, and vegetables were mixed to make dough that was fermented (72 h, 30°C) to make tarhana (Çevik & Ertaş, 2019). *Lactobacillus bulgaricus*, *Streptococcus thermophilus*, and yeast (*Saccharomyces cerevisiae*) were the common strains involved in the fermentation of tarhana dough. Flours of lupine, buckwheat, and quinoa replaced the wheat flour by up to 40% in the tarhana formulations. The sensory and chemical properties of the resulting tarhana samples were studied. Adding the quinoa flour or flours of lupine or buckwheat increased the contents of phytic acid, Ca, Fe, K, Mg, P, Zn, protein, ash, total phenolics, a^* and b^*

Table 12.1 Formulations and properties of diverse food and beverage products containing quinoa and gluten.

Product type	Quinoa form	Formulations	Targeted properties	Major findings	References
Cupcake	Quinoa protein isolate obtained by freeze-drying or heating	Quinoa protein isolate up to 30% replaced wheat flour to make cupcakes containing egg, sugar, vanilla extract, cinnamon, yeast, and oil	Sensory, texture, physicochemical, microbiological properties	Addition of quinoa protein isolate in different forms increased the hardness and chewiness of the cupcakes, while increasing the water activity. The cupcakes containing raw, heat-treated, or freeze-dried quinoa protein isolate samples had higher acceptance of sensory quality and longer microbiological shelf life than the control samples	López-Alarcón et al. (2019)
Tarhana	Flour	Yogurt, wheat flour, yeast, spices and vegetables were mixed to make tarhana (a traditional Turkish food). Flours of lupine, buckwheat, and quinoa replaced wheat flour by up to 40% in the tarhana formulations	Sensory and chemical properties	Adding quinoa flour or flours of lupine or buckwheat increased the contents of phytic acid, Ca, Fe, K, Mg, P, Zn, protein, ash and total phenolics, a^* and b^* of color, and in vitro antioxidant activity of tarhana. Adding quinoa flour gave higher viscosity of the tarhana than adding the other flours. Adding quinoa flour at 20% gave the resulting tarhana the highest sensory scores	Çevik and Ertaş (2019)
Kishk	Seed	Wheat burghul was replaced with quinoa seeds up to 100%. The mixtures were blended with laban rayeb (a fermented milk	Physicochemical, color, sensory and microbiological properties	Increasing quinoa seed addition increased the contents of protein, lipids, fiber, ash, and total phenolics and in vitro antioxidant activity of the kishik, while	Ismail and Rayan (2022)

(*Continued*)

Table 12.1 (Continued)

Product type	Quinoa form	Formulations	Targeted properties	Major findings	References
		product) at a ratio of 1 to 2, before fermentation for 1 day at room temperature. More laban rayeb was added to make the ratio of 1 to 4. The resulting mixtures were shaped and dried at 50°C for 2 days before storage for 3 months		decreasing L^* and b^* of the color. During the storage, counts of total bacteria and lactic acid bacteria decreased for all the samples. The counts of coliform, yeast and mold became under the detection limits. The addition of quinoa seeds at 50% gave the highest sensory scores for acceptance	
Biscuit	Quinoa flour	Quinoa flour replaced wheat flour up to 60% to make biscuit	Effect of quinoa-based biscuit consumption on the biomarkers of cardiovascular diseases in humans using a randomized-controlled, double-blind crossover trial	The quinoa-based biscuit consumption favorably affected the cholesterol levels, body mass index, and body weight in humans, though these changes were relatively small. The risk of developing cardiovascular diseases was decreased by the intake of these biscuits in comparison with that by the consumption of wheat-based biscuits	Pourshahidi et al. (2020)
Cookie	Quinoa flour	Quinoa flour replaced wheat flour by up to 80%	Texture, color, mold-free shelf life	Quinoa flour addition increased the hardness, chewiness and density, microbiological shelf life, yellowness, redness, and darkness of the cookies, while decreasing the specific volume	Wang et al. (2015)

Cookie	Quinoa flour	Quinoa flour replaced wheat flour up to 50%	Chemical composition, shape and sensory properties of cookies	Increasing quinoa flour addition decreased the diameter and spread ratio, the hardness and L^* of the cookies, while increasing the a^* and b^*. The addition of quinoa flour increased the contents of protein, lipids, phenolics, phytic acid, K, Mg, Fe, Ca, and Zn. Quinoa flour addition up to 20% had little effect on the sensory quality of the cookies	Demir and Kılınç (2017)
Cookie	Quinoa flour	Quinoa flour replaced wheat flour by up to 50%	Chemical, physicochemical, textural and sensory properties of the cookies	Adding quinoa flour increased the contents of protein and fiber, water and oil absorption capacities, foaming capacity and bulk density of the flour mixtures, while decreasing the spread ratio of the cookies. Increasing the addition of quinoa flour decreased the overall acceptance of the cookies	Jan et al. (2018)
Bread	Quinoa flour	Quinoa flour replaced wheat flour by up to 20%	Chemical, textural and sensory properties	Adding quinoa flour at 20% decreased the volume and hardness of the breads. Adding quinoa flour increased the concentrations of Zn, Mn, Mg, K, Fe, and phytic acid. Adding quinoa flour at 10% little affected the sensory properties of the breads	Bilgiçli and İbanoğlu (2015)

(*Continued*)

Table 12.1 (Continued)

Product type	Quinoa form	Formulations	Targeted properties	Major findings	References
Bread	Quinoa flour	Quinoa flour replaced wheat flour by up to 80%	Texture, color, mold-free shelf life	Quinoa flour addition increased the hardness, chewiness, and density, microbiological shelf life, yellowness, redness, and darkness of the bread, while decreasing the specific volume	Wang et al. (2015)
Bread	Quinoa flake	Quinoa flakes were added to wheat flour up to 20%	Chemical, physicochemical and textural properties	The addition of quinoa flakes up to 20% significantly increased the contents of protein and tocopherols in the breads. Increasing the addition level of quinoa flakes made the crumb darker, yellower, and less red. The addition decreased the specific volume and firmness without affecting the overall sensory acceptance	Gewehr et al. (2017)
Bread	Flour	Quinoa flour replaced wheat flour by up to 100% to make breads with reduced salt content (0.6%)	Texture, staling and sensory properties	Adding quinoa flour to bread formulations decreased specific volume, number and volume of cells in the crumb, while increasing the hardness. Breads of quinoa flour were perceived as six times more bitter and two times saltier than those of wheat flour. Adding 10% of quinoa flour into wheat breads produced similar sensory quality to breads of wheat flour. Gluten-free breads had faster staling than wheat-based breads	Gostin (2019)

Bread	Flour	Quinoa flours of three varieties replaced wheat flour by 25% to make breads	Physicochemical, textural and sensory properties	Adding quinoa flour increased the staling of the bread crumb. The addition of quinoa flour at 25% increased the concentrations of nutrients including minerals, dietary fiber, proteins, and lipids had minor negative effect on textural and crumb structure, and gave similar sensory scores compared to the wheat control bread	Ballester-Sánchez et al. (2019c)
Bread	Flour	Quinoa flours of three varieties replaced wheat flour by 25% to make breads	Nutritional properties	The addition of quinoa flours increased the contents of linoleic and linolenic acids, dietary fiber, and the percentage of soluble-to-insoluble fiber, reduced the glycemic index and glycemic load, and had little effect on the protein content. The addition increased the contents of Fe and Zn, though the phytic acid content was also increased	Ballester-Sánchez et al. (2019b)
Bread	Flour	Quinoa flours of three varieties (black, red and white) up to 25% were formulated into breads of wheat	Phenolic composition, color, in vitro antioxidant activities	Breads with the addition of black and red quinoa flours had up to five times the in vitro antioxidant activity and two times the contents of both bound and free phenolics. The presence of black and red quinoa flours led to the formation of darker bread crust and crumb. A darker "unconventional" bread is more acceptable by consumers	Ballester-Sánchez et al. (2019a)

(*Continued*)

Table 12.1 (Continued)

Product type	Quinoa form	Formulations	Targeted properties	Major findings	References
Bread	Sprouted quinoa flour	Quinoa seeds were sprouted for 2 days. The flour of the sprouted seeds or normal quinoa flour was added up to 30% to wheat flour for bread formulations (Fig. 12.1)	Chemical, leavening, physicochemical, and textural properties	The addition of sprouted quinoa flour up to 20% had minor effect on the gluten aggregation and pasting behaviors of the flour systems, while increasing the degree of softening and water adsorption of the dough mixing by weakening the gluten network. The sprouted flour contained simple sugars and facilitated better gas production during the leavening than the native quinoa flour. The bread with sprouted flour had less bitterness. The addition of sprouted quinoa flour at 20% gave the bread the lowest crumb hardness and the highest specific volume	Suárez-Estrella et al. (2020)
Pan bread	Quinoa flour	Quinoa flour replaced wheat flour by 5%–30% to make pan bread in the presence of yeast (1% of flour) and salt (1.5%). The dough was fermented at 25°C for 30 min before baking at 250°C for 15 min	Chemical, physicochemical and sensory properties	Adding quinoa flour increased the contents of protein and free amino acids, fibers, vitamins, and minerals, while decreasing that of starch. The quinoa flour addition increased the yellowness, while decreasing the redness and lightness of the pan breads. Overall sensory acceptance of the pan breads was not significantly affected by the quinoa flour addition	El-Sohaimy et al. (2021)

Bread (focaccia)	Quinoa flour liquid sourdough	Quinoa flour was fermented (15 h), using *Weissella cibaria* (a strain with exopolysaccharide producing capacity) to make liquid sourdough. The lipids of focaccia bread formulations were replaced using the liquid sourdough by 20%	Chemical, textural, sensory and glycemic properties	The focaccia breads fortified with quinoa flour liquid sourdough had much reduced lipid content (by ~20%) and increased protein (glutenin/glutelins and albumin) content, lower firmness and glycemic index, and similar sensory profile compared to the control bread	Bavaro et al. (2021)
Tortilla	Flour	Tortillas were made of quinoa flour, oat flour, wheat flour and popped amaranth flour at different ratios. The doughs were prepar by hot-press and baked in an oven of four temperature stages at 200°C, 260°C, 265°C, and 230°C. The tortillas were baked at each temperature for 11 s for each side (22 s in total at each temperature)	Chemical and sensory properties of tortillas	Adding mixtures of oat flour (5%), amaranth flour (20%), and quinoa flour (5%) at 30% to wheat flour increased the contents of protein and soluble dietary fiber and in vitro protein digestibility of the tortillas, decreased the starch content, lightness of color, and the sensory score of odor, and had no effect on the sensory score of texture and rollability	Vázquez-Villegas et al. (2021)

(*Continued*)

Table 12.1 (Continued)

Product type	Quinoa form	Formulations	Targeted properties	Major findings	References
Tortilla	Flour	Quinoa flour was blended into wheat flour obtained using roller mill up to 20% to make tortillas. Stone milled wheat flour (100%) was made into tortillas for comparison	Physicochemical properties of flour mixtures and starch digestibility and sensory properties of tortillas	The tortillas of roller-milled flour and quinoa flour composite had lower starch digestibility, more slowly digestible starch and less rapidly digestible starch as well as comparable sensory quality compared with those of stone milled flour. Quinoa flour fortified tortillas had more protein and fiber than that of wheat flour	Deshpande et al. (2022)
Steamed bread	Quinoa flour	Quinoa flour replaced wheat flour by up to 80%	Texture, color, mold-free shelf life	Quinoa flour addition increased the hardness, chewiness and density, microbiological shelf life, yellowness, redness, and darkness of the color of steamed bread, while decreasing the specific volume	Wang et al. (2015)
Pasta	Raw and fermented flour	Quinoa flour was fermented using *Lactobacillus plantarum* T6B10 and *Lactobacillus rossiae* T0A16. Quinoa flour replaced semolina by up to 20%	Chemical, nutritional and textural properties of the pasta	Adding quinoa flour increased the concentrations of free amino acids, proteins, and fibers of the pasta. Adding fermented flour increased the contents of total phenolics and free amino acids as well as in vitro antioxidant activities by up two times. Adding fermented flour decreased water absorption of pasta during cooking, while increasing the	Lorusso et al. (2017)

				tenacity and elasticity of the cooked pasta. Adding fermented quinoa decreased the expected glycemic index and increased protein nutritional indices and in vitro protein digestibility of the cooked pasta	
Pasta	Raw and germinated (20°C, 2 days, 80%–90% humidity) quinoa flour	Raw or germinated flour replaced wheat semolina up to 30% to make pasta	Chemical, physicochemical, and sensory properties	The germination process significantly increased the contents/values of total phenolics, ash, protein and antioxidant activities, while decreasing that of phytic acid (by 77%). The addition of germinated quinoa flour in the pasta formulations more increased the contents of these nutrients than the native quinoa flour in a dose-dependent manner. At the addition level of 30%, the germinated flour decreased the cooking quality more than the raw quinoa flour. The addition of quinoa flour tended to negatively affect the sensory quality of the pasta	Demir and Bilgiçli (2020)
Beef burger formulation	Flour	Quinoa flour or buckwheat flour replaced beef by up to 50% and bread crumb	Chemical and sensory properties	Burgers containing quinoa flour had more phosphorus, zinc, iron and magnesium, decreased oil adsorption and increased sensory scores	Bahmanyar et al. (2021)

(*Continued*)

Table 12.1 (Continued)

Product type	Quinoa form	Formulations	Targeted properties	Major findings	References
Beer	Un-malted seed and flake	Quinoa seeds/flakes replaced malted barley seeds by up to 40%	Chemical, physicochemical and sensory properties of beer	The types of quinoa being seeds or flakes had little effect on the properties of beer and wort. The beers produced with quinoa samples had less soluble nitrogen-containing compounds than that of the control beers (with malted barley). The foam in the beers produced with quinoa samples was more stable than that in the control beers. The application of un-malted quinoa up to 30% in beer production positively influenced the sensory scores of the beer	Kordialik-Bogacka et al. (2018)

(color parameters) of color, and in vitro antioxidant activity of tarhana. Adding quinoa flour gave higher viscosity of the tarhana than adding the other flours. This may be attributed to the differences in structure and properties of different starches. Adding quinoa flour at 20% gave the highest sensory scores of the resulting tarhana (Çevik & Ertaş, 2019). Kishk is a type of traditional dairy product containing cereals. Wheat burghul was replaced with quinoa seeds by up to 100% for kishk formulations (Ismail & Rayan, 2022). The mixtures were blended with laban rayeb (a fermented milk product) at a ratio of 1 to 2, before fermentation for 1 day at room temperature. More laban rayeb was added to make the ratio of 1 to 4. The resulting mixtures were shaped and dried at 50°C for 2 days before storage for 3 months. Physicochemical, color, sensory, and microbiological properties of the kishk were analyzed. Increasing quinoa seed addition increased the contents of protein, lipids, fiber, ash, and total phenolics and in vitro antioxidant activity of the kishik, while decreasing L and b^* of the color. During the storage, the counts of total bacteria and lactic acid bacteria decreased for all the samples. The counts of coliform bacteria, yeast, and mold became under the detection limits. The addition of quinoa seeds at 50% gave the highest sensory scores of acceptance (Ismail & Rayan, 2022). There are many different types of traditional foods from different regions of the world. It may be feasible to explore the possibility of using quinoa in their formulations for novel food product development.

12.2.3 Biscuit and cookie

Quinoa flour was added to wheat flour of up to 80% to make biscuit and cookies (Pourshahidi et al., 2020; Demir & Kılınç, 2017; Wang et al., 2015; Jan et al., 2018). Wang et al. (2015) replaced quinoa flour with wheat flour by up to 80% to make cookies of which the texture, color, and mold-free shelf life were studied. The quinoa flour addition increased the hardness, chewiness and density, microbiological shelf life, yellowness, redness, and darkness of the cookies, while decreasing the specific volume (Wang et al., 2015). Demir and Kılınç (2017) analyzed the chemical composition, shape, and sensory properties of cookies fortified with quinoa flour by up to 50%. Increasing quinoa flour addition decreased the diameter and spread ratio, the hardness, and L^* of the cookies, while increasing the a^* and b^*. The addition of the quinoa flour increased the contents of protein, lipids, phenolics, phytic acid, K, Mg, Fe, Ca, and Zn. The quinoa flour addition of up to 20% had little effect on the sensory quality of the cookies (Demir & Kılınç, 2017). Jan et al. (2018) studied the chemical, physicochemical, textural, and sensory properties of cookies with quinoa flour replacing wheat flour by up to 50%. Adding the quinoa flour increased the contents of protein and fiber, water and oil absorption capacities, foaming capacity and bulk density of the flour mixtures, while decreasing the spread ratio of the cookies. Increasing the addition of quinoa flour decreased the overall acceptance of the cookies (Jan et al., 2018). Therefore the patterns of changes in the properties of cookies affected by quinoa replacement of wheat flour were seen to be similar among different studies (Demir & Kılınç, 2017; Wang et al., 2015; Jan et al., 2018). The changes in the chemical properties of the cookies due to the quinoa flour addition could be attributed to the different chemical composition of quinoa from that of wheat flour (described in Chapter 2). Importantly, the quinoa flour addition diluted the wheat gluten network, decreasing the water-holding capacity of the dough. This led to decreased hardness of texture, diameter, and spread ratio of the cookies (Wang et al., 2015; Demir & Kılınç, 2017).

Pourshahidi et al. (2020) studied the effect of quinoa-based biscuit consumption on the biomarkers of cardiovascular diseases in humans using a randomized-controlled, double-blind crossover trial. The quinoa-based biscuit consumption favorably affected the cholesterol levels, body mass index, and body weight in humans, though these changes were relatively small. The risk of developing cardiovascular diseases was decreased by the intake of these biscuits in comparison with that by the consumption of wheat-based biscuits (Pourshahidi et al., 2020). Quinoa flour is rich in a range of bioactive components such as fibers, polyphenols, and saponins as described in Chapters 2 and 8. These components have diverse biological functions including preventing/decreasing the occurrences of cardiovascular diseases. Their presence in the biscuits containing quinoa flours contributed to the decreased levels of biomarkers of cardiovascular diseases in humans.

12.2.4 Bread

Quinoa in different forms including flour, flakes, sprouted flour, and liquid sourdough were formulated at different ratios into breadmaking of different types of breads (Bilgiçli & İbanoğlu, 2015; Wang et al., 2015; Gewehr et al., 2017; Ballester-Sánchez et al., 2019a, 2019b, 2019c; Suárez-Estrella et al., 2020; El-Sohaimy et al., 2021; Bavaro et al., 2021; Deshpande et al., 2022). For example, Wang et al. (2015) replaced wheat flour with quinoa flour by up to 80% to make breads of which the texture, color, and mold-free shelf life were studied. The quinoa flour addition increased the hardness, chewiness and density, microbiological shelf life, yellowness, redness, and darkness of the bread, while decreasing the specific volume (Wang et al., 2015). The effect of the addition of three varieties of quinoa flours at 25% on diverse quality attributes of breads was studied by Ballester-Sánchez et al. (2019a, 2019b and 2019c). The addition of quinoa flours increased the contents of linoleic and linolenic acids, dietary fiber and the percentage of soluble-to-insoluble fiber, reduced the glycemic index and glycemic load, and had little effect on the protein content. The addition increased the contents of Fe and Zn, though the phytic acid content was also increased (Ballester-Sánchez et al., 2019b). In another study, Ballester-Sánchez et al. (2019c) showed that adding quinoa flour increased the staling of the bread crumb. The addition of quinoa flour at 25% increased the concentrations of nutrients including minerals, dietary fiber, proteins, and lipids, had minor negative effect on textural and crumb structure, and gave similar sensory scores compared to the wheat control bread (Ballester-Sánchez et al., 2019c). In a third study, Ballester-Sánchez et al. (2019a) showed that breads with the addition of black and red quinoa flours had up to five times the in vitro antioxidant activity and two times the contents of both bound and free phenolics. The presence of black and red quinoa flours led to the formation of darker

	WF	10SQ	20SQ	30SQ	20PQ
Specific volume (mL/g)	3.37±0.11^{b}	3.18±0.04^{a}	3.61±0.11^{c}	3.38±0.10^{b}	2.60±0.10***
Crumb firmness (kg*m/s^2)	9.4±1.0^{c}	8.8±0.9^{c}	5.6±0.7^{a}	6.8±0.9^{b}	13.1±1.8***
Crumb luminosity (L*)	66.9±1.8^{c}	56.7±2.0^{b}	51.4±2.8^{a}	49.7±1.9^{a}	57.4±1.9***
Crumb redness (a*)	2.1±0.2^{a}	2.9±0.2^{b}	3.6±0.3^{c}	4.9±0.3^{c}	2.7±0.2***
Crumb yellowness (b*)	10.8±0.7^{a}	15.1±0.8^{b}	15.6±0.8^{b}	18.5±0.6^{c}	14.0±0.5***

Figure 12.1 Specific volume, crumb firmness (2 h after baking), and crumb color of bread from wheat flour (WF) and with increasing replacement level of sprouted quinoa (SQ) or pearled quinoa (PQ). 10SQ, blend of sprouted quinoa and wheat flour at 10:90 ratio; 20SQ, blend of sprouted quinoa and wheat flour at 20:80 ratio; 30SQ, blend of sprouted quinoa and wheat flour at 30:70 ratio; 20PQ, blend of pearled quinoa and wheat flour at 20:80 ratio. Different letters in the same row indicate a statistically significant difference among samples ($P \leq .05$). The asterisks indicate a statistically significant difference between the mean values of 20SQ and 20PQ (***$P \leq .001$) (Suárez-Estrella et al., 2020).
Source: Reprinted with permission from Elsevier.

bread crust and crumb. A darker "unconventional" bread was more acceptable by consumers.

Suárez-Estrella et al. (2020) used flour of sprouted quinoa seeds or normal quinoa flour of up to 30% to partially replace wheat flour for bread formulations (Fig. 12.1). The addition of the sprouted quinoa flour up to 20% had minor effect on the gluten aggregation and pasting behaviors of the flour systems, while increasing the degree of softening and water adsorption of the dough mixing by weakening the gluten network. The sprouted flour contained simple sugars and facilitated better gas production during the leavening than the native quinoa flour. The bread with the sprouted flour had less bitterness. Indeed, it has been shown that sprouting decreased the saponin content in quinoa seeds, and saponins are largely responsible for the bitter taste of quinoa products (Suárez-Estrella et al., 2021). The addition of the sprouted quinoa flour at 20% gave the lowest crumb hardness and the highest specific volume of the bread (Suárez-Estrella et al., 2020).

Tortillas were made using mixtures of quinoa flour, oat flour, wheat flour, and popped amaranth flour at different ratios (Vázquez-Villegas et al., 2021). The doughs were hot-pressed and baked in an oven at four temperature stages at 200°C, 260°C, 265°C, and 230°C. The tortillas were baked at each temperature for 11 s for each side (22 s in total at each temperature). The chemical and sensory properties of the tortillas were studied. Adding mixtures of oat flour (5%), amaranth flour

(20%), and quinoa flour (5%) at 30% to wheat flour for the formulations increased the contents of protein and soluble dietary fiber and in vitro protein digestibility of the tortillas, decreased the starch content, lightness of color, and the sensory score of odor, and had no effect on the sensory score of texture and rollability (Vázquez-Villegas et al., 2021). In another study, quinoa flour was blended into wheat flour obtained using roller mill of up to 20% to make tortillas (Deshpande et al., 2022). Stone-milled wheat flour (100%) was made into tortillas for comparison. The tortillas of roller-milled flour and quinoa flour composite had lower starch digestibility, more slowly digestible starch, and less rapidly digestible starch as well as comparable sensory quality compared with those of stone-milled flour. Quinoa flour fortified tortillas had more protein and fiber than that of wheat flour (Deshpande et al., 2022). Overall, quinoa flour has the potential for improving the nutritional quality of tortillas without compromising the sensory properties.

Bavaro et al. (2021) studied the feasibility of using quinoa flour liquid sourdough in the formulations of focaccia bread. Quinoa flour was fermented (15 h) using *Weissella cibaria* to make liquid sourdough. This strain had exopolysaccharide-producing capacity. The lipids of focaccia bread formulations were replaced using the quinoa liquid sourdough by 20%. The chemical, textural, sensory, and glycemic properties of the focaccia bread affected by the quinoa liquid sourdough replacement were studied. The focaccia breads fortified with the quinoa flour liquid sourdough had much reduced lipid content (by ~20%) and increased protein (glutenin/glutelins and albumin) content, lower firmness and glycemic index, and similar sensory profile compared to the control bread. The exopolysaccharide produced from *W. cibaria* acted as stabilizer in the dough and interacted with starch and amylases during the hydrolysis, decreasing the glycemic response (Bavaro et al., 2021).

Chinese steamed bread is a popular staple food for Chinese people (Wang et al., 2015). The basic ingredients are water, wheat flour, and yeast. The ingredients are mixed to make dough, which is fermented before steaming. Quinoa flour replaced wheat flour by up to 80% to make steamed breads (Wang et al., 2015). The quinoa flour addition increased the hardness, chewiness, and density, microbiological shelf life, yellowness, redness, and darkness of the steamed bread, while decreasing the specific volume. The changes in the texture could be attributed to the gluten dilution effect by the quinoa flour. The increased microbiological shelf life and changed color could be readily attributed to the special chemical composition of quinoa as described in Chapter 8 of this book.

The results summarized on the effect of quinoa flour addition on different properties of bread products could be explained by the chemical composition and properties of quinoa. The differences in the chemical composition between quinoa and wheat (described in Chapter 2) are largely responsible for the changes observed in the chemical properties of the resulting bread products. Quinoa lacks the gluten type proteins. The quinoa flour addition diluted the gluten matrix of the dough and weakened the gas-holding capacity during the fermentation. This led to decreased volume and altered texture of the resulting breads as reported by different studies (Table 12.1).

12.2.5 Pasta and noodle

Raw, fermented, or sprouted quinoa flours were formulated into pasta (Lorusso et al., 2017; Demir & Bilgiçli, 2020). Chemical, physicochemical, nutritional, sensory, and textural properties of the pasta affected by the addition of the quinoa flours were studied. Lorusso et al. (2017) fermented quinoa flour using *Lactobacillus plantarum* T6B10 and *Lactobacillus rossiae* T0A16. The fermented flour or raw quinoa flour replaced semolina of up to 20% to make pasta. Adding the quinoa flour increased the concentrations of free amino acids, proteins, and fibers of the pasta. Adding the fermented flour increased the contents of total phenolics and free amino acids as well as in vitro antioxidant activities by up to two times. Adding the fermented flour decreased the water absorption of pasta during cooking, while increasing the tenacity and elasticity of the cooked pasta. Adding the fermented quinoa flour also decreased the expected glycemic index and increased protein nutritional indices and in vitro protein digestibility of the cooked pasta (Lorusso et al., 2017). It has been revealed that the fermentation process can affect the chemical composition of quinoa seeds such as increasing the contents of phenolics and amino acids and also improving the biological functions such as increasing antioxidant activity and protein digestibility (Rizzello et al., 2017; Ayyasha et al., 2019). Demir and Bilgiçli (2020) used raw or germinated quinoa flour to replace wheat semolina at 30% to formulate paste. The germination process significantly increased the contents/values of total phenolics, ash, protein, and antioxidant activities, while decreasing that of phytic acid (by 77%). The addition of the germinated quinoa flour in the pasta formulations increased the contents of these nutrients more than the native quinoa flour in a dose-dependent manner. At the addition level of 30%, the germinated flour decreased the cooking quality more than the raw quinoa flour. The addition of quinoa flour tended to negatively affect the sensory quality of the pasta (Demir & Bilgiçli, 2020). Quinoa flour replaced wheat flour by up to 100% to make extruded instant noodles with the addition of 25% of potato starch (Tiga et al., 2021). Increasing quinoa flour content increased the redness (a^*) of color, water absorption, and hardness of the noodles, decreased lightness (L^*) and cohesiveness, and had no effect on the cooking loss. The increased hardness was related to the increased expansion ratio of the noodles (Tiga et al., 2021). The decreased textural and sensory quality of pasta and noodles from the quinoa flour addition may be due to the gluten dilution effect as well as the presence of antinutrients such as saponins. Strategies to improve these quality attributes of pasta with added quinoa flour remain to be developed in detail.

12.2.6 Meat-based product

Quinoa flour or buckwheat flour replaced beef by up to 50% and was formulated into bread crumb (Bahmanyar et al., 2021). Burgers containing quinoa flour had more phosphorus, zinc, iron, and magnesium, decreased oil adsorption, and increased sensory scores. The flavor and aroma profiles of the resulting beef burger should be analyzed to complement the data published by Bahmanyar et al. (2021).

12.2.7 Beer

Unmalted quinoa seeds/flakes replaced malted barley seeds by up to 40% to make beer (Kordialik-Bogacka et al., 2018). The type of quinoa (i.e., seeds or flakes) had little effect on the properties of beer and wort. The beers produced with quinoa samples had less soluble nitrogen-containing compounds than the control beers (made with malted barley). The foam in the beers produced with the quinoa samples was more stable than that in the control beers probably due to the presence of emulsifying compounds such as saponins in quinoa. The application of unmalted quinoa seeds of up to 30% in beer production positively influenced the sensory scores of the beers (Kordialik-Bogacka et al., 2018).

12.3 Quinoa-based gluten-free products

A range of diverse gluten-free food products containing quinoa were developed (Table 12.2). They included extract powder, film for packaging, sprout, infant foods, snacks, biscuit and cookies, breads, pasta, fermented products and yogurt, spreads, beverages, milk substitute, and meat- and fish-based products. Gluten-free bakery products are mostly made from starch-based materials lacking in a range of nutrients. Quinoa addition in formulations may be used to increase the nutritional value of these foods for celiac and gluten-sensitive consumers, although it may impose some hurdles to sensory attributes (Wang & Zhu, 2016).

12.3.1 Intermediate products including composite flour, gel, and extract

Many studies characterized intermediate products of quinoa such as composite flours, gels, and extract powders (Inglett et al., 2015; Romano et al., 2020; Felix et al., 2021). Inglett et al. (2015) mixed quinoa flour and oat products at different ratios and studied the pasting and rheological properties of the mixtures. Increasing the oat product level increased the viscosities during pasting event, water-holding capacity, and elastic modulus (G') and viscous modulus (G'') during dynamic rheological analysis (Inglett et al., 2015). The particle size can influence the physicochemical, chemical, and nutritional properties of quinoa flour (Ahmed et al., 2019). For example, Ahmed et al. (2019) showed that increasing particle size of quinoa flour increased the starch content and decreased that of protein, lipids, dietary fibers, and also the extractability of phenolics. Increasing particle size decreased the sediment volume fraction and water-holding capacity of quinoa flour and increased the final and setback viscosity during pasting (Ahmed et al., 2019). Felix et al. (2021) studied the rheological properties of gels made from quinoa flour suspensions at different concentrations of 15%–30% and different temperatures of 60°C–90°C. The quinoa flour suspensions of high concentrations showed weak gel-like behavior. Gelatinization of starch at ~65°C led to the formation of strong gels. Increasing quinoa concentrations and process temperatures increased the

Table 12.2 Formulations and properties of quinoa-based gluten-free food and beverage products.

Product	Quinoa form	Formulations and processing	Targeted properties	Major findings	References
Extract powder	Quinoa flour	Extract powder obtained using spray-drying	Chemical composition and structure	Increasing drying temperature increased starch swelling and aggregation. Inlet temperature (180 °C) and feed flow speed (7.5 mL/min) were optimized to obtain high yields (85% dry basis) and solid content (0.96), and low water activity (0.047) of the resulting powder	Romano et al. (2020)
Composite flour	Flour	Quinoa flour and oat products were mixed at different ratios	Pasting and rheological properties	Increasing the oat product level increased the viscosities during pasting event, water-holding capacity and elastic modulus (G') and viscous modulus (G'') during dynamic rheological analysis	Inglett et al. (2015)
Gel	Flour	Quinoa flour suspensions at different concentrations of 15%–30% and different temperatures of 60°C–90°C	Rheological properties	The quinoa flour suspensions of high concentrations showed weak gel-like behavior. Gelatinization of starch at ~65°C led to the formation of strong gel. Increasing quinoa concentrations and process temperatures increased the viscoelastic moduli of quinoa systems	Felix et al. (2021)
Film	Flour	Quinoa flour films were fortified with thyme and oregano essential oils up to 2%. Quinoa flour suspension (6%) was heated up to 82°C before the addition of essential oils and plasticizer (1%). The mixtures were homogenized before casting into films	Physical, mechanical, barrier and antimicrobial properties	The addition of the oils imparted the films with antimicrobial capacities against *Staphylococcus aureus* and *Escherichia coli*. Oregano oils were more antimicrobial than thyme oils. *S. aureus* was more sensitive than *E. coli*. The oil type had no effect on physical, mechanical and barrier properties of the films. Increasing essential oil concentration decreased the tensile strength, water vapor permeability solubility and Young's modulus, while increasing the elongation at break	Pagno et al. (2016)

(*Continued*)

Table 12.2 (Continued)

Product	Quinoa form	Formulations and processing	Targeted properties	Major findings	References
Sprout	Sprout	Seeds of 2 varieties were soaked for 3 h before germination for 4 days at 20°C, photoperiod of 24 h, and 70% humidity. The sprouts were packed and stored using modified atmosphere packaging	Chemical, sensory and textural properties of quinoa sprouts	There were differences in germination capacity, sensory property, phenolic content and shelf life between the 2 sprouts. Sprouts stored in passive modified atmosphere packaging had worse texture and more off-flavors than those with active modified atmosphere packaging (20% CO_2 and 5% O_2)	D'ambrosio et al. (2017)
Infant food	Flour	Quinoa flour was precooked with water before drum-drying	Insulin-like growth factor-1 (IGF-1) level in the plasma of undernourished children	Consumption of 200 g of quinoa-based infant foods as supplementary food intake increased the IGF-1 level in the plasma of undernourished children after 15 days	Ruales et al. (2002)
Infant food	Flour	Quinoa flour was mixed with rice flour at different ratios (up to 15% (quinoa) to 48% (rice)). Other ingredients included skimmed milk powder (18%), demineralized whey powder (7%), powder sugar (5%), banana powder (4%), maltodextrin (2%), and sunflower lecithin (1%)	Chemical, rheological, thermal and sensory properties of the infant foods	Increasing quinoa flour addition increased the contents of proteins and fibers of the infant food, while decreasing that of sugar. The quinoa flour addition decreased the loss (G'') and storage (G') modulus of the infant product. A total of 23 aroma active compounds were obtained with the groups of aldehydes, alcohols and ketones being dominant in the infant food. Addition of quinoa flour at 8% gave the infant product the highest sensory acceptance	Ayseli et al. (2020)

Infant weaning food	Extruded or roasted flour	Extruded or roasted quinoa plus millet flour mixtures at a quinoa-to-millet ratio of 7:3 were prepared. A mixture of fructo-oligosaccharide and galacto-oligosaccharide (1 to 1 ratio) was added. The infant weaning food was consisted of the flour mixture (70%), oligosaccharide mixture (3%), skim milk powder (15%), egg yolk powder (8%), minerals (2%) and plant food powder (2%)	Nutritional, prebiotic and physicochemical properties	The protein contents of the quinoa and millet-based weaning foods were higher than that of a commercial millet infant food by ~17%. The weaning foods with extruded flours had higher solubility, water absorption and protein and starch digestibility lower viscosity than those with roasted flours. The digestas of the weaning foods after the simulated digestion promoted the growth of *Lactobacillus delbrueckii* and *Lactobacillus plantarum*	Dong et al. (2021)
Extrudates	Seed	Maize flour was mixed with 20% of quinoa seeds and the mixtures were extruded using a double screw extruder. The barrel heating temperatures were 50°C, 80°C, 100°C, 150°C	Physical properties of extrudes	Adding quinoa seeds increased the expansion rate, smoothness of the surface and decreased the density of the extrudates	Cueto et al. (2015)
Snack product	Flour	Quinoa flour was fortified with the powders of ginger (6%), cayenne pepper (0.7%), or turmeric (6%) and mixed with salt (2%), water (water to solid ratio at 1.2:1) and corn oil (6%) to make doughs. The doughs were heated for 2 min to make snacks (Fig. 12.2)	Sensory quality	Sensory analysis showed that the snacks of quinoa flour had the highest acceptance rate (79%), followed by the samples fortified with cayenne pepper (77%), ginger (73%), and turmeric (70%). Such a range of acceptance rates was high for potential commercial applications	Kahlon et al. (2017)
Snack bar	Flour	Quinoa, flaxseed, dry fruits, brown rice, and honey were used as ingredients. The seeds were dry heated (up to 100°C and 12 min) before formulating with honey (up to 60%). The formulations were optimized for sensory quality	Physicochemical, chemical and sensory properties	Under optimized sensory quality, the snack bar had the contents of moisture (8.5%), ash (1.3%), protein (10.5%), lipids (2.9%), total phenolic (0.51 mg gallic acid equivalent/g), and β-carotene (0.38 μg/g)	Kaur et al. (2018)

(*Continued*)

Table 12.2 (Continued)

Product	Quinoa form	Formulations and processing	Targeted properties	Major findings	References
Snack bar	Puffed quinoa seed	Honey, pollen and propolis were formulated into quinoa with sucrose and agar at different proportions using rotational central composite design	Sensory property and phenolic content	Addition of propolis and pollen decreased the sensory acceptance, while increasing the phenolic content of the snack bars to levels higher than those of common commercial products. Addition of honey increased the acceptance of the bars. Weak astringency and sweetness were the positive sensory attributes, whereas strong yellow color, astringency, bitterness, and pungency were the negative ones	Rios et al. (2018)
Snack bar	Organic flake and puffed seed	Different proportions of grains, seeds, dried fruits, nuts, sunflower oil, salt, agar, panela, honey were formulated into cereal bars	Sensory property, shelf life and peroxide value	Puffed quinoa (9.1%), quinoa flake (12.1%), puffed amaranth (6.1%), chia seeds (5.2%), dried cranberry (12.1%), macadamia nut (15.2%), sunflower oil (12.1%), honey (18.2%), panela (9.1%) and agar (0.8%) were combined to give the highest sensory scores of the resulting cereal bars. The shelf life of the cereal bars 18°C, 25°C, 37°C were 35, 29 and 25 days, respectively	Iuliano et al. (2019)
Snack product	Fermented quinoa product	Fermented quinoa products formulated with/without blackcurrant	Postprandial glycemic property in humans ($n = 26$) affected by sugar intake	Products containing the quinoa product reduced postprandial insulinemia and glycemia in humans	Lappi et al. (2021)
Biscuit	Flour	Quinoa flour replaced wheat flour by up to 60% to make biscuit. The sodium bicarbonate content (0.05%–0.15%) and baking temperatures (160°C–210°C) were variables of the response surface methodology	Acrylamide formation, texture and color	Under optimized formulations (100% quinoa flour and 0.05% sodium-bicarbonate) and processing conditions (baked at 160°C), the resulting biscuit had reduced acrylamide formation and hardness and similar browning compared to the wheat control biscuit	Sazesh and Goli (2020)

Biscuit	Flour	Quinoa flour replaced rice flour up to 100% to make ladyfinger biscuits which were consisted of fresh eggs (55%), sucrose (22%) and rice/ quinoa flour (23%)	Physicochemical, nutritional and sensory properties	Adding quinoa flour decreased the lightness, while increasing the contents of protein, lipids, ash, polyphenols, and flavonoids, in vitro antioxidant activities, amounts of volatile compounds, softness of texture, redness of color and degree of appealing in crust color. Quinoa flour addition of up to 50% gave the biscuits acceptable sensory quality. The addition of >50% of quinoa flour decreased the sensory quality by increasing the bitterness	Cannas et al. (2020)
Biscuit	Flour	Microbial dextran of *Leuconostoc mesenteroides* was added to quinoa flour to make gluten-free biscuits	Production of acrylamide in biscuits	The concentration of acrylamide in biscuits without dextran addition was 2.1 mg/kg biscuit (dry basis) which was significantly higher than the level recommended to be the maximum by the European Commission. Addition of microbial dextran at a suitable level [dextran solution (5%) to water ratio] at 0.880 mL/17.6 mL decreased the production of acrylamide in the biscuits by 89%, probably due to the water retaining capacity of dextran and cross-linking with acrylamide. The sensory properties of the biscuits were not affected by the dextran	Mousa et al. (2022)
Cake	Flour of native and roasted quinoa seed	Seeds were roasted at 177°C up to 45 min before milling into flour. The allergen-free and gluten-free cakes were made of quinoa flour (22%), navy bean (6%), milk (35%), cocoa powder (4%), grape seed oil (6%), maple syrup (10%), agave nectar (15%), baking powder (0.6%) and soda (0.7%), sea salt (0.5%), vanilla extract (0.2%), and apple cider vinegar (0.3%)	Physicochemical and sensory properties	Increasing roasting time increased the final and peak viscosities of the flour during pasting. Cakes made from flour of native seeds had higher sensory scores in color, appearance and texture than the other cakes. A commercial chocolate cake had higher scores in overall acceptability and flavor and lower firmness, moisture content and water activity than quinoa cakes	Rothschild et al. (2015)

(Continued)

Table 12.2 (Continued)

Product	Quinoa form	Formulations and processing	Targeted properties	Major findings	References
Cookie	Flour and flake	Quinoa flour and flakes were mixed with corn starch at different ratios to optimize the quality using experimental mixture design. Other ingredients including water (13%), baking powder (0.55%), egg powder (2.7%), corn oil (13%), salt (0.9%), and sugar (14%) were added to formulate cookies	Sensory and nutritional properties	Increasing levels of quinoa flour and flakes led to the formation of darker cookies. Increasing quinoa flour addition decreased the specific volume of the cookies, and increasing levels of both flour and flakes increased the hardness. A composition of quinoa flour (30%), quinoa flakes (25%) and corn starch (45%) gave the highest sensory scores of the cookies	Brito et al. (2015)
Cake	Flour	Rice flour and potato starch (ratio at 1 to 1) were partially replaced with quinoa flour up to 75%	Rheological, textural and sensory properties	Power law model well described the flow behaviors of the cake batters. Increasing the addition level of quinoa flour increased the storage modulus (G') and loss modulus (G'') of the batter during the frequency sweep from 0.01 to 10 Hz. Temperature sweep test showed that quinoa addition increased mechanical strength of the batter. Increasing the addition level of quinoa flour decreased the cake hardness but increased the volume. An addition level of 50% quinoa flour gave the highest sensory score for the cake	Bozdogan et al. (2019)
Cake	Flour	Cake formulations were modified to have 100% quinoa flour instead of wheat flour in the presence of gum Arabic (up to 1%) and reduced (up to 75%) lipids	Lipid oxidation and acrylamide formation	The addition of gum Arabic (1%) and reduction of lipids (by 50%) reduced the extents of lipid oxidation and acrylamide formation by 64 and 83%, respectively. The gum addition reduced the firmness and increased the lightness of the gluten-free quinoa cakes. The addition of gum Arabic (1%) and reduction of lipids (by 75%) produced similar odor intensity to wheat-based cakes	Mousa et al. (202[illegible])

Bread	Flour	Gluten-free and sucrose free breads were made using quinoa flour or amaranth flour in the presence of sucralose, sucralose–acesulfame-K, and stevia	Sensory properties	Amaranth taste and aroma were related to higher sensory acceptance. In contrast, quinoa taste and aroma as well as brown crust were related to lower sensory acceptance. Breads with amaranth flour and sucralose, sucralose–acesulfame-K and stevia had the highest sensory acceptance	Alencar et al. (2017)
Bread	Flour	Quinoa flour was mixed with buckwheat flour, rice flour, and potato starch at different ratios in the presence of sugar (3%), oil (6%), yeast (3%), xanthan gum (0.5%) and salt (2%)	Batter rheological properties and gluten-free bread properties	The batters of different formulations showed pseudoplastic behavior with more solid-like than liquid-like properties. The percentages of quinoa addition (12.5%–50%) had minor effect on the protein content, specific volume and baking loss of the breads. The quinoa flour addition at 25% gave the softest texture scores and the highest sensory of the breads	Turkut et al. (2016)
Bread	Flour	Rice flour was replaced partially by quinoa flour by up to 25% to make gluten-free breads. Transglutaminase (up to 4.75%), lipase (0.01%) and protease (0.01%) were added to the formulations to improve the quality	Physicochemical, structural and sensory properties	Lipase and protease addition increased the bread volume. The quinoa flour addition at 25% improved the sensory quality and delayed the staling	Azizi et al. (2020)

(*Continued*)

Table 12.2 (Continued)

Product	Quinoa form	Formulations and processing	Targeted properties	Major findings	References
Bread	Flour	Pseudocereal flours including quinoa flour were mixed with rice flour and/or potato starch at different ratios to make gluten-free breads using simplex-centroid design to optimize the sensory quality of the breads	Physicochemical and sensory quality	The sensory quality of gluten-free breads made of 100% pseudocereal flours including quinoa flour was unacceptable. The sensory and physical properties of gluten-free breads made of 50% rice flour and 50% pseudocereal flour were similar to those of white breads. The gluten-free breads of 82% quinoa flour (maximum addition level) and 18% rice flour had acceptable sensory quality, whereas the maximum addition levels of amaranth and buckwheat flours to obtain acceptable gluten-free breads with rice flour were 60 and 85%, respectively	Aguiar et al. (2022)
Bakery product	Flour	Quinoa flour was treated with proteolytic enzymes and transglutaminase	Structure and sensory properties of the bakery product	The transglutaminase treatment affected the microstructure by improving the protein matrix, while increasing the sensory acceptance. Proteases increased the dough rheology but decreased the sensory quality by producing bitter peptides	Romano et al. (2018)
Sourdough	Flour	Quinoa flour was fermented using *Lactobacillus reuteri* R29 and *Lactobacillus brevis* R2Δ with antifungal property, and *L. brevis* L1105 without antifungal property	Chemical properties and microbiological shelf life	Quinoa flour-based sourdough had higher total titratable acidity and organic acid production than rice flour-based sourdough. The former had more types of carboxylic acids at higher concentrations than the latter. *L. reuteri* R29 fermentation produced more carboxylic acids (2-hydroxyisocaproic acid and 3-phenyllactic acid), acetic and lactic acids than the fermentation based on the other strains. Quinoa bread with the addition of *L. reuteri* R29 fermented sourdough had longer shelf life by 2 days than the bread without any acidification. Breads with the addition of *L. reuteri* R29 fermented sourdough had similar shelf life to those chemically acidified breads	Axel et al. (2016)

Pasta	Flour	Corn flour was mixed with 20% of quinoa flour for spaghetti production	Nutritional properties of the pasta	Quinoa flour addition decreased the true protein digestibility and increased the net protein utilization. The quinoa flour addition increased the measuring dialyzable mineral fraction of Fe. The addition of EDTA decreased the sensory quality of the pasta but increased dialyzable mineral fraction of Fe and Zn	Giménez et al. (2016)
Pasta	Flour	Corn flour was replaced with quinoa flour up to 100% to make pasta in the presence of egg and egg white with/without zein	Physicochemical, rheological and textural properties of dough and pasta	Adding quinoa flour decreased the elastic component of the dough, while increasing the protein content and water diffusivity of dough during drying and improving the texture of the cooked product. Addition of zein decreased the structural integrity of protein network in the pasta, while increasing the broken ratio and decreasing the cooking time and cohesiveness of cooked pasta	Sosa et al. (2019)
Pasta	Flour	Egg-free and gluten-free pasta were formulated using native and extruded quinoa flour in combination with other ingredients including tara gum, lupine flour, oxidizing enzyme, potato starch, and pea, potato, and rice proteins (Fig. 12.3)	Physicochemical, nutritional and textural properties	The cooking quality and firmness of the pasta were negatively affected by the addition of tara gum, potato starch and extruded quinoa flour. Increasing the addition level of lupine flour to 12% improved the cooking quality and firmness of the pasta even in the absence of the tara gum. Pea protein in combination with oxidizing enzyme was better in increasing the firmness of the pasta. The optimized formulations of the pasta with the addition of 30% of lupine flour gave high protein (28%) and fiber (15%) contents	Linares-García et al. (2019)
Pasta	Fermented dough	Quinoa flour was fermented with *L. plantarum* CRL 1964 and *L. plantarum* CRL 2107 at 30°C for 24 h before mixing with pregelatinized cassava starch or quinoa flour (up to 20%)	Nutritional properties of preventing deficiency in minerals and vitamins using in vivo mouse model	Feeding mice with the pasta made from quinoa sourdough enhanced the levels of vitamin B2 and B9, Ca^{+2}, Mg^{+2}, Fe^{+2} and P, hematocrit and hemoglobin in the blood as well as the villi length of the small intestine compared to feeding the mice with the control pasta groups	Carrizo et al. (2020)

(*Continued*)

Table 12.2 (Continued)

Product	Quinoa form	Formulations and processing	Targeted properties	Major findings	References
Tempe	Seeds of 3 varieties (white, black and red) fermented under solid state using *Rhizopus oligosporus*	Cooked quinoa seeds were inoculated with *R. oligosporus* under 10^4 spores per g of material to make tempe. The fermentation was done for 2 h at 35°C and then for 28 h at 31°C before steaming and freeze-drying	Chemical and sensory properties	The tempe samples had more free amino acids, protein and soluble fiber than the native seeds. The tempe samples had higher in vitro antioxidant activities than the native seeds. The chemical properties of the tempe samples depended on the quinoa varieties. Tempe of white quinoa seeds had higher antioxidant activities than that of the others, whereas that of black seeds had more fibers than the others	Starzyńska-Janiszewska et al. (2017)
Fermented beverage	Flours of two quinoa varieties	Quinoa flour was gelatinized by cooking and inoculated (1%) with *Lactobacillus plantarum* Q823, *L. lactis* ARH74 and *L. casei* Q11. The samples were fermented for 6 h at 30°C before storage at ~6°C for 4 weeks	Nutritional and microbiological properties	The fermentation was homolactic, while the microbiota of the fermented beverage was stable and viable. One quinoa variety showed advantage of having less saponins, more proteins and lower viscosity decrease from the fermentation process in producing the fermented beverage than the other	Urquizo et al. (2017)
Beer	Unmalted seeds (both raw and gelatinized)	Gluten-free beers of "American pale ale" style were made using the unmalted seeds as brewing adjuncts	The beer making properties of quinoa grains were compared with those of buckwheat, amaranth, millet and sorghum	Partial substitution of malted barley with quinoa seeds had little effect on the alcohol (%v/v) and extract (°P) content of the beers, whereas those with other grains decreased these values. Beers produced using raw seeds of quinoa and also sorghum had acceptable sensory and technological properties. A gluten content of >20 ppm was obtained for all the beers, indicating more work is needed for gluten-free beer making	Cela et al. (2022)

Probiotic drink	Flour	Red quinoa flour suspensions (15%) were pasteurized (80°C, 20 min) before α-amylase hydrolysis. The resulting samples were filtered and inoculated with Bifidobacterium longum or *Lactiplantibacillus plantarum* for fermentation (37°C, 24 h)	Microbiological and chemical properties	The fermentation increased the microbial concentration of the drinks up to 6–8 log CFU/mL. Upon in vitro gastrointestinal digestion, the probiotic viability was significantly retained. The fermentation with *B. longum* after 24 h gave the highest probiotic concentration. *B. longum* was not susceptible to some antibiotics. Fermentation with *Lactiplantibacillus plantarum* gave the samples higher phenolic content and in vitro antioxidant activity after in vitro gastrointestinal digestion than that with *B. longum* and the control	Cerdá-Bernad et al. (2022)
Yogurt	Flour	Quinoa flour up to 2% was added to milk. The mixtures were heated to gelatinize starch. The heated samples were cooled and inoculated with yogurt culture. The fermentation led to a pH of 4.6 of the yogurt	Physicochemical and structural properties	The yogurt samples showed shear thinning behaviors. Increasing quinoa flour addition level decreased the resistance to syneresis. The addition level of 0.6% of quinoa flour gave the yogurt desired viscosity and gel porosity	Codină et al. (2016)
Yogurt	Flour	Quinoa flour suspension (solid-to-water ratio at 1 to 1.5) was autoclaved at 121°C for up to 15 min before skimmed milk (weight to volume at 3–100) and sucrose (6%) addition. The samples were heated and fortified using *Streptococcus thermophiles*, *Lactobacillus delbrueckii* subsp. *bulgaricus*, and *L. rhamnosus* GR-1 for fermentation (38°C, 6 h). The resulting samples were stored at 4°C for 4 weeks	Microbiological and sensory properties	The yogurt with quinoa flour had similar microbial counts to that containing oat, barley, adzuki bean, or mung bean. The count of *L. rhamnosus* GR-1 in the yogurt with quinoa flour (8.2×10^8 CFU/mL) was higher than that in the control sample and the samples containing oat, barley, adzuki bean, or mung bean (1.4–5.7 × 108 CFU/mL). However, the sensory scores of the yogurt with quinoa flour were lower than those of control and the samples containing oat, adzuki bean and barley	Soltani et al. (2018)

(*Continued*)

Table 12.2 (Continued)

Product	Quinoa form	Formulations and processing	Targeted properties	Major findings	References
Yogurt	Flour	Quinoa flour suspension (flour to water ratio at 1:7) was autoclaved (121°C) for 15 min. The cooked samples were subjected to α-amylase (0.5%) treatment (at 50°C) for 20 min, then at 65°C for 90 min. The resulting samples were subjected to high-pressure homogenization (180 MPa). Proteases were added to process the milk. The milk was pasteurized (10 min at 110°C) before inoculating with *Weissella cibaria* MG1 (1 × 107 cfu/mL). The fermentation was done at 30°C for 24 h before storage at 4°C	Microbiological, structural and chemical properties	Treatment of the milk using protease mixtures increased the protein solubility by 55%. Compared to the samples from quinoa milk acidified chemically, the fermented sample (pH 5.2) had higher viscosity and water-holding capacity with high amounts of exopolysaccharide (40 mg/L) and viable cell counts (> 109 cfu/mL). The production of dextran from the fermentation process led to aggregation of macromolecules in the samples. The network of protein and dextran enhanced the yogurt texture	Zanninia et al. (2018)
Yogurt	Extract	Quinoa extracts were added (up to 15%) to goat milk-base for fermentation (45°C) to reach a pH of 4.5	Physicochemical, structural and sensory properties of the yogurt	Adding quinoa extracts decreased the fermentation time, while increasing the apparent viscosity, the lactic acid bacteria viability, and the compactness of the yogurt microstructure. A yogurt fortified with the quinoa extract had high sensory acceptance	El-Shafei et al. (2020)
Yogurt	Germinated seed	Quinoa seed suspensions (sprouted and native) were milled to obtain milk substitute. The quinoa milk or quinoa milk-cow milk mixtures were fermented using *Streptococcus thermophilus* and *Lactobacillus bulgaricus*	In vitro antioxidant activity and physicochemical properties	Yogurt made of the germinated quinoa seeds had higher contents of total phenolics and flavonoids as well as higher antioxidant activities (DPPH, ABTS, and ORAC) than those made of other ingredients	Ujiroghene et al. (2019)

Vegan cheese analog	Quinoa rejuvelac (starter inoculum) (retained liquid of sprouting quinoa seed)	Cashew was blended with the quinoa rejuvelac before fermentation for up to 4 days. Different coloring and flavoring agents were added to obtain different products (Fig. 12.4)	Microbiological and nutritional properties	The cashew-based cheese products had 10^8-10^9 colony forming units/g. The fermentation process little affected the concentrations of vitamins, minerals lipids, and carbohydrates. *Weissella* and *Pediococcus* genera dominated the quinoa rejuvelac. "Blue" and "brie" type products had *Weissella*, *Pediococcus*, and *Lactococcus* as the dominant genera. The fermentation process largely decreased the allergenicity of cashew	Chen et al. (2020)
Ultra-filtration-soft (tallaga) cheese	Flour	Quinoa flour up to 3% was formulated into tallaga cheese. The cheese was stored at 4°C for 30 days	Physicochemical, chemical and sensory properties	Adding quinoa flour increased the contents of fiber and total phenolics, firmness and in vitro antioxidant activity of the soft cheese, while decreasing the syneresis. The soft cheese with up to 2% of quinoa flour addition had similar sensory scores to the control cheese, while the cheese with 3% of the flour addition had decreased sensory acceptance	Abdelmontaleb et al. (2021)
Spoonable vegan product	Flours of 2 varieties	Quinoa flour was fermented using *Lactobacillus plantarum* Q823 before flavoring with date or banana and bilberry	Microbiological, nutritional and sensory properties	Lactic acid bacteria counts in the fermented quinoa products were higher than that of commercial products during storage of 4 weeks. The nutritional quality of the quinoa products was higher than or similar to that of commercial products. Sensory analysis showed that the quinoa products were perceived as being high in fiber, healthy, and novel, with sandy mouthfeel and unpleasant aftertaste	Väkeväinen et al. (2020)
Cream	Organic quinoa flour	Quinoa flour up to 6% was added to organic pumpkin creams. The products were formulated using industrial procedures. The cream products were stored up to 60 days	Physicochemical properties	The addition of quinoa flour increased the pH, color lightness, decreased the yellowness and redness, and had no effect on water activity of the pumpkin cream	Pérez-Álvarez et al. (2021)

(*Continued*)

Table 12.2 (Continued)

Product	Quinoa form	Formulations and processing	Targeted properties	Major findings	References
Porridge	Quinoa seeds of 30 varieties	Saponins were washed off the seeds. The seeds (1.5 g) in water (7.5 mL) were cooked at 100°C for 30 min	Composition and odor activity of volatile compounds in the cooked porridge	A total of 53 volatile compounds were detected in the porridge. Among them, there were 11 compounds varied greatly among different quinoa varieties. The alkanes (23%), acids and esters (44%) were the most abundant. The 30 quinoa varieties may be categorized in eight groups based on the volatile concentrations, or into four groups according to the relative odor activity values of 12 volatile compounds. Hexanal, (E,E)-2,4-decadienal, 1-octen-3-ol, 2-pentylfuran, 6,10-dimethyl-5,9-undecadien-2-one and nonanal were the major aroma contributors	Zhang et al. (2019)
Milk substitute	Quinoa seed	Quinoa seeds were soaked to remove saponins before cooking and homogenization. Amyloglucoside and α-amylase were added to hydrolyze starch. The samples were filtered and mixed with sunflower oil before mixing (Fig. 12.6). The cooking water was also replaced with acidic saline as an alternative process	Chemical, nutritional and sensory properties of the milk	Using acidic saline instead of water for cooking increased the protein content of the resulting milk by threefold. The quinoa milk had lower lipid and sodium contents than cow and rice milk and had an expected glycemic index of 52. The overall sensory profile of quinoa milk was similar to that of rice milk, though additional flavor compounds may be added to improve the sensory quality attributes	Pineli et al. (2015)
Milk substitute	Flour	Flour suspension was treated with amyloglucosidase, α-amylase and protease. Glucose-isomerase was also added to some samples to obtain milk substitute after homogenization and pasteurization. The resulting milk substitute was subjected to fermentation using *Lactobacillus brevis* TR055 and *Leuconostoc citreum* TR116 isolated from sourdough	Microbiological and chemical properties	The two strains were highly viable and significantly reduced the pH of the samples to ∼3.6. Fructose was hydrolyzed into mannitol by the fermentation with the two strains (Fig. 12.7). Glycemic load was significantly reduced by >33%	Jeske et al. (2018b)

Milk substitute	Flour	Flour suspension (50 g in 350 g water) was treated with α-amylase and different proteases at 50°C for 3 h before cooling to 20°C. The samples were subjected to high-pressure homogenization before adjustment of pH to 7	Protein solubility, emulsifying and foaming capacity	Profix 100 L protease treatment degraded the quinoa protein and increased the protein solubility from 48% to 76%. The protease treatments little affected the emulsifying and foaming capacity of the beverage. The selection of suitable proteases is critical to ensure the quality of the milk substitute	Jeske et al. (2018a)
Milk substitute	Flour	Quinoa flour was extruded using a twin-screw extruder (screw and feeding speed were 300 rpm and 30 kg/h). Barrel temperatures were set at 80 to 170°C–180°C. The extruded samples were dried and subjected to α-amylase and amyloglucosidase hydrolysis. The resulting samples were filtered and mixed with lipids and stabilizers (oat β-glucan up to 30% of the quinoa flour) before pH adjustment, high-pressure homogenization (50 MPa, 2 cycles) and sterilization (20 min, 121°C)	Rheological, emulsion, structural and sensory properties	The addition of β-glucan prevented the phase separation of the quinoa milk and increased the stability during storage. Increasing the concentration of β-glucan decreased the range of particle distribution of the milk, had no effect on zeta potential and increased the apparent viscosity. The β-glucan formed network and prevented the phase separation. The addition of β-glucan at 16% gave the highest sensory acceptance of the milk as too high concentration of the β-glucan led to decreased sensory quality	Huang et al. (2021)
Fish pattie	Flour	Fish (*Pseudoplatystoma corruscans*) was skinned, deboned and minced before mixing with quinoa four at different ratios in the presence of butylhydroxyanisole (0.01%) and sodium chloride (1.2%). The fish patties were cooked at 180°C to allow the center to reach a temperature of 78°C	Chemical and sensory properties	The fish patties containing gluten-free flours (including flours of corn, quinoa, rice, and amaranth) had higher cooking yields than control without flour addition. The fish patties containing quinoa or amaranth flours had higher hardness and lower springiness and cohesiveness than those containing rice and corn flour. The addition of flours increased the carbohydrate and mineral contents. However, the overall sensory scores of the fish patties containing different flours were lower than those of control	Romero et al. (2018)

(*Continued*)

Table 12.2 (Continued)

Product	Quinoa form	Formulations and processing	Targeted properties	Major findings	References
Mackerel ingredient	Ethanol extract	Atlantic chub mackerel (*Scomber colias*) was stored in ice-containing quinoa extract for 13 days	Chemical and sensory properties	The presence of quinoa extracts decreased lipid oxidation and hydrolysis in the mackerel. The inhibitory effects became more pronounced toward the end of the storage. The presence of quinoa extracts at a higher addition level (0.05 g/L icing solution) gave the mackerel acceptable sensory quality after the storage, whereas the control mackerel samples became unacceptable for sensory quality	Miranda et al. (2018)
Frozen fatty fish ingredient	Saponin-free ethanol extract	Frozen fatty fish (Atlantic mackerel) was stored at −40°C for 48 h before dipping in glazing solution (0°C, 30 s) containing the quinoa extract (0.45−1.36 g/L)	Chemical and sensory properties and lipid oxidation	The glazing significantly decreased the lipid hydrolysis and free fatty acid formation in the frozen fatty fish due to the presence of antioxidants in the quinoa extracts, while increasing the sensory quality	Trigo et al. (2018)
Meat nugget ingredient	Flour	Quinoa flour was formulated into goat nugget up to 3%	Chemical, rheological, textural and sensory properties	Addition of quinoa flour increased the fiber content and the sensory quality of the nugget at an addition level of 1.5%	Verma et al. (2019)
Fat replacer	Quinoa pastes of three varieties	Quinoa paste of up to 10% was formulated into pork liver pâté	Chemical and textural properties	Adding quinoa paste increased the concentrations/contents of residual nitrite, moisture and ash, while significantly decreasing that of lipids. The quinoa addition increased the hardness of the liver pâté, while having no effect on cohesiveness and springiness. Red quinoa paste addition at 5% gave the highest sensory acceptance. Quinoa addition at 10% gave a lower oxidation of lipids on the samples. Enterobacteriaceae, aerobic bacteria, yeasts, and molds were not found in the samples containing quinoa	Pellegrini et al. (2018)

Fat replacer	Quinoa flour	Quinoa flour and/or teff flour up to 5% were formulated into emulsion type sausages by replacing beef lipids	Physicochemical and textural properties	Addition of quinoa flour increased the emulsion stability, carbohydrate, dietary fiber, protein, and moisture contents and cook yield, decreased the contents of lipids and calories, the hardness, and had little effect on the color	Kerimoğlu et al. (2020)
Ingredients for Bologna-type sausages	Black quinoa seed and fiber-rich fraction	Seeds or fiber-rich fraction up to 2.5% were formulated into sausages which were stored at 4°C for 3 weeks	Physicochemical, chemical and microbiological properties	Adding quinoa products increased the residual nitrite levels, decreased the lipid oxidation, and had little effect on the microbiological safety of the sausages during storage	Fernández-López et al. (2020b)
Binder for Bologna-type sausages	Wet-milling product rich in fibers and whole quinoa seed	Fiber-rich fraction or quinoa seeds up to 3% were added to sausage formulations	Physicochemical and chemical properties	Adding quinoa fiber fraction or seeds significantly increased the dietary fiber and residual nitrite contents of the sausages, improved emulsion stability, and decreased lipid oxidation and water activity	Fernández-Lópe et al. (2020c)
Chicken meatball ingredient	Quinoa starch or seed	Quinoa starch (up to 2.5%), seeds (up to 10%) and potato starch (up to 2.5%) at different ratios were formulated into chicken breast meat (up to 78%) to make chicken meatballs	Chemical, textural and sensory properties of chicken meatballs affected by frozen storage and freeze-thaw cycles	Adding any starch decreased the losses due to cooking, drip, and re-heating, while reducing the chewiness and hardness of the chicken meatball by 50%. The presence of quinoa seeds in the meatballs increased the in vitro antioxidant activity and chewiness and reduced the lipid oxidation of the meatballs during the freeze-thaw cycles. The addition of quinoa starch plus quinoa seeds increased the sensory acceptance of the chicken meatballs compared to that of other combinations	Park et al. (2021)

(*Continued*)

Table 12.2 (Continued)

Product	Quinoa form	Formulations and processing	Targeted properties	Major findings	References
Ingredient for different meat-based products	Flour or seed	Burgers, meatballs, Bolognas, dry-cured sausage	Diverse quality attributes	The review article summarized the applications of different quinoa-based ingredients in the applications and formulations of meat-based products for enhanced nutritional, shelf life, and sensory quality	Fernández-López et al. (2020a)
Fat replacer, binder, functional ingredient	Whole seed, cooked seed, flour, paste, fiber faction	Bolognas, dry-cured sausage, goat nuggets, beef burger and meatball, mortadella, and cooked sausage, pâté were formulated with the addition of quinoa-based ingredients	Diverse quality attributes	The review article summarized the applications of different quinoa-based ingredients in the applications and formulations of meat-based products. Quinoa-based ingredients acted as binder, extender, gelling agent, fat replacer, or filler	Fernández-López et al. (2021)

viscoelastic moduli of quinoa systems (Felix et al., 2021). The studies on the quinoa-based composite flours and gels have not been linked to the properties of any food systems. This kind of analysis is usually not very useful and should be supplemented with the studies of model food systems.

Quinoa flour (10%) was subjected to aqueous extraction at pH of 9 (50°C for 2 h) (Romano et al., 2020). An extract powder was obtained using spray-drying. Increasing drying temperature increased starch swelling and aggregation. The inlet temperature (180°C) and feed flow speed (7.5 mL/min) were optimized to obtain high yields (85% dry basis) and solid content (0.96) and low water activity (0.047) of the resulting powder. The particle size ranged from 9.6 to 20.0 μm (Romano et al., 2020). The extract obtained may be formulated into different food products as functional ingredients; however, this aspect remains to be explored.

12.3.2 Film and coating

Quinoa flour suspension (6%) was heated up to 82°C before the addition of thyme and oregano essential oils (up to 2%) and plasticizer (1%). The mixtures were homogenized before casting into films. The physical, mechanical, barrier, and antimicrobial properties of the films were studied. The addition of the oils imparted the films with antimicrobial capacities against *Staphylococcus aureus* and *Escherichia coli*. The oregano oils were more antimicrobial than the thyme oils. *S. aureus* was more sensitive to the antimicrobial action of the films than *E. coli*. The oil type had no effect on the physical, mechanical, and barrier properties of the films. Increasing essential oil concentration decreased the tensile strength, water vapor permeability solubility, and Young's modulus, while increasing the elongation at break (Pagno et al., 2016). The films remain to be applied in practical food packaging.

12.3.3 Sprout

Seeds of two varieties were soaked for 3 h before germination for 4 days at 20°C with a photoperiod of 24 h and 70% humidity (D'ambrosio et al., 2017). The sprouts were packed and stored using modified atmosphere packaging. The chemical, sensory, and textural properties of the quinoa sprouts were measured. There were differences in germination capacity, sensory property, phenolic content, and shelf life between the sprouts of the two quinoa varieties. The sprouts stored in passive modified atmosphere packaging had worse texture and more off-flavors than those with active modified atmosphere packaging (20% CO_2 and 5% O_2) (D'ambrosio et al., 2017). The sprouting and germination of quinoa seeds by food processing are described in detail in Chapter 11 of the book.

12.3.4 Infant food

Infant foods in different forms were produced using different methods (Ayseli et al., 2020; Dong et al., 2021; Ruales et al., 2002). Quinoa flour was mixed with rice flour at different ratios (up to 15% (quinoa) to 48% (rice)) to make infant foods

(Ayseli et al., 2020). Other ingredients included skimmed milk powder (18%), demineralized whey powder (7%), powder sugar (5%), banana powder (4%), maltodextrin (2%), and sunflower lecithin (1%). Chemical, rheological, thermal, and sensory properties of the infant foods affected by quinoa flour addition were studied. Increasing quinoa flour addition increased the contents of proteins and fibers of the infant food, while decreasing that of sugar. The quinoa flour addition decreased the loss (G'') and storage (G') modulus of the infant product. A total of 23 aroma active compounds were obtained with the groups of aldehydes, alcohols, and ketones being dominant in the infant food. The addition of quinoa flour at 8% gave the highest sensory acceptance of the infant product (Ayseli et al., 2020). Dong et al. (2021) used extruded or roasted quinoa flour to formulate infant weaning food. Extruded or roasted quinoa plus millet flour mixtures at a quinoa-to-millet ratio of 7:3 were prepared. A mixture of fructo-oligosaccharide and galacto-oligosaccharide (1 to 1 ratio) was added. The infant weaning food consisted of the flour mixture (70%), the oligosaccharide mixture (3%), skim milk powder (15%), egg yolk powder (8%), minerals (2%), and plant food powder (2%). Nutritional, prebiotic, and physicochemical properties of the weaning foods were studied. The protein contents of the quinoa- and millet-based weaning foods were higher than that of a commercial millet infant food by ~17%. The weaning foods with the extruded flours showed higher solubility, water absorption, protein, and starch digestibility and lower viscosity than those with the roasted flours. The digestas of the weaning foods after simulated digestion promoted the growth of *Lactobacillus delbrueckii* and *L. plantarum* (Dong et al., 2021). Ruales et al. (2002) precooked quinoa flour with water before drum-drying. Insulin-like growth factor-1 (IGF-1) level in the plasma of undernourished children taking the quinoa infant product was studied. Consumption of 200 g of the quinoa-based infant foods as supplementary food intake increased the IGF-1 level in the plasma of undernourished children after 15 days (Ruales et al., 2002). Such effects could be due to the special chemical composition and nutritional properties of quinoa as described in Chapters 2–9 of this book.

12.3.5 Snack food

Quinoa in different forms were formulated into different snack foods (Cueto et al., 2015; Kahlon et al., 2017; Kaur et al., 2018; Rios et al., 2018; Iuliano et al., 2019; Lappi et al., 2021). For example, Cueto et al. (2015) mixed maize flour with 20% of quinoa seeds. The mixtures were extruded using a double screw extruder. The barrel heating temperatures were set at 50°C, 80°C, 100°C, and 150°C. Adding quinoa seeds increased the expansion rate and smoothness of the surface and decreased the density of the extrudates (Cueto et al., 2015). Kahlon et al. (2017) fortified quinoa flour with the powders of ginger (6%), cayenne pepper (0.7%), or turmeric (6%) in the presence of salt (2%), water (water to solid ratio at 1.2:1), and corn oil (6%) to make doughs. The doughs were heated for 2 min to make snacks (Fig. 12.2). The snacks of pure quinoa flour had the highest sensory acceptance rate (79%), followed by the samples fortified with cayenne pepper (77%), ginger (73%), and turmeric (70%). Such a range of acceptance rates was high with potential

Figure 12.2 Four kinds of cooked gluten-free snacks containing quinoa and other ingredients (Kahlon et al., 2017).
Source: Reprinted with permission under Creative Commons Attribution License.

commercial applications (Kahlon et al., 2017). Lappi et al. (2021) formulated snacks using fermented quinoa product with/without blackcurrant. Postprandial glycemic property in humans (n = 26) affected by sugar and snack intakes was studied. The products containing the quinoa reduced postprandial insulinemia and glycemia in humans (Lappi et al., 2021). The reduced postprandial response may be due to multiple factors including the fibers in quinoa and also the altered formulations of the snacks.

12.3.6 Biscuit and cookie

Biscuits and cookies are considered simple to produce without the presence of gluten-containing flours, though the lack of gluten within formulations commonly leads to less desirable sensory quality (Martínez-Villaluenga et al., 2020).

Quinoa has proven promising in formulations to obtain improved sensory, nutritional, and technological properties of gluten-free biscuits and cookies. Native, roasted quinoa flours or quinoa flakes were formulated into biscuits and cookies (Sazesh & Goli, 2020; Cannas et al., 2020; Brito et al., 2015; Mousa et al., 2022). Quinoa flour replaced rice flour up to 100% to make ladyfinger biscuits that consisted of fresh eggs (55%), sucrose (22%), and rice/quinoa flour (23%) (Cannas et al., 2020). Physicochemical, nutritional, and sensory properties of the biscuits affected by quinoa flour addition were studied. Adding quinoa flour decreased the lightness, while increasing contents of protein, lipids, ash, polyphenols, and flavonoids, in vitro antioxidant activities, amounts of volatile compounds, softness of texture, redness of color, and degree of appeal in crust color. The quinoa flour addition of up to 50% gave the biscuits acceptable sensory quality. The addition of >50% of quinoa flour decreased the sensory quality by increasing the bitterness due to the presence of saponins in quinoa (Cannas et al., 2020). In another study, quinoa flour replaced wheat flour up to 60% to make biscuits (Sazesh & Goli, 2020). The sodium bicarbonate content (0.05%–0.15%) and baking temperatures (160°C–210°C) were variables of the response surface methodology used to optimize the process and formulations in terms of acrylamide formation, texture, and color of the biscuits. Under the optimized formulations (100% quinoa flour and 0.05% sodium-bicarbonate) and processing conditions (baked at 160°C), the resulting biscuit had reduced acrylamide formation and hardness and similar browning compared to the wheat control biscuit (Sazesh & Goli, 2020). Brito et al. (2015) mixed quinoa flour and flakes with corn starch at different ratios to optimize the cookie quality using experimental mixture design. Other ingredients including water (13%), baking powder (0.55%), egg powder (2.7%), corn oil (13%), salt (0.9%), and sugar (14%) were added to formulate cookies. Increasing levels of the quinoa flour and flakes led to the formation of darker cookies. Increasing quinoa flour addition decreased the specific volume of the cookies, and increased the levels of both flour and flakes increased the hardness. A composition of quinoa flour (30%), quinoa flakes (25%), and corn starch (45%) gave the highest sensory scores of the cookies (Brito et al., 2015). Microbial dextran of *Leuconostoc mesenteroides* was added to quinoa flour to make gluten-free biscuits with decreased acrylamide concentration in biscuits (Mousa et al., 2022). The concentration of acrylamide in the biscuits without dextran addition was 2.1 mg/kg biscuit (dry basis), which was significantly higher than the level recommended to be the maximum by the European Commission. The addition of microbial dextran at a suitable level [dextran solution (5%) to water ratio] at 0.880 mL/17.6 mL decreased the production of acrylamide in the biscuits by 89% probably due to the water-retaining capacity of dextran and crosslinking with acrylamide. The sensory properties of the biscuits were not affected by the dextran (Mousa et al., 2022).

12.3.7 Cake

Bozdogan et al. (2019) partially replaced rice flour and potato starch (ratio at 1 to 1) with quinoa flour by up to 75% to make cakes. The rheological properties of the cake batters and the textural and sensory properties of the cakes were studied.

Power law model well described the flow behaviors of the cake batters. Increasing the addition level of the quinoa flour increased the storage modulus (G') and loss modulus (G'') of the batter during the frequency sweep from 0.01 to 10 Hz. Temperature sweep test showed that the quinoa addition increased mechanical strength of the batter. Increasing the addition level of quinoa flour decreased the cake's hardness but increased the volume. An addition level of 50% quinoa flour gave the cake the highest sensory score (Bozdogan et al., 2019). Rothschild et al. (2015) developed allergen-free and gluten-free cakes using flours of native and roasted quinoa seeds. Quinoa seeds were roasted at 177°C up to 45 min before milling into flour. The cakes were made of the quinoa flour (22%), navy bean (6%), milk (35%), cocoa powder (4%), grape seed oil (6%), maple syrup (10%), agave nectar (15%), baking powder (0.6%) and soda (0.7%), sea salt (0.5%), vanilla extract (0.2%), and apple cider vinegar (0.3%). The physicochemical and sensory properties of the cookies were studied. Increasing roasting time increased the final and peak viscosities of the flour during pasting. The cookies made from the flour of native quinoa seeds had higher sensory scores in color, appearance, and texture than the other cakes. A commercial chocolate cake had higher scores in overall acceptability and flavor and lower firmness, moisture content, and water activity than the quinoa cakes (Rothschild et al., 2015). Thus the sensory quality of the quinoa-based cakes remains to be improved for commercial applications. Mousa et al. (2021) used 100% quinoa flour instead of wheat flour in the presence of gum Arabic (up to 1%) and with a reduced amount of (up to 75%) lipids to make cakes. The addition of gum Arabic (1%) and reduction of lipids (by 50%) reduced the extents of lipid oxidation and acrylamide formation in the cakes by 64 and 83%, respectively. The gum addition reduced the firmness and increased the lightness of the gluten-free quinoa cakes. The addition of gum Arabic (1%) and reduction of lipids (by 75%) produced similar odor intensity to that of wheat-based cakes (Mousa et al., 2021).

12.3.8 Bread

Gluten-free breads have been formulated using starch. This kind of breads often has undesirable quality attributes such as crumbly and chalky texture, lower loaf volume, and overall lower sensory acceptance than bread containing gluten (Turkut et al., 2016). Gluten-free breads were developed using quinoa as a major ingredient (Alencar et al., 2017; Turkut et al., 2016; Romano et al., 2018; Aguiar et al., 2022). Turkut et al. (2016) mixed quinoa flour with buckwheat flour, rice flour, and potato starch at different ratios in the presence of sugar (3%), oil (6%), yeast (3%), xanthan gum (0.5%), and salt (2%). The batters of different formulations showed pseudo-plastic behavior with more solid-like than liquid-like properties. The percentages of quinoa addition (12.5%–50%) had minor effect on the protein content, specific volume, and baking loss of the breads. The quinoa flour addition at 25% gave the breads the softest texture and the highest sensory scores (Turkut et al., 2016). Alencar et al. (2017) developed gluten-free and sucrose-free breads using quinoa flour or amaranth flour in the presence of sucralose, sucralose–acesulfame-K, and stevia. Amaranth taste and aroma were related to higher sensory acceptance of the breads. In contrast, quinoa taste and aroma as well as brown crust were related to lower sensory acceptance. The breads with amaranth flour and sucralose, sucralose–acesulfame-K, and stevia had the highest sensory acceptance (Alencar et al., 2017). In a comparison study of Aguiar et al. (2022), pseudocereal flours including quinoa flour were mixed with rice flour and/or potato starch at different ratios to make gluten-free breads using simplex-centroid design to optimize the sensory quality of the breads. The sensory quality of the gluten-free breads made of 100% pseudocereal flours including quinoa flour was unacceptable. The sensory and physical properties of the gluten-free breads made of 50% rice flour and 50% pseudocereal flour were similar to those of white breads. The gluten-free breads of 82% quinoa flour (maximum addition level) and 18% rice flour had acceptable sensory quality, whereas the maximum addition levels of amaranth and buckwheat flours to obtain acceptable gluten-free breads with rice flour were 60% and 85%, respectively (Aguiar et al., 2022).

Overall, methods remain to be developed to improve the sensory quality of quinoa-based breads. Romano et al. (2018) treated quinoa flour with proteolytic enzymes and transglutaminase for breadmaking. The transglutaminase treatment affected the microstructure by improving the protein matrix, while increasing the sensory acceptance. Transglutaminase has crosslinking abilities and improved dough-handling property by increasing the coherence of the protein network. Proteases improved the dough rheology but decreased the sensory quality by producing bitter peptides (Romano et al., 2018). This study again showed that methods remain to be developed to improve the sensory quality of quinoa-based breads. Azizi et al. (2020) studied the effect of lipase and proteases on quinoa flour fortified breads. When the quinoa flour was added to the gluten-free formulation at 15%, the bread volume was increased. The quinoa substitution at 25% delayed bread staling. In sensory evaluation, consumers found the bread with a 25% substitution of quinoa flour formulation to be the most favorable. The lipase addition allowed for the degradation of ester bonds, while protease hydrolyzed the peptide bonds in proteins, helping to improve starch expansion ability and create a more cohesive texture during baking. Protease can help to produce a higher volume and softer crumb to the bread that the quinoa flour may not be able to achieve in the product otherwise, improving the overall sensory attributes of the gluten-free bread (Azizi et al., 2020). One strategy for improving the sensory quality of gluten-free breads is to use sourdough for the formulations. Quinoa flour was fermented using *Lactobacillus reuteri* R29 and *Lactobacillus brevis* R2Δ with antifungal property, and *L. brevis* L1105 without antifungal property (Axel et al., 2016). The quinoa flour-based sourdough had higher total titratable acidity and organic acid production than a rice flour-based sourdough. The former had more types of carboxylic acids at higher concentrations than the latter. Higher protein content and protease activity of the quinoa flour than those of the rice flour related to the higher production of organic acids in quinoa flour fermentation (Axel et al., 2016). The *L. reuteri* R29 fermentation produced more carboxylic acids (2-hydroxyisocaproic acid and 3-phenyllactic acid), acetic and lactic acids than the fermentation based on the other strains. The quinoa bread with the addition of the *L. reuteri* R29 fermented sourdough had longer shelf life by 2 days than the bread without any acidification. The breads with the addition of *L. reuteri* R29 fermented sourdough had similar shelf life to those chemically acidified breads (Axel et al., 2016). Gluten-free breads based on quinoa sourdough remain to be developed and diversified.

12.3.9 Pasta

Raw and fermented quinoa flours were formulated into pasta of different shapes such as penne, tagliatelle, and spaghetti (Giménez et al., 2016; Sosa et al., 2019; Carrizo et al., 2020; Linares-García et al., 2019; Schoenlechner et al., 2010). Methods to apply quinoa in the formulations with other ingredients have been found effective in making gluten-free and nutrient-denser pasta similar to the traditional pasta made from wheat flour (Wang & Zhu, 2016). For example, corn flour was replaced with quinoa flour up to 100% to make pasta in the presence of egg and egg white with/without zein (Sosa et al., 2019). Adding quinoa flour decreased the elastic component of the dough, while increasing its protein content and water diffusivity during drying, as well as improving the texture of the cooked product. The addition of zein decreased the structural integrity of protein network in the pasta, while increasing the broken ratio and decreasing the cooking time and cohesiveness of the cooked pasta (Sosa et al., 2019). Linares-García et al. (2019) developed egg-free and gluten-free pasta using native and extruded quinoa flour in combination with other ingredients including tara gum, lupine flour, oxidizing enzyme, potato starch, and pea, potato, and rice proteins (Fig. 12.3). The cooking quality and firmness of the pasta were negatively affected by the addition of tara gum, potato starch, and extruded quinoa flour. Increasing the addition level of lupine flour to 12% improved the cooking quality and firmness of the pasta even in the absence of the tara gum. Pea protein in combination with oxidizing enzyme was more efficient in increasing the firmness of the pasta. The optimized formulations of the pasta with

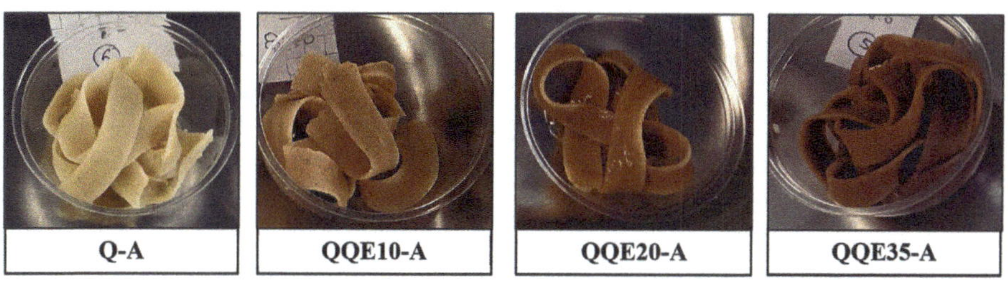

Figure 12.3 Development of gluten-free and egg-free pasta based on quinoa with the addition of lupine flour, vegetable proteins, and the oxidizing enzyme Pox; Q–A (quinoa flour), QQE10-A (quinoa flour, extruded quinoa flour 10%), QQE20-A (quinoa flour, extruded quinoa flour 20%), and QQE35-A (quinoa flour, extruded quinoa flour 35%) (Linares-García et al., 2019).
Source: Reprinted under the terms of the Creative Commons Attribution License.

the addition of 30% of lupine flour gave high protein (28%) and fiber (15%) contents (Linares-García et al., 2019). The sensory properties of the pasta should be studied and optimized in relation to the formulations. Pasta was made using fermented quinoa flour and the nutritional properties were studied in vivo (Carrizo et al., 2020). Quinoa flour was fermented with *L. plantarum* CRL 1964 and *L. plantarum* CRL 2107 at 30°C for 24 h before mixing with pregelatinized cassava starch or the quinoa flour (up to 20%) to make pasta. Feeding mice with the pasta made from the quinoa sourdough enhanced the levels of vitamin B2 and B9, Ca^{+2}, Mg^{+2}, Fe^{+2}, P, hematocrit, and hemoglobin in the blood as well as the villi length of the small intestine compared to feeding the mice with the control pasta groups (Carrizo et al., 2020).

12.3.10 Tempe

Tempe or tempeh is a traditional fermented food product originated in Indonesia. The cooked quinoa seeds of three varieties (white, black, and red) were inoculated with *Rhizopus oligosporus* under 104 spores per g of material to make tempe (Starzyńska-Janiszewska et al., 2017). The fermentation was done for 2 h at 35°C and then for 28 h at 31°C before steaming and freeze-drying. The tempe samples had more free amino acids, protein, and soluble fiber than the native seeds. The tempe samples had higher in vitro antioxidant activities than the native seeds. The chemical properties of the tempe samples depended on the quinoa varieties. The tempe of white quinoa seeds had higher antioxidant activities than that of the others, whereas that of black seeds had more fiber than the others (Starzyńska-Janiszewska et al., 2017). The sensory properties of the quinoa-based tempe remain to be studied.

12.3.11 Probiotic and fermented drinks and yogurt

Urquizo et al. (2017) developed fermented beverage using flours of two quinoa varieties. Quinoa flour was gelatinized by cooking and inoculated (1%) with *Lactobacillus plantarum* Q823, *L. lactis* ARH74, and *L. casei* Q11. The samples were fermented for 6 h at 30°C before storage at ~6°C for 4 weeks. The fermentation was homolactic, while the microbiota of the fermented beverage was stable and viable. One quinoa variety had the advantage of having less saponins, more proteins, and lower decrease in viscosity from the fermentation process in producing the fermented beverage than the other (Urquizo et al., 2017). Cerdá-Bernad et al. (2022) developed probiotic drink using flour of a red quinoa variety. The flour suspensions (15%) were pasteurized (80°C, 20 min) before α-amylase hydrolysis. The resulting samples were filtered and inoculated with *Bifidobacterium longum* or *Lactiplantibacillus plantarum* for fermentation (37°C, 24 h). The fermentation increased the microbial concentration of the drinks up to 6–8 log CFU (colony forming units)/mL. Upon in vitro gastrointestinal digestion, the probiotic viability was significantly retained. The fermentation with *B. longum* after 24 h gave the highest probiotic concentration. *B. longum* was not susceptible to some antibiotics.

The fermentation with *Lactiplantibacillus plantarum* gave the samples higher phenolic content and in vitro antioxidant activity after the in vitro gastrointestinal digestion than that with *B. longum* and the control (Cerdá-Bernad et al., 2022). The above studies have not analyzed the sensory properties of the fermented products.

Yogurt samples were developed using different forms of quinoa products including raw flour, extracts, and germinated seeds (Codină et al., 2016; Soltani et al., 2018; Zanninia et al., 2018; El-Shafei et al., 2020; Ujiroghene et al., 2019). For example, Soltani et al. (2018) developed quinoa-based yogurt by first autoclaving flour suspensions (solid-to-water ratio at 1−1.5) at 121°C for up to 15 min. Then, skimmed milk (weight to volume at 3−100) and sucrose (6%) were added to the autoclaved samples. The samples were heated and fortified using *Streptococcus thermophiles*, *Lactobacillus delbrueckii* subsp., *Bulgaricus*, and *L. rhamnosus* GR-1 for fermentation (38°C, 6 h). The resulting samples were stored at 4°C for 4 weeks. Comparison studies showed that the yogurt containing quinoa flour had similar microbial counts to those containing oat, barley, adzuki bean, or mung bean. The count of *L. rhamnosus* GR-1 in the yogurt containing quinoa flour (8.2×10^8 CFU/mL) was higher than that in the control sample and the samples containing oat, barley, adzuki bean, or mung bean ($1.4-5.7 \times 10^8$ CFU/mL). However, the sensory scores of the yogurt containing quinoa flour were lower than those of control and the samples containing oat, adzuki bean, and barley (Soltani et al., 2018). Thus methods to improve the sensory quality of yogurt containing quinoa flour should be developed. Zanninia et al. (2018) studied the feasibility of using *Weissella cibaria* MG1 for the fermentation of α-amylase hydrolyzed quinoa samples and also the use of proteases to treat the samples. This strain has the capacity of producing exopolysaccharide. The treatment of the milk using protease mixtures increased the protein solubility by 55%. Compared to the samples from quinoa milk acidified chemically, the fermented sample (pH 5.2) had higher viscosity and water-holding capacity with high amounts of exopolysaccharide (40 mg/L) and viable cell counts ($> 10^9$ CFU/mL). The production of dextran from the fermentation process led to aggregation of macromolecules in the samples. The network of protein and dextran enhanced the yogurt texture (Soltani et al., 2018). El-Shafei et al. (2020) added quinoa extracts (up to 15%) to goat milk for fermentation (45°C) to reach a pH of 4.5. Adding quinoa extracts decreased the fermentation time, while increasing the apparent viscosity, the lactic acid bacteria viability, and the compactness of the yogurt microstructure. The yogurt fortified with the quinoa extract had high sensory acceptance (El-Shafei et al., 2020).

12.3.12 Cheese and cream analog

Vegan cheese and cream products were developed using quinoa and related products as an ingredient (Chen et al., 2020; Abdelmontaleb et al., 2021; Väkeväinen et al., 2020; Pérez-Álvarez et al., 2021). Chen et al. (2020) obtained quinoa rejuvelac as the retained liquid of sprouting quinoa seeds. Cashew was blended with the quinoa rejuvelac as starter inoculum before fermentation for up to 4 days. Different coloring and flavoring agents were added to obtain different products (Fig. 12.4).

Figure 12.4 Five flavors of fermented cashew products using a quinoa-based rejuvelac starter culture. From left to right: "brie," red, herb, "cheddar," and "blue" cashew cheese-like products (Chen et al., 2020).
Source: Reprinted with permission under the terms of Creative Commons Attribution.

The cashew-based cheese products had 108–109 CFU/g. The fermentation process little affected the concentrations of vitamins, minerals lipids, and carbohydrates. *Weissella* and *Pediococcus* genera dominated the quinoa rejuvelac. "Blue" and "brie" type products had *Weissella*, *Pediococcus*, and *Lactococcus* as the dominant genera. The fermentation process largely decreased the allergenicity of cashew (Chen et al., 2020). Väkeväinen et al. (2020) made spoonable vegan product by fermenting quinoa flour using *Lactobacillus plantarum* Q823 before flavoring with date or banana and bilberry (Fig. 12.5). Lactic acid bacteria counts in the fermented quinoa products were higher than those of commercial products during storage of 4 weeks. The nutritional quality of the quinoa products was higher than or similar to that of commercial products. Sensory analysis showed that the quinoa products were perceived as being high in fiber, healthy, and novel, with sandy mouthfeel and unpleasant aftertaste (Väkeväinen et al., 2020). The texture and taste quality of the resulting spoonable vegan products containing quinoa should be improved. Organic quinoa flour up to 6% was added to organic pumpkin creams to make cream (Pérez-Álvarez et al., 2021). The products were formulated using industrial procedures. The cream products were stored up to 60 days. The addition of the quinoa flour increased the pH and color lightness, decreased the yellowness and redness, and had no effect on water activity of the pumpkin cream (Pérez-Álvarez et al., 2021). However, the sensory property of the resulting creams has not been studied.

12.3.13 Milk substitute and porridge

Zhang et al. (2019) studied the composition and odor activity of volatile compounds in the cooked porridge samples of quinoa seeds of 30 varieties. A total of 53

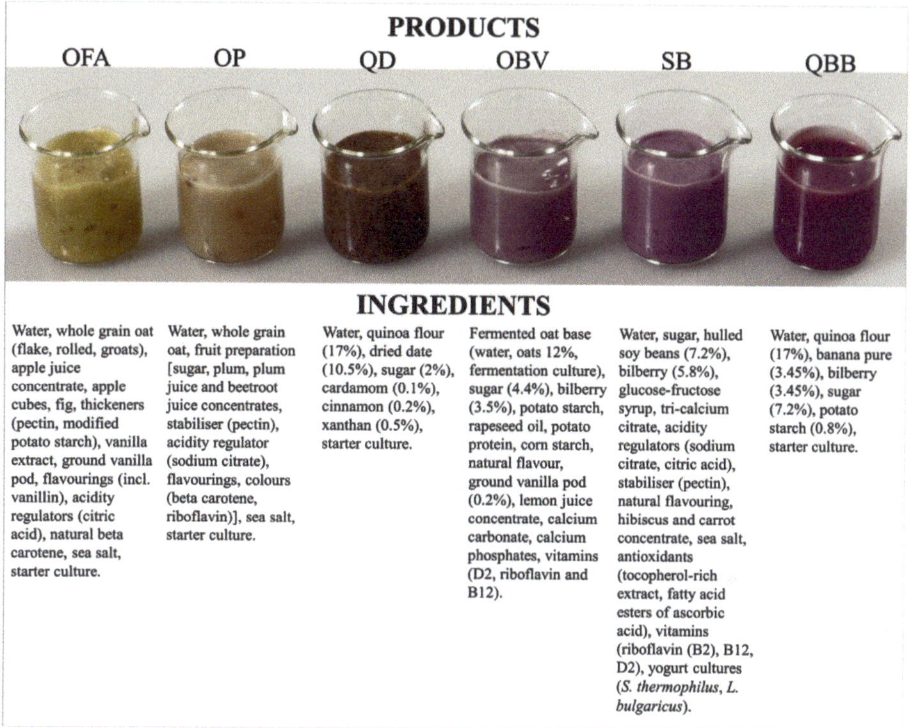

Figure 12.5 The experimental quinoa products and commercial vegan spoonable products and their ingredients. OFA: Yosa Break fig apple, OP: Yosa plum, QD: Quinoa date, OBV: Oatly bilberry vanilla, SB: Alpro bilberry, and QBB: Quinoa bilberry banana. Yosa, Oatly and Alpro are commercial brand names (Väkeväinen et al., 2020).
Source: Reprinted with permission from Elsevier.

volatile compounds were detected in the porridge samples. Among them, 11 compounds varied greatly among different quinoa varieties. The alkanes (23%) and acids and esters (44%) were the most abundant. The 30 quinoa varieties may be categorized in eight groups based on the volatile concentrations, or into four groups according to the relative odor activity values of 12 volatile compounds. Hexanal, (E,E)-2,4-decadienal, 1-octen-3-ol, 2-pentylfuran, and 6,10-dimethyl-5,9-undecadien-2-one and nonanal were the major aroma contributors (Zhang et al., 2019). The sensory properties of the quinoa porridge samples remain to be studied.

There has been increasing interest in the production and consumption of plant-based milk including quinoa-based products (Silva et al., 2020). Quinoa seeds were soaked to remove saponins before cooking and homogenization (Pineli et al., 2015). Amyloglucoside and α-amylase were added to hydrolyze starch. The samples were filtered and mixed with sunflower oil before mixing to make milk substitute (Fig. 12.6). The cooking water was also replaced with acidic saline as an alternative process. Using acidic saline instead of water for cooking increased the protein

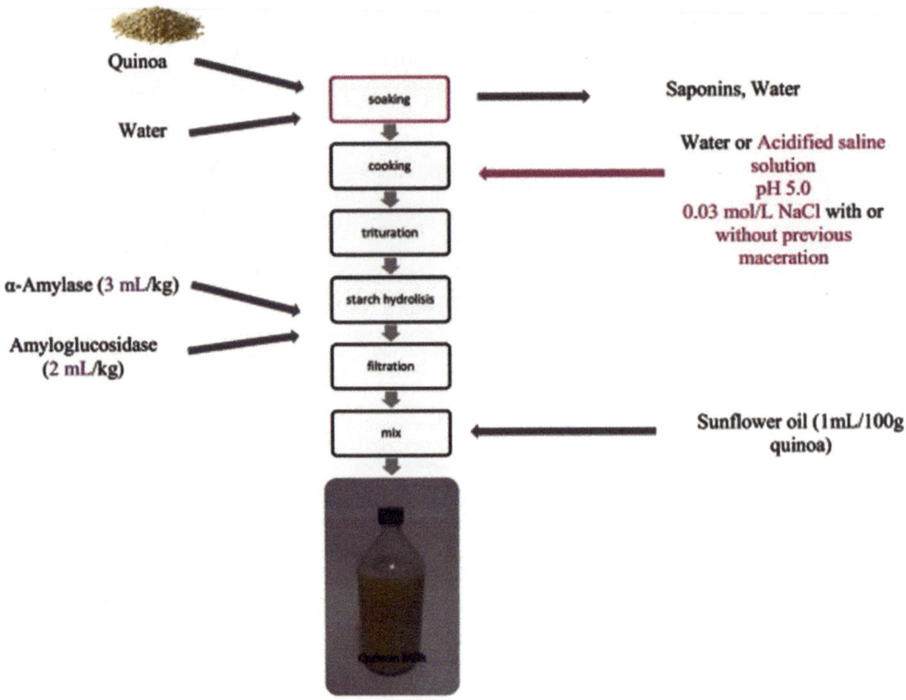

Figure 12.6 Quinoa milk process (operations in pink color are part of alternative flowchart, with increased protein content in quinoa milk) (Pineli et al., 2015).
Source: Reprinted with permission from Elsevier.

content of the resulting milk by threefold. The quinoa milk had lower lipid and sodium contents than cow and rice milk and had an expected glycemic index of 52. The overall sensory profile of the quinoa milk was similar to that of rice-based milk substitute; though additional flavor compounds may be added to improve the sensory quality attributes (Pineli et al., 2015). In the studies of Jeske et al. (2018a and 2018b), quinoa flour suspensions were treated with amyloglucosidase, α-amylase, and protease. Glucose-isomerase was also added to some samples to obtain milk substitute after homogenization and pasteurization. The resulting milk substitute was subjected to fermentation using *Lactobacillus brevis* TR055 and *Leuconostoc citreum* TR116 isolated from sourdough. The two strains were highly viable and significantly reduced the pH of the samples to ~3.6. Fructose was hydrolyzed into mannitol by the fermentation with the two strains (Fig. 12.7). Glycemic load was significantly reduced by >33% (Jeske et al., 2018b). Profix 100 L protease treatment degraded the quinoa protein and increased the protein solubility from 48% to 76%. The protease treatments little affected the emulsifying and foaming capacity of the beverage. The selection of suitable proteases was critical to ensure the quality of the milk substitute (Jeske et al., 2018a). The sensory quality of the milk substitute produced remains to be studied. Huang et al. (2021)

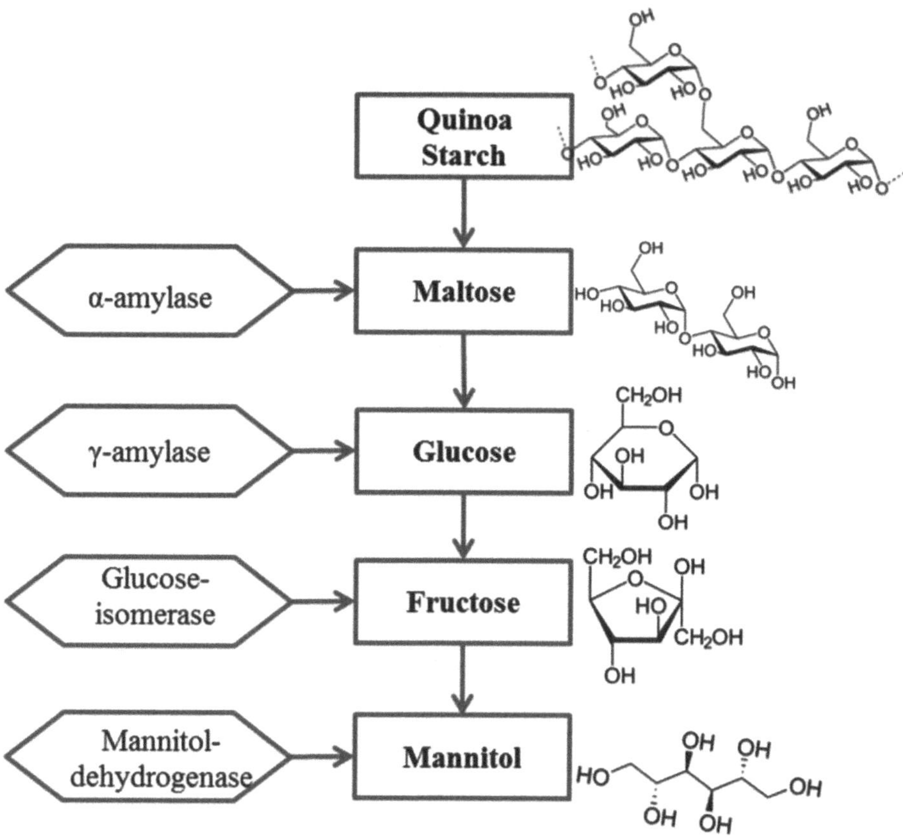

Figure 12.7 Enzymatic processing of quinoa starch with exogenous enzymes (α-amylase, γ-amylase, and glucose-isomerase), and endogenous enzymes, secreted by lactic acid bacteria (mannitol-dehydrogenase) (Jeske et al., 2018b).
Source: Reprinted with permission from Elsevier.

used a different process to make quinoa milk substitute containing oat β-glucan. Quinoa flour was extruded using a twin-screw extruder (screw and feeding speed were 300 rpm and 30 kg/h). Barrel temperatures were set at 80°C to 170°C–180°C. The extruded samples were dried and subjected to α-amylase and amyloglucosidase hydrolysis. The resulting samples were filtered and mixed with lipids and stabilizers (oat β-glucan up to 30% of the quinoa flour) before pH adjustment, high-pressure homogenization (50 MPa, 2 cycles), and sterilization (20 min, 121°C). The addition of β-glucan prevented the phase separation of the quinoa milk and increased the stability during storage. Increasing the concentration of β-glucan decreased the range of particle distribution of the milk, had no effect on zeta potential, and increased the apparent viscosity. The β-glucan formed network and prevented the phase separation. The addition of β-glucan at 16% gave the milk the highest sensory acceptance as too high concentrations of the β-glucan led to decreased sensory quality.

12.3.14 Fish- and meat-based products

Quinoa products in different forms have been used as ingredients in different fish- and meat-based products (Miranda et al., 2018; Romero et al., 2018; Verma et al., 2019; Pellegrini et al., 2018; Kerimoğlu et al., 2020; Fernández-López et al., 2020b, 2020c, 2021; Park et al., 2021; Trigo et al., 2018). Romero et al. (2018) formulated quinoa flour into fish patties. Fish (*Pseudoplatystoma corruscans*) was skinned, deboned, and minced before mixing with quinoa flour at different ratios in the presence of butylhydroxyanisole (0.01%) and sodium chloride (1.2%). The fish patties were cooked at 180°C to allow the center to reach a temperature of 78°C. The fish patties containing gluten-free flours (including flours of corn, quinoa, rice, and amaranth) had higher cooking yields than the control without flour addition. The fish patties containing quinoa or amaranth flours had higher hardness and lower springiness and cohesiveness than those containing rice and corn flour. The addition of the flours increased the carbohydrate and mineral contents. However, the overall sensory scores of the fish patties containing different flours were lower than those of the control (Romero et al., 2018). Quinoa extracts have been used in protection systems to better preserve mackerel and frozen fatty fish (Miranda et al., 2018; Trigo et al., 2018). Atlantic chub mackerel (*Scomber colias*) was stored in ice-containing quinoa ethanol extract for 13 days. The presence of quinoa extracts decreased lipid oxidation and hydrolysis in the mackerel. The inhibitory effects became more pronounced toward the end of the storage. The presence of quinoa extracts at a higher addition level (0.05 g/L icing solution) gave the mackerel acceptable sensory quality after the storage, whereas the control mackerel samples became unacceptable for sensory quality (Miranda et al., 2018). In another study, frozen Atlantic mackerel was stored at −40°C for 48 h before dipping in glazing solution (0°C, 30 s) containing the quinoa extract (0.45−1.36 g/L) (Trigo et al., 2018). The glazing significantly decreased the lipid hydrolysis and free fatty acid formation in the frozen fatty fish due to the presence of antioxidants in the quinoa extracts, while increasing the sensory quality (Trigo et al., 2018).

Fernández-López et al. (2020a and 2021) reviewed the recent advances in the applications of quinoa products in the formulations of meat-based products for improved quality. The quinoa products used included flour, raw and cooked seeds, paste, and fiber faction. The meat-based products included burgers, meatballs, Bolognas, dry-cured sausage, goat nuggets, beef burger and meatball, mortadella, cooked sausage, and pâté. Diverse quality attributes of these meat products affected by quinoa product addition were studied. The quinoa-based ingredients acted as binder, extender, gelling agent, fat replacer, or filler for enhanced nutritional, shelf life, and sensory quality (Fernández-López et al., 2020a, 2021). A few examples of detailed studies from recent years are given here. Quinoa pastes of three varieties of up to 10% as fat replacer was formulated into pork liver pâté (Pellegrini et al., 2018). Adding quinoa paste increased the concentrations/contents of residual nitrite moisture and ash of the pork liver pâté, while significantly decreasing that of lipids. The quinoa addition increased the hardness of the liver pâté, while having no effect on cohesiveness and springiness. The red quinoa paste addition at 5% gave the highest sensory acceptance. The quinoa addition at 10% gave lower oxidation of lipids on the samples. Enterobacteriaceae, aerobic bacteria, yeasts, and molds were not found in the samples containing quinoa (Pellegrini et al., 2018). Fiber-rich fraction from wet milling or quinoa seeds up to 3% were added to sausage formulations as binder (Fernández-Lópe et al., 2020c). Adding quinoa fiber fraction or seeds significantly increased the dietary fiber and residual nitrite contents of the sausages, improved emulsion stability, and decreased lipid oxidation and water activity (Fernández-Lópe et al., 2020c). Quinoa starch (up to 2.5%), seeds (up to 10%), and potato starch (up to 2.5%) at different ratios were formulated into chicken breast meat (up to 78%) to make chicken meatballs (Park et al., 2021). Adding any starch decreased the losses due to cooking, drip, and reheating during the frozen storage, while reducing the chewiness and hardness of the chicken meatball by 50%. The presence of quinoa seeds in the meatballs increased the in vitro antioxidant activity and chewiness and reduced the lipid oxidation of the meatballs during the freeze-thaw cycles. The addition of quinoa starch plus quinoa seeds increased the sensory acceptance of the chicken meatballs compared to that of other combinations (Park et al., 2021).

12.3.15 Beer

Effects of gluten-free grains on sensory properties of beer were summarized by Yang and Gao (2022). Although the use of quinoa in beer production is possible, it produced variable flavors, body, and foaming stability compared to that of beer made from barley due to its differences in chemical properties. The high fructose and maltose contents of quinoa make it suitable for producing malt, but the low enzymatic activity and carbohydrate content in comparison to barley limit its ability for malting (Yang and Gao, 2022). This issue of quinoa for use in beer formulations was also highlighted by the study of Kerpes et al. (2017). The lower levels of carbohydrates and lower enzymatic ability from α-amylase and β-amylase of quinoa than barley caused insufficient saccharification of quinoa starch. This process is essential in breaking down starch into sugars for high wort extraction and for fermentation to take place. Deželak et al. (2014) showed that the saccharification process of quinoa starch took longer than that of barley. Both Kerpes et al. (2017) and Deželak et al. (2014) suggested that this may be improved by supplementary addition of exogenous enzymes such as various starch-degrading amylases used for malting processes of rice and sorghum, in order to improve the body and flavor of quinoa beer. Germinating grains is a method to increase their nutritional value (e.g., increased polyphenol content and antioxidant activity) and decrease the concentrations of antinutrients. The malting process using quinoa requires modifying the germination time for optimal grain malting and kilning temperatures to preserve enzymes. Successful use of quinoa to produce gluten-free beer required alkaline steeping for 11–33 h at 8°C–20°C, followed by germination for 3–7 days at 8°C–15°C, then kilning for 5–48 h at 40°C–74°C. This vastly improved the quantities of α-amylase, β-amylase, protease, and α-glucosidase for improved brewing performance (Yang & Gao, 2022).

Compared to barley for beer production, quinoa has higher levels of total protein and fermentable amino nitrogen, which produces unique volatile compounds (Yang & Gao, 2022). These may not be desirable as quinoa may produce flavorings too different from common beer. Deželak et al. (2014) showed that gluten-free beer of 100% quinoa was not as accepted based on its sensory characteristics in comparison to beer made from 100% barley. Quinoa should be better studied for beer production when used in formulations alongside other gluten-free grains to improve nutritional value and functional properties and to create a familiar beer without an entirely altered flavor composition. Deželak et al. (2014) showed that beer formulated from 100% quinoa had significantly higher levels of iron (1.95 mg/L) and copper (0.33 mg/L) than barley in the wort. These higher concentrations potentially impaired yeast growth and fermentation, reducing the quality of brews and yield. Means in which these micronutrients may be reduced from the quinoa grain prior to beer production may help to improve the body, taste, and color of gluten-free beer through improving fermentation outcomes. Foam stability was found to be positively affected by gluten components, in which hordein may improve both the quality and stability of foam in beer. Quinoa does not contain these hordeins as it is gluten-free, but quinoa beer may still have good foaming stability caused by the high levels of soluble nitrogen and fermentable amino nitrogen (Yang & Gao, 2022). Greater protein hydrolysis in grain malt also causes poor foam stability, suggesting that methods such as addition of raw quinoa with unmodified proteins into gluten-free wort formulation may be valid for improving foam formation and stability of the final beer. In a recent study by Cela et al. (2022), gluten-free beer of "American pale ale" style was made using unmalted seeds as brewing adjuncts. The beer-making properties of quinoa grains were compared with those of buckwheat, amaranth, millet, and sorghum (both raw and gelatinized). Partial substitution of malted barley with quinoa seeds had little effect on the alcohol (%v/v) and extract (°P) content of the beers, whereas those with the other grains decreased these values. Beers produced using raw seeds of quinoa and also sorghum had acceptable sensory and technological properties. A gluten content of >20 ppm was obtained for all the beers, indicating more work is needed for gluten-free beer making (Cela et al., 2022).

12.4 Labeling and increasing price issues of quinoa-based gluten-free products

Development of gluten-free food and beverage products using quinoa grains should strictly follow food standard regulations in *Codex Alimentarius*. In order to label a food or beverage as "gluten-free," the gluten level in the products should be less than 20 ppm (Yang & Gao, 2022; Cela et al., 2022). This provides a challenge to gluten-free food and beverage production in that any additional substances or ingredients apart from quinoa used in formulations to enhance or alter nutritional and sensory properties must not contain gluten. Cross-contamination in the production facilities could be an issue. Research into testing of gluten presence could be used to ensure the safety of consumption in products for those with celiac disease or gluten sensitivity, while meeting standards of labeling requirements (Osorio et al., 2019; Cela et al., 2022).

A major concern in the production of gluten-free food and beverage products formulated from quinoa is cost. With the increase in popularity and demand of the quinoa, farm-gate price has increased in recent years (Arendt & Zannini, 2013; Hirich et al., 2021). This price increase has the potential to make quinoa food products and related gluten-free alternatives less accessible to many consumers worldwide. This could be improved upon by researching into quinoa processing methods to reduce costs from the production end.

Most of newly developed food/beverage products from the laboratory fail to enter the market for sales. Among those which enter the marketplace, most of them fail to perform well commercially and are removed from the shelves soon after their launches. Therefore, quinoa based food product development should be rigorously tested for any viable commercial applications in the future.

12.5 Conclusions

Quinoa has special chemical and nutritional properties. It has high potential in the development of diverse gluten-containing and gluten-free foods and beverages including both intermediate and finished products. Quinoa in various forms has been formulated into cakes, biscuits and cookies, breads, pastas, yogurt, cheese analogs, meat and fish products, fermented beverages and beers, and a range of traditional foods. Overall, the quinoa addition tends to increase the nutritional value of the products (especially the gluten-free ones), while reducing the sensory quality. Methods such as the addition of enzymes, emulsifiers, and texturizers and the application of special processes such as fermentation have shown to be beneficial in improving the sensory properties, along with optimizing formulations to balance the nutritional value. Potential issues in the production may include its high price due to the popularity of quinoa, as well as regulations around gluten-free labeling to adhere to food standards. The functions of the bioactive and other minor components of quinoa in relation to food and beverage production are not yet fully understood. The potential enhancement of quinoa genetic varieties may be useful to reduce unfavorable compounds, which may improve the overall acceptability of quinoa in food and beverage products, as well as provide further improvement in its nutritional value. Selection of suitable quinoa varieties such as sweet or bitter ones is important for food product formulations. In many of the studies summarized in this chapter, some technical problems such as low sensory acceptance were identified but not solved. The methods used to evaluate the quality of the developed products were not comprehensive. In future studies, the chemical, physicochemical, nutritional, sensory, and shelf-life properties of the products developed from quinoa should be analyzed comprehensively to provide a holistic picture. Environmental impact and carbon emissions of quinoa based product development should also be taken into consideration.

References

Abdelmontaleb, H. S., Othman, F. A., Degheidi, M. A., & Abbas, K. A. (2021). The influence of quinoa flour addition on the physicochemical, antioxidant activity, textural, and sensory characteristics of UF-soft cheese during refrigerated storage. *Journal of Food Processing and Preservation*, *45*, e15494.

Aguiar, E. V., Santos, F. G., Centeno, A. C. L. S., & Capriles, V. D. (2022). Defining amaranth, buckwheat and quinoa flour levels in gluten-free bread: A simultaneous improvement on physical properties, acceptability and nutrient composition through mixture design. *Foods*, *11*, 848.

Ahmed, J., Thomas, L., & Arfat, Y. A. (2019). Functional, rheological, microstructural and antioxidant properties of quinoa flour in dispersions as influenced by particle size. *Food Research International*, *116*, 302–311.

Alencar, N. M. M., de Morais, E. C., Steel, C. J., & Bolini, H. M. A. (2017). Sensory characterisation of gluten-free bread with addition of quinoa, amaranth flour and sweeteners as an alternative for coeliac patients. *International Journal of Food Science and Technology*, *52*, 872–879.

Arendt, E. K., & Zannini, E. (2013). *Cereal grains for the food and beverage industries* (1st ed.). Woodhead Publishing.

Axel, C., Brosnan, B., Zannini, E., Furey, A., Coffey, A., & Arendt, E. K. (2016). Antifungal sourdough lactic acid bacteria as biopreservation tool in quinoa and rice bread. *International Journal of Food Microbiology*, *239*, 86–94.

Ayseli, M. T., Yilmaz, M. T., Cebi, N., Sagdic, O., Ozmen, D., & Capanoglu, E. (2020). Physicochemical, rheological, molecular, thermal and sensory evaluation of newly developed complementary infant (6–24 months old) foods prepared with quinoa (*Chenopodium quinoa* Willd.) flour. *Food Chemistry*, *315*, 126208.

Ayyasha, M., Johnson, S. K., Liu, S. Q., Mesmari, N., Dahmani, S., Dhaheri, A. S. A., & Kizhakkayil, J. (2019). In vitro investigation of bioactivities of solid-state fermented lupin, quinoa and wheat using *Lactobacillus* spp. *Food Chemistry*, *275*, 50–58.

Azizi, S., Azizi, M. H., Moogouei, R., & Rajaei, P. (2020). The effect of quinoa flour and enzymes on the quality of gluten-free bread. *Food Science & Nutrition*, *8*, 2373–2382.

Bahmanyar, F., Hosseini, S. M., Mirmoghtadaie, L., & Shojaee-Aliabadi, S. (2021). Effects of replacing soy protein and bread crumb with quinoa and buckwheat flour in functional beef burger formulation. *Meat Science*, *172*, 108305.

Ballester-Sánchez, J., Gil, J. V., Haros, C. M., & Fernández-Espinar, M. T. (2019a). Effect of incorporating white, red or black quinoa flours on free and bound polyphenol content, antioxidant activity and color of bread. *Plant Foods for Human Nutrition*, *74*, 185–191.

Ballester-Sánchez, J., Millán-Linares, M. C., Fernández-Espinar, M. T., & Haros, C. M. (2019b). Development of healthy, nutritious bakery products by incorporation of quinoa. *Foods*, *8*, 379.

Ballester-Sánchez, J., Yalcin, E., Fernández-Espinar, M. T., & Haros, C. M. (2019c). Rheological and thermal properties of royal quinoa and wheat flour blends for breadmaking. *European Food Research and Technology*, *245*, 1571–1582.

Bavaro, A. R., Biase, M. D., Caputo, A. C. S. L. L. L., Cedola, A., Nobile, M. A. D., Logrieco, A. F., & Valerio, P. L. F. (2021). *Weissella cibaria* short-fermented liquid sourdoughs based on quinoa or amaranth flours as fat replacer in focaccia bread formulation. *International Journal of Food Science and Technology*, *56*, 3197–3208.

Bilgiçli, N., & İbanoğlu, Ş. (2015). Effect of pseudo cereal flours on some physical, chemical and sensory properties of bread. *Journal of Food Science and Technology*, *52*, 7525–7529.

Bozdogan, N., Kumcuoglu, S., & Tavman, S. (2019). Investigation of the effects of using quinoa flour on gluten-free cake batters and cake properties. *Journal of Food Science and Technology*, *56*, 683–694.

Brito, I. L., de Souza, E. L., Felex, S. S. S., Madruga, M. S., Yamashita, F., & Magnani, M. (2015). Nutritional and sensory characteristics of gluten-free quinoa (*Chenopodium quinoa* Willd)-based cookies development using an experimental mixture design. *Journal of Food Science and Technology*, *52*, 5866–5873.

Cannas, M., Pulina, S., Conte, P., Del Caro, A., Urgeghe, P. P., Piga, A., & Fadda, C. (2020). Effect of substitution of rice flour with quinoa flour on the chemical-physical, nutritional, volatile and sensory parameters of gluten-free ladyfinger biscuits. *Foods*, *9*, 808.

Carrizo, S. L., de LeBlanc, A. D. M., LeBlanc, J. G., & Rollán, G. C. (2020). Quinoa pasta fermented with lactic acid bacteria prevents nutritional deficiencies in mice. *Food Research International*, *127*, 108735.

Cela, N., Galgano, F., Perretti, G., Di Cairano, M., Tolve, R., & Condellia, N. (2022). Assessment of brewing attitude of unmalted cereals and pseudocereals for gluten free beer production. *Food Chemistry*, *384*, 132621.

Cerdá-Bernad, D., Valero-Cases, E., Pastor, J. J., Frutos, M. J., & Pérez-Llamas, F. (2022). Probiotic red quinoa drinks for celiacs and lactose intolerant people: Study of functional, physicochemical and probiotic properties during fermentation and gastrointestinal digestion. *International Journal of Food Sciences and Nutrition*, *73*, 49–59.

Chen, J. M., Al, K. F., Craven, L. J., Seney, S., Coons, M., McCormick, H., Reid, G., O'Connor, C., & Burton, J. P. (2020). Nutritional, microbial, and allergenic changes during the fermentation of cashew 'cheese' product using a quinoa-based rejuvelac starter culture. *Nutrients*, *12*, 648.

Codină, G. G., Franciuc, S. G., & Mironeasa, S. (2016). Rheological characteristics and microstructure of milk yogurt as influenced by quinoa flour addition. *Journal of Food Quality*, *39*, 559–566.

Cueto, M., Porras-Saavedra, J., Farroni, A., Alamilla-Beltrán, L., Schöenlechner, R., Schleining, G., & Buera, P. (2015). Physical and mechanical properties of maize extrudates as affected by the addition of chia and quinoa seeds and antioxidants. *Journal of Food Engineering*, *167*, 139–146.

D'ambrosio, T., Amodio, M. L., Pastore, D., De Santis, G., & Colelli, G. (2017). Chemical, physical and sensorial characterization of fresh quinoa sprouts (*Chenopodium quinoa* Willd.) and effects of modified atmosphere packaging on quality during cold storage. *Food Packaging and Shelf Life*, *14*, 52–58.

Demir, B., & Bilgiçli, N. (2020). Changes in chemical and anti-nutritional properties of pasta enriched with raw and germinated quinoa (*Chenopodium quinoa* Willd.) flours. *Journal of Food Science and Technology*, *57*, 3884–3892.

Demir, M. K., & Kılınç, M. (2017). Utilization of quinoa flour in cookie production. *International Food Research Journal*, *24*, 2394–2401.

Deshpande, R. H., Kumar, A., Katare, M., Sakhare, S. D., & Inamdar, A. A. (2022). Effect of grinding techniques and supplementation of quinoa on the carbohydrate profile of tortillas. *Journal of Food Science and Technology*, *59*, 3600–3608.

Deželak, M., Zarnkow, M., Becker, T., & Košir, I. J. (2014). Processing of bottom-fermented gluten-free beer-like beverages based on buckwheat and quinoa malt with chemical and sensory characterization. *Journal of the Institute of Brewing*, *120*, 360–370.

Dong, G. M., Dong, J. L., Zhu, Y. Y., Shen, R. L., & Qu, L. B. (2021). Development of weaning food with prebiotic effects based on roasted or extruded quinoa and millet flour. *Journal of Food Science*, *86*, 1089–1096.

El-Shafei, S. M. S., Sakr, S. S., & Abou-Soliman, N. H. I. (2020). The impact of supplementing goats' milk with quinoa extract on some properties of yoghurt. *International Journal of Dairy Technology*, *73*, 126–133.

El-Sohaimy, A. S., Shehata, G. M., Djapparovec, T. A., Mehany, T., Zeitoun, M. A., & Zeitoun, A. M. (2021). Development and characterization of functional pan bread supplemented with quinoa flour. *Journal of Food Processing and Preservation*, *45*, e15180.

Felix, M., Camacho-Ocaña, Z., López-Castejón, M. L., & Ruiz-Domínguez, M. (2021). Rheological properties of quinoa-based gels. An alternative for vegan diets. *Food Hydrocolloids*, *120*, 106827.

Fernández-López, J., Lucas-González, R., Roldán-Verdú, A., Viuda-Martos, M., Sayas-Barberá, E., Ballester-Sánchez, J., Haros, C. M., & Pérez-Álvarez, J. A. (2020b). Effects of black quinoa wet-milling coproducts on the quality properties of bologna-type sausages during cold storage. *Foods*, *9*, 274.

Fernández-López, J., Lucas-González, R., Viuda-Martos, M., Sayas-Barberá, E., Ballester-Sánchez, J., Haros, C. M., Martínez-Mayoral, A., & Pérez-Álvarez, J. A. (2020c). Chemical and technological properties of bologna-type sausages with added black quinoa wet-milling coproducts as binder replacer. *Food Chemistry*, *310*, 125936.

Fernández-López, J., Viuda-Martos, M., & Pérez-Alvarez, J. A. (2021). Quinoa and chia products as ingredients for healthier processed meat products: technological strategies for their application and effects on the final product. *Current Opinion in Food Science*, *40*, 26–32.

Fernández-López, J., Viuda-Martos, M., Sayas-Barberá, M. E., de Vera, C. N. R., Lucas-González, R., Roldán-Verdú, A., Botella-Martínez, C., & Pérez-Alvare, J. A. (2020a). Chia, quinoa, and their coproducts as potential antioxidants for the meat industry. *Plants*, *9*, 1359.

Franco, W., Evert, K., & Van Nieuwenhove, C. (2021). Quinoa flour, the germinated grain flour, and sourdough as alternative sources for gluten-free bread formulation: Impact on chemical, textural and sensorial characteristics. *Fermentation*, *7*, 115.

Gewehr, M. F., Danelli, D., Melo, L. M. D., Flöres, S. H., & Jong, E. V. D. (2017). Nutritional and technological evaluation of bread made with quinoa flakes (*Chenopodium quinoa* Willd). *Journal of Food Processing and Preservation*, *41*, e12803.

Giménez, M. A., Drago, S. R., Bassett, M. N., Lobo, M. O., & Sammán, N. C. (2016). Nutritional improvement of corn pasta-like product with broad bean (*Vicia faba*) and quinoa (*Chenopodium quinoa*). *Food Chemistry*, *199*, 150–156.

Gostin, A. I. (2019). Effects of substituting refined wheat flour with wholemeal and quinoa flour on the technological and sensory characteristics of salt-reduced breads. *LWT-Food Science and Technology*, *114*, 108412.

Guo, H., Hao, Y., Yang, X., Ren, G., & Richel, A. (2022). Exploration on bioactive properties of quinoa protein hydrolysate and peptides: A review. *Critical Reviews in Food Science and Nutrition*, *28*, 1–14.

Hirich, A., Rafik, S., Rahmani, M., Fetouab, A., Azaykou, F., Filali, K., Ahmadzai, H., Jnaoui, Y., Soulaimani, A., Moussafir, M., El Gharous, M., Karboune, s., Sbai, A., & Choukr-Allah, R. (2021). Development of quinoa value chain to improve food and nutritional security in rural communities in Rehamna, Morocco: Lessons learned and perspectives. *Plants*, *10*, article no. 301.

Huang, K., Zhang, S., Guan, X., Li, C., Li, S., Liu, Y., & Shi, J. (2021). Effect of the oat β-glucan on the development of functional quinoa (*Chenopodium quinoa* wild) milk. *Food Chemistry*, *349*, 129201.

Inglett, G. E., Chen, D., & Liu, S. X. (2015). Pasting and rheological properties of quinoa-oat composites. *International Journal of Food Science and Technology*, *50*, 878–884.

Ismail, H. A., & Rayan, A. M. (2022). Preparation and evaluation of quinoa-kishk as a novel functional fermented dairy product. *Journal of Food Science and Technology*, *59*, 1063–1074.

Iuliano, L., González, G., Casas, N., Moncayo, D., & Cote, S. (2019). Development of an organic quinoa bar with amaranth and chia. *Ciência e Tecnologia de Alimentos*, *39*, 218–224.

Jan, K. N., Panesar, P. S., & Singh, S. (2018). Textural, in vitro antioxidant activity and sensory characteristics of cookies made from blends of wheat-quinoa grown in India. *Journal of Food Processing and Preservation*, *42*, e13542.

Jeske, S., Zannini, E., Cronin, M. F., & Arendt, E. K. (2018a). Impact of protease and amylase treatment on proteins and the product quality of a quinoa-based milk substitute. *Food & Function*, *9*, 3500.

Jeske, S., Zannini, E., Lynch, K. M., Coffey, A., & Arendt, E. K. (2018b). Polyol-producing lactic acid bacteria isolated from sourdough and their application to reduce sugar in a quinoa-based milk substitute. *International Journal of Food Microbiology*, *286*, 31–36.

Kahlon, T. S., Avena-Bustillos, R. J., & Chiu, M. C. M. (2017). Sensory evaluation of gluten-free quinoa whole grain snacks. *Heliyon*, *3*, e00213.

Kaur, R., Ahluwalia, P., Sachdev, P. A., & Kaur, A. (2018). Development of gluten-free cereal bar for gluten intolerant population by using quinoa as major ingredient. *Journal of Food Science and Technology*, *55*, 3584–3591.

Kerimoğlu, B. Ö., Kavuşan, H. S., Tabak, D., & Serdaroğlu, M. (2020). Formulating reduced fat sausages with quinoa or teff flours: Effects on emulsion characteristics and product quality. *Food Science of Animal Resources*, *40*, 710–721.

Kerpes, R., Fischer, S., & Becker, T. (2017). The production of gluten-free beer: Degradation of hordeins during malting and brewing and the application of modern process technology focusing on endogenous malt peptidases. *Trends in Food Science & Technology*, *67*, 129–138.

Kordialik-Bogacka, E., Bogdan, P., Pielech-Przybylska, K., & Michałowska, D. (2018). Suitability of unmalted quinoa for beer production. *Journal of the Science of Food and Agriculture*, *98*, 5027–5036.

Lappi, J., Raninen, K., Väkeväinen, K., Kårlund, A., Törrönen, R., & Kolehmainen, M. (2021). Blackcurrant (*Ribes nigrum*) lowers sugar-induced postprandial glycaemia independently and in a product with fermented quinoa: A randomised crossover trial. *British Journal of Nutrition*, *126*, 708–717.

Li, G., & Zhu, F. (2018). Quinoa starch: Structure, properties, and applications. *Carbohydrate Polymers*, *181*, 851–861.

Linares-García, L., Repo-Carrasco-Valencia, R., Paulet, P. G., & Schoenlechner, R. (2019). Development of gluten-free and egg-free pasta based on quinoa (*Chenopdium quinoa* Willd) with addition of lupine flour, vegetable proteins and the oxidizing enzyme Pox. *European Food Research and Technology*, *245*, 2147–2156.

Lorusso, A., Verni, M., Montemurro, M., Coda, R., Gobbetti, M., & Rizzello, C. G. (2017). Use of fermented quinoa flour for pasta making and evaluation of the technological and nutritional features. *LWT-Food Science and Technology*, *78*, 215–221.

López-Alarcón, C. A., Cerdán-Leal, M. A., Beristain, C. I., Pascual-Pineda, L. A., Azuara, E., & Jiménez-Fernández, M. (2019). The potential use of modified quinoa protein isolates in cupcakes: Physicochemical properties, structure and stability of cupcakes. *Food & Function, 10*, 4432.

Martínez-Villaluenga, C., Peñas, E., & Hernández-Ledesma, B. (2020). Pseudocereal grains: Nutritional value, health benefits and current applications for the development of gluten-free foods. *Food and Chemical Toxicology, 137*, 111178.

Miranda, J. M., Carrera, M., Pasten, A., Vega-Gálvez, A., Barros-Velázquez, J., & Aubourg, S. P. (2018). The impact of quinoa (*Chenopodium quinoa* Willd.) ethanolic extracts in the icing medium on quality loss of Atlantic chub mackerel (*Scomber colias*) under chilling storage. *European Journal of Lipid Science and Technology, 120*, 1800280.

Mousa, R. M. A. (2021). Simultaneous inhibition of acrylamide formation and fat oxidation in quinoa cakes using gum Arabic supplementation coupled with fat reduction. *International Journal of Food Properties, 24*, 749–763.

Mousa, R. M. A. (2022). Inhibition of acrylamide in gluten-free quinoa biscuits by supplementation with microbial dextran. *International Journal of Food Properties, 25*, 11–23.

Navruz-Varli, S., & Sanlier, N. (2016). Nutritional and health benefits of quinoa (*Chenopodium quinoa* Willd.). *Journal of Cereal Science, 69*, 371–376.

Osorio, C. E., Mejías, J. H., & Rustgi, S. (2019). Gluten detection methods and their critical role in assuring safe diets for celiac patients. *Nutrients, 11*, 2920.

Pagno, C. H., Klug, T. V., Costa, T. M. H., de Oliveira Rios, A., & Flôres, S. H. (2016). Physical and antimicrobial properties of quinoa flour-based films incorporated with essential oil. *Journal of Applied Polymer Science, 133*, 43311.

Park, J. H., Lee, Y. J., Lim, J. G., Jeon, J. H., & Yoon, K. S. (2021). Effect of quinoa (*Chenopodium quinoa* Willd.) starch and seeds on the physicochemical and textural and sensory properties of chicken meatballs during frozen storage. *Foods, 10*, 1601.

Pellegrini, M., Lucas-Gonzalez, R., Sayas-Barberá, E., Fernández-López, J., Pérez-Álvarez, J. A., & Viuda-Martos, M. (2018). Quinoa (*Chenopodium quinoa* Willd) paste as partial fat replacer in the development of reduced fat cooked meat product type pâté: Effect on quality and safety. *CyTA - Journal of Food, 16*, 1079–1088.

Pineli, L. D. L. D. O., Botelho, R. B. A., Zandonadi, R. P., Solorzano, J. L., Oliveira, G. T. D., Reis, C. E. G., & Teixeira, D. D. S. (2015). Low glycemic index and increased protein content in a novel quinoa milk. *LWT-Food Science and Technology, 63*, 1261–1267.

Pourshahidi, L. K., Caballero, E., Osses, A., Hyland, B. W., Ternan, N. G., & Gill, C. I. R. (2020). Modest improvement in CVD risk markers in older adults following quinoa (*Chenopodium quinoa* Willd.) consumption: a randomized-controlled crossover study with a novel food product. *European Journal of Nutrition, 59*, 3313–3323.

Pérez-Álvarez, J. Á., Botella-Martínez, C. M., de Vera, C. N. R., Sayas-Barberá, E., Viuda-Martos, M., Fernández-López, J., & Sánchez-Zapata, E. (2021). A Preliminary study on the incorporation of quinoa flour in organic pumpkin creams: Effect on the physicochemical properties. *Proceedings, 70*, 71.

Rios, F., Lobo, M., & Samman, N. (2018). Acceptability of beehive products as ingredients in quinoa bars. *Journal of the Science of Food and Agriculture, 98*, 174–182.

Rizzello, C. G., Lorusso, A., Russo, V., Pinto, D., Marzani, B., & Gobbetti, M. (2017). Improving the antioxidant properties of quinoa flour through fermentation with selected autochthonous lactic acid bacteria. *International Journal of Food Microbiology, 241*, 252–261.

Romano, A., Masi, P., Bracciale, A., Aiello, A., Nicolai, M. A., & Ferranti, P. (2018). Effect of added enzymes and quinoa flour on dough characteristics and sensory quality of a gluten-free bakery product. *European Food Research and Technology, 244*, 1595–1604.

Romano, N., Ureta, M. M., Guerrero-Sánchez, M., & Gómez-Zavaglia, A. (2020). Nutritional and technological properties of a quinoa (*Chenopodium quinoa* Willd.) spray-dried powdered extract. *Food Research International, 129*, 108884.

Romero, M. C., Fogar, R. A., Rolhaiser, F., Clavero, V. V., Romero, A. M., & Judis, M. A. (2018). Development of gluten-free fish (*Pseudoplatystoma corruscans*) patties by response surface methodology. *Journal of Food Science and Technology, 55*, 1889–1902.

Rothschild, J., Rosentrater, K. A., Onwulata, C., Singh, M., Menutti, L., Jambazian, P., & Omary, M. B. (2015). Influence of quinoa roasting on sensory and physicochemical properties of allergen-free, gluten-free cakes. *International Journal of Food Science and Technology, 50*, 1873–1881.

Ruales, J., de Grijalva, Y., Lopez-Jaramillo, P., & Nair, B. M. (2002). The nutritional quality of an infant food from quinoa and its effect on the plasma level of insulin-like growth factor-1 (IGF-1) in undernourished children. *International Journal of Food Sciences and Nutrition, 53*, 143–154.

Sazesh, B., & Goli, M. (2020). Quinoa as a wheat substitute to improve the textural properties and minimize the carcinogenic acrylamide content of the biscuit. *Journal of Food Processing and Preservation, 44*, e14563.

Schoenlechner, R., Drausinger, J., Ottenschlaeger, V., Jurackova, K., & Berghofer, E. (2010). Functional properties of gluten-free pasta produced from amaranth, quinoa and buckwheat. *Plant Foods for Human Nutrition, 65*, 339–349.

Silva, A. R. A., Silva, M., M. N., & Ribeiro, B. D. (2020). Health issues and technological aspects of plant-based alternative milk. *Food Research International, 131*, 108972.

Soltani, M., Hekmat, S., & Ahmadi, L. (2018). Microbial and sensory evaluation of probiotic yoghurt supplemented with cereal/pseudo-cereal grains and legumes. *International Journal of Dairy Technology, 71*, 141–148.

Sosa, M., Califano, A., & Lorenzo, G. (2019). Influence of quinoa and zein content on the structural, rheological, and textural properties of gluten-free pasta. *European Food Research and Technology, 245*, 343–353.

Starzyńska-Janiszewska, A., Bączkowicz, M., Sabat, R., Stodolak, B., & Witkowicz, R. (2017). Quinoa tempe as a value-added food: sensory, nutritional, and bioactive parameters of products from white, red, and black seeds. *Cereal Chemistry, 94*, 491–496.

Suárez-Estrella, D., Borgonovo, G., Buratti, S., Ferranti, P., Accardo, F., Pagani, M. A., & Martim, A. (2021). Sprouting of quinoa (*Chenopodium quinoa* Willd.): Effect on saponin content and relation to the taste and astringency assessed by electronic tongue. *LWT-Food Science and Technology, 144*, 111234.

Suárez-Estrella, D., Cardone, G., Buratti, S., Pagani, M. A., & Marti, A. (2020). Sprouting as a pre-processing for producing quinoa-enriched bread. *Journal of Cereal Science, 96*, 103111.

Tiga, B. H., Kumcuoglu, S., Vatansever, M., & Tavman, S. (2021). Thermal and pasting properties of quinoa—wheat flour blends and their effects on production of extruded instant noodles. *Journal of Cereal Science, 97*, 103120.

Trigo, M., Rodríguez, A., Dovale, G., Pastén, A., Vega-Gálvez, A., & Aubourg, S. P. (2018). The effect of glazing based on saponin-free quinoa (*Chenopodium quinoa*) extract on the lipid quality of frozen fatty fish. *LWT-Food Science and Technology, 98*, 231–236.

Turkut, G. M., Cakmak, H., Kumcuoglu, S., & Tavman, S. (2016). Effect of quinoa flour on gluten-free bread batter rheology and bread quality. *Journal of Cereal Science, 69*, 174–181.

Ujiroghene, O. J., Liu, L., Zhang, S., Lu, J., Zhang, C., Lv, J., Pang, X., & Zhang, M. (2019). Antioxidant capacity of germinated quinoa-based yoghurt and concomitant effect of sprouting on its functional properties. *LWT-Food Science and Technology, 116*, 108592.

Urquizo, F. E. L., Torres, S. M. G., Tolonen, T., Jaakkola, M., Pena-Niebuhr, M. G., von Wright, A., et al. (2017). Development of a fermented quinoa-based beverage. *Food Science & Nutrition*, *5*, 602–608.

Verma, A. K., Rajkumar, V., & Kumar, S. (2019). Effect of amaranth and quinoa seed flour on rheological and physicochemical properties of goat meat nuggets. *Journal of Food Science and Technology*, *56*, 5027–5035.

Vázquez-Villegas, P., Perez-Carrillo, E., García, C. P., & Camacho, M. C. (2021). Effect of wheat flour substitution and popped amaranth flour content on the rheological, physicochemical and textural properties of hot-press wheat-oat-quinoa-amaranth composite flour tortillas. *CyTA-Journal of Food*, *19*, 571–578.

Väkeväinen, K., Ludena-Urquizo, F., Korkala, E., Lapveteläinen, A., Peräniemi, S., von Wright, A., & Plumed-Ferrer, C. (2020). Potential of quinoa in the development of fermented spoonable vegan products. *LWT-Food Science and Technology*, *120*, 108912.

Wang, S., Opassathavorn, A., & Zhu, F. (2015). Influence of quinoa flour on quality characteristics of cookie, bread and Chinese steamed bread. *Journal of Texture Studies*, *46*, 281–292.

Wang, S., & Zhu, F. (2016). Formulation and quality attributes of quinoa food products. *Food and Bioprocess Technology*, *9*, 49–68.

Yang, D., & Gao, X. (2022). Progress of the use of alternatives to malt in the production of gluten-free beer. *Critical Reviews in Food Science and Nutrition*, *62*, 2820–2835.

Zanninia, E., Jeskea, S., Lyncha, K. M., & Arendt, E. K. (2018). Development of novel quinoa-based yoghurt fermented with dextran producer *Weissella cibaria* MG1. *International Journal of Food Microbiology*, *268*, 19–26.

Zhang, Y., Zhang, S., Fan, W., Duan, M., Han, Y., & Li, H. (2019). Identification of volatile compounds and odour activity values in quinoa porridge by gas chromatography-mass spectrometry. *Journal of the Science of Food and Agriculture*, *99*, 3957–3966.

Zoumpopoulou, G., & Tsakalidou, E. (2019). Chapter 8 – Gluten-free products. In C. M. Galanakis (Ed.), *The role of alternative and innovative food ingredients and products in consumer wellness* (pp. 213–237). Academic Press.

Çevik, A., & Ertaş, N. (2019). Effect of quinoa, buckwheat and lupine on nutritional properties and consumer preferences of tarhana. *Quality Assurance and Safety of Crops & Foods*, *11*, 145–155.

Food and nonfood applications of quinoa fractionated products

13

13.1 Introduction

Value-added processing of quinoa is important for its commercial success. Like other food grains, quinoa seeds have been fractionated into different products rich in certain components or into pure components such as starch and protein (Duran et al., 2018; Lotfy et al., 2021; Zuo et al., 2022). Overall, the diverse components of quinoa have the potential to be used for human nutrition applications and functional food formulations (Graf et al., 2015; Ng & Wang, 2021). Compared with major cereals such as maize and wheat, the fractionation and value-added processing of quinoa grains have been much less conducted both in the laboratory and in industries. More importantly, the technical advantage of using quinoa-based fractions for food and nonfood applications should be revealed in comparison with those of other major food grains such as maize and wheat. This is important since the cost of quinoa grains is much higher than that of those major grains. Even similar technical applicability and capacity of quinoa-based products to those of major crops do not imply the commercial viability of quinoa for applications. A cost-effective approach is essential for its industrial applications.

The properties and structure of individual components in quinoa seeds have been described in Chapters 4–8 of the book. The chemical components have special properties. For example, quinoa starch granules are very small compared to the majority of other starch types. Small starch granules can be chemically modified with octenyl succinic anhydride to stabilize Pickering emulsions (Li & Zhu, 2018). Quinoa protein isolate has film-forming property and could be formulated into edible films and coatings for food packaging (Zade et al., 2021). Quinoa has also been used in nonfood areas. For example, quinoa powder was used for corrosion inhibition in carbon steel in strong acidic condition (Mobtaker et al., 2021). The seed coat of some quinoa varieties is rich in saponins. Saponins are biosurfactants and can be used in the formulations of cosmetics and fossil oil refinery processes (nonfood applications) (Góral & Wojciechowski, 2020; Norouzpour et al., 2022).

The fractionated products of quinoa seeds of current interest include starch, protein, dietary fiber, and saponin- and polyphenol-enriched extracts. These products are either pure components or rich in certain components such as saponins. They have been used in a range of different food and nonfood applications. In this chapter, representative studies were selected from a vast number of publications by focusing on the literature from the past 3 years. Food and nonfood applications of quinoa-fractionated products are described.

Quinoa. DOI: https://doi.org/10.1016/B978-0-323-99909-0.00003-9

13.2 Food and nutraceutical applications of quinoa protein

13.2.1 Functional food ingredient and food component interactions

Quinoa protein isolate was used in different food systems and formulations (López-Alarcón et al., 2019; Duran et al., 2018; Lotfy et al., 2021) (Table 13.1). Quinoa protein isolate was modified by freeze-drying or heating (López-Alarcón et al., 2019). The modified protein of up to 30% replaced wheat flour to make cupcakes containing egg, sugar, vanilla extract, cinnamon, yeast, and oil. The addition of the quinoa protein isolate in different forms increased the hardness and chewiness of the cupcakes, while increasing the water activity. The cupcakes containing the raw, heat-treated, or freeze-dried quinoa protein isolate samples had higher acceptance of sensory quality and longer microbiological shelf life than the control samples (López-Alarcón et al., 2019). During food formulations and processing, food component interactions are critical for ensuring food quality. Interactions of quinoa proteins with carbohydrates including ι-carragennan and xylose were studied under different conditions (Duran et al., 2018; Lotfy et al., 2021). ι-Carragennan and quinoa protein isolate interacted in solutions at different pH, which were studied computationally and experimentally (Duran et al., 2018). At pH of $1-2.9$, carragennan and the quinoa protein associated. At pH of >5.5, the charges of carragennan and the quinoa protein became negative and their interactions showed Coulombic repulsion. At pH between 2.9 and 5.5, the two molecules were attractive as the charge regulation was dominant over electrostatic repulsion. These results were confirmed by computational simulation studies (Duran et al., 2018). Quinoa protein hydrolysates (alcalase and flavourzyme) and xylose were subjected to Maillard reactions (2 h, 140°C) at a range of initial pH (5–9) (Lotfy et al., 2021). Decreasing the initial pH decreased the browning intensity, while increasing the in vitro antioxidant activity and contents of free amino acids of the Maillard reaction products. The Maillard reactions at a lower pH tended to produce more furanones, furans, and cyclopentanones. Pyrazines were the major products of the Maillard reactions at pH 9 (Lotfy et al., 2021).

13.2.2 Nutraceutical ingredients

Quinoa protein isolate was used as nutraceutical ingredient due to their biological functions, which were tested in vitro or in vivo (Vilcacundo et al., 2017; Capraro et al., 2020; Guo et al., 2021; Pompeu et al., 2022). For example, in the study of Vilcacundo et al. (2017), quinoa protein isolate was subjected to in vitro gastrointestinal digestion at different pH of 1.2–3.2 related to the gastric phase. The inhibitory activities of the digestas against lipid peroxidation were studied using Zebrafish larva model. The digesta at the end of the digestion process had inhibitory activity against lipid peroxidation comparable to butylated hydroxytoluene. The digestion process of duodenal phase increased the inhibitory activity. The digesta obtained using pH of 1.2 had the highest inhibitory activity

Table 13.1 Diverse applications of quinoa protein fraction.

Application	Experimental	Major findings	Reference
Functional ingredient	Quinoa protein isolate was modified by freeze-drying or heating. The modified protein up to 30% replaced wheat flour to make cupcakes containing egg, sugar, vanilla extract, cinnamon, yeast, and oil	Addition of quinoa protein isolate in different forms increased the hardness and chewiness of the cupcakes, while increasing the water activity. The cupcakes containing raw, heat-treated, or freeze-dried quinoa protein isolate samples had higher acceptance of sensory quality and longer microbiological shelf life than the control samples	López-Alarcón et al. (2019)
Food component interaction	ι-Carragennan and quinoa protein isolate interacted in solutions of different pH, which were studied computationally and experimentally	At pH of 1 − 2.9, carragennan and quinoa protein associated. At pH of >5.5, the charges of carragennan and quinoa protein became negative and their interactions showed Coulombic repulsion. At pH between 2.9 and 5.5, the two molecules were attractive as the charge regulation was dominant over electrostatic repulsion. These results were confirmed by computational simulation studies	Duran et al. (2018)
Food component interaction	Quinoa protein hydrolysates (alcalase and flavourzyme) and xylose were subjected to Maillard reactions (2 h, 140°C) at a range of initial pH (5–9)	Decreasing the initial pH decreased the browning intensity, while increasing the in vitro antioxidant activity and content of free amino acids of the Maillard reaction products. The Maillard reactions at a lower pH tended to produce more furanones, furans, and cyclopentanones. Pyrazines were the major products of the Maillard reactions at pH 9	Lotfy et al. (2021)

(Continued)

Table 13.1 (Continued)

Application	Experimental	Major findings	Reference
Nutraceutical ingredient	Quinoa protein isolate was subjected to in vitro gastrointestinal digestion at different pH of 1.2–3.2 for the gastric phase. The inhibitory activities of the digestas against lipid peroxidation were studied using Zebrafish larva model	The digesta at the end of the digestion process had inhibitory activity against lipid peroxidation which was comparable to butylated hydroxytoluene. The digestion process of duodenal phase increased the inhibitory activity. The digesta obtained using pH of 1.2 had the highest inhibitory activity	Vilcacundo et al. (2017)
Nutraceutical ingredient	Caco-2 cells were used as model system to study the antiinflammatory of chenopodin isoforms	Both of lowly charged and highly charged chenopodin isoforms showed antiinflammatory capacity with different molecular mechanisms of action	Capraro et al. (2020)
Nutraceutical ingredient	Spontaneously hypertensive rats were fed with a diet containing quinoa protein isolate for 5 weeks	Upon gastrointestinal, peptide precursors were released. Quinoa protein intake decreased the blood pressure, while changing the composition of gut microbiota in the hypertensive rats to be more similar to that of normal rats	Guo et al. (2021)
Nutraceutical ingredient	Chenopodin was tested in vivo for antiinflammatory activity using mouse model. Mice were treated with carrageenan to induce neutrophil recruitment and paw edema. The mice were also fed with a diet containing chenopodin isolated from quinoa seeds	Chenopodin intake decreased the neutrophil recruitment and paw edema in the mice, being antiinflammatory	Pompeu et al. (2022)
Nutraceutical ingredient and emulsifier	Quinoa protein concentrate was hydrolyzed using pancreatin. Emulsifying and in vitro antioxidant activity of the protein hydrolysates were studied	A degree of hydrolysis of 19% was obtained after 3 h of hydrolysis. The protein concentrate had lower in vitro antioxidant, foaming, and emulsifying capacities than the protein hydrolysate, whereas the latter had lower stabilizing capacity for the foams and emulsions	Daliri et al. (2021)

Encapsulation	Flavonoids (quercetin, resveratrol, curcumin, and luteolin) and nanomicelles of quinoa protein interacted and assembled	The flavonoids and the protein formed assembly through hydrophobic interactions and hydrogen bonding. The encapsulation into the nanomicelles increased the stability and solubility of the flavonoids	Liu et al. (2021)
Encapsulation	Quinoa protein and gum Arabic were combined to form coacervates which were used to encapsulate eugenol. The microcapsules were freeze-dried and were used to preserve minced pork	The coacervates were optimally formed at pH 4 with a gum to protein ratio of 1:4 by electrostatic interactions to have a mean size of 30 μm and an encapsulation efficiency of 89%. Eugenol was encapsulated in the coacervates via hydrogen bond. When applied in minced pork, the microcapsules reduced the increases in total bacterial counts, total volatile basic nitrogen level, and pH compared to pure eugenol. The application of the eugenol-containing microcapsules increased the shelf life of minced pork (4°C, 9 days)	Chen et al. (2021)
Emulsion stabilizer	Soluble proteins were obtained using alkaline solubilization and acid precipitation from quinoa seeds and were used to stabilize oil (rapeseeds)-in-water emulsions which were stored at 30°C for 1 week	The emulsions stabilized with quinoa protein were less stable against physical aggregation and lipid oxidation than those stabilized with Tween20 and those stabilized with amaranth protein. Cooxidation of the protein and lipids in the emulsions were obtained	Gürbüz et al. (2018)
Emulsion stabilizer	The interfacial properties of quinoa protein concentrate affected by the pH (9 or 11) of extraction solution were studied	The protein extracted using pH 11 formed more sustained films on the interface and smaller oil droplets of O/W emulsions. The protein extracted using pH 9 gave more compact interfacial structure and lower tension. Increasing protein concentration increased the emulsion stability and decreased oil droplet size	López-Castejón et al. (2020)

(*Continued*)

Table 13.1 (Continued)

Application	Experimental	Major findings	Reference
Emulsion stabilizer	Quinoa protein nanoparticles were produced from quinoa protein isolate using ultrasound treatment (100 W, 20 min) at different NaCl concentrations (0 – 500 mM). The protein nanoparticles were used to stabilize Pickering emulsions which were subjected to freeze-thaw cycles	Increasing NaCl concentration increased the hydrophobic interactions and the percentage of β-sheet units of the resulting nanoparticles and increased freeze-thaw stability of the emulsions stabilized by these nanoparticles. The NaCl-treated nanoparticles gave higher stability of the Pickering emulsions than those without NaCl treatment	Qin et al. (2018)
Emulsion stabilizer	High internal phase emulsions were stabilized using quinoa protein nanoparticles formed using ultrasound	Increasing ultrasonic density decreased the size of the protein nanoparticles (down to 130 nm), which were spherical. The high internal phase emulsions stabilized by the quinoa protein nanoparticles were stable against oxidation and aggregation, and had internal phases at up to 89% (by volume). Smaller protein nanoparticles gave higher shear thinning and elastic behaviors of the emulsions (Fig. 13.2)	Zhang et al. (2021a)
Emulsion stabilizer	Quinoa protein isolate was subjected to high pressure homogenization. The resulting protein samples (up to 5%) were used to stabilize high internal phase emulsions	Viscoelastic gel-like property was obtained for the high internal phase emulsions. Increasing protein concentration in the emulsions significantly increased the elastic modulus and decreased the oil droplet size. The emulsions had high stability against oil coalescence, phase separation, and creaming under heating (80°C, 0.5 h) or storage (30 days). Heating significantly increased the strength of the emulsion gels. The storage increased the elastic modulus without affecting the oil droplet size	Zhang et al. (2021b)

Emulsion stabilizer	Quinoa protein isolate was subjected to ultrasound (360 W, 20 kHz) at 5.0 kJ/mL. The sonicated protein was used to stabilize high internal phase emulsions at different pH (3–9). The physicochemical properties of the emulsions were studied	High internal phase emulsions had smaller droplets, higher stability, and smoother appearance by the sonicated protein at pH 7 and 9 than at pH 3 and 5. Increasing concentration of protein and decreasing pH increased the solid component of the emulsions. At pH 3 and 5, the proteins aggregated to fill the space between droplets, whereas at pH 7 and 9, the proteins tended to be distributed at oil-water interface	Zuo et al. (2022)
Emulsion stabilizer	Plant protein-based emulsions were formed using mixtures of lentil and quinoa protein at ratios of 40:60 and 0:100 and homogenization. Quinoa protein was in the form of flour rich in protein (33%)	Combination of lentil and quinoa proteins increased the physical stability of the protein-based emulsions with more balanced amino acid profiles	Alonso-Miravall-es et al. (2022)
Film	Coatings of thymol-containing nanoemulsion, quinoa protein, and chitosan were created to coat strawberries (5°C, up to 16 days)	The sensory quality of aroma and flavor of the coated strawberries was negatively affected initially as compared to the control. After 5 days of storage, the sensory quality of the coated strawberries became similar to that of the control. After 12 days of storage, the aroma of the coated strawberries was higher than that of the control. The coating with the thymol-containing nanoemulsion, quinoa protein, and chitosan increased the microbiological shelf life, reduced the weight loss of the strawberries without affecting their content of soluble solids, pH, and titrable acidity	Robledo et al. (2018)

(Continued)

Table 13.1 (Continued)

Application	Experimental	Major findings	Reference
Film	Edible films were made of quinoa protein, chitosan and chitosan thymol nanoparticles via ionic gelation (Fig. 13.3). They were used for the packaging of blueberries and tomatoes during storage	The addition of nanoparticles decreased the water vapor permeability of the films, and the effects of the nanoparticles were more efficient than those of sunflower oil. The application of the antimicrobial films as coating for the fruits decreased the weight loss during storage	Medina et al. (2019)
Film	Chitosan and quinoa protein (ratio of 1:10) mixtures were treated with transglutaminase to form suspensions with film-forming property. Thyme or rosemary essential oils (up to 2%) were added to the suspensions for antimicrobial applications	For inhibition on *Salmonella* sp., minimum inhibitory concentrations of rosemary and thyme essential oils were 1% and 0.5%, respectively. The addition of the essential oils increased the particle sizes of the suspensions	Escamilla-García et al. (2020)
Film	Edible films were made of quinoa protein, chitosan-thymol nanoparticles and chitosan. The combined effect of transglutaminase and high-intensity ultrasound on the film properties was studied	The application of transglutaminase combined with high-intensity ultrasound increased the thermal stability and water vapor permeability, tensile strength, and thickness of the composite films, while decreasing the percentage of elongation at break. Ultrasound treatment alone increased the mechanical tensile strength of the films	Vera et al. (2020)
Film	Quinoa protein isolate (up to 80%) in combination with cellulose nanocrystals (up to 7%) and starch (up to 30%) were formulated into films	Adding the nanocrystals decreased water vapor permeability. Adding starch and nanocrystals enhanced the mechanical properties of the films. Decreasing the amount of the nanocrystals added increased the film transparency. The films with the lowest water solubility and water vapor permeability and the highest tensile strength and transparency were obtained with 3.2% nanocrystals, 18% starch, and 79% protein	Zade et al. (2021)

Film	Quinoa protein-based films were fortified with gum extracts of *Mesona procumbens* (hsian-tsao) by up to 50% (glycerol (20%) as plasticizer)	Electrostatic interactions and hydrogen bonding between quinoa protein and gum extracts contributed to the system compatibility. Films with 10% gum had more compacted structure with improved tensile strength, water barrier, and elongation at break. The films became cracked on surface and decreased in functionalities with more gum addition. Gum addition significantly reduced the oxidative rancidity of vegetable oils by increasing the antioxidant activities	Zhao et al. (2022)

(Vilcacundo et al., 2017). In the study of Guo et al. (2021), spontaneously hypertensive rats were fed with a diet containing quinoa protein isolate for 5 weeks. Upon gastrointestinal hydrolysis, peptide precursors were released. The quinoa protein intake decreased the blood pressure, while changing the composition of gut microbiome in the hypertensive rats to be more similar to that of normal rats. The quinoa protein intake lowered blood pressure, while reducing gut dysbiosis induced by hypertension (Guo et al., 2021). Pompeu et al. (2022) tested the antiinflammatory activity of chenopodin in vivo using mouse model. Mice were treated with carrageenan to induce neutrophil recruitment and paw edema. The mice were also fed with a diet containing chenopodin isolated from quinoa seeds. The chenopodin intake decreased the neutrophil recruitment and paw edema in the mice, being antiinflammatory (Pompeu et al., 2022). Quinoa protein isolate was hydrolyzed using pancreatin. Emulsifying and in vitro antioxidant activity of the protein hydrolysates were studied (Daliri et al., 2021). A degree of hydrolysis of 19% was obtained after 3 h of hydrolysis. The protein concentrate had lower in vitro antioxidant, foaming, and emulsifying capacities than the protein hydrolysate, whereas the latter had lower stabilizing capacity for the foams and emulsions (Daliri et al., 2021).

13.2.3 Encapsulation of bioactive compounds

Food bioactive compounds may have limited bioavailability and need to be encapsulated for controlled release and targeted delivery (Liu et al., 2021; Chen et al., 2021). Liu et al. (2021) studied the interactions between flavonoids (quercetin, resveratrol, curcumin, and luteolin) and nanomicelles of quinoa protein for encapsulation. The flavonoids and the protein formed assembly through hydrophobic interactions and hydrogen bonding. The encapsulation into the nanomicelles increased the stability and solubility of the flavonoids (Liu et al., 2021). Chen et al. (2021) combined quinoa protein and gum Arabic to form coacervates that were used to encapsulate eugenol (Fig. 13.1). The microcapsules were freeze-dried and used to preserve minced pork. The coacervates were optimally formed at pH 4 with a gum to protein ratio of 1:4 via electrostatic interactions to have a mean size of 30 μm and an encapsulation efficiency of 89%. Eugenol was encapsulated in the coacervates via hydrogen bond. When applied in minced pork, the microcapsules reduced the increases in total bacterial counts, total volatile basic nitrogen level, and pH compared to pure eugenol. The application of the eugenol-containing microcapsules increased the shelf life of minced pork (4°C, 9 days) (Chen et al., 2021).

13.2.4 Emulsion stabilizers

Quinoa proteins in different forms were used to stabilize different emulsions (Gürbüz et al., 2018; López-Castejón et al., 2020; Qin et al., 2018; Zhang et al., 2021a,b; Zuo et al., 2022; Alonso-Miravalles et al., 2022). For example, Gürbüz et al. (2018) obtained soluble proteins from quinoa seeds using alkaline solubilization and acid precipitation. The soluble proteins were used to stabilize oil

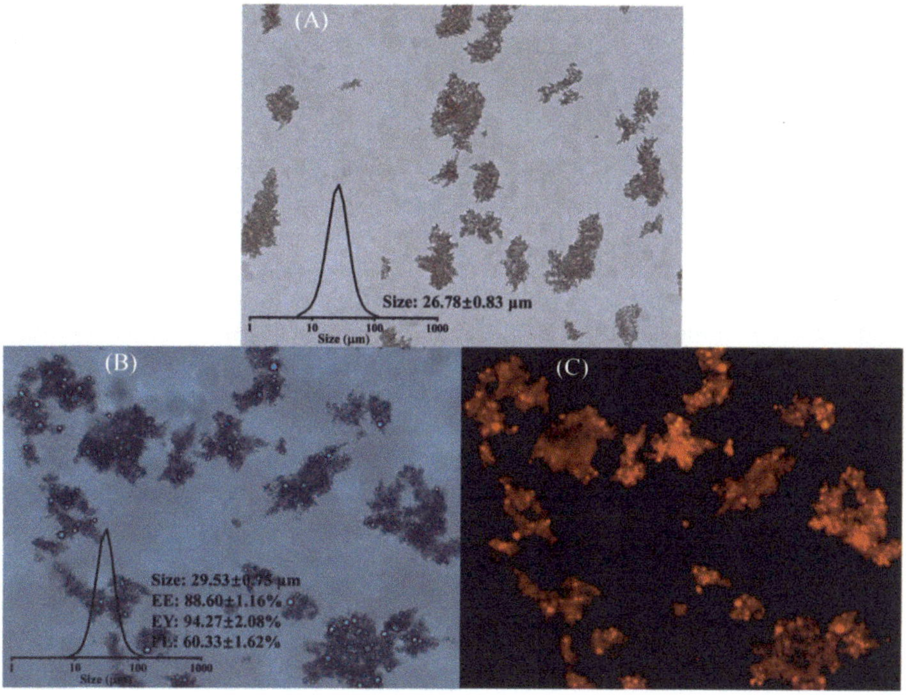

Figure 13.1 Inverted fluorescence microscopic images of quinoa protein-gum Arabic coacervates, eugenol encapsulated microcapsules at the magnification of $\times 200$ and their size distribution, encapsulation yield (EY), encapsulation efficiency (EE), and payload (PL) (inserted). Quinoa protein-gum Arabic coacervates under bright field (A); eugenol encapsulated microcapsules under bright field (B); eugenol encapsulated microcapsules under fluorescence field (C) (Chen et al., 2021).
Source: Reprinted with permission from Elsevier.

(rapeseeds)-in-water emulsions that were stored at 30°C for 1 week. The emulsions stabilized with the quinoa protein were less stable against physical aggregation and lipid oxidation than those stabilized with Tween20 and also those stabilized with amaranth protein. Cooxidation of the protein and lipids in the emulsions was obtained (Gürbüz et al., 2018). Quinoa protein nanoparticles were produced from quinoa protein isolate using ultrasound treatment (100 W, 20 min) at different NaCl concentrations (0 – 500 mM) (Qin et al., 2018). The protein nanoparticles were used to stabilize Pickering emulsions that were subjected to freeze-thaw cycles. Increasing NaCl concentration increased the hydrophobic interactions, the percentage of β-sheet units of the resulting nanoparticles, and the freeze-thaw stability of the emulsions stabilized by these nanoparticles. The NaCl-treated nanoparticles gave higher stability of the Pickering emulsions than those without NaCl treatment (Qin et al., 2018). The internal phase volume fractions of high internal phase emulsions were >74% (the close packing limit) (Gao et al., 2021). These emulsions were semisolid. In the study of Zhang et al. (2021b), quinoa protein isolate was subjected to high-pressure homogenization. The resulting protein samples (up to 5%) were used to stabilize high internal phase emulsions. Viscoelastic gel-like property was obtained for the high internal phase emulsions. Increasing protein concentration in the emulsions significantly increased the elastic modulus and decreased the oil droplet size. The emulsions had high stability against oil coalescence, phase separation, and creaming under heating (80°C, 0.5 h) or storage (30 days). Heating significantly increased the strength of the emulsion gels. The storage increased the elastic modulus without affecting the oil droplet size (Zhang et al., 2021b). High internal phase emulsions may be used as replacement of partially hydrogenated oils for healthy product formulations (Zhang et al., 2021a). In the study of Zhang et al. (2021a), high internal phase emulsions were stabilized using quinoa protein nanoparticles formed via ultrasound processing. Increasing ultrasonic density decreased the size of the protein nanoparticles (down to 130 nm), which were spherical in shape. The high internal phase emulsions stabilized by the quinoa protein nanoparticles were stable against oxidation and aggregation and had internal phases at up to 89% (by volume). Smaller protein nanoparticles gave the emulsions higher shear-thinning and elastic behaviors (Fig. 13.2) (Zhang et al., 2021a). Similarly, sonicated quinoa protein isolate was used to stabilize high internal phase emulsions at different pH (3–9) (Zuo et al., 2022). The high internal phase emulsions had smaller droplets, higher stability, and smoother appearance with the sonicated protein at pH 7 and 9 than at pH 3 and 5. Increasing concentration of protein and decreasing pH increased the solid component of the emulsions. At pH 3 and 5, the proteins aggregated to fill the space between the oil droplets, whereas at pH 7 and 9, the proteins tended to be distributed at the oil-water interface (Zuo et al., 2022). Alonso-Miravalles et al. (2022) found that a combination of lentil and quinoa proteins increased the physical stability of the protein-based emulsions with more balanced amino acid profiles.

13.2.5 Films and coatings

Quinoa protein was used to formulate edible films and coatings for food packaging applications (Robledo et al., 2018; Medina et al., 2019; Escamilla-García et al., 2020; Vera et al., 2020; Zade et al., 2021; Zhao et al., 2022). For example, chitosan and quinoa protein (ratio of 1:10) mixtures were treated with transglutaminase to form suspensions with film-forming property (Escamilla-García et al., 2020). Thyme or rosemary essential oils (up to 2%) were added to the suspensions for antimicrobial applications. For inhibition on *Salmonella* sp., the minimum inhibitory concentrations of rosemary and thyme essential oils were 1% and 0.5%, respectively. The addition of the essential oils increased the particle sizes of the suspensions (Escamilla-García et al., 2020). In the study of Medina et al. (2019), edible films were made of quinoa protein, chitosan, and chitosan-thymol nanoparticles via ionic gelation (Fig. 13.3). They were used for the packaging of blueberries and tomatoes during storage. The addition of the nanoparticles decreased the water vapor permeability of the films, and the effects of the nanoparticles were more

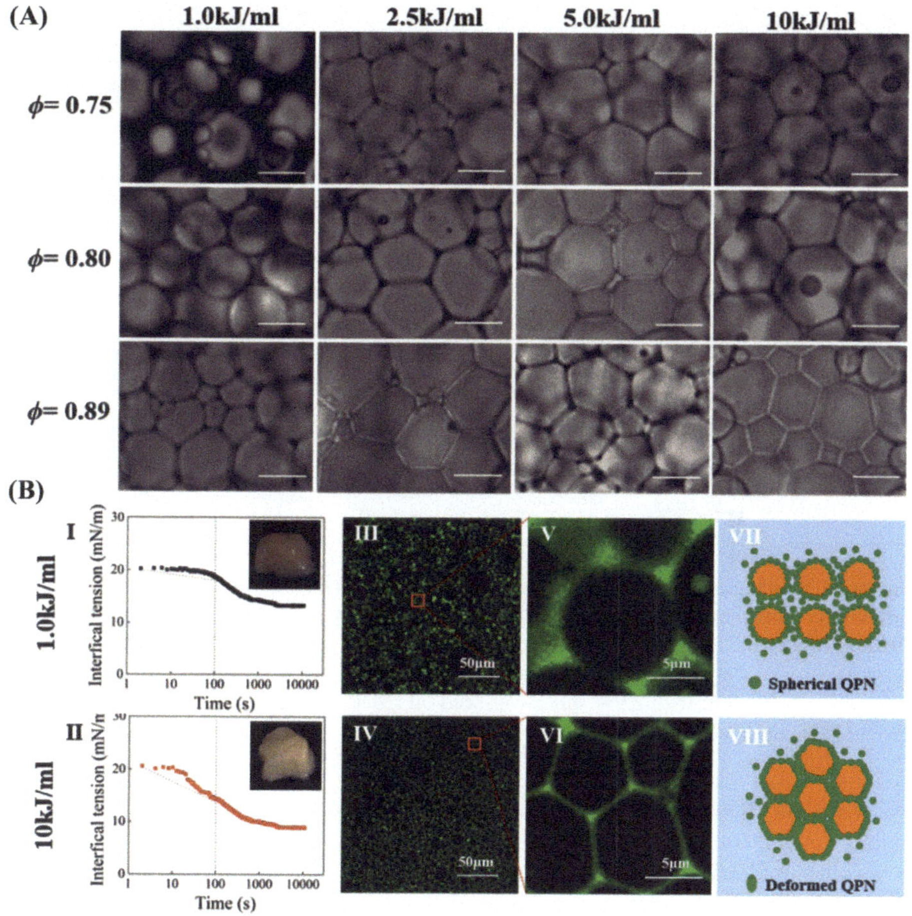

Figure 13.2 Microstructure of high internal phase emulsions stabilized by quinoa protein nanoparticles. (A) Optical microscopic images of fresh high internal phase emulsions stabilized by quinoa protein nanoparticles at varying φ values in the range of 0.6–0.9. The emulsion pictures were taken immediately after preparation. The quinoa protein nanoparticles were prepared with different ultrasonic densities (1.0, 2.5, 5.0, and 10 kJ/mL). The scale bar is 10 μm. (B, I–II) Dynamic interfacial tension (Inset: visual appearance) of fresh high internal phase emulsions ($\varphi = 0.8$) stabilized by quinoa protein nanoparticles (c = 4.0%); (B, III–VI) Confocal laser scanning microscopy of fresh high internal phase emulsions ($\varphi = 0.8$) stabilized by quinoa protein nanoparticles (c = 4.0%). (B, VII–VIII) Schematic illustration for the microstructure of high internal phase emulsions stabilized by quinoa protein nanoparticles (Zhang et al., 2021a).
Source: Reprinted with permission from Elsevier.

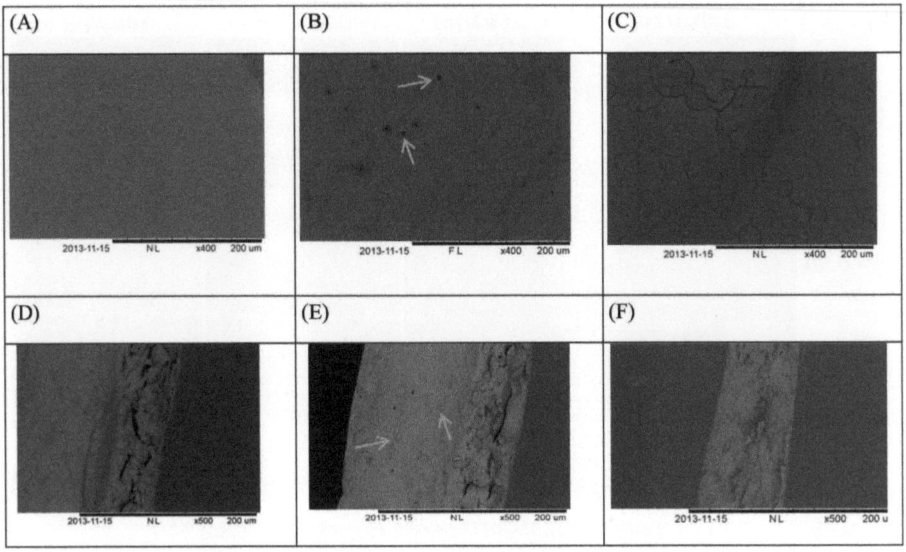

Figure 13.3 Scanning electron micrographs of surfaces of chitosan-quinoa protein films: (A) film without chitosan nanoparticles, (B) film with 5% chitosan nanoparticles, (C) film with 2.9% sunflower seed oil. Scanning electron micrographs of cross section of quinoa-chitosan blend films: (D) film without chitosan nanoparticles, (E) film with 5% chitosan nanoparticles, (F) film with 2.9% sunflower seed oil (Medina et al., 2019).
Source: Reprinted with permission from Elsevier.

efficient than those of sunflower oil. The application of the antimicrobial films as coating for the fruit decreased the weight loss during storage (Medina et al., 2019). In the study of Zade et al. (2021), quinoa protein isolate (up to 80%) in combination with cellulose nanocrystals (up to 7%) and starch (up to 30%) were formulated into films. Adding the nanocrystals decreased water vapor permeability. Adding starch and nanocrystals enhanced the mechanical properties of the films. Decreasing the amount of the nanocrystals added increased the film transparency. The films with the lowest water solubility and water vapor permeability and the highest tensile strength and transparency were obtained with 3.2% nanocrystals, 18% starch, and 79% protein (Zade et al., 2021). Plant extracts are becoming a popular ingredient for the fortification of films and coatings (Zhu, 2021). Quinoa protein-based films were fortified with gum extracts of *Mesona procumbens* (hsian-tsao or xian-cao) by up to 50% with the addition of glycerol (20%) as plasticizer (Zhao et al., 2022). Electrostatic interactions and hydrogen bonding between the quinoa protein and the gum extracts contributed to the system compatibility. The films with 10% gum had more compacted structure with improved tensile strength, water barrier, and elongation at break. The films became cracked on surface and decreased in functionalities with more gum addition. The gum addition significantly reduced the oxidative rancidity of vegetable oils by increasing the in vitro antioxidant activities (Zhao et al., 2022) (Table 13.1).

13.3 Food and nonfood applications of quinoa starch

Quinoa starch has special properties partially due to its small granule size (Lindeboom et al., 2004; Li & Zhu, 2018). Previous studies on quinoa starch dated before 2018 were focused on formulations of films and coatings and also on Pickering emulsions (Li & Zhu, 2018). The majority of recent studies from the last few years have continued to focus on the film and coating formulations, encapsulation, and formation of emulsions (Table 13.2).

13.3.1 Films and coatings

Many polysaccharides including starch have film-forming property (Zhu, 2021). Starches of different quinoa varieties were made into edible films and coatings. The films were compared with those based on potato, lentil, and pumpkin (Fig. 13.4) (Pająk et al., 2019). Quinoa starch-based films showed different physicochemical properties compared to those based on other starches. Variations in the properties of the films made from the starches of different quinoa varieties were obtained. Quinoa starch-based films had higher water vapor permeability than potato starch-based films. They also had low percentages of elongation at break. This showed that the functional performance of these films remains to be improved (Pająk et al., 2019). In the study of Alak et al. (2019), quinoa starch and lemon or sage essential oils (2%) were formulated into edible films that were used to coat rainbow trout fillets for packaging application and storage (15 days, 4°C) with a degree of success.

13.3.2 Applications in meat products

Quinoa starch, corn starch (2.5%), or quinoa seeds (10%) were formulated into chicken meatballs that were subjected to five freeze (−18°C)–thaw (4°C) cycles (Park et al., 2021). Adding quinoa or corn starch decreased the reheating, cooking and drip losses during the freeze-thaw cycles in the meatballs, while reducing the hardness and chewiness by ~50%. Adding starch and quinoa seeds reduced the values of thiobarbituric acid reactive substance and total volatile basic nitrogen. Adding quinoa starch and quinoa seeds increased the sensory scores of overall acceptability, texture, and taste (Park et al., 2021). It may be expected that other meat products such as pork-based processed foods can be formulated with quinoa starch for improved quality. The potential of quinoa protein in the formulation of meat analogues and vegan meat products remains to be tested.

13.3.3 Production of nanoparticles

Quinoa starch was hydrolyzed using H_2SO_4 at 30°C − 40°C and waxy maize starch at 40°C to make nanoparticles (Velásquez-Castillo et al., 2020). Increasing temperature increased the hydrolysis extents of the quinoa starch from 63% at 30°C to 91% at 40°C (on the fifth day). Increasing temperature decreased the thermal stability of the nanoparticles, while decreasing the zeta potential. The hydrolysis extent of

Table 13.2 Food and nonfood applications of quinoa starch.

Application	Experimental	Major findings	Reference
Film	Starches of quinoa different varieties were made into edible film. The films were compared with that based on potato, lentil and pumpkin (Fig. 13.4)	Quinoa starch-based films showed different physicochemical properties compared to those based on other starches. Variations in the properties of the films made from the starches of different quinoa varieties were obtained. Quinoa starch-based films had higher water vapor permeability than potato starch-based films. They also had lower percentages of elongation at break. This indicated that the functional performance of these films remains to be improved	Pająk et al. (2019)
Film	Quinoa starch and lemon or sage essential oils (2%) were formulated into edible films that were used to coat rainbow trout fillets for packaging application and storage (15 days, 4°C)	The film of quinoa starch with sage essential oils was more antimicrobial in the protection of the fish fillets than that with lemon essential oils. The films of quinoa starch plus lemon essential oils were the most effective in the prevention against the lipid oxidation of the fish fillets	Alak et al. (2019)
Chicken meatball ingredient	Quinoa starch, corn starch (2.5%) or quinoa seeds (10%) were formulated into chicken meatballs which were subjected to 5 freeze (−18°C)–thaw (4°C) cycles	Adding quinoa or corn starch decreased the input of reheating and drip losses during the freeze-thaw cycles in the meatballs, while reducing the hardness and chewiness by ~50%. Adding starch and quinoa seeds reduced the values of thiobarbituric acid reactive substance and total volatile basic nitrogen. Adding quinoa starch and quinoa seeds increased the sensory scores of overall acceptability, texture, and taste	Park et al. (2021)

Nanoparticle production	Quinoa starch was hydrolyzed using H_2SO_4 at 30°C – 40°C and waxy maize starch at 40°C to make nanoparticles	Increasing temperature increased the hydrolysis extents of quinoa starch from 63% at 30°C to 91% at 40°C (fifth day). Increasing temperature decreased the thermal stability of the nanoparticles, while decreasing the zeta potential. The hydrolysis extent of quinoa was higher than that of the maize starch. The particles produced at 40°C and 35°C were nanosized and conical-shaped, whereas those at 30°C were microsized. The yield of waxy maize starch nanoparticles was higher than that of quinoa starch nanoparticles, whereas the size of the former was larger than that of the latter	Velásquez-Castillo et al. (2020)
Encapsulation	Piroxicam was encapsulated into quinoa starch nanoparticles using nanoprecipitation technique	The particle size of the nanoparticles ranged from 282–870 nm, depending on the concentrations of the starch and piroxicam. A total of 99% of the drug was released in 2 h of dissolution. The drug encapsulated in the nanoparticles had higher in vitro antiinflammatory activity than pure piroxicam	Bhatia and Rohilla (2020)
Encapsulation	Quinoa and maize starch nanoparticles were produced using antisolvent method to encapsulate rutin	The starch nanoparticles had sizes of 107 (quinoa) and 222 (maize) nm and loading efficiency of 27 (quinoa) and 23% (maize). The encapsulated samples were subjected to in vitro digestion. The rutin bioavailability was higher in quinoa starch-based system than in maize-based one	Remanan and Zhu (2021)
Encapsulation	Quinoa and maize starch nanoparticles were produced using antisolvent method to encapsulate quercetin	Full factorial design was used to produce the smallest particle size of the nanoparticiples by optimizing the solvent composition and ultrasound treatment conditions. Maize starch nanoparticles were larger with lower loading capacity than quinoa starch nanoparticles (166 nm of size and 26% of loading capacity). The former was less protective on the encapsulated quercetin than the latter	Jiang et al. (2022)

(*Continued*)

Table 13.2 (Continued)

Application	Experimental	Major findings	Reference
Emulsion	Quinoa starch granules were treated with α-amylase up to 9 h to obtain smaller starch particles which were used to stabilize Pickering emulsions	Increasing hydrolysis time decreased the particle size of the starch, which led to increased contact angle and decreased droplet size of oil in the emulsions. The highest emulsion capacity of the starch was obtained after hydrolysis of 9 h. The resulting starch particles had lower emulsifying capacity at an oil volume fraction of 20% than at 50%	Zhang et al. (2021)
Emulsion	Quinoa starch granules were modified using octenyl succinic anhydride. They were used to stabilize Pickering emulsions. The effects of oil volumefraction and degree of substitution on the properties of the emulsions were studied	Octenylsuccinylation enhanced the contact angle of the droplets from 36 to 69. Pickering emulsion gels were obtained with the starch octenylsuccinate at a degree of substitution of 0.0286. The oil volume fractions varied from 50% to 70%. Increasing DS or oil volume fraction of the starch enhanced storage modulus and apparent viscosity of the emulsion gels	Li et al. (2019a)
Emulsion	Quinoa starch granules were modified with dodecenyl succinic anhydride (DDSA) to various degrees of substitution (0.0023–0.0095). The modified starch was used to stabilize Pickering emulsions	Increasing degrees of substitution of the DDSA-modified starch reduced the size of the oil droplets in the emulsions stabilized by the starch and increased the capacity in stabilizing the Pickering emulsions during storage (10 days)	Li et al. (2019b)
Emulsion	Octenylsuccinate quinoa starch was prepared and used to stabilize Pickering emulsion gels containing lutein at varying oil volume fractions (30%–60%)	Increasing oil volume fractions increased the gel-like feature of the emulsions by enhancing the apparent viscosity, storage modulus (G'), and droplet size through aggregation. The retention index of lutein was 55% with the half-life increased to 41 days (from 21 days) after 31 days of storage	Li et al. (2020)

Emulsion	Quinoa starch granules were modified by octenyl succinic anhydride. The modified starch was used (200 and 600 mg/mL oil) to stabilize oil-in-water Pickering emulsions which were made using a high-pressure homogenizer and were then stored at 6°C for 8 years	The Pickering emulsions were stable over 8 years of storage. Increasing starch concentrations decreased the size of droplets. Dissociation of aggregated droplets and the subsequent release of trapped starch granules resulted in decreased droplet size during the storage	Marefati and Rayner (2020)
Emulsion	Quinoa starch granules were modified with nonenyl succinic anhydride (NSA) to various degrees of substitution (0.008–0.0548). The modified starch was used to stabilize Pickering emulsions	The NSA-modified starch with medium degrees of substitution gave the most stable Pickering emulsions. The size of oil droplets of the emulsions decreased before increasing with increasing degrees of NSA substitution of quinoa starch granules. The NSA-modified starch could be an alternative to the commonly used octenyl succinic anhydride-modified starch	Li and Zhu (2021)
Emulsion and encapsulation	OSA-modified (0.6%) quinoa starch granules were used to stabilize Pickering emulsions to encapsulate curcumin. The emulsions were heated and subjected to simulated digestion	The encapsulation efficiency of curcumin in the emulsions was 80%. Heating the emulsions increased the stability of the encapsulation during storage (24 h) and in vitro oral and intestinal digestion process, but had no effect on that during in vitro gastric digestion. The presence of bile salts decreased the physical stability of the emulsions during the in vitro digestion process	Marefati et al. (2017)

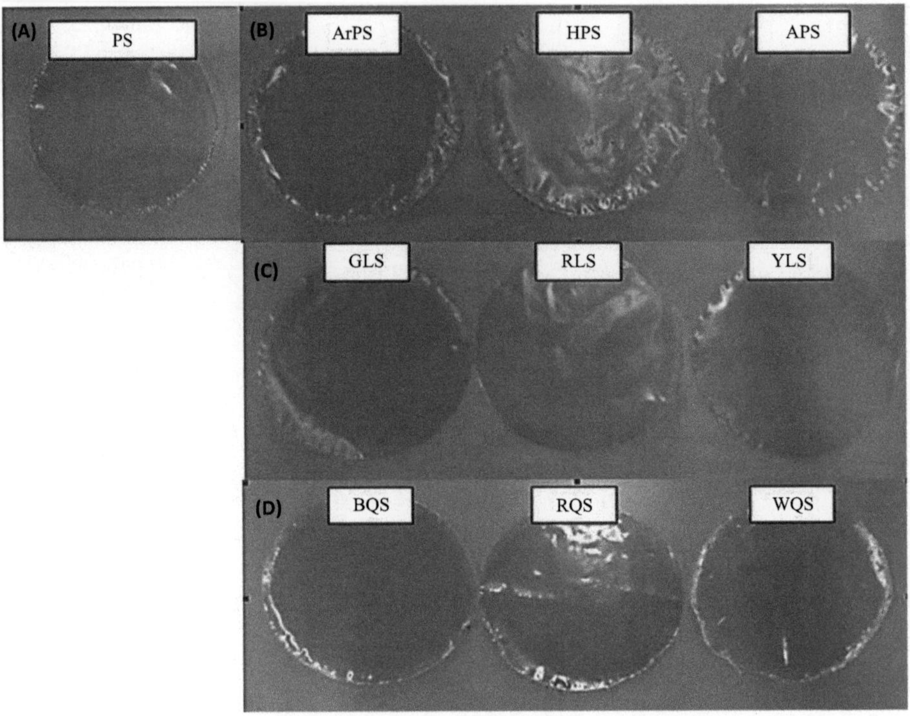

Figure 13.4 Appearance of edible films based on different starches and starch of different varieties: (A) potato starch—PS, (B) pumpkin starch (Ambar—ArPS, Hokkaido—HPS and Amazonka—APS), (C) lentil starch (green—GLS, red—RLS and yellow—YLS), (D) quinoa starch (black—BQS, red—RQS and white—WQS) (Pająk et al., 2019).
Source: Reprinted with permission from Elsevier.

quinoa starch was higher than that of the maize starch. The particles produced at 40°C and 35°C were nanosized and conical-shaped, whereas those at 30°C were microsized. The yield of waxy maize starch nanoparticles was higher than that of quinoa starch nanoparticles, whereas the size of the former was larger than that of the latter (Velásquez-Castillo et al., 2020). Starch nanoparticles have different applications such as in the formulations of edible films and coatings. Quinoa starch nanoparticles remain to be used for different food and nonfood applications.

13.3.4 Encapsulation of bioactive compounds

Food bioactive compounds may have limited bioavailability and need to be encapsulated for controlled release and targeted delivery (Bhatia & Rohilla, 2020; Li et al., 2020; Remanan & Zhu, 2021; Jiang et al., 2022). In the study of Remanan and Zhu (2021), quinoa and maize starch nanoparticles were produced using anti-solvent method to encapsulate rutin (a flavonoid). The starch nanoparticles had average sizes of 107 (quinoa) and 222 (maize) nm and loading efficiency of 27 (quinoa) and 23% (maize). The encapsulated samples were subjected to in vitro digestion. The rutin bioavailability was higher in quinoa starch-based system than in maize starch-based one (Remanan & Zhu, 2021). Similarly, quinoa and maize starch nanoparticles were produced using antisolvent method to encapsulate quercetin by Jiang et al. (2022). Full factorial design was used to produce the smallest particle size of the nanoparticiples by optimizing the solvent composition and ultrasound treatment conditions. The maize starch nanoparticles were larger with lower loading capacity than the quinoa starch nanoparticles (166 nm of size and 26% of loading capacity). The former was less protective on the encapsulated quercetin than the latter (Jiang et al., 2022). Octenylsuccinate quinoa starch was prepared and used to stabilize Pickering emulsion gels containing lutein at varying oil volume fractions (30%–60%) (Li et al., 2020) (Fig. 13.5). Increasing oil volume fractions increased the gel-like feature of the emulsions by enhancing the apparent viscosity, storage modulus (G'), and droplet size through aggregation. The retention

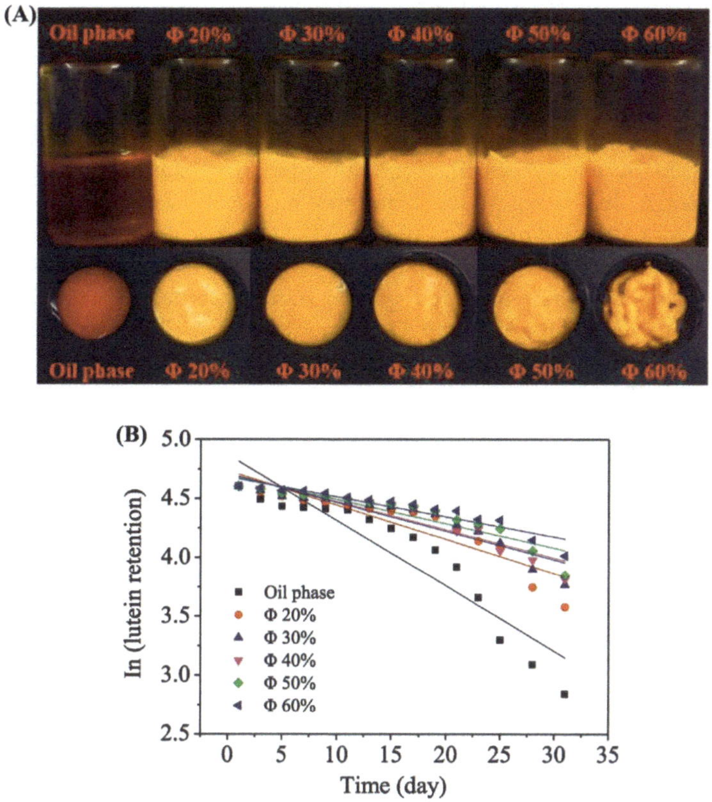

Figure 13.5 Photographs (A) and storage stability (B) of lutein-loaded oil phase and Pickering emulsions stabilized by octenylsuccinate quinoa starch granules as a function of Φ (oil volume fraction) (Li et al., 2020).
Source: Reprinted with permission from Elsevier.

index of lutein was 55% with the half-life increased to 41 days (from 21 days) after 31 days of storage (Li et al., 2020). In vivo studies remain to be done to confirm these results obtained using in vitro studies.

13.3.5 Emulsions

Studies have been focused on the preparation of Pickering emulsions using quinoa starch-based particles (Zhang et al., 2021; Li et al., 2019a,b, 2020; Marefati & Rayner, 2020; Li & Zhu, 2021; Marefati et al., 2017). For example, novel modified quinoa starches including those modified with dodecenyl succinic anhydride (DDSA) and nonenyl succinic anhydride (NSA) were obtained at various degrees of substitution (Li & Zhu, 2021; Li et al., 2019b). Increasing degrees of substitution of the DDSA-modified starch reduced the size of the oil droplets in the emulsions stabilized by the starch and increased the capacity in stabilizing the Pickering emulsions during storage (10 days) (Li et al., 2019b). The NSA-modified starch with medium degrees of substitution gave the most stable Pickering emulsions. The size of oil droplets of the emulsions decreased before increasing with increasing degrees of NSA substitution of quinoa starch granules. The NSA-modified starch could be an alternative to the commonly used octenyl succinic anhydride (OSA)-modified starch (Li et al., 2022; Li & Zhu, 2021). Marefati and Rayner (2020) showed that the Pickering emulsions stabilized by octenyl succinic anhydride-modified starch were stable over 8 years of storage. Increasing starch concentrations decreased the size of droplets. Dissociation of aggregated droplets and the subsequent release of trapped starch granules resulted in decreased droplet size during the storage (Marefati & Rayner, 2020). Octenyl succinic anhydride-modified quinoa starch stabilized Pickering emulsions were used to encapsulate curcumin (Marefati et al., 2017). The emulsions were heated and subjected to simulated digestion. The encapsulation efficiency of curcumin in the emulsions was 80%. Heating the emulsions increased the stability of the encapsulation during storage (24 h) and in vitro oral and intestinal digestion process, but had no effect on that during in vitro gastric digestion. The presence of bile salts decreased the physical stability of the emulsions during the in vitro digestion process (Marefati et al., 2017).

13.4 Food, nutraceutical, and nonfood applications of quinoa dietary fiber and nonstarch polysaccharides

Zhu (2020) summarized the literature published up to 2020 on biological activities and food uses of quinoa dietary fiber polysaccharides. Fiber polysaccharide-rich fractions of quinoa seeds showed diverse health effects such as antioxidant, anticancer, and immunomodulatory activities. The fiber fraction was incorporated into bakery and meat-based products for improved quality (Zhu, 2020). In recent years, fiber polysaccharide-rich fractions of quinoa seeds were used as functional ingredients for food and nutraceutical formulations (Table 13.3).

Table 13.3 Applications of quinoa dietary fiber and nonstarch polysaccharides.

Application	Experimental	Major findings	Reference
Food and nutraceutical applications			
Meat based product	The review article summarized related papers on the use of different quinoa-based ingredients including fiber-rich fraction on meat product formulations	Different quinoa-based ingredients including fiber-rich fractions were feasible in the applications and formulations of meat-based products. Quinoa-based ingredients acted as binder, extender, gelling agent, fat replacer, or filler to improve the technical and nutritional properties of the meat-based products	Fernández-López et al. (2021)
Bakery ingredient	Fiber-rich fractions were obtained from quinoa seeds using dry and wet milling methods. The fiber fractions were added to bread formulations	The addition of the fiber fraction from the dry-milling process gave softer crumb and higher specific volume of the loaf. The addition of the fiber fractions into the breads enhanced in vitro antioxidant capacity and dietary fiber content. Wet-milling process gave higher fiber content (72%) of the fraction than dry-milling (59%)	Ballester-Sánchez et al. (2020)
Nutraceutical ingredient	Rats were fed with a high-fat diet and also soluble polysaccharide (up to 600 mg/kg/day) for 8 weeks. The soluble fiber fraction was tested for gut microbiota regulation and hypolipidemia	The levels/activities of serum low density lipoprotein cholesterol, total triglycerides, malondialdehyde, hepatic lipid accumulation, total glutamic pyruvic transaminase, and glutamic-oxaloacetic transaminase in the rats fed with high fat diet were decreased by the polysaccharide fraction. The high-fat diet disrupted the gut microbiota composition. The quinoa polysaccharide intake increased the microbial diversity, and decreased the percentages of *Proteobacteria*, *Desulfovibrio* and *Allobaculum* and the ratio of *Firmicutes* to *Bacteroides*	Cao et al. (2020)

(Continued)

Table 13.3 (Continued)

Application	Experimental	Major findings	Reference
Nutraceutical ingredient	Five fractions of polysaccharide extract were obtained using gradient ethanol precipitation and tested for the inhibitory activities against DPPH and ABTS free radicals and α-amylase and α-glucosidase	The five fractions of polysaccharide extract showed antioxidant and antidiabetic activities to different extents	Tan et al. (2020)
Nutraceutical ingredient	Three fractions of polysaccharide extract were obtained using microfiltration and ultrafiltration. They were tested for the inhibitory activities against DPPH and ABTS radicals and α-amylase and α-glucosidase and antiinflammatory activity using RAW264.7 cells	The three fractions of polysaccharide extract showed antioxidant and antidiabetic activities to different extents in a dose-dependent manner. They also showed immune-regulatory activity	Tan et al. (2021b)
Nutraceutical ingredient	Human fecal microbiota was used to ferment raw, cooked quinoa or quinoa polysaccharide extract after in vitro human digestion	Anaerobic fermentation of the samples for 24 h increased contents of short-chain fatty acids and decreased the pH, while increasing the amounts of *Bacteroides* and *Prevotella* species. In particular, quinoa polysaccharide fraction increased the amounts of *Collinsella* and *Bifidobacterium* species with prebiotic function	Zeyneb et al. (2021)
Nutraceutical ingredient	Polysaccharide fraction was isolated from quinoa seeds using column chromatography. The effect of the fraction on adipocyte (3T3-L1) differentiation was studied in vitro	The fraction inhibited 3T3-L1 differentiation. The expressions of peroxisome proliferator-activated receptor gamma, enhancer-binding protein α, β and δ, sterol regulatory element-binding protein-1c and antiadipocyte protein 2 were inhibited by the polysaccharide fraction. The polysaccharide fraction had antiobese function	Teng et al. (2020)

Nutraceutical ingredient	Polysaccharide fraction was isolated from quinoa seeds using alkaline solution and column chromatography. The in vitro antioxidant activity of the fraction was studied using different chemical assays	The polysaccharide fraction showed antioxidant activities according to different antioxidant measurement assays (scavenging capacities against ABTS, DPPH, superoxide, and hydroxyl free radicals, and ferric reducing capacity)	Teng et al. (2021)
Nonfood applications			
Waste water treatment	Aerogels of quinoa polysaccharide and chitosan (ratio at 1:1) were synthesized using green method. The aerogel was used to adsorb methylene blue and Congo red in wastewater	The aerogels obtained from the green synthesis were loose, mesoporous hydrophobic, and flexible. The process and formulations of the aerogels were optimized to obtain maximum adsorption loads of methylene blue (58 mg/g) and Congo red (340 mg/g) in waste water	Tan et al. (2021a)

A review article of Fernández-López et al. (2021) summarized related papers on the use of different quinoa-based ingredients including fiber-rich fraction on meat product formulations. Quinoa-based ingredients acted as binder, extender, gelling agent, fat replacer, or filler to improve the technical and nutritional properties of the meat-based products (Fernández-López et al., 2021). Ballester-Sánchez et al. (2020) obtained fiber-rich fractions from quinoa seeds using dry and wet milling methods. The fiber fractions were added to bread formulations. The addition of the fiber fraction from the dry-milling process gave the loaf softer crumb and higher specific volume. The addition of the fiber fractions into the breads enhanced in vitro antioxidant capacity and dietary fiber content. The wet-milling process gave higher fiber content (72%) of the fraction than the dry-milling (59%) (Ballester-Sánchez et al., 2020).

Various biological functions of fiber polysaccharide-rich fractions of quinoa seeds confirmed their potential role as nutraceutical ingredient (Cao et al., 2020; Tan et al., 2020, 2021b; Zeyneb et al., 2021; Teng et al., 2021). For example, Tan et al. (2021b, 2020) obtained different fractions of polysaccharide extract from quinoa seeds using gradient ethanol precipitation or microfiltration and ultrafiltration. These fractions were tested for the inhibitory activities against 2,2-diphenyl-1-picrylhydrazyl (DPPH) and 2,2'-azino-bis(3-ethylbenzothiazoline-6-sulfonic acid) (ABTS) radicals, α-amylase and α-glucosidase, and inflammation in vitro. The fractions of polysaccharide extract showed antioxidant, antidiabetic, and immune-regulatory activities to different extents (Tan et al., 2021b, 2020). Zeyneb et al. (2021) used human fecal microbiota to ferment raw, cooked quinoa, or quinoa polysaccharide extracts after in vitro human digestion. The anaerobic fermentation of the samples for 24 h increased the contents of short-chain fatty acids and decreased the pH, while increasing the amounts of *Bacteroides* and *Prevotella* species. In particular, the quinoa polysaccharide fraction with prebiotic function increased the amounts of *Collinsella* and *Bifidobacterium* species (Zeyneb et al., 2021). Similarly, Cao et al. (2020) showed that soluble polysaccharide fraction of quinoa seeds had gut microbiota regulation and hypolipidemia properties using in vivo rat model. These studies confirmed that quinoa fiber polysaccharides have the potential to be used as nutraceutical ingredient for functional food development.

Nonfood applications of quinoa nonstarch polysaccharide were reported. Aerogels of quinoa polysaccharide and chitosan (ratio at 1:1) were synthesized using a green method (Tan et al., 2021a). The aerogel was used to adsorb methylene blue and Congo red in waste water. The aerogel obtained from the green synthesis was loose, mesoporous hydrophobic, and flexible. The process parameters and formulations of the aerogels were optimized to obtain maximum adsorption loads of methylene blue (58 mg/g) and Congo red (340 mg/g) in the wastewater (Tan et al., 2021a).

13.5 Food and nonfood applications of quinoa lipids

There have been relatively few studies on the application of lipid fractions from quinoa seeds. This could be due to the relatively small amount of the lipids in the seeds, though quinoa oils were studied as described in Chapter 7 of this book. Camillo-Andrade et al. (2020) subjected quinoa seed oils to transesterification to obtain bioesters. The effect of the bioesters on human reconstructed epidermis tridimensional cell cultures after exposing for up to 2 days was studied. Proteomics analysis showed that the bioesters prevented protection against oxidative damage and UV radiation, premature aging, tissue regeneration, and homeostasis. They had potential in skin care product formulations (Camillo-Andrade et al., 2020).

13.6 Food, nutraceutical and nonfood applications of quinoa saponin, polyphenol, and other extracts

The relative contents of saponins and polyphenols in quinoa seeds are low compared to that of starch, protein, and lipids. However, the functional and biological properties of these minor components render them useful in a range of food and nonfood applications (Table 13.4).

13.6.1 Food and nutraceutical applications of saponins

There have been a few reports on the food and nutraceutical applications of saponins from quinoa (Miranda et al., 2018; Del Hierro et al., 2021; Ceylan & Basturk, 2022). Atlantic chub mackerel (*Scomber colias*) was stored in ice containing quinoa ethanol extract for 13 days (Miranda et al., 2018). The presence of the quinoa extracts decreased lipid oxidation and hydrolysis in the mackerel. The inhibitory effects became more pronounced toward the end of the storage. The presence of quinoa extracts at a higher addition level (0.05 g/L icing solution) gave the mackerel acceptable sensory quality after the storage, whereas the control mackerel samples became unacceptable for sensory quality (Miranda et al., 2018). Saponin-rich extracts were tested for in vitro inhibitory activity on pancreatic lipase and hypocholesterolemia capacity (Del Hierro et al., 2021). The extracts from quinoa, uckun, propolis, or BHT (1000–1500 ppm) were added to palm olein oil for deep frying of potatoes (Ceylan & Basturk, 2022). The thermal oxidation properties of the palm oils were measured in terms of free fatty acidity, viscosity, total polar matter, peroxide value, total oxidation value, *p*-anisidine value, conjugated diene, and triene. The extracts of quinoa were not as effective as that of propolis to retard oil oxidation and the antioxidant effect of the latter was close to that of BHT (Ceylan & Basturk, 2022).

13.6.2 Nonfood applications of saponins

13.6.2.1 Cosmetics

Saponins from quinoa were tested in different aspects for potential applications in cosmetic formulations (Jurek et al., 2019; Góral & Wojciechowski, 2020;

Table 13.4 Food and nonfood applications of saponin, polyphenol and other extracts.

Applications	Experimental	Major findings	Reference
Saponins			
Different food and nonfood uses	This article reviewed the extraction, chemistry and biological activities of saponins from quinoa covering the literatures from the last three decades or so	The review of literature showed that saponins from quinoa were structurally diverse and had antifungal, hemolytic, molluscicidal, anti-inflammatory, and cytotoxic activities	El Hazzam et al. (2020)
Functional ingredient for food formulations	Atlantic chub mackerel (*Scomber colias*) was stored in ice containing quinoa ethanol extract for 13 days	The presence of quinoa extracts decreased lipid oxidation and hydrolysis in the mackerel. The inhibitory effects became more pronounced toward the end of the storage. The presence of quinoa extracts at a higher addition level (0.05 g/L icing solution) gave the mackerel acceptable sensory quality after the storage, whereas the control mackerel samples became unacceptable for sensory quality	Miranda et al. (2018)
Nutraceutical ingredient	Saponin-rich extracts were tested for in vitro inhibitory activity on pancreatic lipase and hypocholesterolemia capacity	Only after acid hydrolysis, the extracts had increased lipase inhibition activity of quinoa extract, and significantly reduced the bioaccessible cholesterol. The hydrolyzed extracts had similar hypocholesterolemia effects compared to β-sitosterol	Del Hierro et al. (2021)

Cosmetics or pharmaceutics	The capacity of quinoa saponins as skin penetration enhancer was tested in vitro. An intercellular lipid mixture mimicking the outermost layer of skin (stratum corneum) was created and tested against saponin-rich extracts from quinoa	Saponins got incorporated into the lipid monolayer and could be used as enhancers for active components to penetrate into skin for pharmaceutical or cosmetics applications	Jurek et al. (2019)
Cosmetics	Extracts containing saponins were obtained using water from 45 different plants including quinoa seeds	Analysis of the surface activity and foaming properties showed that the surface dilational response of the extracts was mostly contributed by the elasticity. Quinoa extract rich in saponins was among the best surfactant materials for household product and cosmetics applications	Góral and Wojciechowski (2020)
Cosmetics	Extracts containing saponins were obtained from quinoa seeds using optimized conditions (temperature and solvent type and concentration). The emulsification property of the extracts was studied	The extract rich in triterpene saponins decreased the water surface tension from 72 to 31. It has potential use in cosmetics formulations	Bezerra et al. (2021)
Biopriming agent for crop growing	Saponin-rich extracts were obtained from quinoa hull meal. Soybean seedlings were primed with saponins extract at 5%–10%. The primed soybean seedlings were grown under salt stress	Soybean seedlings primed at 5% of saponins reduced the decreases in chlorophyll and carotenoid synthesis and crop growth due to the salt stress. The priming increased antioxidant enzyme activities, membrane stability, water content and the production of ascorbic acid, glutathione, sugars, proline, and glycine betaine, and decreased lipid peroxidation	Soliman et al. (2020)

(Continued)

Table 13.4 (Continued)

Applications	Experimental	Major findings	Reference
Silver nanoparticle synthesis	Saponin-rich extract from quinoa husks to make silver nanoparticles by reducing silver ions	The silver nanoparticles were stable and had sizes ranging from 5 and 50 nm. The nanoparticles had no phytotoxicity and were phytostimulant on radish seeds for germination in a dose-dependent manner	Segura et al. (2020)
Pesticide	Ethanol extracts of quinoa seed coats containing saponins were obtained. The acaricidal effect on the carmine spider mite (*Tetranychus urticae* Koch) of different growing stages was studied	Increasing concentration of the extracts applied on female adults to 9.1% increased the mortality to 93%. For the nymphs of the mite, the lethal effect was obtained after 24 h of applying the extract. The mortality reached 99% after 24 h of applying the extract at 9.1%. Applying the extract at 9.1% to the eggs led to a hatch rate of 76% for the larvae after 120 h. A sublethal effect of the quinoa extracts was obtained on the offspring	Harder et al. (2016)
Pesticide	n-Hexane extract of quinoa inflorescence was tested for pesticidal activity	9,12-Octadecadienoic acid(z,z), n-hexadecanoic acid and 9,12-octadecadienoic acid (z,z)-methyl ester were dominant compounds among 14 compounds identified in the extract revealed by GC-MS. Methyl tetradecanoate was nematicidal. 9,12-Octadecadienoic acid(z,z) and E,Z-1,3,12-nonadecatriene were antimicrobial	Khan et al. (2018)

Pesticide	Aquatic toxicity of saponin-rich extract from quinoa seed coat was tested against algae, plants, worms, arthropods, mollusks, bacteria, and fish embryo	The highest concentration of protecting a total of 95% of the aquatic species for quinoa saponin-rich extract was 23 mg/L, which was higher than that of the saponin-rich extracts from tea seed coat (0.22 mg/L) and quillaja bark (2.9 mg/L). The analysis of the predicted environmental concentrations showed that the quinoa extract could be a risk to the aquatic environment	Jiang et al. (2018)
Molluscicide	Saponin extracts were obtained from quinoa seed husk and were treated with alkaline before neutralizing and drying to obtain the powder. The powder contained 36% of saponins. The molluscicidal effect of the powder against *Pomacea maculate* was tested in lab and in field	In the lab, 7 ppm of saponins was enough to kill 100% of the *Pomacea maculate* after 3 days. In the field in Spain, 5 − 7.7 ppm of saponins were enough to kill 88 − 93% of *P. maculate* after 4 days. Based on the tests of rabbits, rats, and guinea pigs, the product had no acute oral toxicity, slight acute dermal toxicity, and acute eye irritation. It was a nonskin sensitizer with no acute dermal irritation. The saponins-based product was highly safe to algae (*Selenastrum capricornutum*), crustaceans (*Daphnia*), and common carp	Castillo-Ruiz et al. (2018)
Oil recovery for fossil fuel	Saponin extract from quinoa was used to increase oil recovery for fossil fuel applications	The extract was a nonionic surfactant. The extract was more efficient in changing the wettability of oil-water mixtures and had higher emulsifying capacity than NX-610 and TR-880 (two commercial ones). The extract can increase the oil recovery to 24% in original oil in place	Norouzpour et al. (2022)

(*Continued*)

Table 13.4 (Continued)

Applications	Experimental	Major findings	Reference
Polyphenol or saponin-rich extract			
Nutraceutical ingredient (antioxidants)	Extracts of quinoa seeds containing saponins or without saponins were formulated into marine oil for inhibitory activity against lipid oxidation	Extracts containing no saponins had lower in vitro antioxidant activity and less tocopherols and phenolics than those containing saponins, though the two types of extracts had similar inhibitory activity against lipid oxidation in the marine-oil models	Miranda et al. (2017)
Antioxidation against oil oxidation	Extracts from quinoa, uckun, propolis or BHT (1000–1500 ppm) were added to palm olein oil for deep frying of potatoes	The thermal oxidation properties of palm oils were measured in terms of free fatty acidity, viscosity, total polar matter, peroxide value, total oxidation value, *p*-anisidine value, conjugated diene, and triene. The extracts of quinoa were not as effective as that of propolis to retard oil oxidation and the antioxidant effect of latter was close to that of BHT	Ceylan and Basturk (2022)
Nutraceutical ingredient (hypolipidemic)	Ethanol extracts of quinoa seeds were prepared. Rats were treated with Triton X-100 (intraperitoneal injection) to induce hyperlipidemia. The rats also received quinoa extracts for 40 days	The feeding of extract to the hyperlipidemic rats had hepato-renal protective effect, while decreasing lipid peroxidation, enhancing growth performance and antioxidant activity, and regulating the levels of leptin and adiponectin hormone. The exact compounds responsible for the biological functions were not clear	Hashem et al. (2021)

Nutraceutical ingredient (prevention of cardiovascular disease and steatohepatitis)	Ethanol extract of quinoa seeds of two varieties were prepared. Rats were fed with a high fructose-high fat diet to induce cardiovascular disease and steatohepatitis. These rats were also fed with the extracts	The high fructose and high fat diet increased the levels of liver lipids, plasma malondialdehyde, low-density lipoprotein cholesterol, total cholesterol, and triglycerides and decreased the levels of total antioxidants and high-density lipoprotein cholesterol in the rats. The quinoa extracts reduced the extents of these changes in the rats, while improving their glucose tolerance. The extract of the quinoa variety hualhuas had more effect in reducing the risk of the diseases than that of the variety quinoa 1	Al-Okbi et al. (2021)
Functional ingredient for food formulations	Saponin-free ethanol extract of quinoa seeds was obtained. Frozen fatty fish (Atlantic mackerel) was stored at −40°C for 48 h before dipping in glazing solution (0°C, 30 s) containing the quinoa extract (0.45 − 1.36 g/L)	The glazing significantly decreased the lipid hydrolysis and free fatty acid formation in the frozen fatty fish due to the presence of antioxidants in the quinoa extracts, while increasing the sensory quality	Trigo et al. (2018)

Bezerra et al., 2021). The capacity of quinoa saponins as skin penetration enhancer was tested in vitro. An intercellular lipid mixture mimicking the outermost layer of skin (stratum corneum) was created and tested against saponin-rich extracts from quinoa (Jurek et al., 2019). Saponins got incorporated into the lipid monolayer and could be used as enhancer for active components to penetrate into skin for pharmaceutical or cosmetics applications (Jurek et al., 2019). In a comparative study, extracts containing saponins were obtained using water from 45 different plants including quinoa seeds (Góral & Wojciechowski, 2020). Analysis of the surface activity and foaming properties showed that the surface dilational response of the extracts was mostly contributed by the elasticity. Quinoa extract rich in saponins was among the best surfactant materials for household product and cosmetics applications. Bezerra et al. (2021) obtained extracts containing saponins from quinoa seeds using optimized conditions (temperature and solvent type and concentration). The emulsification property of the extracts was studied. The extract rich in triterpene saponins decreased the water surface tension from 72 to 31. It has potential use in cosmetics formulations (Bezerra et al., 2021).

13.6.2.2 Agricultural applications

Saponins were used as biopriming agent for crop growing and for silver nanoparticle synthesis (Soliman et al., 2020; Segura et al., 2020). Soybean seedlings were primed with saponin-rich extract from quinoa hull meal at 5%–10% (Soliman et al., 2020). The primed soybean seedlings were grown under salt stress. The seedlings primed at 5% of saponins reduced the decreases in chlorophyll and carotenoid synthesis and crop growth due to the salt stress. The priming increased antioxidant enzyme activities, membrane stability, water content, and the production of ascorbic acid, glutathione, sugars, proline, and glycine betaine, while decreasing lipid peroxidation. The priming of the soybean seedlings with saponins increased its capacity to tolerate salt stress (Soliman et al., 2020). Silver nanoparticles have different applications as sensors and phytostimulants (Segura et al., 2020; Alberti et al., 2021). Saponin-rich extract from quinoa husks was used to make silver nanoparticles by reducing silver ions (Segura et al., 2020). The silver nanoparticles were stable and had sizes ranging from 5 and 50 nm. The nanoparticles had no phytotoxicity and were phytostimulant on radish seeds for germination in a dose-dependent manner (Segura et al., 2020).

13.6.2.3 Pesticide

Saponins from quinoa seeds and inflorescence were used as pesticides (Harder et al., 2016; Jiang et al., 2018; Castillo-Ruiz et al., 2018). For example, ethanol extracts of quinoa seed coats containing saponins were obtained (Harder et al., 2016). The acaricidal effect on the carmine spider mite (*Tetranychus urticae* Koch) of different growing stages was studied. Increasing concentration of the extracts applied on the female adults to 9.1% increased the mortality to 93%. For the nymphs of the mite, the lethal effect was obtained after 24 h of applying the extract.

The mortality reached 99% after 24 h of applying the extract at 9.1%. Applying the extract at 9.1% to the eggs led to a hatch rate of 76% for the larvae after 120 h. A sublethal effect of the quinoa extracts was obtained on the offspring (Harder et al., 2016). In the study of Castillo-Ruiz et al. (2018), saponin extracts were obtained from quinoa seed husk and were treated with alkaline solution before neutralizing and drying to obtain the powder. The powder contained 36% of saponins. The molluscicidal effect of the powder against *Pomacea maculate* was tested in laboratory and in field (Castillo-Ruiz et al., 2018). In the laboratory, 7 ppm of saponins were enough to kill 100% of the *P. maculate* after 3 days. In the field in Spain, 5 − 7.7 ppm of saponins were enough to kill 88 − 93% of *P. maculate* after 4 days. Based on the tests of rabbits, rats, and guinea pigs, the product had no acute oral toxicity, slight acute dermal toxicity, and acute eye irritation. It was a nonskin sensitizer with no acute dermal irritation. The saponins-based product was highly safe to algae (*Selenastrum capricornutum*), crustaceans (*Daphnia*), and common carp (Castillo-Ruiz et al., 2018). Khan et al. (2018) obtained n-hexane extract of quinoa inflorescence for pesticidal activity. 9,12-Octadecadienoic acid (z,z), n-hexadecanoic acid, and 9,12-octadecadienoic acid (z,z)-methyl ester were dominant compounds among 14 compounds identified in the extract revealed by gas chromatography−mass spectrometry (GC-MS). Methyl tetradecanoate was found to be nematicidal. 9,12-Octadecadienoic acid (z,z) and E,Z-1,3,12-nonadecatriene were antimicrobial (Khan et al., 2018).

13.6.2.4 Oil recovery for fossil fuel

Saponin extract from quinoa was used to increase oil recovery for fossil fuel applications (Norouzpour et al., 2022). The extract was a nonionic surfactant. The extract was more efficient in changing the wettability of oil-water mixtures and had higher emulsifying capacity than NX-610 and TR-880 (two commercial ones). The extract can increase the oil recovery to 24% in original oil in place (Norouzpour et al., 2022).

13.6.3 Applications of compounds other than saponins in quinoa

Extracts of quinoa seeds may contain non-saponin compounds such as polyphenols that have biological and functional properties (Table 13.4). In the study of Miranda et al. (2017), extracts of quinoa seeds containing saponins or without saponins were formulated into marine-oil models for inhibition against lipid oxidation. The extracts containing no saponins had lower in vitro antioxidant activity and less tocopherols and phenolics than those containing saponins, though the two types of extracts had similar inhibitory activity against lipid oxidation in the marine-oil models (Miranda et al., 2017). A saponin-free ethanol extract of quinoa seeds was obtained (Trigo et al., 2018). Frozen fatty fish (Atlantic mackerel) was stored at −40°C for 48 h before dipping in glazing solution (0°C, 30 s) containing the quinoa extract (0.45 − 1.36 g/L) for retarding the lipid oxidation in the fish

(Trigo et al., 2018). Al-Okbi et al. (2021) prepared ethanol extracts of quinoa seeds of two varieties. Rats were fed with a high fructose-high fat diet to induce cardiovascular disease and steatohepatitis. These rats were also fed with the extracts. The high fructose and high-fat diet induced increases in the levels of liver lipids, plasma malondialdehyde, low-density lipoprotein cholesterol, total cholesterol, and triglycerides and decreases in the levels of total antioxidants and high-density lipoprotein cholesterol in the rats. The quinoa extracts reduced the extents of these changes in the rats, while improving their glucose tolerance. The extract of the quinoa variety hualhuas had more effect in reducing the risk of the diseases than that of the variety quinoa 1. These biological effects were largely attributed to the presence of polyphenols in the quinoa (Al-Okbi et al., 2021). Comparisons of the results between Sections 13.6.1 and 13.6.3 (also within Chapter 8) showed that there were overlapping biological functions of polyphenols and saponins.

13.7 Lack of comparative studies between quinoa-based fractionated products and products from other food grains

The cost of quinoa grain is much higher than that of common cereals such as maize, wheat, and rice (Rafik et al., 2021). In order to commercialize the fractionated products of quinoa grain, the technical advantages and special properties of quinoa-based fractionated products should be illustrated against those of common cereals and commercial products. Overall, the special properties of quinoa starch and saponins have been largely revealed. The former is small in size, rendering it useful in some applications as outlined in Section 13.3 above. For example, modified quinoa starch granules are suitable to stabilize Pickering emulsions (Table 13.2). On the other hand, a range of starchy crops such as amaranth, taro, and rice also produce small starch granules. Rice starch can be readily available due to the large-scale production of the crop. Similar to quinoa starch granules, rice starch granules can be modified with octenyl succinic anhydride to efficiently stabilize Pickering emulsions (Wang et al., 2020). Another unique property of quinoa is that it contains saponins that are not found in most of the cereals and other pseudocereals. Saponins have a range of food and nonfood applications (El Hazzam et al., 2020). On the other hand, many other plants also contain saponins that could be extracted for different applications (Góral & Wojciechowski, 2020). The questions remain regarding the uniqueness of quinoa-based saponin extracts using cost-effective approaches.

The applications of quinoa proteins are seen to be similar to those of proteins from other pseudocereals, cereals, and different sources. For example, nanoparticles may be produced from zein, soy protein, whey protein, and gelatin to efficiently stabilize Pickering emulsions (Yan et al., 2020). It is obvious that these proteins can be readily available due to the industrial and large scale of the production of the source materials. Unless there are obvious technical advantages of quinoa protein in specific applications, the high cost of quinoa grain will make the large scale commercial applications of the protein fractions unrealistic. Similar comments can be applied to the potential applications of other fractions of quinoa including dietary fiber-, lipid-, and polyphenol-rich fractions. Unfortunately, our research community continues to ignore such practical and economic aspects of food technology, while producing a large amount of publications out of academic common practice. In this regard, too many of the published articles are rather misleading.

13.8 Lack of studies on the impact of quinoa fractionation on the environment

The value-added processing of quinoa may generate high-value products with high demands from the market. For example, different patents have been published on the applications (e.g., formulation of skin care products) of quinoa-fractionated products in recent years (Raskin & Graf, 2019; Lee et al., 2018; Dejmek et al., 2018). The market demands for these products may drive the production of quinoa crop and the fractionation of the seeds. However, the impact of the increasing quinoa production and the subsequent fractionation processes on the environment has largely been ignored. It is well known that extensive farming and cultivation of quinoa on the highlands of South America damage the ecosystem (Magrach & Sanz, 2020). The wet-milling and extraction processes of quinoa seeds to obtain certain fractions tend to generate waste including wastewater (Asgharnejad et al., 2021), which should be studied in terms of their environmental impact. Life cycle assessment should be done on the chain processes of quinoa production to fractionation. However, this aspect of quinoa processing remains largely under-studied. The research scenario on quinoa has remained very unbalanced so far. Many studies have been done on the fractionation of quinoa seeds for specific applications. However, there have been few studies on the impact of the whole process to obtain a large picture of sustainability.

13.9 Conclusions

There is great potential to develop quinoa for value-added processing and products. Quinoa seeds have been fractionated into starch, protein, dietary fiber, lipid, saponins, or polyphenol-rich products. Quinoa starch in different forms have been used in formulations of edible films and coatings, meat-based products, encapsulation of food ingredients, and stabilizing emulsions. Quinoa protein isolate in different forms have been used as bakery and food ingredients, nutraceutical ingredients with biological functions, encapsulation of ingredients, stabilizing emulsions, and formulations of edible films and coatings. Dietary fiber-rich fractions have been used in the formulations of bakery and meat-based products and nutraceutical ingredients with biological functions. Quinoa lipid fraction has been transesterified for skin care product formulations. Saponin-rich fractions have been used as nutraceutical ingredients for formulations of foods, cosmetics, pharmaceutics, pesticides, bio-priming agents for crop growing, and for silver nanoparticle synthesis. Polyphenol-rich fractions have the potential for nutraceutical formulations. The applications of quinoa-based fractions should be compared with those of fractions based on major cereals and crops. The impact of the fractionation, application, and cultivation of quinoa on the environment remains to be studied.

References

Alak, G., Guler, K., Ucar, A., Parlak, V., Kocaman, E. M., Yanık, T., & Atamanalp, M. (2019). Quinoa as polymer in edible films with essential oil: Effects on rainbow trout fillets shelf life. *Journal of Food Processing and Preservation*, *43*, e14268.

Alberti, G., Zanoni, C., Magnaghi, L. R., & Biesuz, R. (2021). Gold and silver nanoparticle-based colorimetric sensors: New trends and applications. *Chemosensors*, *9*, 305.

Al-Okbi, S. Y., Hamed, T. E., Elewa, T. A., Ramadan, A. A., El-Karamany, M. F., & Bakry, B. A. (2021). Role of polar extracts from two quinoa varieties in prevention of steatohepatitis and cardiovascular diseases and improving glucose tolerance in rats. *Journal of Herbmed Pharmacology*, *10*, 93–101.

Alonso-Miravalles, L., Zannini, E., Bez, J., Arendt, E. K., & O'Mahony, J. A. (2022). Formation and thermal and colloidal stability of oil-in-water emulsions stabilized using quinoa and lentil protein blends. *Journal of the Science of Food and Agriculture*, *102*, 5077–5085.

Asgharnejad, H., Nazloo, E. K., Larijani, M. M., Hajinajaf, N., & Rashidi, H. (2021). Comprehensive review of water management and wastewater treatment in food processing industries in the framework of water-food-environment nexus. *Comprehensive Reviews in Food Science and Food Safety*, *20*, 4779–4815.

Ballester-Sánchez, J., Fernández-Espinar, M. T., & Haros, C. M. (2020). Isolation of red quinoa fibre by wet and dry milling and application as a potential functional bakery ingredient. *Food Hydrocolloids*, *101*, 105513.

Bezerra, K. G. O., Silva, I. G. S., Almeida, F. C. G., Rufino, R. D., & Sarubbo, L. A. (2021). Plant-derived biosurfactants: Extraction, characteristics and properties for application in cosmetics. *Biocatalysis and Agricultural Biotechnology*, *34*, 102036.

Bhatia, M., & Rohilla, S. (2020). Formulation and optimization of quinoa starch nanoparticles: Quality by design approach for solubility enhancement of piroxicam. *Saudi Pharmaceutical Journal*, *28*, 927–935.

Camillo-Andrade, A. C., Santos, M. D. M., Fischer, J. S. G., Swinka, B. B., Bosquetti, B., Schuck, D. C., Pincerati, M. R., Lorencini, M., & Carvalho, P. C. (2020). Proteomics reveals that quinoa bioester promotes replenishing effects in epidermal tissue. *Scientific Reports*, *10*, 19392.

Cao, Y., Zou, L., Li, W., Song, Y., Zhao, G., & Hu, Y. (2020). Dietary quinoa (*Chenopodium quinoa* Willd.) polysaccharides ameliorate high-fat diet-induced hyperlipidemia and modulate gut microbiota. *International Journal of Biological Macromolecules*, *163*, 55–65.

Capraro, J., De Benedetti, S., DiDio, M., Bona, E., Abate, A., Corsetto, P. A., & Scarafoni, A. (2020). Characterization of chenopodin isoforms from quinoa seeds and assessment of their potential anti-inflammatory activity in Caco-2 cells. *Biomolecules, 10*, 795.

Castillo-Ruiz, M., Cañon-Jones, H., Schlotterbeck, T., Lopez, M. A., Tomas, Á., & San Martín, R. (2018). Safety and efficacy of quinoa (*Chenopodium quinoa*) saponins derived molluscicide to control of *Pomacea maculata* in rice fields in the Ebro Delta, Spain. *Crop Protection, 111*, 42–49.

Ceylan, M. M., & Basturk, A. (2022). Investigation of the effects of uckun (*Rheum ribes* L.), quinoa (*Chenopodium quinoa* Willd.), and propolis extracts on the thermal oxidation of palm olein oil during the deep-frying process. *Journal of Food Processing and Preservation, 46*, e16210.

Chen, K., Zhang, M., Mujumdar, A. S., & Wang, H. (2021). Quinoa protein-gum Arabic complex coacervates as a novel carrier for eugenol: Preparation, characterization and application for minced pork preservation. *Food Hydrocolloids, 120*, 106915.

Daliri, H., Ahmadi, R., Pezeshki, A., Hamishehkar, H., Mohammadi, M., Beyrami, H., Heshmati, M. K., & Ghorbani, M. (2021). Quinoa bioactive protein hydrolysate produced by pancreatin enzyme-functional and antioxidant properties. *LWT-Food Science and Technology, 150*, 111853.

Dejmek, P., Timgren, A., SJöö, M., & Rayner, M. (2018). *New particle stabilized emulsions and foams*. European Patent EP2651243B1.

Del Hierro, J. N., Casado-Hidalgo, G., Reglero, G., & Martin, D. (2021). The hydrolysis of saponin-rich extracts from fenugreek and quinoa improves their pancreatic lipase inhibitory activity and hypocholesterolemic effect. *Food Chemistry, 338*, 128113.

Duran, N. M., Spelzini, D., Wayllace, N., Boeris, V., & da Silva, F. L. B. (2018). A combined experimental and molecular simulation study of factors influencing interaction of quinoa proteins-carrageenan. *International Journal of Biological Macromolecules, 107*, 949–956.

El Hazzam, K., Hafsa, J., Sobeh, M., Mhada, M., Taourirte, M., EL Kacimi, K., & Yasri, A. (2020). An insight into saponins from quinoa (*Chenopodium quinoa* Willd): A review. *Molecules, 25*, 1059.

Escamilla-García, M., Ríos-Romo, R. A., Melgarejo-Mancilla, A., Díaz-Ramírez, M., Hernández-Hernández, H. M., Amaro-Reyes, A., Pierro, P. D., & Regalado-González, C. (2020). Rheological and antimicrobial properties of chitosan and quinoa protein filmogenic suspensions with thyme and rosemary essential oils. *Foods, 9*, 1616.

Fernández-López, J., Viuda-Martos, M., & Pérez-Alvarez, J. A. (2021). Quinoa and chia products as ingredients for healthier processed meat products: Technological strategies for their application and effects on the final product. *Current Opinion in Food Science, 40*, 26–32.

Gao, H., Ma, L., Cheng, C., Liu, J., Liang, R., Zou, L., Liu, W., & McClements, D. J. (2021). Review of recent advances in the preparation, properties, and applications of high internal phase emulsions. *Trends in Food Science & Technology, 112*, 36–49.

Graf, B. L., Rojas-Silva, P., Rojo, L. E., Delatorre-Herrera, J., Baldeón, M. E., & Raskin, I. (2015). Innovations in health value and functional food development of quinoa (*Chenopodium quinoa* Willd.). *Comprehensive Reviews in Food Science and Food Safety, 14*, 431–445.

Guo, H., Hao, Y., Fan, X., Richel, A., Everaert, N., Yang, X., & Ren, G. (2021). Administration with quinoa protein reduces the blood pressure in spontaneously hypertensive rats and modifies the fecal microbiota. *Nutrients, 13*, 2446.

Góral, I., & Wojciechowski, K. (2020). Surface activity and foaming properties of saponin-rich plants extracts. *Advances in Colloid and Interface Science, 279*, 102145.

Gürbüz, G., Kauntola, V., Diaz, J. M. R., Jouppila, K., & Heinonen, M. (2018). Oxidative and physical stability of oil-in-water emulsions prepared with quinoa and amaranth proteins. *European Food Research and Technology*, *244*, 469–479.

Harder, M. J., Tello, V. E., & Giliomee, J. H. (2016). The acaricidal effect of ethanolic extracts of *Chenopodium quinoa* Willd. on *Tetranychus urticae* Koch (Acari: Tetranychidae). *African Entomology*, *24*, 50–60.

Hashem, M. A., Mahmoud, E. A., & Abd-Allah, N. A. (2021). Hypolipidemic activity of an ethanolic extract of quinoa seeds in Triton X-100-induced hyperlipidemic rats. *Comparative Clinical Pathology*, *30*, 473–482.

Jiang, F., Du, C., Zhao, N., Jiang, W., Yu, X., & Du, S. K. (2022). Preparation and characterization of quinoa starch nanoparticles as quercetin carriers. *Food Chemistry*, *369*, 130895.

Jiang, X., Hansen, H. C. B., Strobel, B. W., & Cedergreen, N. (2018). What is the aquatic toxicity of saponin-rich plant extracts used as biopesticides? *Environmental Pollution*, *236*, 416–424.

Jurek, I., Góral, I., Gęsiński, K., & Wojciechowski, K. (2019). Effect of saponins from quinoa on a skin-mimetic lipid monolayer containing cholesterol. *Steroids*, *147*, 52–57.

Khan, I. H., Javaid, A., Ahmed, D., & Khan, U. (2018). Pesticidal constituents in n-hexane inflorescence extract of *Chenopodium quinoa*. *Mycopath*, *16*, 43–46.

Lee, J. Y., Lee, S. H., Kwon, L.K., Paik, B. R., & Lee, H. K. (2018). *Cosmetic composition for skin moisturizing containing minerals and quinoa extract*. US Patent 10,028,901.

Li, G., Hemar, Y., & Zhu, F. (2022). Supramolecular structure of quinoa starch affected by nonenyl succinic anhydride (NSA) substitution. *International Journal of Biological Macromolecules*, *218*, 181–189.

Li, G., Xu, X., & Zhu, F. (2019b). Physicochemical properties of dodecenyl succinic anhydride (DDSA) modified quinoa starch. *Food Chemistry*, *300*, 125201.

Li, G., & Zhu, F. (2018). Quinoa starch: Structure, properties, and applications. *Carbohydrate Polymers*, *181*, 851–861.

Li, G., & Zhu, F. (2021). Physicochemical, rheological, and emulsification properties of nonenyl succinic anhydride (NSA) modified quinoa starch. *International Journal of Biological Macromolecules*, *193*, 1371–1378.

Li, S., Zhang, B., Li, C., Fu, X., & Huang, Q. (2020). Pickering emulsion gel stabilized by octenylsuccinate quinoa starch granule as lutein carrier: Role of the gel network. *Food Chemistry*, *305*, 125476.

Li, S., Zhang, B., Tan, C. P., Li, C., Fu, X., & Huang, Q. (2019a). Octenylsuccinate quinoa starch granule-stabilized Pickering emulsion gels: Preparation, microstructure and gelling mechanism. *Food Hydrocolloids*, *91*, 40–47.

Lindeboom, N., Chang, P. R., & Tyler, R. T. (2004). Analytical, biochemical and physicochemical aspects of starch granule size, with emphasis on small granule starches: A review. *Starch/Stärke*, *56*, 89–99.

Liu, K., Zha, X. Q., Li, Q. M., Pan, L. H., & Luo, J. P. (2021). Hydrophobic interaction and hydrogen bonding driving the self-assembling of quinoa protein and flavonoids. *Food Hydrocolloids*, *118*, 106807.

Lotfy, S. N., Saad, R., El-Massrey, K. F., & Fadel, H. H. M. (2021). Effects of pH on headspace volatiles and properties of Maillard reaction products derived from enzymatically hydrolyzed quinoa protein-xylose model system. *LWT-Food Science and Technology*, *145*, 111328.

López-Alarcón, C. A., Cerdán-Leal, M. A., Beristain, C. I., Pascual-Pineda, L. A., Azuara, E., & Jiménez-Fernández, M. (2019). The potential use of modified quinoa protein isolates in cupcakes: Physicochemical properties, structure and stability of cupcakes. *Food & Function*, *10*, 4432.

López-Castejón, M. L., Bengoechea, C., Díaz-Franco, J., & Carrera, C. (2020). Interfacial and emulsifying properties of quinoa protein concentrates. *Food Biophysics*, *15*, 122–132.

Magrach, A., & Sanz, M. J. (2020). Environmental and social consequences of the increase in the demand for 'superfoods' world-wide. *People and Nature*, *2*, 267–278.

Marefati, A., Bertrand, M., Sjöö, M., Dejmek, P., & Rayner, M. (2017). Storage and digestion stability of encapsulated curcumin in emulsions based on starch granule Pickering stabilization. *Food Hydrocolloids*, *63*, 309–320.

Marefati, A., & Rayner, M. (2020). Starch granule stabilized Pickering emulsions: An 8-year stability study. *Journal of the Science of Food and Agriculture*, *100*, 2807–2811.

Medina, E., Caro, N., Abugoch, L., Gamboa, A., Díaz-Dosque, M., & Tapia, C. (2019). Chitosan thymol nanoparticles improve the antimicrobial effect and the water vapour barrier of chitosan-quinoa protein films. *Journal of Food Engineering*, *240*, 191–198.

Miranda, J. M., Carrera, M., Pasten, A., Vega-Gálvez, A., Barros-Velázquez, J., & Aubourg, S. P. (2018). The impact of quinoa (*Chenopodium quinoa* Willd.) ethanolic extracts in the icing medium on quality loss of Atlantic chub mackerel (*Scomber colias*) under chilling storage. *European Journal of Lipid Science and Technology*, *120*, 1800280.

Miranda, M., Barbosa, R. G., Trigo, M., Uribe, E., Vega-Galvez, A., & Aubourg, S. P. (2017). Enhancement of the rancidity stability in a marine-oil model by addition of a saponin-free quinoa (*Chenopodium quinoa* Willd.) ethanol extract. *European Journal of Lipid Science and Technology*, *119*, 1600291.

Mobtaker, H., Azadi, M., Hassani, N., Neek-Amal, M., Rassouli, M., & Bidi, M. A. (2021). The inhibition performance of quinoa seed on corrosion behavior of carbon steel in the HCl solution; theoretical and experimental evaluations. *Journal of Molecular Liquids*, *335*, 116183.

Ng, C. Y., & Wang, M. (2021). The functional ingredients of quinoa (*Chenopodium quinoa*) and physiological effects of consuming quinoa: A review. *Food Frontiers*, *2*, 329–356.

Norouzpour, M., Nabipour, M., Azdarpour, A., Akhondzadeh, H., Santos, R. M., & Keshavarz, A. (2022). Experimental investigation of the effect of a quinoa-derived saponin-based green natural surfactant on enhanced oil recovery. *Fuel*, *318*, 123652.

Pająk, P., Przetaczek-Rożnowska, I., & Juszczak, L. (2019). Development and physicochemical, thermal and mechanical properties of edible films based on pumpkin, lentil and quinoa starches. *International Journal of Biological Macromolecules*, *138*, 441–449.

Park, J. H., Lee, Y. J., Lim, J. G., Jeon, J. H., & Yoon, K. S. (2021). Effect of quinoa (*Chenopodium quinoa* Willd.) starch and seeds on the physicochemical and textural and sensory properties of chicken meatballs during frozen storage. *Foods*, *10*, 1601.

Pompeu, D. G., Cordeiro, H. G., Cristina, F., Tonelli, P., Godin, A. M., de Melo, I. S. F., Matsui, T. C., et al. (2022). Chenopodin as an anti-inflammatory compound. *Natural Product Research*. Available from https://doi.org/10.1080/14786419.2021.1980791.

Qin, X. S., Luo, Z. G., Peng, X. C., Lu, X. X., & Zou, Y. X. (2018). Fabrication and characterization of quinoa protein nanoparticle-stabilized food-grade Pickering emulsions with ultrasound treatment: Effect of ionic strength on the freeze – thaw stability. *Journal of Agricultural and Food Chemistry*, *66*, 8363–8370.

Rafik, S., Chaoui, M., Assabban, Y., Salima Jazi., Choukr-Allah, R., El Gharouss, M., & Hirich, A. (2021). Quinoa value chain, adoption, and market assessment in Morocco. *Environmental Science and Pollution Research*, *28*, 46692–46703.

Raskin, I., & Graf, B. (2019). *Methods of making quinoa leachates and uses thereof.* US Patent 10,441,594.

Remanan, M. K., & Zhu, F. (2021). Encapsulation of rutin using quinoa and maize starch nanoparticles. *Food Chemistry*, *353*, 128534.

Robledo, N., López, L., Bunger, A., Tapia, C., & Abugoch, L. (2018). Effects of antimicrobial edible coating of thymol nanoemulsion/quinoa protein/chitosan on the safety, sensorial properties, and quality of refrigerated strawberries (*Fragaria* × *ananassa*) under commercial storage environment. *Food and Bioprocess Technology*, *11*, 1566–1574.

Segura, R., Vásquez, G., Colson, E., Gerbaux, P., Frischmon, C., Nesic, A., García, D. E., & Cabrera-Barjas, G. (2020). Phytostimulant properties of highly stable silver nanoparticles obtained with saponin extract from *Chenopodium quinoa*. *Journal of the Science of Food and Agriculture*, *100*, 4987–4994.

Soliman, M. H., Abdulmajeed, A. M., Alhaithloul, H., Alharbi, B. M., El-Esawi, M. A., Hasanuzzaman, M., & Elkelish, A. (2020). Saponin biopriming positively stimulates antioxidants defense, osmolytes metabolism and ionic status to confer salt stress tolerance in soybean. *Acta Physiologiae Plantarum*, *42*, 114.

Tan, M., Chang, S., Liu, J., Li, H., Xu, P., Wang, P., Wang, X., Zhao, M., Zhao, B., Wang, L., & Zhao, Q. (2020). Physicochemical properties, antioxidant and antidiabetic activities of polysaccharides from quinoa (*Chenopodium quinoa* Willd.) seeds. *Molecules*, *25*, 3840.

Tan, M., Zhao, Q., & Zhao, B. (2021b). Physicochemical properties, structural characterization and biological activities of polysaccharides from quinoa (*Chenopodium quinoa* Willd.) seeds. *International Journal of Biological Macromolecules*, *193*, 1635–1644.

Tan, M., Zheng, S., Lv, H., Wang, B., Zhao, Q., & Zhao, B. (2021a). Rational design and synthesis of chitosan–quinoa polysaccharide composite aerogel and its adsorption properties for Congo red and methylene blue. *New Journal of Chemistry*, *45*, 9829.

Teng, C., Qin, P., Shi, Z., Zhang, W., Yang, X., Yao, Y., & Ren, G. (2021). Structural characterization and antioxidant activity of alkali-extracted polysaccharides from quinoa. *Food Hydrocolloids*, *113*, 106392.

Teng, C., Shi, Z., Yao, Y., & Ren, G. (2020). Structural characterization of quinoa polysaccharide and its inhibitory effects on 3T3-L1 adipocyte differentiation. *Foods*, *9*, 1511.

Trigo, M., Rodríguez, A., Dovale, G., Pastén, A., Vega-Gálvez, A., & Aubourg, S. P. (2018). The effect of glazing based on saponin-free quinoa (*Chenopodium quinoa*) extract on the lipid quality of frozen fatty fish. *LWT-Food Science and Technology*, *98*, 231–236.

Velásquez-Castillo, L. E., Leite, M. A., Ditchfield, C., Sobral, P. J. D. A. I., & Moraes, C. F. (2020). Quinoa starch nanocrystals production by acid hydrolysis: Kinetics and properties. *International Journal of Biological Macromolecules*, *143*, 93–101.

Vera, A., Tapia, C., & Abugoch, L. (2020). Effect of high-intensity ultrasound treatment in combination with transglutaminase and nanoparticles on structural, mechanical, and physicochemical properties of quinoa proteins/chitosan edible films. *International Journal of Biological Macromolecules*, *144*, 536–543.

Vilcacundo, R., Barrio, D., Carpio, C., García-Ruiz, A., Rúales, J., Hernández-Ledesma, B., & Carrillo, W. (2017). Digestibility of quinoa (*Chenopodium quinoa* Willd.) protein concentrate and its potential to inhibit lipid peroxidation in the zebrafish model. *Plant Foods for Human Nutrition*, *72*, 294–300.

Wang, P. P., Luo, Z. G., Chen, C., Fu, X., & Tamer, T. M. (2020). Effects of octenyl succinic anhydride groups distribution on the storage and shear stability of Pickering emulsions formulated by modified rice starch. *Carbohydrate Polymers*, *228*, 115389.

Yan, X., Ma, C., Cui, F., McClements, D. J., Liu, X., & Liu, F. (2020). Protein-stabilized Pickering emulsions: Formation, stability, properties, and applications in foods. *Trends in Food Science & Technology*, *103*, 293–303.

Zade, M. S. R., Aghkhani, M. H., Abbaspour-Fard, M. H., Hosseini, F., & Qazanfarzadeh, Z. (2021). Optimization of nanocomposite films based on quinoa protein isolate incorporated with cellulose nanocrystal and starch. *Journal of Food Processing and Preservation*, *45*, e15926.

Zeyneb, H., Pei, H., Cao, X., Wang, Y., Win, Y., & Gong, L. (2021). In vitro study of the effect of quinoa and quinoa polysaccharides on human gut microbiota. *Food Science & Nutrition*, *9*, 5735–5745.

Zhang, L., Xiong, T., Wang, X. F., Chen, D. L., He, X. D., et al. (2021). Pickering emulsifiers based on enzymatically modified quinoa starches: Preparation, microstructures, hydrophilic property and emulsifying property. *International Journal of Biological Macromolecules*, *190*, 130–140.

Zhang, X., Zuo, Z., Ma, W., Yu, P., Li, T., & Wang, L. (2021a). Assemble behavior of ultrasound-induced quinoa protein nanoparticles and their roles on rheological properties and stability of high internal phase emulsions. *Food Hydrocolloids*, *117*, 106748.

Zhang, R., Cheng, L., Luo, L., Hemar, Y., & Yang, Z. (2021b). Formation and characterisation of high-internal-phase emulsions stabilised by high-pressure homogenised quinoa protein isolate. *Colloids and Surfaces A: Physicochemical and Engineering Aspects*, *631*, 127688.

Zhao, Y., Chen, F., Gao, C., Feng, X., & Tang, X. (2022). Structure, physical and antioxidant properties of quinoa protein/hsian-tsao gum composite biodegradable active films. *LWT-Food Science and Technology*, *155*, 112985.

Zhu, F. (2020). Dietary fiber polysaccharides of amaranth, buckwheat and quinoa grains: A review of chemical structure, biological functions and food uses. *Carbohydrate Polymers*, *248*, 116819.

Zhu, F. (2021). Polysaccharide based films and coatings for food packaging: Effect of added polyphenols. *Food Chemistry*, *359*, 129871.

Zuo, Z., Zhang, X., Li, T., Zhou, J., Yang, Y., Bian, X., & Wang, L. (2022). High internal phase emulsions stabilized solely by sonicated quinoa protein isolate at various pH values and concentrations. *Food Chemistry*, *378*, 132011.

Index

Note: Page numbers followed by "*f*" and "*t*" refer to figures and tables, respectively.

A

Adipogenesis inhibition, 129
Agronomic analysis, 11
Alkylresorcinols, 196–197
Amaranth grains, 159
Amino acid composition of proteins, 25*t*
Amylopectin, 79–80
  retrogradation, 86
Amylose, 76–78
Antibacterial activity, 129, 193
Anticancer, 195
Anticancer capacity, 130–131
Anticancer property, 110
Antidiabetic capacity, 130
Antihaemolytic activity, 128–129
Antihypertension, 129, 245
Antiinflammatory activity, 131
Antiobesity and antidiabetes, 232–245
Antioxidant activity, 128
Arabinoxylans, 98
Aroma, 274

B

Baking and roasting, 270
Betalains, 182–183, 196–197
Bioactive compounds, 169
Bioprocessing
  fermentation, 295–309
  germination, 310
  malting, 309–310
Boiling, 269

C

Carbohydrates of quinoa, 27–28
Carotenoids, 32
Column chromatography, 98
Consumer preferences, 267–268
Contaminated quinoa, 256
COVID-19 pandemic, 10–11
Crude fiber measurement, 94–95

D

Dairy and gluten-free products, 317–318
Dietary fiber of quinoa seeds
  anticancer property, 110
  antidiabetic activity, 109
  antioxidant activity, 104
  biological properties, 93–94
  cholesterol and bile acid adsorption, 109–110
  fermentative properties and gut microbiota, 111
  fiber fractions and polysaccharides
    biological properties of, 104–111
    chemical structure, 97–98
    dry and wet milling of, 97*f*
    enzyme/ultrasound assisted extraction, 96*f*
    physical structure, 99
  fiber-rich fractions, 93
  gastro-protection, 110
  "healthy" and "novel" food products, 93
  immunomodulation, 110–111
  inhibition on food digestion, 104–109
  insoluble and soluble fiber fractions, 94
  physicochemical properties of
    glucose, cholesterol, and bile acid adsorption capacity, 103–104
    thermal and rheological properties, 99–103
    water- and oil-holding capacities, 103
  resistant starch, 94
Diferulic acid oligosaccharides, 98
Differential scanning calorimetry (DSC), 159–160

Domestic cooking
  baking and roasting, 270
  boiling, 269
  steaming, 269–270
Drying, 198
Dry milling, 40–57, 57*f*, 63–65
Dynamic oscillatory and flow analysis, 83–86

E
Ecuador, 1–2
Electron beam irradiation, 162, 209–210

F
Fermentation, 198–206
Fiber-rich fraction, 95
Flavonoids, 198
Food digestion, 104–109

G
Gas chromatography (GC), 156
Gastro-protection, 110
Gelatinization parameters, 80
Gel textural property, 87
Genetic diversity, 165
Germination, 161, 198–206
Global Collaborative Network on Quinoa, 5–6
Glycemic control, 249–250
Gut microbiota regulation, 193–195

H
Heating and drying, 139–142
Hepatoprotection, 247
High hydrostatic pressure, 209–210
High-pressure homogenization, 139
Home cooking methods, 207–209
Human endothelial NCI-H716 cells, 109
Hypercholesterolemia, 232
Hypolipidemia, 247

I
Immunoadjuvant activity, 195
Immunomodulation, 110–111
Immunonutritional property, 245–246
Insertion/Deletion markers, 2–4
International Quinoa Research Symposium, 5–6
Intersimple sequence repeat (ISSR), 2–4
Ion-exchange chromatography, 95–97

L
Life Cycle Assessment, 7–8
Linear polysaccharides, 70
Lipid profile and immune regulating, 245
Long terminal repeat (LTR), 2–4

M
Membrane and skin permeability, 195
Metabolic improvement, 249
5-Methyltetrahydrofolate, 31–32
Microwave-assisted hydrolysis, 210
Minerals, 28–30

N
Neuroprotection, 247–248
Non-LTR retrotransposons, 2–4
Nonstarch polysaccharides, 27–28
Nonthermal processing
  cold plasma modification, 284
  electron beam irradiation, 285
  high hydrostatic pressure treatment, 283–284
  ultrasound modification, 284–285
  washing, 283

P
Pancreatic lipase inhibition and hypocholesterolemia, 195
Pectin sidechains, 98
Perisperm, 19
Phenolics, 112
Phospholipids, 156–157
Phytic acid and organic acids, 30–31
Phytoecdysteroids, 182–183, 196–197
Plasma lipid profile improvement, 250
Polymorphism, 76
Polyphenols, 94
Procambium, 19
Protein, in quinoa
  biological properties of
    adipogenesis inhibition, 129
    antibacterial activity, 129
    anticancer capacity, 130–131
    antidiabetic capacity, 130
    antihaemolytic activity, 128–129

antihypertension property, 129
antiinflammatory, 131
antioxidant activity, 128
quinoa protein, hydrolysates, and peptides, 131
chemical and enzymatic treatments
enzyme hydrolysis, 143
oxidation, 142–143
quinoa protein and peptides, 143–144
sprouting, 143
composition, fractionation, and structure of, 118
functional properties of, 117
digestibility, 123
emulsifying property, 121
foaming property, 121–122
gelling property, 120–121
solubility, 120
thermal property, 122
water and oil-binding property, 122
immunonutritional, antihypertensive, and gut microbiota modulating properties, 246–247
modifications and processing, 131–144
physical treatments
fractionation, 132–138
heating and drying, 139–142
high-pressure homogenization, 139
ultrasound, 139
physicochemical properties, 117
quinoa protein isolates, 118–119
Proteins and amino acids, 24–26

Q

Quinoa
alkylresorcinols, 196–197
antihypertension, 245
antiobesity and antidiabetes, 232–245
applications of compounds other than saponins, 411–412
betalains, 182–183, 196–197
biochemical and chemical changes, 267
biological activities of, 197–210
biological processing of, 296*t*
bioprocessing of
fermentation, 295–309
germination, 310
malting, 309–310
carbohydrates, 27–28
carotenoids, 32
chemical composition of, 21–23, 198–210
climate conditions, 1
combined cooking methods, 270–271
component-biofunction relationships of, 251–252
consumer preference for, 268
consumption and utilization of, 169
cooked quinoa, sensory properties of
aroma, 274
sensory attributes of, 275–276
taste, 271–274
cuisines and commercial food products, 267
cultivation, in different environmental conditions, 4–5
development of, 6–7
domestic cooking
baking and roasting, 270
boiling, 269
steaming, 269–270
ecological and socioeconomic concerns of, 10–11
epidemiological studies, 317–318
extracts of
hepatoprotection, 247
hypolipidemia, 247
neuroprotection, 247–248
fermentation, 267
for food and nutritional applications, 279
genetic diversity of, 1–2
germination, 267
grain structure, 17–20
hepatoprotection, 232
high-pressure assisted cooking, 270
human trials on health and nutritional effects
cardiovascular disease risk, 250–251
glycemic control, 249–250
metabolic improvement, 249
plasma lipid profile improvement, 250
hypercholesterolemia, 232
immunonutritional property, 245–246
in vivo studies on health and nutritional effects, 232–249
interpretation and data comparability, 23–24
lipid profile and immune regulating, 245

Quinoa (*Continued*)
lipids, 26–27
minerals, 28–30
mitigating chemical toxicity, 246
negative health effects of
allergy revealed by clinical reports, 252–256
contaminated quinoa, 256
intestinal permeability, 257
nonthermal processing, 279, 280*t*
cold plasma modification, 284
electron beam irradiation, 285
high hydrostatic pressure treatment, 283–284
ultrasound modification, 284–285
washing, 283
nutritional composition and physicochemical properties of, 279–283
nutritional properties, 1
organic quinoa, 7–8
in outer space, 9
phenolics-related biological activities
anticancer, 191–192
antiobesity and antihypoglycemia, 192
antioxidant activity, 183–191
liver protection, 192–193
phytic acid and organic acids, 30–31
phytoecdysteroids, 182–183, 196–197, 248–249
polyphenols and saponins in
flavonoids and derivatives in, 178–179
phenolic acids and derivatives in, 170–178
phenolics extraction, 179–182
polysaccharide fraction
gut microbiota and hyperlipidemia regulation, 248
immunity-improving, 248
proteins and amino acids, 24–26
quinoa greens, 8–9
quinoa seed storage, 9–10
research gaps on health effects of, 257–258
saponins, 17
saponins-related biological activities
antibacterial activity, 193
anticancer, 195
gut microbiota regulation, 193–195
immunoadjuvant activity, 195
membrane and skin permeability, 195
pancreatic lipase inhibition and hypocholesterolemia, 195
toxicity, 195–196
sensory and nutritional properties of, 267
thermal processes, 279, 286*t*
boiling, steaming, and roasting, 292–294
drying, 294
extrusion, 285–292
heating under pressure, 294
microwave heating, 294–295
popping, 295
vitamins, 31–32
Quinoa-based fractionated products, 412–413
Quinoa-based gluten-free products, 317–318, 334–368
beer, 334, 367–368
biscuit and cookie, 329–330, 355–356
bread, 330–332, 357–359
cake, 356–357
cheese and cream analog, 361–362
cupcakes, 318
film and coating, 353
fish and meat-based products, 366–367
infant food, 353–354
intermediate products, 334–353
labeling and increasing price issues of, 368–369
meat-based products, 333
milk substitute and porridge, 362–365
pasta, 359–360
pasta and noodle, 333
probiotic and fermented drinks and yogurt, 360–361
snack food, 354–355
sprout, 353
tarhana and kishk, 318–329
tempe, 360
Quinoa dietary fiber and nonstarch polysaccharides
food, nutraceutical, and nonfood applications of, 398–402
Quinoa fractionation, 132–138, 413
Quinoa Germinate Platform, 6
Quinoa germ oil, 165
Quinoa husk, 210

Quinoa insoluble fiber, 103
Quinoa lipids, 26–27
aging and oxidation, 160–161
electron beam irradiation, 162
extraction of, 151–156
fatty acid composition, 157–159
food and nonfood applications of, 402–403
germination, 161
lipid content and fatty acid composition, 156
melting properties of, 159–160
neutral and polar lipids, 151, 156–157
oils of quinoa leaves, 164
physicochemical properties, 162
quinoa seed oil, 162–164
thermal processing, 160
triacylglycerol composition, 157
unsaponifiable lipidic fraction, 159
Quinoa milling fractions, 60–61
Quinoa milling process, 209
alternatives to conventional processes, 60
devices, 61
dry milling, 40–57, 57*f*, 63–65
nutritional and chemical composition and particle size, 61–63
physicochemical and functional properties, 63
of quinoa grains, 41*t*
quinoa milling fractions, 60–61
wet milling, 40, 57–60, 57*f*, 63–65
Quinoa protein fraction
food and nutraceutical applications
emulsion stabilizers, 386–388
encapsulation of bioactive compounds, 386
films and coatings, 388–390
functional food ingredient and food component interactions, 378
nutraceutical ingredients, 378–386
Quinoa soluble fiber, 103
Quinoa starch
amylopectin and amylose molecules, 69
chemical structure
amylopectin, 79–80
amylose, 76–78
composition, 70–73
food and nonfood applications of, 391–398
emulsions, 398
encapsulation of bioactive compounds, 396–398
films and coatings, 391
meat products, 391
production of nanoparticles, 391–396
gelatinization and retrogradation properties, 69
isolation, 69–70
physicochemical and functional characteristics of
rheological properties, 83–86
swelling and solubility, 80–83
thermal property, 80
polymorphism, 76
shape and size of starch granules, 73–75

R

Rapid amplified polymorphic DNA (RAPD) markers, 2–4
RAW 264.7 cells, 110–111

S

Saponins, 17, 39, 169
antibacterial activity, 193
anticancer property, 195
food and nutraceutical applications of, 403–412
gut microbiota regulation, 193–195
immunoadjuvant activity, 195
membrane and skin permeability, 195
nonfood applications of
agricultural applications, 410
cosmetics, 403–410
oil recovery for fossil fuel, 411
pesticide, 410–411
pancreatic lipase inhibition and hypocholesterolemia, 195
toxicity, 195–196
Short chain fatty acid (SCFA), 111
Simple sugars, 28
Simulated gastrointestinal digestion, 198
Soxhlet extraction, 151–156
Sprouted quinoa-based yogurt, 109
Squalene, 159, 162–163
Starch, 27
Starch digestibility, 87–88
Starch gels, 87
Starch granules, 19

Steaming, 269–270
Stone and ultrafine milling, 61
Super-long amylopectin chains, 79–80
Swelling power (SP) values, 80–83

T
Tartary buckwheat grains, 61
Taste, 271–274
Texture, 274–275
Thermal processes, 279, 286*t*
  boiling, steaming, and roasting, 292–294
  drying, 294
  extrusion, 285–292
  heating under pressure, 294
  microwave heating, 294–295
  popping, 295

U
Ultrasound, 139, 209–210
Ultrasound-assisted extraction, 210
Ultrasound/microwave processing, 95
United States Department of Agriculture (USDA), 6–7
UV-B radiation, 9

V
Vitamins, 31–32

W
Water solubility index (WSI), 80–83
Wet milling, 40, 57–60, 57*f*, 63–65
Worldwide Consortium on Evolutionary Participatory Breeding in Quinoa, 5–6

FSC
www.fsc.org
MIX
Papier | Fördert
gute Waldnutzung
FSC® C083411

Kolibri 360 GmbH
Ferdinand-Jühlke-Straße 7
99095 Erfurt, Deutschland
produktsicherheit@kolibri360.de